Environmental Footprints and Eco-design of Products and Processes

Series editor

Subramanian Senthilkannan Muthu, SGS Hong Kong Limited,
Hong Kong, Hong Kong SAR

More information about this series at http://www.springer.com/series/13340

Miguel Angel Gardetti
Editor

Sustainable Management of Luxury

Springer

Editor
Miguel Angel Gardetti
Center for Studies on Sustainable Luxury
Buenos Aires
Argentina

ISSN 2345-7651 ISSN 2345-766X (electronic)
Environmental Footprints and Eco-design of Products and Processes
ISBN 978-981-10-2916-5 ISBN 978-981-10-2917-2 (eBook)
DOI 10.1007/978-981-10-2917-2

Library of Congress Control Number: 2016960193

Printed on acid-free paper

This Springer imprint is published by Springer Nature
The registered company is Springer Nature Singapore Pte Ltd.
The registered company address is: 152 Beach Road, #21-01/04 Gateway East, Singapore 189721, Singapore

Preface

Christopher L. Berry in his work "The idea of Luxury" from 1994—one of the most interesting and comprehensive pieces on the concept of luxury, particularly its intellectual history—establishes that luxury has changed throughout time and that it reflects social norms and aspirations. Therefore, luxury depends on cultural, economic, or regional contexts which transform luxury into an ambiguous concept. Also, according to Ricca and Robins (2012), luxury is a source of inspiration, controversy, admiration, and considerable economic success. And, along this controversial line, already in 1999 Robert H. Frank stated in his book "Luxury Fever—Weighing the Cost of Excess" the need to minimize the culture of "excess" to restore the true values of life. And this is in line with the World Commission on Environment and Development (WCED 1987) report, Our Common Future, also known as the Brundtland Report, which defines sustainable development as the development model that allows us to meet present needs, without compromising the ability of future generations to meet their own needs. Luxury—according to Kleanthous (2011)—is becoming less exclusive and less wasteful and more about helping people to express their deepest values. So, sustainable luxury would not only be the vehicle for more respect for the environment and social development, but it will also be synonym of culture, art, and innovation of different nationalities, maintaining the legacy of local craftsmanship (Gardetti 2011).

For a better understanding, this book has been divided into six parts, as shown below:

- **Part 1**: Sustainability and Business Management in Luxury Goods Companies
- **Part 2**: Sustainable Practices Within the Luxury Universe
- **Part 3**: Sustainable Luxury Business Practices: Different Aspects
- **Part 4**: Sustainable Management in the Luxury Value Chain
- **Part 5**: Sustainable Luxury, Entrepreneurship (Including Family Business) and Craftsmanship
- **Part 6**: Sustainable Luxury and Consumption

Part I—**Sustainability and Business Management in Luxury Goods Companies**—begins with a paper by Gulen Hashmi called "Redefining the Essence of Sustainable Luxury Management: The SLOW Value Creation Model," where the author provides the luxury industry and academia with a value creation model, the SLOW (**S**ustainable **L**uxury for **O**verall **W**ell-being) that represents different types of sustainability engagement relevant to the luxury industry dynamics. In the next chapter, "The Strategic Role of Authenticity in the Luxury Business," by Patricia Anna Hitzler and Günter Müller-Stewens, the authors argue that authenticity is considered to be the major challenge for the luxury segment of our time. Moreover, authenticity is a dynamic construct with an etymology that can be traced back to the 16th century, which, in the course of its usage, has repeatedly reformatted itself. If seen from an economic point of view, authenticity first and foremost requires a match between the internal identity of an organization, as it is lived by its members, and the external image, as it is perceived by outsiders. In an empirical study, the authors analyzed the main features of the construct authenticity with experts from six companies in the luxury sector. Besides, they linked the research on authenticity with the different activities of the strategic value chain of a luxury goods company. Therefore, this chapter gives an overview of drivers and features of authenticity integrated into a luxury-specific value chain that can be used as guidance for the management.

In line with this first part, Misha Pinskarov and Joshi Nair developed "Creating a Culture of Shared Value Through Luxury Branding," a chapter where they describe the three key ideas below:

> The pursuit of shared value relies on the individual as the common element between business and society to reconcile these two institutions' competing goals.
>
> Brands consist of people, so they must align their values with those of society, define their purpose as citizens, and shape the behavior of their people to suit.
>
> Luxury brands are not just citizens but leaders, which gives them the ability to raise their stature by influencing the wider community.

Completing the first part, France Riguelle and Didier Van Caille present the chapter "Sustainable Performance Management in High-End Luxury Goods Firms: The Use of the "Reputation-Clock" Model." This work builds on the specific case of a small Belgian firm in the luxury industry. This paper describes the motivations and content of a reputational performance management tool, the so-called "Reputation-Clock" Model, which incorporates different key success factors that a firm from the high-luxury goods industry has to leverage in order to master its strategic performance from a reputation point of view.

Part II—**Sustainable Practices Within the Luxury Universe**—begins with the work titled "Sustainable Luxury and Fashion: From Global Standardisation to Critical Customisation" written by Patrizia Calefato. In her work, the author explains that fashion and luxury can create a new critical and ethical paradigm, so the chapter focuses on this new paradigm of luxury concerning fashion. The word "critical" refers, above all, to the concept of "choice." One can choose his/her clothes and can stand out by taking care of the senses, of the materials, of the time. The next chapter called "Identifying the Luxury Sustainability Paradox: Three Steps Toward a Solution" was developed by Oliver P. Heil and Daniel André Langer.

They state that many luxury products seem less sustainable than they could, and probably should, these days bringing what they call the "Luxury Sustainability Paradox" to surface. In essence, luxury product superiority appears to not span across all attributes and, paradoxically, to do so may be very challenging if not impossible. In short, being at the very top, along with certain performance attributes, may paradoxically preclude top scores on sustainability.

Moving on to the second part, Hakan Karaosman, Gustavo Morales-Alonso, and Alessandro Brunr developed "Strike a Pose: Luxury for Sustainability," a chapter where they introduce sustainability and fashion paradox. They also research into how luxury could positively change the vicious cycle of Desire—Speed—Consume—Dispose, and how this study brings a novel perspective by explaining the degree to which luxury fashion could indeed have a positive impact on sustainability through craftsmanship, heritage, and artisan skills.

The chapter called "Sustainable Luxury in Hotels and Resorts: Is It Possible?" written by Gianna Moscardo shows how the increasing affluence of countries such as India, China, and Brazil supports a growing global middle-class interested in travel, consumption, and luxury purchases. This trend presents some serious challenges for addressing a range of sustainability issues in tourism. Rising awareness of these sustainability issues has resulted in a number of luxury brands and businesses adopting sustainability practices. This chapter begins by describing this response in tourism using an approach adapted from the general sustainability analysis—Drivers, Pressures, State, Impacts, and Responses (DPSIR) framework—to understand the context for the development of the sustainable luxury tourism concept. It then examines three cases of sustainable luxury tourism experiences exploring the extent and nature of the sustainability strategies adopted. In turn, Thomaï Serdari, in his paper "Diamonds, Development, and Dollars: A Case Study on a Newly Founded Gemological Lab in Botswana Aiming at Sustainable Community Empowerment" presents the development of a model gemological lab in Botswana. This work poses questions on the nature of diamonds mining and its challenging relationship with natural resources and showcases the best practices in reinforcing sustainable value chain management by focusing on the following: production processes, employee relationships, employee training, and other related organizational issues.

And, closing this Part II, the paper titled "More than Skin Deep: A Service Design Approach to Making the Luxury Personal Care Industry More Sustainable" developed by Rachael Wakefield-Rann presents a case study of luxury personal care company LUSH, and examines how its innovative approach to service design could provide a genuinely sustainable model for luxury personal care companies, and potentially the broader industry. The core elements of this model include local production, "naked" products, short expiry dates, and innovative retail design.

Part III of this volume dealing with **Sustainable Luxury Business Practices: Different Aspects** presents only one paper titled "Promoting Sustainable Management: World Wildlife Fund's Hybrid Strategy to Change the Luxury Industry," where the author, Graham Bullock, explores the *Deeper Luxury* report (WWF) used to engage the luxury industry in 2007. This chapter analyzes the report descriptive and injunctive normative statements that positively engage the industry and its controversial rating system that directly confronts the top ten luxury firms. The chapter also examines the sustainability-related documents of the ten rated companies and finds that their responses to WWF's normative demands varied significantly, with L'Oreal, LVMH, and PPR demonstrating the greatest responsiveness. It concludes that this hybrid approach had mixed results in promoting sustainable management within the luxury industry and may require collaboration among organizations with different skills and reputations.

Part IV of the book—**Sustainable Management in the Luxury Value Chain**—begins with a paper by John Armitage, Joanne Roberts, and Yasmin Sekhon called "Luxury Products and Services and the Sustainable Value Chain: Six Management Lessons from Gucci," where the authors show how *Gucci*'s sustainable luxury products encourage the "reduction," "reuse," and, crucially, "recycling" of luxury products, transforming luxury "waste materials" into reusable luxury materials to avert misuse, reducing the consumption of new luxury materials, energy use, air and water pollution by reducing the need for established waste disposal and lower greenhouse gas production. They also explore *Gucci*'s approach to sustainability, an approach that is determined by its holding group's (Kering) use of the "Environmental Profit and Loss" (EP&L) tool. The next chapter, "Luxury Fashion Brand Sustainability and Flagship Store Design. The Case of 'Smart Sustainable Stores'" by Elisa Arrigo, shows—for the first time—how sustainable design and management of the flagship store help to minimize the environmental impact of the store's operations with an enhanced and efficient use of energy and to make stakeholders more aware of the luxury fashion brand's commitment toward environmental sustainability by safeguarding and improving its reputation.

In line with this part, Hakan Karaosman, Alessandro Brun, and Gustavo Morales-Alonso developed "Vogue or Vague: Sustainability Performance Appraisal in Luxury Fashion Supply Chains." It deals with a study called 360 Degrees Performance Appraisal model, and the main contribution is to be able to evaluate the impact of supply chain (SC) operations on sustainability. Hence, this chapter provides an understanding of (i) how SC of a luxury goods company must be configured toward sustainability, (ii) how sustainability performance must be assessed with the incorporation of a wide range of stakeholders, and (iii) how sustainability could be further advanced in luxury fashion supply chains (FSCs).

Completing the fourth part, Kyung Eun Lee presents the chapter "Application of Digital Enterprise Technology (DET) for Green Made-to-Measure in Korean Luxury Fashion Industry." This study attempted to examine Samsung Fashion Division's DET usage status and its effectiveness toward the company's green luxury business practice. More specifically, this research is about the following: (a) Samsung's DET system implementation and business process design, (b) green luxury apparel product development, (c) company's CSR store operation, and (d) DET system's impact to maintain businesses in environmentally sustainable approaches.

Part V—**Sustainable Luxury, Entrepreneurship (Including Family Business) and Craftsmanship**—begins with the chapter written by Miguel Angel Gardetti and Rachida Justo called "Sustainable Luxury Fashion: The Entrepreneurs' Vision." This paper shows the results and analysis of a study conducted for the authors to fifteen entrepreneurs of different nationalities with sustainable projects in the luxury fashion and fashion accessory sector. All these entrepreneurs participated in the IE Award for Sustainability in the Premium and Luxury Sectors. Some of them won in the category "fashion and accessories" and others reached the final stages of the award. The purpose of this study is not only to get to know what entrepreneurs understand by sustainable luxury and, particularly sustainable luxury fashion, but also to learn their vision about the future of this sector.

Following, Elisa Giacosa's work "The Family Business Phenomenon as a Means for a Sustainable Business in the Clothing Luxury Business" focuses on the family business model in the clothing luxury industry. The main goal of this chapter was to analyze how the innovation strategy implemented by family businesses operating in the clothing context could represent a means for sustainable development in the society, in terms of economic, social, and cultural aspects. Indeed, thanks to its history and reputation, the family could increase the integrity of its business, generating a positive impact on both its turnover and employment, as well as on society, at large.

Closing this part, the next chapter called "Craft's Path to the Luxury Market: Sustaining Cultures and Communities Along the Way" was developed by Jana M. Hawley and Judy Frater. The authors explain that as the luxury market became more and more democratized, it began to lose its exclusivity—an essential aspect of luxury. Crafts, those items made by human hands, are (1) unique (2) authentic to the maker, (3) sustainable to the maker and community, and (4) tactile thus providing intimacy of touch. Exceptional craftsmanship for the luxury consumer is a compilation of time, skill, passion, and ideas. Luxury consumers want to show themselves as people of good taste and style and they can accomplish this with craft products that are instilled with the human traits of personality, knowledge, traditions, and memory. Traditional crafts can capitalize the bespoke element and bring exclusivity in a new form.

The last part dealing with **Sustainable Luxury and Consumption** consists of five papers, beginning with the one by authors Vertica Bhardwaj and Sergio C. Bedford, titled "(Not) Made in Italy: Can Sustainability and Luxury Co-exist?," where the authors outline a conceptual framework that explores factors that influence how consumption of sustainable luxury is perceived. They built upon the idea that luxury brands acts as a psychological meaning of self-expression and symbolic association that individuals aspire to create or maintain. The ultimate goal of this chapter was to explore whether consumers are increasingly becoming conscious of the adverse effects of social, moral, and environmental imbalances created by luxury brands.

The next chapter, "Luxury, Sustainability, and Corporate Social Responsibility: Insights from Fashion Luxury Case Studies and Consumers' Perceptions," by Marco D'Anolfo, Cesare Amatulli, and Matteo De Angelis, delves into a currently neglected area of research—namely consumers' perceptions of the relationship between luxury, sustainability, and CSR strategies that major luxury companies adopt to counter said perceptions. To analyze luxury brands' initiatives, the authors conducted a qualitative study on three well-known brands—Gucci, Tod's, and Bulgari—and examined their eco-sustainable collections and approaches. They also supplemented this research with a survey of consumer perceptions regarding luxury, sustainability, and CSR specifically by interviewing 200 consumers via an online questionnaire.

Moving on to this last part, Nadine Hennigs, Evmorfia Karampournioti, and Klaus-Peter Wiedmann developed "Anything Worth Doing: The Ambiguity of Values in Sustainable Luxury," a chapter in which they research into the relationship between the set of core values and norms guiding consumer behavior defined as "*desirable goals, varying in importance, that serve as guiding principles in peoples' lives*" (Schwartz 1994, p. 21) with special focus on environmental orientation, on the one hand, and dimensions of customer perceived value understood as the "*consumer's overall assessment of the utility of a product (or service) based on perceptions of what is received and what is given*" (Zeithaml 1988, p. 14), on the other. In turn, Eirini Koronaki and George Panigyrakis prepared a paper titled "Lust for Our Better Versions—The Meta Modern Reality's Roots and Implications." The connection between ideal selves and ethical luxury is examined, and they ask themselves: Can luxury brands provide their customers with socially-oriented benefits? Which activities are more efficient in that direction?

Completing the last part, Duane Windsor presents the chapter "Moral Responsibility for Sustainable Luxury Consumption and Production." This chapter develops the proposition that luxury consumption and production generate special moral responsibility for sustainability—responsibility that holds for both buyers and sellers. The focus is on the responsibility imposed by wealth to make luxury consumption and production as environmentally friendly as possible.

It is important to highlight that all of these diverse contributions represent a great step forward in expanding the insights in the field of sustainable management of luxury. It is certainly the most comprehensive collection of writings on this subject to date. Note that this initiative has received a large international response, and it is expected to continue to stimulate further debate.

Buenos Aires, Argentina | Miguel Angel Gardetti

Bibliography

Berry CJ (1994) The idea of luxury—a conceptual and historical investigation. Cambridge University Press, New York

Gardetti MA (2011) Sustainable luxury in Latin America. In: Conference delivered at the seminar sustainable luxury & design within the framework of IE, Instituto de Empresa, Business School MBA, Madrid, Spain

Kleanthous A (2011) Simple the best is no longer simple—*Raconteur on* sustainable luxury, July, 2011:3. Source: http://theraconteur.co.uk/category/sustainability/sustainable-luxury/. Accessed Dec 2012

Ricca M, Robins R (2012) Meta-luxury—brands and the culture of excellence. Palgrave Macmillan, New York

Schwartz SH (1994) Are there universal aspects in the structure and contents of human values? J Soc Issues 50(4):19–45

World Commission on Environment and Development—WCED (1987) Our common future. Oxford University Press, Oxford

Zeithaml VA (1988) Consumer perceptions of price, quality, and value: a means-end model and synthesis of evidence. J Mark 52(3):2–22

Contents

Part I
Sustainability and Business Management in Luxury Goods Companies

Redefining the Essence of Sustainable Luxury Management: The Slow Value Creation Model

Gulen Hashmi

Abstract This chapter aims to provide the luxury industry and academia with a value creation model, the SLOW (Sustainable Luxury for Overall Well-being) that represents different types of sustainability engagement relevant to luxury industry dynamics. The SLOW model focuses on sustainability strategies and sustainable practices of luxury goods/services companies based on a value creation model that takes into consideration both shareholder value and social value. The vertical axis is built on the luxury goods/services company's need to manage the business with added shareholder value while the horizontal axis is built on the company's need to create social value, as necessitated by today's societal and environmental challenges. The four plots that represent different types of sustainability engagement in the luxury industry are Self-indulgent Luxury, Refined Luxury, Thoughtful Luxury, and Sustainable Luxury for Overall Well-being, respectively. The SLOW value creation model is a theoretical contribution that helps understand and differentiate between four different types of sustainability engagement that lead to varying degrees of shareholder value and social value. The model sets the stage for further empirical research in the sustainable management of luxury.

Keywords Sustainable luxury · Value creation · Value redistribution · Shareholder value · Social value · Sustainability engagement · Self-indulgent luxury · Refined luxury · Thoughtful luxury · Sustainable luxury for overall well-being

1 Introduction

Luxury and sustainability are two words one may not particularly expect to find in the same sentence at first glance. To many people, "sustainable luxury" is a term that might best be found in the dictionary under the entry for oxymoron. The word

G. Hashmi (✉)
Business School Lausanne, Lausanne, Switzerland
e-mail: gulen.hashmi@bsl-lausanne.ch

M.A. Gardetti (ed.), *Sustainable Management of Luxury*,
Environmental Footprints and Eco-design of Products and Processes,
DOI 10.1007/978-981-10-2917-2_1

luxury derives from the Latin word "luxus" and often carries with it connotations of excess and waste, and it is associated with fashion, an industry prone to fads that change quickly. Traditionally, luxury may be defined as an inessential, desirable item that is expensive or difficult to obtain. Thus, luxury goods are defined by reference to great disparities in income or wealth that is in itself a pervasive symptom of unsustainability. On the other hand, according to the Brundtland Report, sustainable development is, by definition: "*meeting the needs of the current generation without compromising the ability of future generations to meet theirs*" (WCED 1987). Indeed, the two concepts seem utterly paradoxical. Luxury is argued to be the polar opposite of sustainability since it is superfluous, conspicuous, and excessive devoid of any utilitarian use (Guercini and Ranfagni 2013). These facts make the term notoriously difficult to define (Doran 2013).

It's true that the luxury industry has been criticized by sustainability advocates for the opaque traceability of most of the gold, diamonds, rubies and other gemstones coming from under-developed countries; for taking advantage of poor labor conditions in developing countries where some luxury brands have their products made; for water use in cotton production; or for resort maintenance of luxury hotels. The reason for this is that luxury brands have always been addressing more desires than necessities, and in this sense, it has looked somehow hypocritical for the industry to engage in sustainability (Friedman 2010). In a 2014 survey of 1618 consumers by Salon, consumers put the luxury industry last in a ranking of industries associated with sustainable commitments, ranking lower than the financial and petrol sectors.

Yet, a paradigm shift is currently taking place in the domain of luxury as consumers in all social classes are increasingly concerned about social and environmental issues, and prefer ethical and green products that reflect their own values and beliefs. Evidence shows that 57% of wealthy Americans are willing to pay more for socially responsible and environmentally friendly goods (Lochard and Murat 2011), and this trend is also valid for two-thirds of Chinese luxury clients (Lau 2010). Thus, there have recently been positive efforts within the sustainability movement to redefine luxury as something that embodies social and environmental credentials of a product or service. More strikingly, a 2010 Havas Media sustainable consumption research carried out with 20,000 consumers in ten markets covering more than fifty brands across eight different sectors, revealed that 86% of people considered sustainability issues when deciding what to buy and that with the same conditions of price and quality, 80% of people under the age of 35 would choose the sustainable option. This simply shows that people expect large companies to be involved in social issues and their quality of life.

If we go beyond appearances and return to the very definition of luxury, we can easily say that sustainability is part of the luxury DNA. Sustainable products also share the same "essential" qualities of luxury goods. They call for extraordinary creativity and design, they need to be made from exceptional materials, with good quality and they are rare and expensive. It seems that the two concepts are not that far apart from each other. This awareness has led to the concept of "sustainable luxury—a return to the essence of luxury with its ancestral meaning, to the thoughtful purchase, to the artisan manufacturing, to the beauty of materials in the broadest sense, and to the respect for social and environmental issues" (Girón 2009). And the discussion of sustainable luxury has finally moved on from creating feel-good initiatives to recognizing the potential for creating business value, highlighting the need for sustainable management of luxury. A good example is the Sustainable Luxury Working Group made up of leading luxury brands that are committed to advancing good social, environmental, and animal-welfare practices in their business operations, including sustainable sourcing practices. This can be considered as a tipping point for the luxury industry. Luxury is now in the spotlight of sustainable development.

Since luxury brands create desire through innovative design, influence consumption processes, and embody preciousness in experiences rather than products, they can become leaders in sustainability. A 2013 Leverhulme International Network research on luxury found that millennials aspire to justify their luxury purchasing in terms of investment in health, well-being and security, seeking the value beyond the product itself. It appears that the concept of luxury across society has transitioned from the material to immaterial. The idea of "sustainable luxury" is becoming an integral part of leading luxury brands' brand image in the sense it is associated with the possibility for people to express their deepest values in a justifiable way with social and ecological concerns and some degree of positive social value creation. This is clearly reflected in the WWF report *Deeper Luxury* (Bendell and Kleanthous 2007) where authentic luxury brands are defined as "*those that provide the greatest positive contribution to all affected by their creation and that identify their consumers as having the means and motivation to respect both people and planet*".

The luxury industry, also because it is so profitable, has a special obligation to harness its potential to create the most environmentally and socially responsible products. Thus, as a high potential, promising sector, the luxury industry has a bigger responsibility towards society in the creation of environmental and social value. Over the past five years, the global fashion and luxury industry has delivered extremely strong value creation, due to improving consumer confidence, an emerging middle class in developing markets, and a willingness among shoppers to trade up to higher-priced goods. According to a 2014 Boston Consulting Group annual study of Total Shareholder Return (TSR)[1] among 1620 publicly traded companies in 26 industry sectors, the fashion and luxury companies in the sample returned an annualized TSR of 35%, making the sector the first among the five groups of consumer companies studied—and first among all 26 industry sectors (The BCG 2014 Consumer Value Creators Series 2014). This reflects the strong rebound in consumer confidence since the depths of the financial crisis, and it shows that investors have high expectations for the continued performance of fashion and luxury companies in the future.

[1]Total Shareholder Return (TSR) is a metric that encompasses all sources of value that accrue to shareholders, such as changes in sales, margins, and valuation multiples, along with all sources of free cash flow to investors and debt holders, such as changes in dividends, net debt, and the number of shares outstanding.

Furthermore, it shows that luxury companies have a major impact on consumer behavior, as well as a considerable transformative potential in other industries (Bendell and Kleanthous 2007). Their powerful current impact and potential could be harnessed to promote environmentally and socially responsible consumption behavior patterns both to luxury and mass consumers, and stimulate the offer of eco-products and environmentally friendly design and conception if those are unique and desirable in the mass market. Thus, as a trendsetter and strong value creator, the fashion and luxury industry, compared to other industries, can create a larger impact on educating consumers to be less wasteful.

The conceptualized value creation model, the SLOW (Sustainable Luxury for Overall Well-being), aims to contribute to the literature on sustainable management of luxury in the context of value creation, drawing attention to different sustainability engagements of the luxury industry with regard to shareholder value and social value. In the quest for sustainable management of luxury, the industry needs to first start with sustainability challenges and create social value in the form of social and environmental excellence while managing the business to deliver shareholder value at the same time. Although some frameworks and models have taken shareholder value into consideration in corporate sustainability (Elkington 1997; Orsato 2006; Nidumolu et al. 2009; Porter and Kramer 2011), none of these theoretical works except for the S-WELL Grid (Hashmi and Muff 2015) has looked into types of corporate sustainability in light of both shareholder value and social value. As such, the SLOW model might be a useful instrument for luxury brand managers to successfully create and sustainably manage luxury brands that also contribute to social value creation with well-being in mind.

The remainder of the chapter continues with a section on the holistic concept of "well-being" and its crucial role in the creation of social value. Then, the SLOW (Sustainable Luxury for Overall Well-being) value creation model is introduced, highlighting four different types of sustainability engagement for value creation, widely seen in the luxury industry. The section gives examples of famous luxury brands that portray these different types of sustainability engagement, highlighting the need for "slow yet meaningful value creation" that would better align with the timelessness and quality the luxury industry is built upon. The chapter concludes with implications of the proposed model for the luxury industry.

2 From a Mechanistic to a Holistic Worldview Through "Well-Being"

> Two concepts will determine the course of civilization in the next millennium: Relativity and Holism.
>
> Albert Einstein

Capitalism as practiced today is in question for good reason. Fullerton (2015) explains this as the following: "As the mechanistic, equilibrium-based theoretical construct for today's dominant form of capitalism, neoliberal economics assumes that the primacy of the individual and societal prosperity can be achieved through the operations of unrestricted free markets that efficiently allocate resources, presumably maximizing what economists call the 'utility' of the participants in the system". These unquestioned beliefs rest on the assumption that the economy is separate from the biosphere and the environment, and that profit maximization for shareholders, growing GDP, and optimization of consumer material utility all lead to wealth. However, the idea poses a clear and present danger to the health of human communities and all life on Earth because a firm, with its customers and employees, depends on the greater whole of society as well as on the biosphere and environment.

Contrastingly, today's newly emerging scientific worldview—originally called holism and now referred to under titles ranging from systems theory and complexity to the ecological or living systems worldview—assumes that whole systems can only be understood through the dynamic relationships among all parts (Fullerton 2015). The concept of holism was first defined in 1926 as "the tendency in nature to form wholes that are greater than the sum of the parts through creative evolution" (Smuts 1926). The environment is us, or more accurately, we are a part of the greater whole that is the environment and beyond (Hashmi and Muff 2014). So, economy, literally the "management of the household" is inseparable from "ecology" which means "the study of the household". This demands that we shift from a competitive, mechanistic worldview to a more collaborative, socio-ecological worldview, just like life in mature and sustainable forests, which is collaborative at its core. Thus, a key issue to become a sustainable business is the integration of different actions and sectors of society, economy and environment, taking a holistic view and overcoming barriers between disciplines (Hashmi and Muff 2014). It is indeed this interconnectedness between planetary resources, human systems and the goal of human well-being that requires holism, and makes it difficult to manage towards sustainability (Thanh and Bosch 2010).

A purpose of 'well-being" possesses almost all aspects of a holistic worldview. Well-being is best thought of as a dynamic process, emerging from the way in which people interact with the world around them (Foresight Mental Capital & Well-being Project 2008). It is captured in four major aspects: overall satisfaction with life, education, health, and social justice (DEFRA 2010). As such, well-being goes hand in hand with the concept of sustainability, which takes into account the broader world, namely the environmental and social impacts of businesses. As well-being can be measured based on what people value presently in contrast to the intergenerational influences of sustainability, it may be a healthy indicator of future well-being. A 2011 research by the UK ONS found the key issues of societal well-being to be social relationships, environment, security and health (physical and mental), which can be associated with the 3P's of sustainability—profit, people and planet. In this regard, viewing sustainable management of luxury as a means of achieving overall well-being, rather than an end itself, could be easier to conceive.

While academic debate continues on how "well-being" should be defined, UK's New Economics Foundation (NEF) defines it as "the extent to which people experience happiness and satisfaction, and are functioning well" (NEF 2011). Thus, well-being is most usefully thought of as the dynamic process that gives people a sense of how their lives are going, through the interaction between their circumstances, activities and psychological resources or "mental capital" (ONS 2011). The concept of "well-being" fits well within the luxury industry, mainly because luxury is rooted in social norms and aspirations—the intangible, and is believed to be a sign of prosperity, power, social status and well-being. Recent research shows that among luxury buyers, differentiation and class will now come from environmentally friendly and socially responsible purchases (Kapferer and Michaut-Denizeau 2015). In this respect, the four aspects of "well-being" could serve luxury buyers well, as buying products with additional environmental and social value that contribute to well-being could help justify their expensive purchases and offset their guilt. Luxury buyers who perceive the functional value of their luxury purchases higher than their high price would certainly feel happy, in light of their overarching assumption that, because of their high prices, luxury brands must have taken care of sustainability already.

The luxury industry appears to be promising in the creation of social value for overall well-being. In the context of overall satisfaction with life, namely, sense of pleasure and feeling of happiness, luxury brands offer people fulfillment of their individual desires by introducing better consumer choices—rather than making them feel guilty about themselves. Luxury brands have the ability to make sustainable behavior an aspirational mainstream lifestyle choice, which is a much more effective way for creating social value in the form of social change than simply using guilt to make people feel bad about the environment. Contrary to the popular belief that luxury purchases are exclusively based around the notions of reputation, image or social standing, various studies have shown that these types of purchases can also act as gestures that translate authenticity and caring into existing relationships (Miller 2001: 229; Thorpe 2010), as well as engagement into new relationships (Sundie et al. 2011). Thus, forward-thinking luxury companies operate not only on "face value", but also on a social platform on which they build their sustainable communication strategies (Fowles 1996: 167).

In the context of education, luxury brands offer opportunities for artisans of the world to regain their dignity; contribute to creating awareness of responsible products through sharing of know-how; and help develop respect for traditional knowledge. This would mean contribution to culture, art and innovation of different nationalities, and build upon the legacy of local craftsmanship (Gardetti 2011). In the context of health, the health impact of toxic residues present in many textiles and cosmetics are becoming more and more of an issue for consumers that are increasingly willing to pay more for eco-friendly and ethical products. This presents opportunities for luxury brands to develop methodologies for well-being, develop products made from species of plants that can be used for healing while protecting biodiversity, and contribute to healthier working conditions in their supply chains through health programs. "Slow food", for instance, has emerged from a drive for health and well-being and is the fastest growing segment that has led to a new luxury food category (Altagamma Fondazione 2012). Finally, in the context of social justice, luxury brands can offer long-term contracts and fair wages to indigenous people, thus bring back dignity to the trade. They can, through various partnerships with NGOs and policy-makers, help create jobs amongst under-developed countries' most impoverished women—single mothers, widows and HIV/Aids victims. Thus, forward-thinking luxury companies that prioritize social value creation and well-being are highly likely to contribute positively to society, and further increase sustainable business value.

Deriving from a holistic worldview with the ultimate purpose of creating both social value and business value in service of well-being, the SLOW value creation model in the following section, aims to serve as a practical guide for the luxury industry that aspire to sustainably manage value creation. As such, the term "sustainable luxury for overall well-being" indicates luxury brands that sustainably manage luxury for the well-being of all stakeholders.

3 The "Slow" Value Creation Model

The SLOW value creation model builds on the True Business Sustainability Typology (Dyllick and Muff 2013), which is based on concerns (inputs), organizational perspective (processes), and the type of value created (outputs) as typical elements of a business process. The typology highlights that while today's neo-liberal economics requires one-sided economic concerns such as turnover, market share, profit and shareholder value, sustainability requires a socio-ecological mindset that takes into account social, environmental and economic concerns. The authors assert that true business sustainability starts with sustainability challenges such as human rights to water and sanitation, nutrition, climate change, social justice and gender inequality; and works its way towards the capabilities and resources of the firm in an effort to align these challenges to organization's sustainability goals. Thus, true business sustainability possesses an outside-in business perspective, as a differentiator to other types of business sustainability. Finally, the authors highlight the need for active contribution to positive value creation, rather than reactive contribution in the form of negative impact minimization. The typology serves as a relevant framework to clarify the drivers and purpose of business sustainability (Hashmi and Muff 2015).

The SLOW value creation model further expands the True Business Sustainability Typology on one of its three elements of the business process model: type of value created. The model redefines sustainable management of luxury with regard to both shareholder value and social value, shareholder value on one axis and social value on the other axis. The vertical axis is built on the luxury companies' need to create shareholder value while the horizontal axis is built on the luxury companies' need to create social value for overall well-being. Taking into consideration the nature and the current sustainability practices of the luxury sector, the four plots of the SLOW model distinguish between "self-indulgent luxury", "refined luxury", "thoughtful luxury" and "sustainability luxury for overall well-being (SLOW)", which represent different types of sustainability engagement relevant to

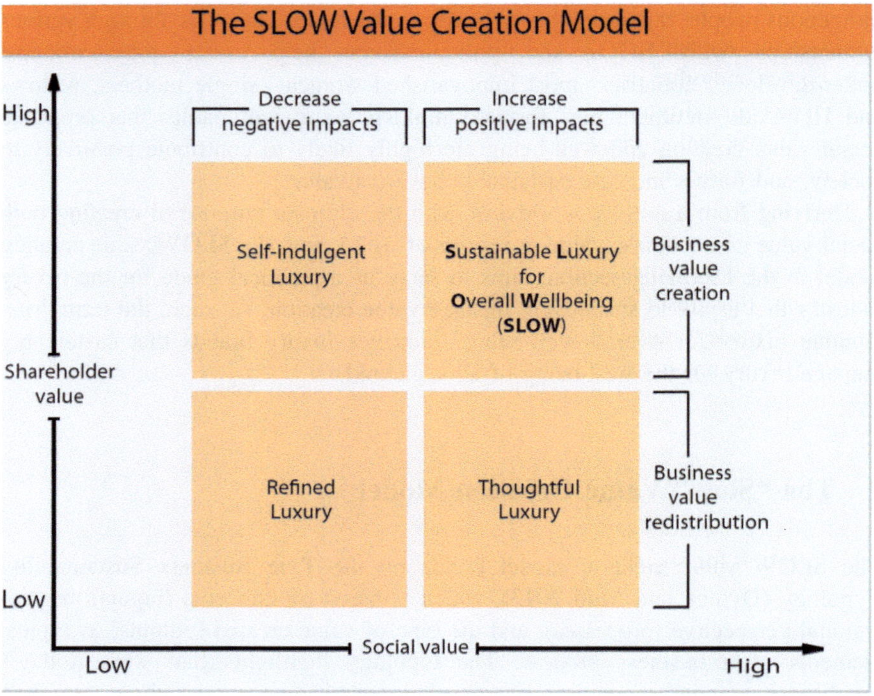

Fig. 1 Sustainable Luxury for Overall Well-being (SLOW) value creation model, highlighting different types of sustainability engagement in the luxury sector

the luxury industry (Fig. 1). While "self-indulgent luxury" and "refined luxury" largely focus on decreasing negative impacts of business, "thoughtful luxury" and "sustainable luxury for overall well-being" focus on increasing positive impacts of business operations on social value. The model also distinguishes between business value creation and business value redistribution. While "self-indulgent luxury" and "sustainable luxury for overall well-being" largely contribute to business value creation, "refined luxury" and "thoughtful luxury" largely contribute to business value redistribution.

Plot "Self-indulgent luxury" implies "The business of business is business" Friedman (1970) with some reaction to social and ecological concerns, yet shareholder value creation remains the only priority. Plot "refined luxury" characterizes sustainability engagement that goes beyond the one-sided goal of shareholder value creation and creates simultaneous social value through implementation of sustainability strategies and plans. Plot 'thoughtful luxury' is where donations to worthy social causes and volunteering are largely at play, with little regard to shareholder value creation. Finally, plot "sustainable luxury for overall well-being (SLOW)" is where relatively bigger social value and bigger shareholder value are created in a slow yet meaningful way. This is where luxury companies that sustainably manage luxury address social issues without forgetting that, at some point, they have to become profitable and self-sufficient. Luxury companies that fall into this strategic sustainability engagement type form various collaborations with NGOs, government bodies and other private companies to change the rules of the game in solving specific sustainability challenges.

The SLOW value creation model refers to 'slow value creation', not only because value creation process is slow due to various collaborations that take time to change the rules of the game for significant impact, but also because it is connected to the "slow food" movement that originated in Italy. This type of sustainability engagement emphasizes creation of value largely for society's overall well-being, thus rests on the principle that each element of the value creation process is made regionally with value created to improve local craftsmanship, preserve local resources, and develop local clusters through collaborations for capacity-building of the luxury industry, industry-related infrastructure improvements, etc.

Now, let's walk through each type of sustainability engagement represented in the SLOW plots. The plots challenge luxury companies to think about which sustainability practices and strategies they should pursue with regard to sustainable value creation for society and for themselves.

3.1 The First SLOW Plot: Self-indulgent Luxury

This type of sustainability engagement starts with the assumption that economy is separate from the biosphere and the environment, and that maximizing profits for shareholders, growing GDP and optimizing customer utility all lead to prosperity for everyone. Another assumption behind this engagement is that when markets are fair and transparent, they will efficiently allocate scarce resources, irrespective of temporal considerations. Indeed, both of these assumptions are grounded in a mechanistic worldview that ignores the harm done to other parts of society and the biosphere. Today, this is the case with many self-indulgent luxury businesses, majority of which are well-established international brands. These brands rarely go beyond incremental innovation since they are anchored to the usual neo-liberal economic mindset. Indeed, these established brands' progress towards sustainable management of luxury is slow (Bendell and Kleanthous 2007; Doyle and Bendell 2011). In a study of 20 luxury brands, which were assessed under a series of criteria such as environment (gold and diamonds, furs, testing on animals, etc.), rights of workers and activity in politics, it was found that, in general, the luxury industry reacts to what market and consumers demand, rather than act proactively to sustainability challenges (Gardetti and Girón 2014).

Enjoying high profit margins, self-indulgent luxury brands believe that optimizing near-term shareholder value should be a firm's primary goal (Friedman 1970), and societal or ecological concerns must not distract private companies. Such shareholder primacy mindset derives from the view that non-shareholders are the "means to the ends". Luxury brands with this type of engagement react to some social and ecological concerns through compliance, although economic objectives remain the overarching priority. They recognize that sustainability management can help them save costs and reduce risks, yet they do not go far enough to harness sustainability to prove their attractiveness as an employer, differentiate them from their competitors, innovate new products or markets, and change rules of the game in the industry. They very much need the motivation to create disruptive type of change to advance to transformational behavior. Such a cost and risk reduction approach to profit maximization and increased shareholder value with secondary regard to sustainability value as social value can be considered a reactive approach towards management of sustainability (Dyllick and Muff 2013).

Competitiveness through cost and risk reduction derives from the resource-based view of business, which is built on the argument that business will engage in sustainability practices for inimitable competitive advantage (Hart 1995). A cost and risk reduction approach leads to eco-efficiency strategies, which reduce the use of resources that ultimately result in shareholder earnings, and reduction in potential losses (Hart and Milstein 2003). Eco-efficiency within and beyond the firm is a common environmental strategy used with this type of sustainability engagement, if any. As McDonough and Braungart (2000) assert, companies with such eco-efficiency orientation try to reduce the negative effects on the environment through use of resource reduction such as energy use, water and emissions, as well as reduction of landfill waste and toxic releases. Yet, reducing and recycling do not create value for society, let alone stop environmental degradation and destruction. Rather, radical innovation and product redesign are crucial strategies for long-term sustainability (McDonough and Braungart 2000).

The luxury giant Louis Vuitton Moët Hennessy (LVMH), a corporation comprising over sixty luxury brands, showcases a good example of this type of sustainability engagement, in that LVMH has had an environmental charter since 2001, and has been auditing its carbon footprint since 2004. More recently, in July 2013, the company acquired an 80% stake in the Italian business family Loro Piana, to strategically be able to control a century-old, prestigious brand over vicuna wool, a valuable and scarce resource. LVMH was fully aware that continued supply of raw materials for the luxury industry depends on well-managed populations of plant and animal species, both in the wild and elsewhere, such as farming operations. The overarching driver of this acquisition was to minimize resource risks and the desire to create strategic business value for their brand, with some regard to creating social value for all stakeholders involved. Similarly, Burberry's decision to close factories in Britain in 2007 and relocate manufacturing to China was driven by cost reductions for profit maximization.

Interestingly, numerous luxury brands such as Prada, Hugo Boss and Dolce and Gabbana have been highlighted in a recent Clean Clothes Campaign report on conditions in the "Euro-Mediterranean textile cluster". One striking finding of the report was that the gap between legal minimum wages (the legal minimum a person can be paid) and minimum estimates for living wages (the minimum pay rate needed for a person to lead a decent life) tended to be even larger in Europe's cheap labor countries such as eastern Europe and Turkey, than in Asia. Understandably, profit depends on companies maximizing the difference between a garment's sale price and its production cost, yet some accredited luxury brands can currently manage to exist outside of this system by paying higher living wages for happier staff who feel valued, which ultimately leads to lower staff turnover and cost savings on training (Living Wage Foundation 2014).

High shareholder value with some interest in decreasing negative impacts is strikingly at play here. Thus, self-indulgent luxury brands strongly believe that addressing societal or ecological concerns is government's responsibility, and compliance with government rules and regulations would be provided should they be imposed on them. Yet, it is more desirable for businesses to have less regulation in order to have more freedom in decision-making to be able to meet market and social factors (Bronn and Vidaver-Cohen 2009). Going beyond compliance can indeed give businesses huge opportunities for out-designing the competition (McDonough and Braungart 2000). Some luxury brands have already sensed the business opportunities inherent in social and environmental issues, which are, when coupled with discerning customers, good margins and international visibility, make sustainable management of luxury aspirational. Thus, luxury brands have started becoming more proactive and indeed more vocal about sustainability initiatives that extend beyond run-of-the-mill environmental impact reduction. Some luxury brands have already started shifting from pure compliance to decrease negative impacts, to the integration of sustainability as a cornerstone of their operations.

This brings us to another type of sustainability engagement "refined luxury", which is also characterized by reduction of negative impacts, yet largely through redistribution of business value among economic, environmental and social value types, rather than creation of business value only.

3.2 The Second SLOW Plot: Refined Luxury

Plot "refined luxury" represents another type of sustainability engagement whereby luxury brands rethink their one-sided goal of profit maximization and pursue a Triple Bottom Line (TBL) approach. TBL value creation is based on the strategic thinking that a company's ultimate success or well-being can and should be measured not just by its business value, but also by its social and environmental performance (Norman and MacDonald 2004). As such, the concept of TBL measures the multidimensional business contributions to sustainability (Elkington 1997). Luxury brands with this type of engagement are not only concerned about economic goals but also social and environmental goals as opportunities for creating business value. As such, a higher mark-up that is a characteristic of "self-indulgent luxury" sustainability engagement, gives way to higher standards with this type of engagement, which are costly to implement. While environmental goals relate to minimizing negative impacts to a great extent, social goals most often relate to extra philanthropic CSR activities that fall outside of the luxury brands' core business with the expected benefit of image improvement and other reputation impacts.

They often differentiate and innovate within the boundaries of their business models, harnessing the distinguishing principle of sustainable luxury products "buy less and pay more". Sustainability actually represents an opportunity for differentiation and innovation within the luxury industry. As Hart and Milstein (2003) assert, sustainability is a quest for creating synergistic value through sustainable practices, with focus on differentiation and innovation. Such environmental product differentiation would create greater environmental benefits or impose smaller environmental costs, compared to similar products (Orsato 2006). Orsato (2006) asserts on his theory of eco-branding that this environmental differentiation in terms of products and services could further provide competitive advantage. This supports Nidumolu et al.'s (2009) assertion that "becoming sustainable" is an outcome of developing sustainable offerings or redesigning existing ones to become eco-friendly.

Some luxury brands with "refined luxury" engagement are even envisaged to go further to develop new business models, to create value that will change the basis of competition. Although raised operating costs would be the case, such differentiation could justify a company's price premium (Reinhardt 1998), by simply satisfying luxury customers' willingness to pay a premium for sustainable products (Blanco et al. 2009). Bendell and Kleanthous (2007) argue that the investment required to improve sustainability performance will be repaid because of increasing demand from luxury consumers for more sustainable options.

Luxury brands with "refined luxury" engagement embed TBL concerns of profit, people and planet into their supply chain and establish sustainability strategies and plans through sustainability management systems and proper sustainability reporting. Programs and activities are managed, measured and reported through integration of sustainability goals into management and governance structures such as cross-functional committees, policies and guidelines. Internal reporting includes differentiated TBL activities and results with improved TBL value creation. An evaluation of 56 academic studies by Deutsche Bank revealed that companies with high ratings for ESG (environmental, social and governance) factors have a lower cost of debt and equity. In 89% of the studies reviewed, companies with high ESG ratings outperformed the market in the medium (three to five years) and long (five to ten years) term (Fulton 2012). This implies that integrating sustainability into business operations contributes to increased shareholder value.

Value creation goes beyond shareholders and takes into account other stakeholders who are also indirectly involved in the business itself. This stakeholder perspective is built on the premise that interests of many non-shareholders are also viewed as "ends", rather than "means". Value is largely created by clean traceability, reaching beyond the tangible beauty. Luxury brands with this type of sustainability engagement use traceability as a starting point for creating value along their supply chain. For these luxury brands, sustainable luxury means clear traceability. Traceability can be defined as the journey to the past to understand where a product comes from, its components, raw materials and their extraction (Gardetti and Girón 2014). For Elvis and Kresse and Bottletop, for instance, their approach to traceability is to use waste material from fire hoses and ring pulls from drink cans, respectively. For We are Knitters, Big Blue Bike and Pachacuti, traceability is about using wool, Panama straw or alpaca fibres, produced or grown sustainably whereby local communities are involved to a great extent. The launch of the first fully traceable luxury eco-bag by Gucci in 2013 is another initiative in this regard. The luxury eco-bag is accompanied by a passport that certifies the leather's provenance from ranches in Brazil that do not cause deforestation, the work of craftsmen in Gucci's factory in Florence, and the origins of the handle made from fast-growing bamboo (Armstrong 2013).

These luxury brands start out from the assumption that if they give visibility to the supply chain of their products, they thereby reveal their strongest values (Hockerts and Wustenhagen 2009). For that reason they revalue the local culture and indigenous people, and respect both direct and indirect stakeholders. Indeed, the role of values is very important in that idealistic values regarding social and environment goals can be translated into value of economic assets (Dixon and Clifford 2007). As such, there have been some attempts to set minimum standards for the sourcing of raw materials in the luxury industry, such as The Sustainable Luxury Working Group's industry-wide animal sourcing principles developed by its members such as Burberry Ltd., Cartier, Gucci and LVMH (BSR 2011), leading to supply chain transparency of some luxury brands.

However, the sector still has significant ground to make up in the field of sustainability. A 2013 PricewaterhouseCoopers (PWC) sustainability report on the ASEAN region, which accounts for a large share of wildlife-sourced products used in the luxury industry, shows that the luxury industry still has a lot to cover regarding sustainable management of luxury. The report highlighted that sustainability is still not measured in the supply chain of 80% of companies in the report (PWC 2013: 5). Similarly, according to a 2012 Swiss-based private Bank Sarasin report, the challenge of luxury brands is to develop sustainable supply chain management systems for traceability and more transparency (Ashida 2012). Yet, no single company is able to tackle such a multi-dimensional challenge alone, in isolation from its environmental and social context—namely, without a holistic approach (Hashmi and Muff 2014).

Thus, the relational view of a firm comes into play with this type of sustainability engagement. The theory of the relational view highlights the importance of relational networks, network resources, and dynamic capabilities as these play a crucial role in creating competitive advantage. In this concept, collaborative actions are considered key determinants of competitive advantage, and these collaborations are required to be adopted internally and externally at inter-firm level (Daveu 2014; Jackson 2011). These collaborations, because they are considered specialized knowledge, lead to inimitable capabilities, thus possible sources of competitive advantage (Conner and Prahalad 1996; Dyer and Singh 1998; Barney 1999; Duschek 2004). The luxury menswear brand Ermenegildo Zegna's collaboration with Loro Piana—the luxury textile mill, is a good example of such inter-firm synergy creation. In this collaboration, both companies have committed resources by supporting Peruvian shepherds in raising Vicuna, an endangered species animal that produces wool of exceptional quality. Thus, they simultaneously contribute to the sustainability of diverse species and shepherds' livelihoods, as well as secure the supply of wool for their suiting clothes (Bendell 2012; Kahn 2009), creating shared value (CSV).

The concept of creating shared value can be considered a recent development of TBL value creation strategy that utilizes the relational view of a firm to a great extent. Creating shared value is achieved through reconceiving products and markets, redefining productivity in the value chain and enabling local cluster development in the form of local, industry and inter-industry collaborations (Porter and Kramer 2011). CSV rests on the principle of creating economic value in a way that also creates value for society by addressing its needs and challenges (ibid). Thus, optimization of economic, social and environmental value is achieved linking corporations to society at large, which is the polar opposite of shareholder value management (Friedman 1970; Rappaport 1998). This balanced approach to sustainability recalls the wisdom of one ancient philosopher, Aristotle, in his description of and his desirability for achieving the "Golden Mean"—nothing to excess (Aristotle, n.d.).

Although additional value is created to some extent, by offering sustainable products and services and communicating them in ways that empower and educate consumers, which in turn strengthens the brand and the company's reputation (Porter and Kramer 2006; Ottman 2011), this additional value is created through the issues and concerns that emphasize "the business case for sustainability", thus business value (Dyllick and Muff 2013). In this framing, issues such as global warming, food scarcity, decreasing biodiversity, a health care crisis, and just about every other environmental and social issue are considered through the profit-making calculus, rather than as a matter of corporate or shared responsibility. Such an inside-out perspective limits value creation in the sense that tackling social and environmental issues becomes confined to an organization's resource capabilities and its business case for sustainability.

Kering can be considered a leading best practice in terms of "refined luxury" sustainability engagement. Kering's five-year social and environmental plan for its key luxury brands is, as yet, the most ambitious. As part of the group's overall Profit and Loss account, Kering's multi-tiered action plan focuses on the reduction of carbon-dioxide emissions, waste and water, sourcing of raw materials, hazardous chemicals and materials, and paper and packaging. Kering's code of business practice at the group level includes guidelines for leather, exotic skins and fur. Specifically, the group aims to phase out PVC entirely, source 100% of its paper from certified sustainably managed forests (sourcing a minimum 50% recycled content) and evaluate key suppliers once every two years, by 2016. Kering aims to eliminate all hazardous chemicals from its production by 2020.

In terms of social value creation, Kering is engaged in both philanthropic CSR and strategic CSR. Within Kering's luxury portfolio, social value integration is more advanced at some brands than it is at others, but the group's overall CSR objectives look cohesive overall. Gucci, for instance, has emerged as one of the leaders, through its philanthropic activities such as charity, donations and sponsorships. In terms of strategic CSR, Kering contributes positively to capacity building of the luxury industry, thus creates social value to some extent. The luxury conglomerate has signed a five-year partnership with the London College of Fashion lending support to fashion education on sustainability issues. It has set up a laboratory in order to develop sustainable solutions regarding materials and manufacturing.

With this type of sustainability engagement, motivation for increased business value still overweighs desire for increased social and environmental values. Thus, business value is largely redistributed, rather than created solely through sustainability challenges. Although luxury brands with "refined luxury" work up and down their value chain and collaborate closely with external stakeholders, social value is yet to be created in a holistic way—without distinguishing between environmental, social and economic values separately, and with an external focus of seeing the organization as an inter-independent figure in the economy.

3.3 The Third SLOW Plot: Thoughtful Luxury

In "Thoughtful luxury", Corporate Social Responsibility (CSR) projects play a dominant role in the creation of social value. Having reached some operational efficiency through compliance, luxury brands with this type of sustainability engagement direct their attention to what society wants from them and what social value they can add to society through voluntary CSR initiatives. Luxury brands start to delve into giving back to society, specifically on the social issues they prefer. This is largely because CSR is voluntary compliance rather than a well-planned solution to social challenges. Most luxury brands with this type of sustainability engagement portray philanthropic CSR activities, which consist of donations to worthy social causes, charity and volunteering. In most cases, these activities fall outside of the company's core business rather than enlarge core business or develop new business. Such philanthropic CSR engagement creates only marginal social impact because activities most often do not strategically align with the company's core business (Porter and Kramer 2006). Thus, the engagement remains reactive and non-strategic, as the underlying purpose is to seek reputational benefits and legitimization for business concerns. To sustain economic viability and create business value, CSR needs to be strategic (ibid). Since CSR is about business value appropriation through giving financial resources with no positive repercussion (Aupperle et al. 1985), it can be considered a source of competitive disadvantage.

Luxury brands with high CSR orientation seek societal legitimization in the eyes of stakeholders, which, once achieved, may lead to enhanced reputation and growth of shareholder value (Hashmi and Muff 2015). However, there is still lack of empirical research on the potential association between CSR and shareholder value. Although search for increased shareholder value is largely rooted in the belief that there is a relationship between CSR and CFP (corporate financial performance) (Carroll and Shabana 2010), various studies have shown positive, neutral and negative associations in the relationship between CSR and CFP, making extant research inconclusive (Margolis and Walsh 2001). Hart and Milstein (2003) assert that as the theory of legitimization extends to numerous external stakeholders in the value chain, stakeholders' interests are likely to affect or can be affected by the achievement of an organization's objectives.

When it comes to philanthropic CSR, Gucci can be considered a leading best practice. Gucci is particularly active in philanthropy relating to women's empowerment and girls' education. To date, Gucci has contributed over $10 million to UNICEF during an eight-year partnership, through a mix of charitable donations and by creating "UNICEF" products, whereby 25% of the retail proceeds are passed on to the charity. In 2013–2014, Gucci's donations from sales of the Gucci Nice shopper bag has benefited UNICEF's "Schools for Africa" and "Schools for Asia" initiatives, which aim to provide access to quality education for millions of the most vulnerable children. Chime for Change, a campaign launched in 2013 to support women and girls' education, health services and justice have perhaps been Gucci's most visible CSR initiative. Gucci actively supports the Foundation for Women's Dignity and Rights, which combats violence against women and promotes the empowerment by supporting community-based projects and encouraging employee engagement to sustain women's causes around the world. The company has also established two women's film awards with the Tribeca Film Institute and the Gucci Award for Women in Cinema with the Venice Film Festival. Last but not least, Gucci was one of the first companies in the luxury industry to voluntarily initiate a CSR certification process—SA 8000, for its entire production cycle. According to the brand, this certification—relating to the leather goods, jewelry, footwear and clothing supply chains and to the logistics hub—stands for values such as business ethics, respect for human rights, workers' health and safety, and equal opportunities.

Most luxury brands that display "Thoughtful luxury" react positively to societal issues without focus for business value creation. However, some brands perform CSR activities close to their existing core business, and target environmental and social performance of existing business operations. They expect improvements to environmental and social aspects of their core business in the form of cost and risk reduction and reputational improvement. Few luxury brands start getting strategic with their CSR activities and integrate them into their core business operations. Since reputation is the main competitive advantage of luxury companies, their CSR strategy most often aims to preserve and diffuse the image of a responsible business. Luxury brands with integrative CSR engagement (Halme and Laurila 2009) prefer to adhere to mechanisms to develop and regulate practices in the regulation of human rights, such as the Responsible Jewellery Council, which is committed to the regulation of social and environmental practices, and human rights in the gold and diamond supply chain (Gomes and Yarime 2015). They also support NGOs in connecting with local communities or reviving traditional skills to apprentices.

The iconic Fifth Avenue jeweler—Tiffany and Co., demonstrates a good example of such integrative CSR engagement. Tiffany's overarching contribution to social value creation is the creation of awareness on threatened coral reefs due to destructive fishing techniques, particularly trawling. Emphasizing that coral reefs generate a great deal of marine life where fish spawn and young fish find protection,

Tiffany and Co. has not only removed coral jewelry from its product range, but it continuously draws attention and gives visibility to this environmental issue by devoting its shop window space to informing consumers. Although its decision not to use coral has a marginal impact on the process of destruction of coral reefs, the company's positive contribution in the form of educating consumers on this environmental challenge has a significant impact on society. Tiffany and Co. also collaborates with organizations such as Oceana to seek a change in legislation to restrict the practice of trawling.

Hermes and Mulberry are other luxury brands that preserve their reputation and brand integrity by acting as an ethical and responsible company, with CSR activities close to its existing core business. Hermes, for instance, has offered consumers the chance to experience the skills associated with the brand through workshops or craft days. Festival Des Metiers at the Saatchi Gallery in 2013 was an insight into the craftsmanship of Hermes, whereby visitors queued for hours to see a silk scarf screen-printed. The exhibition put the brand in a gallery context and boosted its cultural credentials. Similarly, Mulberry Loves Craft tent at Wilderness festival in 2014 was a hugely popular workshop where festivalgoers could customize a Mulberry leather armband.

Other examples of integrative CSR initiatives are from the timepiece industry such as: The International Watch Company (IWC)'s partnership with the Charles Darwin Foundation to conserve the endemic flora and fauna of the Galapagos Islands, Omega's support of the GoodPlanet initiative to raise awareness about environmental protection, and Audemars Piguet's contribution to the conservation of the world's forests since 1992. Clearly, CSR activities of these luxury brands contribute to social value creation as a positive side effect of being environmentally friendly, rather than as a well-planned solution to societal challenges.

3.4 The Fourth SLOW Plot: Sustainable Luxury for Overall Well-Being

Luxury brands with this type of sustainability engagement start with a holistic purpose such as well-being, and create relatively slow yet meaningful value for society, improving aspects of well-being, such as quality of life, education, health and social justice. They address social issues without forgetting that at some point, they have to become profitable and self-sufficient. They create social value as well as shareholder value by developing alliances as well as disruptive products and services aimed at improving overall well-being of society. This is in accordance with an extension of the resource-based view, which is called dynamic capabilities where competitive advantage may accrue in changing market conditions. It is all about fusing changing market conditions as external resources, with a firm's internal capabilities and competences through a holistic approach. Indeed, companies can create both social and business value by quickly responding to changing market conditions (Eisenhardt and Martin 2000; O'Regan and Ghobadian 2004). In view of today's diminishing planetary boundaries that demand resource efficiency, minimization of negative business impacts is a must-have in crafting and managing sustainability strategies. Yet, what is more important is worsening societal well-being that demands eco-effectiveness and socio-effectiveness from businesses, which necessitates positive contribution to society at large.

Luxury brands demonstrating "sustainable luxury for overall well-being" sustainability engagement are very much aware of changing market conditions such as sustainability challenges, which they see as potential added value to their business value. Thus, they start with sustainability challenges and positive contribution to society in mind, and work towards their resources, competences and business model, to improve the well-being of the planet and society through economically viable projects that ultimately lead to increased business value also. This is why they are generous to share their knowledge and experience with those who may contribute to their dream with success and speed. These brands are good at systems building, often through interdependent collaborations between many disparate organizations that create positive impacts on the society and the planet. Indeed, research findings suggest that companies can start with societal challenges, adapt to them, profit from them and enhance overall well-being (Network for Business Sustainability 2012). According to a 2013 UN Global Compact report, 84% of the 1000 global CEOs surveyed agreed that business should lead efforts to define and deliver new goals on current sustainability challenges.

Luxury brands with this type of sustainability engagement perform their sustainability activities holistically based on the underlying principle of "do good by doing new things with others". They deliberately choose to contribute to social value creation in one or some of the four well-being aspects identified previously: quality of life, education, health and social equity, which largely align with their values. These brands then aspire to deliver better products and services to customers around the world, with focus on differentiation and innovation that further lead to business value creation. This is different from "refined luxury" engagement, which derives from well-defined TBL values of people, planet and profit for the ultimate purpose of business value creation. With "sustainable luxury for overall well-being", social value and business value are considered equally important, yet social value creation is given priority in deploying organizational resources and competences.

Eden Diodati can be considered a good example of a luxury brand that appears to be leading "sustainable luxury for overall well-being" sustainability engagement. Launched in 2012, Eden Diodati chose to contribute to social equity and quality of life aspects of well-being with an outside-in perspective that currently shapes its overarching business strategy. Thus the brand committed to a fashion ethos based on four principles: global cultural collaborations for stunning designs, an ethical manufacturing with artisans to re-create their futures, a vision to offer women the opportunity to experience true beauty through compassion and love for others, and finally a commitment to donate 10% of shareholder dividends to Medecins sans Frontieres, an international humanitarian-aid NGO and Nobel Peace Prize laureate, best known for its projects in war-torn regions and developing countries facing endemic diseases. The brand manufactures a sizeable portion of its garments through a non-profit cooperative in Italy, where disadvantaged women on the fringes of society are given the chance to learn skills and rebuild their lives through their work with Eden Diodati. Thus, the brand contributes to social equity by integrating disadvantaged women in its business model and empowering them. Eden Diodati has been recognized by the Ethical Fashion Forum, the industry body for sustainable fashion, and through its Fellowship 500, unites fashion innovators around the globe in a network supporting sustainability "from field to final product", namely, along their supply chain (www.ethicalfashionforum.com). Eden Diodati also justifiably contributes to quality of life in war-torn and vulnerable areas of the world by linking beauty to a broader value, namely love and compassion for others, vulnerable people in particular.

Another best practice is Elvis and Kresse, a certified B-Corp and a Positive Luxury trusted brand whose products are stocked in Harrods and are renowned for a stylish range of high-end bags, wallets, belts and cases from decommissioned fire-brigade hoses that would normally end up in landfill. For Elvis and Kresse, design begins with a sustainability challenge, rather than with an idea. Elvis and Kresse's business was established in 2005 to solve environmental and social issues, waste problems in particular. Having started with fire-hose, the company now manages to recycle all the fire hoses in the UK, which would otherwise be thrown in landfills each year (Gardetti and Girón 2014). The company currently reclaims more than 10 different materials, and further engages their material partners and their key stakeholders to share and see if more good could be done with the surplus of an already good business. Thus, they're looking for franchisees around the world to implement the methodology they have successfully developed in the UK, further extending into the business of sustainability to increase their social impact in other countries.

For Elvis and Kresse, proving value for society, changing perception and respecting resources is the challenge. The company re-distributes up to 50% of profits to various projects and charities related to the unique materials they reclaim. For instance, 50% of the profits from their fire hose range are donated to the Fire Fighters Charity. Elvis and Kresse see success not only as the bottom line, but also in two additional measures that they consider equally important: amount of waste diverted from landfill, and amount of money redistributed to charities. The brand creates social value while also creating business value that is further redistributed as social value in the form of donations. Elvis and Kresse contributes to the health of the planet with its focus on resolving the waste issue, as well as educating its stakeholders by sharing know-how and involving them in their social entrepreneurship business model; thus the company clearly contributes to the health and education aspects of well-being for society. Although this may look like creating shared value at first glance, it is indeed different, in the sense that Elvis and Kresse starts with a sustainability challenge such as waste and develops its business model around it with an outside-in perspective, rather than start with the conventional neo-liberal value creation mindset and then create environmental, social and economic value with the overarching "business case for sustainability" must-have in mind.

As we can see, sustainability engagement of the aforementioned two best-practices, is built upon collaborations with charities or NGOs for increased social impact. Creation of next-practice platforms in the form of business models, synthesizing technologies and regulations in different industries would require becoming a network of alliances either across the value chain or in partnership with other companies, or policy-makers (Nidumolu et al. 2009). Similarly, Porter and Kramer (2011) emphasize the importance of enabling local cluster development as an essential element of creating shared value. Finally, Larson et al. (2000) draw attention to the importance of forming alliances to be truly sustainable and solve sustainability challenges. It can easily inferred that, modification of organizational structures around well-being issues and integration of well-being metrics into sustainability reporting, in consultation with relevant societal representatives such as policymakers and NGO representatives, is the case with this type of engagement.

One major challenge with well-being focused sustainability engagement is the difficulty of using a common denominator to understand social value creation in financial terms on the same valuation basis (Gray et al. 1996). As aspects of well-being such as education, health, social justice and quality of life require economic and social metrics that are qualitative in nature, it becomes challenging, if not possible, for companies to make up poor social performance on one well-being issue by doing better on another one (Hashmi and Muff 2015). Social value measures, as compared to environmental value measures, are relatively more qualitative in nature, and their qualitative nature presents difficulties for standard setting and benchmarking (Wöber 2002). Indeed, measurement with the same or at least comparable methodology for different elements of sustainability is still a challenge (Beckett and Jonker 2002), and yet to be improved in academic research. Forward-thinking luxury brands, however, would almost always aspire and commit to well-being issues regardless of challenges and keeping the following quote in mind:

> Capitalism does millions of things better than the alternatives…. However, it is totally ill equipped to deal with a small handful of issues unfortunately. They are the issues that are absolutely central to our long-term well-being and even survival.

(Jeremy Grantham)

4 Conclusions

The globalization and exponential growth of the luxury industry, fostered by strong demand from China, has led to greater social and environmental impacts, raising the issue of sustainable management of luxury: the sourcing of raw materials such as gold and gems, the health impacts of toxic residues present in the textile supply chain, attendant labor conditions due to outsourcing to developing countries, and animal welfare issues for luxury leather goods manufacturers. This lays the ground for the luxury industry to embrace new environmental and labor standards and sustainably manage their brands. While the luxury industry remains relatively far from sustainable in the pure sense of the word, steps by luxury brands towards sustainable management of luxury still present a positive change, regardless of how big or small. Yet, the duty of luxury becomes not only to act, but also to mobilize, to have a large impact in educating customers to be less wasteful, mainly because of its role as a trendsetter.

Luxury goes hand in hand with sustainability. A sensible starting point is that luxury can be considered as placing an importance on durability in line with the concept of fewer but better. Luxury brands, unlike premium or mass brands, sell products that are rare, and are thus resource-dependent, which increases their need for sustainability. Indeed, luxury is, by its very nature, timeless, durable and long lasting. Luxury goods have then implicit sustainability built in as they do not go out of fashion and are long-life products. Moreover, luxury is very close to sustainable preoccupations as it is founded on respect for fine natural resources and craftsmanship that result in a rare and beautiful product charged with meaning, heritage and a story. Luxury companies can promote specialized skills and train employees, thus ensuring that such skills are sustained and enhanced. To this end, luxury brands are potentially less sensitive to cost reduction compared to fast fashion, and more with creating value. Furthermore, exclusive goods and services with sound social and environmental credentials as well as positive social and business value sound more meaningful, and the link to sustainability for overall well-being becomes less disagreeable.

Yet, the luxury industry's approach to sustainable management of luxury is quite fragmented with differing motivations, types of sustainability engagement and measures of value creation. The Sustainable Luxury for Overall Well-being (SLOW) value creation model, in this sense, aims to provide luxury industry professionals and academia with a tool to understand and differentiate between various types of sustainability engagement that lead to varying degrees of shareholder value and social value. Due to their high potential and high profile to make a relatively higher positive impact as compared to other industries, luxury brands have a bigger responsibility to solve sustainability challenges and prioritize social value over shareholder value. This requires a radical paradigm shift from the mechanistic industrial mind to the holistic socio-ecological mind, which in turn, necessitates a broader purpose such as "overall well-being". Indeed, the holistic concept of well-being breeds within itself solutions to solve social and environmental issues such as social equity, public health, poverty, education, quality of life, energy and water. Thus, linking the conventional purpose of business to happiness or well-being of society would be a worthwhile attempt in sustainability management of luxury companies.

The SLOW value creation model is a proposed theoretical contribution that sets the foundation for further empirical research in the sustainable management of luxury. Although the SLOW model was elaborated with reference to assessments of some top most valuable luxury brands' sustainability engagements, the adoptability of SLOW, yet, needs to be applied to other luxury brands in the industry. SLOW offers important insights to luxury brand managers and sustainability experts who aspire to resolve today's social challenges with business acumen. Last but not least, the SLOW value creation model can be considered an important contribution to transcending to the social entrepreneurship mindset in these times when there's growing need for awareness of "*it's not how big companies grow, but it's how they grow big*", to improve today's current economic paradigm.

References

Altagamma Fondazione (2012) Worldwide markets monitor. Fondazione Altagamma, Milan

Aristotle (n.d.) A brief description of Aristotle's Golden Mean. http://www.anus.com/zine/articles/draugdur/golden_mean/. Accessed 6 Mar 2016

Armstrong L (2013) Eco-fashion: why green is the new black for Gucci. Telegraph

Ashida M (2012) The quest for authenticity: can luxury brands justify a premium price? (Basel. Bank Sarasin, Switzerland

Aupperle KE, Carroll AB, Hatfield JD (1985) An empirical examination of the relationship between corporate social responsibility and profitability. Acad Manage J 28(2):446–463

Barney JB (1999) How a firm's capabilities affect boundary decisions. Sloan Manag Rev 137–145

BCG (Boston Consulting Group) (2014) Consumer value creator series. https://www.bcgperspectives.com/content/articles/consumer_products_value_creation_strategy_fashion_luxury_riding_strong_recovery/. Accessed 14 Mar 2016

Beckett R, Jonker J (2002) Accountability 1000: a new social standard for building sustainability. Manag Auditing J 17(36):36–42

Bendell J (2012) Elegant disruption: how luxury and society can change each-other for good. Griffith University, Queensland

Bendell J, Kleanthous A (2007) Deeper Luxury. WWF-UK, Godalming

Blanco E, Lozano J, Ray-Maquieira J (2009) A dynamic approach to voluntary environmental contributions in tourism. Ecol Econ 69(1):104–114

Bronn PS, Vidaver-Cohen D (2009) Corporate motives for social initiative: legitimacy, sustainability or the bottom line? J Bus Ethics 87:91–109

BSR (2011) Sustainable luxury working group: animal sourcing principles, 8 Feb 2011. http://www.bsr.org/files/SLWG_Animal_Sourcing_Principles.pdf. Accessed 14 Mar 2016

Carroll AB, Shabana KM (2010) The business case for corporate social responsibility: a review of concepts, research and practice. Int J Manag Rev 12(1):85–105

Conner KR, Prahalad CK (1996) A resource-based theory of the firm: knowledge versus opportunitism. Organ Sci 7:477–501

Daveu MC (2014) Placing sustainability at the heart of Kering. Business of fashion Ltd., London. http://www.businessoffashion.com/2014/10/placing-sustainability-heart-kering.html. Accessed 12 Feb 2016

DEFRA (2010). Measuring progress: sustainable development indicators 2010. National Statistics Compendium Publication, London. http://www.sd.defra.gov.uk/progress/national/annual-review. Last accessed 27 Dec 2015

Dixon SEA, Clifford A (2007) Ecopreneurship: a new approach to managing the triple bottom line. J Organ Change Manag 20(3):32645

Doran S (2013) The new luxury is luxury for all, suggests Jean-Noel Kapferer. Luxury Society. http://luxurysociety.com/articles/2013/04/the-new-luxury-is-luxury-for-all-suggests-jean-noel-kapferer. Accessed 10 Feb 2016

Doyle I, Bendell J (2011) Uplifting the earth: the ethical performance of luxury jewellery brands. http://www.lifeworth.com/consult/2011/06/uplifting. Accessed Feb 2016
Duschek S (2004) Inter-firm resources and sustained competitive advantage. Manag Revue 15:53–73
Dyer JH, Singh H (1998) The relational view: cooperative strategy and sources of interorganizational competitive advantage. Acad Manag Rev 23:660–679
Dyllick T, Muff K (2013) Clarifying the meaning of sustainable business: Introducing a typology from business-as-usual to true business sustainability. http://ssrn.com/abstract=2368735. Accessed 01 June 2014
Eisenhardt KM, Martin JA (2000) Dynamic capabilities: what are they? Strateg Manag J 21: 1105–1121
Elkington J (1997) Cannibals with forks: the triple bottom line of 21st century business. Capstone, Oxford
Ethical Fashion Forum. http://source.ethicalfashionforum.com/article/about-the-fellowship-500. Accessed 5 Mar 2016
Foresight Mental Capital & Well-being Project (2008). https://www.gov.uk/government/uploads/system/uploads/attachment_data/file/292453/mental-capital-wellbeing-summary.pdf. Accessed 25 Mar 2016
Fowles J (1996) Advertising and popular culture. Sage, London
Friedman M (1970) A Friedman doctrine: the social responsibility of business is to increase its profits. NY Times Mag 13:33
Friedman, V (2010) Sustainable fashion: what does green mean? Financial Times, London. http://www.ft.com/cms/s/2/2b27447e-11e4–11df-b6e3-00144feab49a.html. Accessed 6 Jan 2016
Fullerton J (2015) Regenerative capitalism: how universal principles and patterns will shape our new economy. Capital Institute. http://fieldguide.capitalinstitute.org. Accessed 25 Mar 2016
Fulton M (2012) Sustainable investing: establishing long-term value and performance. DB Climate Change Advisors, Deutsche Bank Group. http://www.dbadvisors.com. Accessed 08 Aug 2014
Gardetti MA (2011) Sustainable luxury in Latin America. Conference delivered at the seminar sustainable luxury and design within the framework of the MBA of IE – Instituto de Empresa – Business School, Madrid, Spain
Gardetti MA, Girón ME (2014) Sustainable luxury and social entrepreneurship: stories from the pioneers (Sheffield. Greenleaf Publishing), UK
Girón ME (2009) Inside luxury. Editorial LID, Madrid
Gomes CP, Yarime M (2015) Diversity of human capital as a driver for corporate responsibility engagement: the case of the luxury industry. In: Gardetti MA, Torres AL (eds) Sustainability luxury: managing social and environmental performance in iconic brands. Greenleaf Publishing, Sheffield
Gray R, Owen D, Adams C (1996) Accounting and accountability: changes and challenges in corporate social and environmental reporting. Prentice Hall, Hemel Hempstead
Guercini S, Ranfagni S (2013) Sustainability and luxury: the Italian case of a supply chain based on native woods. J Corp Citiz 52:76–79
Halme M, Laurila J (2009) Philanthropy, integration or innovation? Exploring the financial and societal outcomes of different types of corporate responsibility. J Bus Ethics 84(3):325–339
Hart SL (1995) A natural-resource-based view of the firm. Acad Manage Rev 20(4):986–1014
Hart SL, Milstein MB (2003) Creating sustainable value. Acad Manag Executive 17(2):13
Hashmi ZG, Muff K (2014) Rethinking corporate sustainability within the 'Well-Being' context: the 'COGWHEEL' sustainability framework. Building Sustain Legacies 2:67–96
Hashmi ZG, Muff K (2015) Evolving towards truly sustainable hotels through a "well-being" lens: the S-WELL sustainability grid. In: Gardetti MA, Torres AL (eds) Sustainability in hospitality: how innovative hotels are transforming the industry. Greenleaf Publishing, Sheffield, pp 117–135
Hockerts K, Wustenhagen R (2009) Emerging Davids versus greening Goliaths (Frederiksberg, Denmark: CBS Center for Corporate Social Responsibility)

Jackson F (2011) Giving the environment a sporting chance. The Times, London. http://www.sportanddev.org/?1072/Giving-the-environment-a-Sporting-Chance. Accessed 6 Jan 2016

Kahn J (2009) Luxury-goods makers embrace sustainability. NY Times A11

Kapferer JM, Michaut-Denizeau A (2015) Are luxury purchasers insensitive to sustainable development? New insights from research. In: Gardetti MA, Torres LA (eds) Sustainable luxury. Greenleaf Publishing Limited, Sheffield

Larson AL, Teisberg EO, Johnson RR (2000) Sustainable business: opportunity and value creation. Interfaces 30(3):1–12

Lau J (2010) Foreign luxury companies take note: Chinese consumers favor CSR. http://blogs.ft.com/beyond-brics/2010/07/21/67251/. Accessed 4 Feb 2016

Living Wage Foundation (2014) What are the benefits? http://www.livingwage.org.uk/what-are-benefits. Accessed 23 Feb 2016

Lochard C, Murat A (2011) Luxe et Développement Durable: La Nouvelle Alliance. Eyrolles, Paris

Margolis JD, Walsh J (2001) People and profits? The search for a link between a company's social and financial performance. Lawrence Erlbaum Associates, Mahwah, NJ

McDonough W, Braungart M (2000) A world of abundance. Interfaces 30(3):55–65

Miller D (2001) The poverty of morality. J Consum Culture 1(2):225–243

NEF (2011) Measuring our progress. http://www.neweconomics.org/publications/measuring-our-progress. Accessed 04 Jan 2016

Network for Business Sustainability (NBS) (2012) Innovating for sustainability: a guide for executives'. http://www.nbs.net. Accessed 28 Feb 2016

Nidumolu R, Prahalad C, Rangaswami M (2009) Why sustainability is now the key driver of innovation. Harvard Bus Rev 57–64

Norman W, MacDonald C (2004) Getting to the bottom of "Triple Bottom Line". Bus Ethics Q 14(2):243–262

Office for National Statistics (ONS). (2011) Measuring what matters: national statistician's reflection's on the national debate on measuring national well-being. The national archives: London. http://www.ons.gov.uk/ons/rel/wellbeing/measuring-national-well-being/discussion-paper-on-domains-and-measures/measuring-national-well-being—discussion-paper-on-domains-and-measures.html. Accessed 16 Mar 2016

O'Regan N, Ghobadian A (2004) The importance of capabilities for strategic direction and performance. Manag Decis 42:292–312

Orsato RJ (2006) What does it pay to be green? Calif Manag Rev 48(2):127–143

Ottman, JA (2011) The new rules of green marketing: strategies, tools and inspiration for sustainable brands. Berrett-Koehler

Porter ME, Kramer MR (2006) The link between competitive advantage and corporate social responsibility. Harvard Bus Rev 11:78–92

Porter M, Kramer M (2011) Creating shared value: how to reinvent capitalism and unleash a wave of innovation and growth. Harvard Bus Rev 62–77

PWC (2013) Going beyond philanthropy? Pulse-check on sustainability. PWC, Kuala Lumpur

Rappaport A (1986) Creating shareholder value. Free Press, New York (1998)

Reinhardt F (1998) Environmental product differentiation: implications for corporate strategy. Calif Manag Rev 4(4):43–73

Smuts J (1926) Holism and evolution. Macmillan & Company Limited, New York

Sundie JM, Kenrick DT, Griskevicius V, Tybur JM, Vohs KD, Beal DJ (2011) Peacocks, Porsches, and Thorstein Veblen: conspicuous consumption as a sexual signaling system. J Pers Soc Psychol 100(4):664–680

Thanh VM, Bosch OJH (2010) Systems thinking approach as a unique tool for sustainable tourism development: a case study in the cat ba biosphere reserve of Vietnam. International Society for Systems Sciences, Wilfrid Laurier University, Waterloo, ON, Canada, pp 18–23

The UN Global Compact (2013) Accenture CEO study on sustainability 2013: architects of a better world. Accenture and United Nations Global Compact. http://www.unglobalcompact.org. Accessed 12 Feb 2016

Thorpe A (2010) Design's role in sustainable consumption. Des Issues 26(2):3–16

WCED (World Commission on Environment and Development) (1987) Our common future oxford. Oxford University Press, Oxford, p 400

Wöber K (2002) Benchmarking in tourism and hospitality industries: the selection of benchmarking partners. CABI, Wallingford

The Strategic Role of Authenticity in the Luxury Business

Patricia Anna Hitzler and Günter Müller-Stewens

Abstract Authenticity is considered to be one of the major challenges for the luxury segment of our time. Crises and scandals in the economic world change behaviors of luxury consumption: Today, luxury clients are looking for authenticity, purpose and sustainability. Furthermore, the progressive democratization of luxury and the increasing imitation of a luxury-specific strategy by mass product manufacturers lead to an erosion of the notion of "luxury" and its mysticism. Luxury becomes commonplace.

Keywords Authentic employees · Authentic luxury · Authentic strategy · Authenticity · Authenticity concept · Craftsmanship · Differentiation · Drivers of authenticity · Heritage · Identity · Key bearers of authenticity · Luxury · Luxury business · Management of luxury · Mission · Passion · Positioning tool · Public image · Self-image · Spirit · Values

The luxury industry sustainably tries to defend its exclusive status. It attempts to escape from the dilution with mass producers and now refers to itself as "authentic luxury" or as an "authentically operating luxury segment." If, however, we take a closer look at the notion of "authenticity", we find that it is an elusive concept and that it is difficult and vague to provide the strategic positioning of a luxury product with sufficient advice. What does authenticity precisely mean nowadays? What does it mean in the contemporary economic environment? What features and criteria have to be taken into consideration in order to act authentically as a company in the luxury business and to be perceived by stakeholders as a real authentic enterprise?

In an empirical study, we analyzed the main features of the construct authenticity with experts from six companies of the luxury industry. To be sufficiently concrete with our results for decision-makers, we linked our research on authenticity with the

P.A. Hitzler (✉) · G. Müller-Stewens
University of St. Gallen, St. Gallen, Switzerland
e-mail: patricia.hitzler@oliverhitzler.de

G. Müller-Stewens
e-mail: guenter.mueller-stewens@unisg.ch

M.A. Gardetti (ed.), *Sustainable Management of Luxury*,
Environmental Footprints and Eco-design of Products and Processes,
DOI 10.1007/978-981-10-2917-2_2

different activities of the strategic value chain of a luxury goods company. Therefore, this paper gives an overview of the drivers and features of authenticity integrated in a luxury-specific value chain that can be used as guidance for management. It also provides a tool that may be applied for individual implementation of a strategy based on authenticity. With the choice of and concentration on the core criteria, this authenticity concept can, as a differentiation and positioning tool, protect the luxury segment in a sustainable way.

1 The Luxury Segment: Risk of Dilution

Luxury is omnipresent today. Many products and services are declared as "luxury", "premium", "exquisite" and in similar terms. Discounters, for instance, offer "gourmet" and "deluxe" product lines. Drugstore chains publish advertising magazines under the name "Luxury" and the music industry promotes CD albums as limited "deluxe" editions. Fashion companies that mass-produce their textiles in the Far East are considered high quality premium providers and low price chains avail themselves of renowned designers to profit from the symbolic added value of the luxury industry. In recent years, the manufactures of mass products have increasingly adopted the strategy of the luxury goods segment and use the same communication tactics to place their products with the consumer (Kapferer and Bastien 2012, pp. 39–40). They attempt to escape the image of a commodity product as a means of differentiating themselves and to be in a position to realize a price premium. The term "luxury" therefore loses its meaningfulness and mysticism—it becomes commonplace. The exclusive status of the luxury industry starts to fade.

But some luxury goods manufacturers contribute to this dilution themselves. Examples are expensive perfumes or bags dumped onto the market. And some manufacturers of luxury brands have meanwhile opened one of its stores "on every corner" and the uniqueness is lost. Then the luxury products are simply expensive, but anything other than special.

This leads to confusion and skepticism among luxury consumers (Bellaiche et al. 2010, pp. 1–5). The advance of democratization of luxury and changing consumer behavior compound this dilemma. Is it not more elegant for a liveried chauffeur from Uber to drive you in a Bentley than to own and drive one yourself? Soon everyone will have access to one of the numerous luxury product categories. The social strata that used to elevate true luxury consumers from others are becoming diluted and are partly vanishing (Kapferer and Bastien 2012, p. 11).

Furthermore, the crises and scandals in the business world have also changed customer behavior. "Understatement" has been propagated since the 2007/2008 recession (Wittig et al. 2014, pp. 91–94). The luxury clientele has developed from a "consumer society" to a society that longs for authenticity, purpose and sustainability (Albers 2014, p. 118).

The Internet, and especially social media and blogs, also offer a high degree of transparency. This can be used for the benefit of enterprises through word-of-mouth or may ruin their reputation within a few hours in the case of failures, e.g. uncovering unethical behavior.

Demographic change accompanied by the replacement of the often satiated "baby boomers" as the highest-spending consumer group with "Generation Y", leads to changes in the luxury segment (Bain & Company 2014). In 2016, this generation will already represent the main consumer group for luxury goods (The Economist 2014); a generation that has grown up with digitalization and views may things skeptically. They expect transparency, authenticity, ethical working conditions and practices from luxury goods manufacturers (Young 2014, p. 107).

All these changes in the framework conditions and new customer requirements lead to new challenges in the luxury segment. This sector "that ideally sells staged stories and perfect illusions" (Albers 2014, p. 118) also has to rethink and reposition itself.

The luxury industry is already attempting to respond to these external influences, to sustainably protect its exclusive status and to escape dilution with the manufacturers of mass products. Some call themselves "authentic luxury", as an "authentically operating luxury segment". Yet if the notion of "authenticity" is examined in more detail, hardly anyone can put it into words or express a strategic recommendation for management.

Authenticity is considered *the* challenge for the luxury segment of our time. If the familiar world starts to falter with doubts and mistrust arising, authenticity offers people safety and confidence (Peterson 2005, p. 1083). Authenticity can be deployed in enterprise management as a positioning and differentiation tool. Where this is appropriately applied in the enterprise, i.e. such that authentic leadership is also considered, it can strengthen the market position, generate value, ensure customer retention and loyalty, as well as stronger identification with the company, its brands and products (Napoli et al. 2014, p. 1096).

But what exactly is understood by "authenticity" today? What does it mean in enterprise management? Which features and criteria have to be considered in order to be in a position to respond authentically as an organization in the luxury segment and to be perceived as authentic by one's stakeholders?

2 Authenticity—Definition and Characteristics

The origin of authenticity is to be found in ancient philosophy (Harter 2002, p. 382ff). But the construct has undergone synonymization and abstraction with terms from several fields (Knaller 2006, pp. 17–18).

2.1 Origin and Development

Today's noun "authenticity" has its origins in the Greek adjective "authentikós" and the late Latin adjective "authenticus", which means "vouched", "true" (Pandel

2009, p. 30), “reliable”, “correct” or “in relation to the originator” (Knaller 2006, p. 18).

In the German-speaking region, the terms “recognized, rightful and binding” that describe today’s adjective “authentic” date back to the 16th century. Knaller (2006) explains that the noun first appeared in the mid 18th century in German, French and English usage. However, the normative concept of value and validity meaning “truthful, actual, direct, undisguised, unadulterated” only arises in the 20th century with the fusion of the preparatory terms (original, true, truthful, immediate) with the original meaning (authorship, sense of being bound to tradition and vouched for) (pp. 31–32).

Firstly used in author authentication in the literature (end of the 18th century), the word was then used in the context of originality and genuineness in the arts scene, and thereafter (mid 20th century) as the original or in relation to the originator in the music world, it now still has the invariable meaning. Etymologically, “authenticity” is anchored in semantics (Knaller 2006, pp. 20–31). If you enter “authenticity” in the search for reference books from a well-known online provider, literature results from over 20 different disciplines appear, such as philosophy, linguistics, fine arts, sociology, social sciences, educational theory, business, psychology, architecture, mathematics, medicine, chemistry, physics, astronomy and many more besides.

In summary, authenticity can be described as a dynamic concept with a broad range of associations (Burmann and Schallehn 2008, p. 20). Because with the permanent changes in the environment (new areas of research, change in society), new concept and synonym formations arise (Knaller 2006, pp. 21–31).

2.2 *Authenticity in Research and Science*

The dynamics of authenticity are also noticeable in the attempts at definition in research. The sociologist Wang (1999) clarifies that authenticity cannot be captured in a simple definition. She writes:

> […] authenticity is not a matter of black or white, but rather involves a much wider spectrum, rich in ambiguous colors (p. 353);

and illustrates that authenticity has many facets. It may be understood as a multifaceted, ambiguous concept (Gilmore and Pine 2007), which is defined in research over a wide range as (Fig. 1):

- Original and no imitation (cf. Jones et al. 2005, p. 893; Gillmore and Pine 2007, pp. 49–56; Kennick 1985, pp. 4–5; Schallehn 2012a, p. 77),
- something “truthful and unadulterated” (Kennick 1985, pp. 4–5), or
- something real and undisguised (Schallehn 2012a, p. 77),
- sincerity and honesty (cf. Beverland 2005, pp. 1025–1027; Boyle 2004, pp. 16–22; Fine 2003, pp. 160–162; Holt 2002, pp. 25–31; Peterson 2005, pp. 1086–1090),

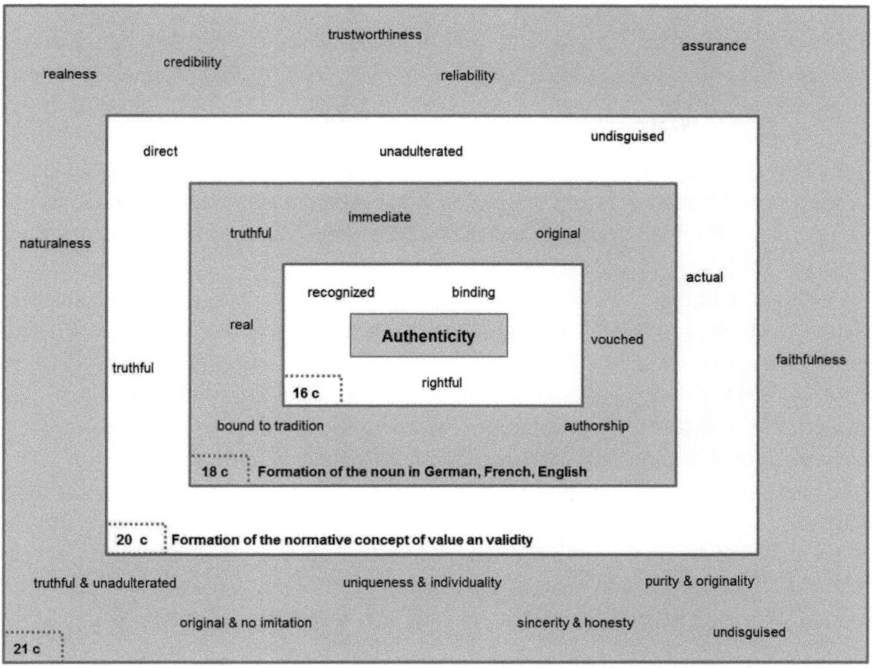

Fig. 1 Etymology of the term authenticity (*Source* Authors, own illustration)

- naturalness (cf. Boyle 2004, pp. 16–22; Gilmore and Pine 2007, pp. 49–56),
- purity and originality (cf. Fine 2003, pp. 160–162; Holt 2002, pp. 25–31; Gilmore and Pine 2007, pp. 49–56),
- uniqueness and individuality (cf. Boyle 2004, p. 285; Schallehn 2012a, p. 77),
- faithfulness (Peterson 2005, pp. 1086–1090).

Authenticity may be related to subjects and objects (Knaller 2006, pp. 21–22). *Object authenticity* is based on a relationship to the originator and is considered as evidence or a guarantee of genuineness (Grayson and Martinec 2004, p. 310), which can be confirmed by institutions and experts (Knaller 2006, pp. 21–22). In anthropology, the object is placed in relation to a social or cultural identity (Schallehn 2012a, p. 25). Here the authenticity of the object is not questioned, but rather whether it corresponds with the perceived social or cultural identity (Schallehn 2012a, p. 25).

Subject authenticity is described in anthropology as “an internal […] state to be strived for, which is considered to be ideal, which includes the return to or arrival at a place of origin” (Schallehn 2012a, p. 28). For instance, in times of crisis and upheaval people long for a safe, authentic place of retreat (Huyssen 2006, S. 235), where personal goals, such as control, connection, belonging and virtuousness, can be fulfilled by authenticity (Beverland and Farrely 2010, pp. 841–848). In psychology, the individual and his self-realization is at the focus of attention (Schallehn

2012a, p. 27). Here "self-images and emotional states [...] are expressed in a subjective coherent way" (Schallehn 2012b, p. 56), even if these leave no positive impression. In sociology, authenticity builds upon truthful and real self-portrayal, i.e. an individual presents himself externally in a way that is consistent with his or her own identity (Wang, 1999, pp. 359–360). The social psychologist Fine (2003) warns that authenticity can be lost if, e.g. artists allow themselves to be seduced by the market and no longer work according to their own conviction, imagination and artistic vein, but rather create objects the market wants, so as to be able to sell them better (p. 166).

However, authenticity can be both a theoretical, as well as a created construct (Napoli et al. 2014, p. 1096). It is a matter of discussion whether it may be "staged" and "constructed" (Peterson 2005) or it has to be real and not artificial, not arranged (cf. Fine 2003, p. 166; Jones et al. 2005, pp. 897–898; Schallehn 2012a, p. 77). Furthermore, the construct as a mixture of true and invented elements is subject to discussion (cf. Grayson and Martinec 2004; Rose and Wood 2005).

In their study, Grayson and Martinec (2004), elicited, for instance, that both reality and fiction can be perceived as authentic, as long as it makes sense for the recipient. It has to correspond with the individual's power of imagination, experience and belief (p. 310), as well as the cultural symbolism (Culler 1981), and must be internally honest and consistent and fit the respective context (Wang 1999, pp. 355–357).

It is important to realize that people may perceive things and situations differently because of their cultural background, their environment, their experience. Hence, the evaluation of truth and genuineness along with the associated evaluation of authenticity may turn out differently (Grayson and Martinec 2004, pp. 296–297). Consequently, it must be stressed that the assessment of authenticity is dependent on individual judgement, one's own identity and the given circumstances (Leigh 2006, p. 482).

2.3 Authenticity in Enterprise Management

The identity of an enterprise also plays a key role from a business point of view. Here, as in sociology, authenticity is understood to be the agreement of the self-portrayal of the corporation with its own identity. The identity of an organization is shaped by predefined values, as well as through its competences, performance, personality, origin and vision (Burmann et al. 2003, pp. 12–25).

As presented graphically in Fig. 2 for an enterprise it means that the behavior of an organization must correspond with its identity. All activities and actions have to confirm the enterprise's identity. The self-image of the company must be firmly anchored in the heads of the various stakeholders (Schallehn 2012b, pp. 55–58), such that comparison between the image created and the behavior of the organization produces the same picture and hence may be perceived as authentic. In a nutshell: "Authenticity is about doing what is said" (Eggers et al. 2013, p. 346).

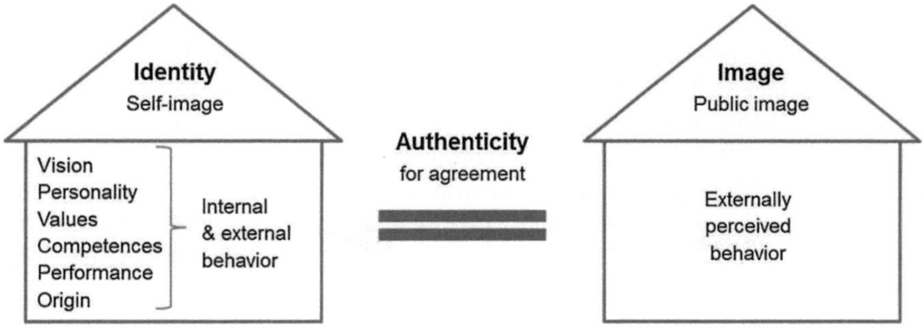

Fig. 2 Authenticity in enterprise management (*Source* Authors, own illustration)

In the sense of the *subject authenticity* mentioned above, the managers should also embody authenticity in their management style as far as possible. In this way, authenticity can relate to managers and their management style. Authentic leadership is called for (Gardner et al. 2011). According to today's view (Walumba et al. 2008), authentic leaders are firstly characterized by their self-awareness. The key is their ability and willingness to engage in self-perception and self-reflection. They possess extensive knowledge of themselves, of their own strengths and weaknesses, but also their feelings, attitudes and motives. The basis is formed for the capability of self-regulation. This includes (a) that they are in a position to openly discuss their considerations and positions with their colleagues free of emotions (relational transparency). In return they are (b) proactively open to other positions and integrate them in their own decision-making (balanced processing). And they make these decisions (c) on the basis of a well-founded system of values, which is also formative for the way in which they conduct their relationships (internalized moral perspective). The focus is not only on the leader, but "followership" also matters, i.e. the relationships between the managers and the managed.

2.4 Authenticity from the Customer Perspective

But what is the luxury clientele searching for? What does the luxury consumer expect from a luxury brand and the enterprise behind it? What do they expect from their products and the associated processes? And who is the true luxury consumer?

Firstly, it is important to point out that every single consumer of luxury goods strives to differentiate or elevate themselves from other individuals (Chevalier and Mazzalovo 2012, pp. 118–136). Interpersonal goals are of paramount importance, such as conspicuousness, uniqueness and the search for confirmation, as well as personal motives like reward, fun and pleasure or pursuing top quality (Vigneron and Johnson 1999). The clientele's wealth and status play a key role in this respect.

The establishment of longstanding customer relationships is likely to be only possible in the customer group with high status and considerable wealth (Wittig et al. 2014, p. 93). Here is where the true luxury customer is to be found. This customer no longer needs to display his high standing and his wealth, but prefers to be among his equals (Kapferer and Bastien 2012, pp. 114–126). His often long experience with luxury products leads to immaterial needs: he yearns for discretion, a sense of purpose and self-fulfillment (Wittig et al. 2014, pp. 92–99). Not seldom, does he also identify himself relatively closely with a brand and prefers to cover his consumption as fully as possible with this brand.

It is necessary to stress that the demands of these luxury customers are changing. Today they are more selective and search for the "true value" behind the product (Anders 2014, pp. 103–108). They are searching for discrete and authentic products (Kapferer and Bastien, pp. 114–126). They scrutinize where and how these products are made and are e.g. more willing to invest in custom-made shoes from an authentic niche provider than to buy insubstantial products that are perishable (Tungate 2009, pp. 25–32).

One can therefore conclude that the consumer behavior of the true luxury customers can be shaped by an authentic enterprise with authentic managers, with an authentic brand and authentic products and processes. But what behavior favors positive perception of authenticity? Which criteria must be considered in the luxury industry to be perceived as an authentic enterprise with an authentic brand that does not only appear to be created by communication and advertising campaigns?

3 On the Path towards Integrated Management of Authenticity in the Luxury Segment

The question firstly arises, which are the drivers of authenticity in the luxury segment as viewed from the existing literature in order to then search for ways in which they can be put into use for a corporation in this segment (Fig. 3).

3.1 Drivers of Authenticity

Building on the importance of authenticity in enterprise management, an organization firstly has to know its identity to be in a position to pursue a successful authenticity-driven strategy.

In his empirical study, Schallehn (2012a) elicits the authenticity determinants *consistency, continuity* and *individuality* (pp. 77, 143–153). The so-called "levers to influence the perception of [...] authenticity" (p. 170) support consumers in perceiving the identity of an enterprise as authentic and consequently to build up its brand and its products (Eggers et al. 2013, p. 346).

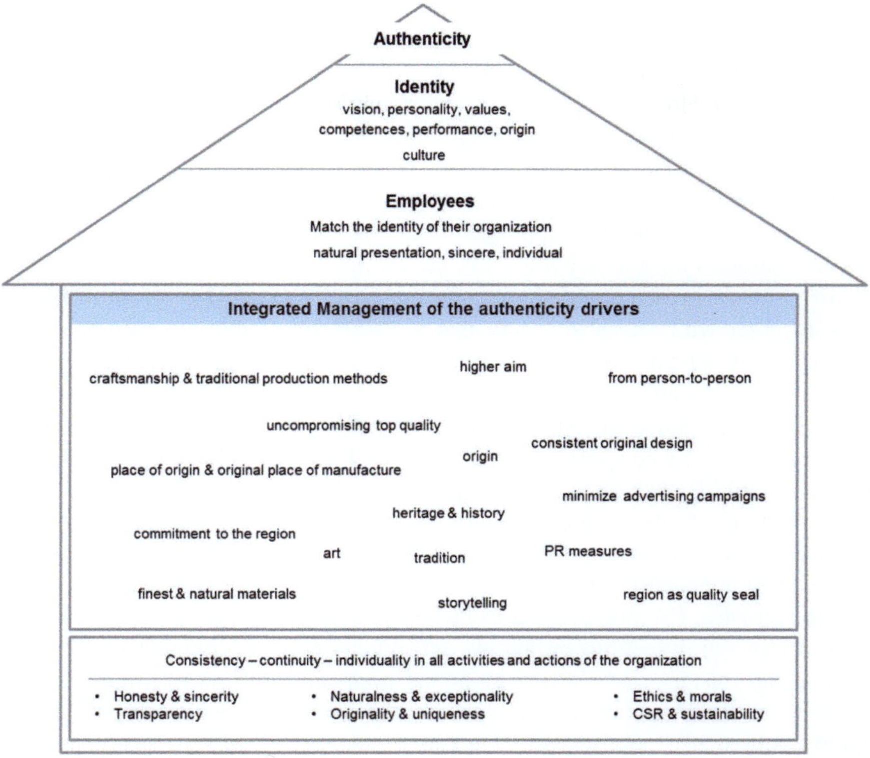

Fig. 3 Overview of authenticity drivers (*Source* Authors, own illustration)

Moreover, consistency, together with congruence between the identity of a company or its values and its action, raises the assessment of authenticity. Communication, i.e. what is said, must be followed by actions (Eggers et al. 2013, p. 346), because *honesty* and *sincerity* form the foundation for authentic enterprise management (Grayson and Martinec 2004, p. 310).

An organization can only be perceived as authentic if it demonstrates honesty and sincerity in its actions and activities (Arthur Page 2007, pp. 16–18). Besides keeping promises, a high degree of *transparency* also counts (Eggers et al. 2013, p. 346). In the information age, with the rapid propagation of news in social media, as well as the strong position of bloggers, enterprises have to behave in an open, correct and honest way (Pedraza 2014). This not only includes self-evident *ethical* and *moral behavior* (Boyle 2004, p. 16) from the organization, as well as considered and active action in the sense of *sustainability* and *CSR concepts* (Boyle 2004, pp. 19–22; Gilmore and Pine 2007, pp. 56, 72–77), but especially an open information culture and admission of mistakes before third parties can spread any shortcomings in the media (Boyle 2004, pp. 271–276).

Generally, authentically operating luxury corporations are recommended to pursue a *higher goal* than a purely entrepreneurial profit philosophy. According to the empirical study results in the luxury wine industry from Beverland (2005), this criterion represents a key driver of authenticity. Other scientists also share this opinion (e.g. Boyle 2004, pp. 270–278; Eggers et al. 2013, p. 341; Gilmore and Pine 2007, pp. 72–77; Holt 2002, pp. 25–31) and recommend acting sustainably and being committed to social issues. *Art* plays an important role in this context. By its very nature, art is authentic (Kapferer 2014, pp. 377–376): an artist with his very special life history and identity creates a work of art out of inner conviction, an original, something unique (Fine 2003, pp. 166–172). Accordingly, the synthesis of luxury and art reinforces the external impact of the enterprise producing luxury products.

For instance, in the course of searching for creative ideas, it is recommended to work together with artists in a way that is externally visible. Or the artist even appears on one's own products, as is practiced in the fashion industry. Or art can be integrated in communication campaigns, because, besides a certain altruism as part of the corporate objective, luxury-driving organizations are advised to *minimize classical advertising campaigns* and to shift their focus (Holt 2002, pp. 25–31). The reason for this is that firstly, aggressive marketing of products contradicts distancing oneself from commercial motives and secondly, people are generally skeptical towards marketing campaigns (Eggers et al. 2013, p. 341).

It is recommended to pursue sponsoring, event and promotion activities and to focus on classical PR measures instead (Beverland 2005, pp. 1023–1025). Here the entire bandwidth of features promoting authenticity can be communicated, e.g. the heritage, the history and the origin of the company with storytelling. Because tradition and the past play an central role in the concept of authenticity (Chevalier and Mazzalovo 2012, pp. 18–19): A link has to be established between the past and today. For instance, the history and origin of a luxury enterprise and therefore by association its cultural impact should be used as references (Napoli et al. 2014, p. 1091). Above all, this includes a strong connection with the region (Beverland 2005. pp. 1015–1019), which is ideally shown by a seal of approval issued by a profession association, as well as the place of origin or original place of manufacture (Newman and Dhar 2014). In their research, Newman and Dhar (2014) ascertained that the original production site is especially relevant for the perception of authenticity. Here the volunteers from their study perceived luxury products produced at the original site to be more authentic than those from out-sourced production. The history of the original manufactory, which is transferred to the product, serves as the explanation for this. A bag from Paris appears especially valuable, because a better story is associated and can be communicated with the French capital.

Beside the commitment to the region, a *consistent original design* has to be pursued (Beverland 2006, pp. 253–254). Gilmore and Pine (2007) speak in this context of "original authenticity" and refer to products that in their design and in esthetics reflect the uniqueness and exceptional nature of the enterprise and the brand. *Nostalgia* also plays a relevant role in this context (pp. 57–62). Apart from the presentation of their products, the assurance of *uncompromising dedication to premium quality* is the main prerequisite for a luxury producing company (Beverland 2006, pp. 253–254). The use of the finest and most natural materials contributes just as much (Gilmore and Pine 2007, pp. 49–56), as practicing traditional production methods and hence the association with *true craftsmanship* (cf. Carroll and Saminathan 2000). But not only the craftsman, as the guarantor for quality and mediator between the past and today, plays a key role in the authentic concept.

Personality, humanity and primarily *person-to-person contact* is the driving force of authenticity (Boyle 2004, pp. 270–271). Therefore, it is recommended to maximize face-to-face contact—even though digital means of interaction are expanded in parallel. The condition for successful implementation of this driver is that the employees, especially those in public positions and are in customer contact, should fit the identity of the enterprise. Ward and Dahl (2014) determined in their study that the personnel can increasingly confirm the client's image of the brand and the company, provided it fits the image of the organization and the personnel identifies with the organization. These employees send stronger signals to their environment than those who only make an effort and fail to match the identity of the enterprise.

Furthermore, the investigation from Ward and Dahl (2014) suggests that an arrogant stance from the sales personnel that fits the image of the brand or the image of the luxury industry can have a positive impact on sales. Here it should be enhanced that this is only of a short-term nature, however. Over the course of time, the arrogance of the employee will be evaluated as negative. Other studies have shown that sales personnel with a friendly, natural manner who meet the highest standards are desirable for the luxury clientele (Maloney 2012, pp. 139–143). Gilmore and Pine (2007, pp. 49–50) speak in this context of "exceptional authenticity" and refer to extraordinarily good service that is offered individually, sincerely and caringly.

Fulfilling these criteria and therefore being authentic and perceived as authentic is easier for smaller enterprises with fewer employees than for large ones (Boyle 2004, pp. 271–277). In their study, Carroll and Swaminathan (2000) recognized that demonstrating authenticity is dependent on the size of the company. Small organizations are perceived as more authentic, as the criteria associated are intuitively attributed by consumers, such as simplicity and reputation for craftsmanship. Small niche providers, such as a shoemaker, stand for traditional production methods and the use of natural ingredients, which, in turn, suggest high quality and product features (Tungate 2009, pp. 25–32).

Organizations in the luxury segment, which, over the course of their development, are no longer able to emit these intuitive authenticity signals in a truly natural way, should strategically help improve the perception of authenticity. The drivers and features of authenticity presented here should be anchored in all areas and all activities of the corporation to be in a position to present them to the outside world as a matter of course and thus to be perceived as authentic.

In summary, it has to be stressed that an authenticity-based strategy in a large enterprise requires *integrated management of the authenticity drivers* (Schallehn 2012a, pp. 169–178).

3.2 *Integration of Authenticity Drivers in the Value Creation Chain*

All activities of an organization in the luxury segment are subject to the identity of the brand (Müller-Stewens and Berghaus 2014). The primary goal of the luxury industry is to establish a strong brand on the market and to protect its identity and its exclusive status. The authenticity concept makes a great contribution to its protection and to strengthening its positioning towards imitators of the luxury-specific strategy (Schallehn 2012b, pp. 53–55). A price structure appropriate to the luxury industry, consistent quality leadership and pursuance of the features of authenticity presented here may not only assure a high status (Beverland 2005), but going beyond this—its implementation in enterprise management has a positive effect on the organization's sales (Eggers et al. 2013, p. 344).

To assist the implementation of integrated management, the criteria that promote authenticity in the key activities of the value creation chain of a luxury goods enterprise (design, procurement, production, communication, distribution and the human factor) are now considered.

3.2.1 Design

The authenticity determinants of consistency and individuality play a key role in the value creation stage of design. The extraordinary design of luxury goods and their quality leadership are assured in this area (Beverland 2005, pp. 1025–1027). The strength and originality of the brand remain protected in the long-term (Kapferer and Bastien 2012, pp. 211–214). An example of this is the use of historical objects and processes as a source of inspiration (Gilmore and Pine 2007, p. 62). Parts of old products can flow into new ones, classics can be rolled out again. Nostalgia is a popular theme, especially in times of uncertainty, and offers a connection to a safe world (Brown et al. 2003, p. 4).

Consequently, the development team should always keep an eye on origin, history and culture, as well as the tradition of the company, and, against this backdrop, develop new products or services that deviate from the standard. New trends should arise, innovations. Art, for instance, can flow into new product series. Cooperation agreements can be established with artists, e.g. for the development of new fabric motifs in the textile field or for furnishing and equipping new facilities in the hotel trade.

Furthermore, at the very beginning of product development, the highest quality standard must be considered and included in calculations. The prices of the products are determined at this value creation stage. A true luxury market strategy is based on pricing power. A lower limit should be considered to be perceived as true luxury; an upper limit does not exist according to Kapferer and Bastien (2012). Instead, they recommend increasing prices as soon as demand grows to protect the exclusivity of the brand.

The possibility of exerting individual influence, such as allowing one's own creative impulses to flow into the object ordered, allows the customer base to blithely accept the high prices (Tungate 2009, p. 31). This holds a major advantage for small authentic trade firms. They can cater for the individual taste of their customers as early as in the design process without any great effort. This inclusion offers great appeal and helps retain customer (Anders 2014, p. 104). It is a testimony to individuality and authentic quality.

For large organizations too, it is of particular importance to be able to act with agility. Their manufactories should operate closely the creative teams, not only to realize individual production (Kapferer and Bastien 2012, p. 214). The process of developing a new product is very complex and requires intensive, interpersonal, personal and interdisciplinary communication. Because often the best product ideas arise in the exchange between production (e.g. craftsmen) and development (e.g. designers) (Kapferer and Bastien 2012, p. 83).

3.2.2 Procurement

For the luxury industry, it is especially important to be certain where the raw materials come from, under which conditions they are obtained and processed, and whether the necessary operations and work steps are sustainable, ethical and morally defensible (Tungate 2009).

For a jewelry manufacturer, it is of particular importance to be able to document where the diamonds to be used in the end product originate. The enterprise should be able to exclude that they are blood or conflict diamonds whose sale finances weapons for violent rebel movements. The Antwerp World Diamond Centre (AWDC), the world's most important center for processing and trading diamonds, for example, has worked towards a certification process (Kimberley Process 2015) that monitors the origin of the gemstone. Diamonds from the conflict regions in Africa cannot be traded in Antwerp without this certificate. This process monitors each individual step from extraction in the mine through to delivery in Antwerp. Furthermore, the accompanying documents and certificates are again checked for genuineness on receipt of the goods in Antwerp.

Another example is the use and processing of real furs, as is often discussed in the luxury textile industry. Kapferer and Bastien (2012) explain that it is, of course, unacceptable to use rare animal species threatened with extinction or have to die under wretched circumstances and recommend backward integration. Breeding rare races can achieve a relativized effect. The luxury company could build up a new business field in the organization. This would give it the certainty that the animals are correctly treated, on the one hand, and it would create jobs, on the other. This way, the enterprise could publicize its responsibility towards nature and society (p. 103). The company Hermès already practices this approach. The manufacturer of luxury goods owns an alligator farm in Florida and is therefore self-sufficient in its manual production with this exotic material (Thomas 2007, p. 175). However, these types of measures don't protect the luxury goods companies against negative headlines in the long-term. In 2015, PETA, the animal rights group, e.g. purported in a video, that the crocodiles shown, which were skinned under cruel conditions, were used for the legendary "Birkin Bag". Although the company Hermès denied the allegations and stressed that neither the farm in the video belonged to the company, nor were the croc skins that were supplied used in Birkin bags, a negative impression persists.

It is therefore all the more important to select the right external partners and suppliers, as well as to operate and to check on-site plants, which are partly scattered around the world. Generally, many organizations in the luxury segment are backwards integrated, as it represents a major challenge to find external partners that can fulfil the high demands. All the features that safeguard the exclusivity and authenticity of the luxury corporation should also be ensured by its suppliers (Chevalier and Mazzalovo 2012, p. 236).Top quality, creativity, uniqueness and innovation also play a key role in procurement. And natural, original ingredients are the main constituents of an authentic luxury product (Gilmore and Pine 2007, pp. 49–56).

3.2.3 Production

As in the "procurement" value creation stage, ethics and sustainability play a special role here too. They form a fundamental prerequisite that not only applies for pursuing an authentic corporate strategy, but generally for the luxury industry (Coignet 2014, pp. 191–196). When buying luxury products, a luxury consumer assumes that the correct behavior is ensured by the organization in its creation. This is not questioned, but is assumed. The high price implies the appropriate payment and fair treatment of the employees, and the use of natural materials and no use of harmful substances (Coignet 2014, pp. 191–196).

Moreover, the creation of luxury products is associated with the highest level of craftsmanship—even if this is in combination with modern technology. Craftsmen who manufacture products for their clientele with a love for detail is the image of every luxury customer. In earlier times, this concept was largely fulfilled by the producers. Every luxury product was manufactured individually by hand at the customer's request. However, today only a few individual parts are still classically handcrafted (Thomas 2007). The rest are produced industrially, in many cases from subcontractors all around the world (Chevalier and Mazzalovo 2012, p. 28) and in the ideal case finished by hand before they are offered via the individual distribution channels (Kapferer and Bastien 2012, p. 211).

The expectations and standards of the luxury clientele, which are attributed to the history and communication of the luxury industry, present the production of the luxury goods industry with two basic challenges: The decision between manual or industrial manufacture and weighting to the two manufacturing processes. And the choice between internal and external production and, closely associated with this, shifting production to other countries.

Given a shift of production to cheaper regions, not only the question arises how the highest quality can be assured, but also how the connection to history, to the origin and the identity of the organization are to be maintained over the long term. Take a leather processor in France or a watchmaker in Switzerland who are not just average craftsmen. They are highly trained and have known their products for many years or even generations (Tungate 2009, pp. 29–32). They pay the necessary attention to luxury-oriented details and subtleties and are therefore shaped by their culture and nationality (Thomas 2007, p. 200).

Generally, many of the European production sites are considered as the seal of approval for the entire luxury industry and as the main pillars of their business model (ECCIA 2014, p. 33). The manufacture seal "Made in …" should be seen as an important competitive advantage and sustainably protected. It strengthens the authenticity and the aura of luxury brands produced in the European countries known to be guarantors for quality. Furthermore, in the case of outsourcing, it is advisable to enter into and nurture long-term partnerships (ECCIA 2014, p. 6). The chosen subcontractors should pursue the same quality standards and be the best in their field (Thomas 2007, pp. 218–221). According to the authenticity concept, these subcontractors should pursue a comparable philosophy and be located in the regions in which their profession is recognized as a specialist area (Chevalier and Mazzalovo 2012, p. 236).

The second challenge in this main activity lies in judging and weighing up between craftsmanship and automated production methods. As many steps as possible should indeed be performed industrially to be able to remain cost-efficient and competitive. Yet, the highest qualitative and individual craftsmanship must be upheld to avoid destroying the foundation of luxury and to preserve its authenticity. It is therefore advisable to protect the traditional, manual part of production in the long term. As consistent with an authentic strategic orientation, this should be preserved at the original locations (Kapferer and Bastien 2012, pp. 211–212), which connect the history and originality of the brand and the enterprise. Preservation of the well-known location offers the necessary continuity, can be used as a reference and thus enhances the perception of an authentic company (Newman and Dhar 2014, p. 383). Furthermore, the human hand confers the product a soul (Tungate 2009, p. 83). The products themselves are easy to copy, a soul is not.

The importance of the criteria promoting authenticity in the field of production is known to luxury enterprises: Their communication very much builds on commitment to location, tradition and craftsmanship. However, the accusation has been made that their externally communicated handcrafted manufacturing methods and original production sites do not match the internal activities (cf. Thomas 2007; Tungate 2009). It would be desirable to offer customers more transparency and honesty. In consideration of the authenticity aspect, the luxury industry should find a way to communicate industrial manufacture and production outsourced to external partners in a more open way (Coignet 2014, pp. 191–196). Because if "untruths" are uncovered and discussed publicly, the brand and the company could suffer severe damage (Pedraza 2014, p. 132).

3.2.4 Communication

Traditional marketing in the luxury industry has served its time and is being superseded by a strong brand and communication strategy. The focus is on a strong brand and its identity, as well as the corresponding communication.

For the luxury industry, this means subordinating all marketing activities in favor of the brand and advancing its prominence and fascination. The brand has to embody all features specific to luxury and authenticity in order to be perceived as authentic (Beverland 2005, pp. 1013, 1017). Together with the determinants of consistency, individuality and continuity, these features play a major part at this value creation stage. Through communication, all the required features can be conveyed to the outside world and form the image of the company and brand for the recipient. The luxury brand is then considered as something exclusive and as a bearer of good taste. Accordingly, communication has to consistently and continuously convey this outstanding taste (Molleda 2010, p. 232–233). The brand must be unique, of the highest quality and be represented in the most exclusive circles (Kapferer and Bastien 2012, pp. 256–258).

Corporations in the luxury industry increasingly use storytelling, print advertising and events as classical means of transmitting these brand-specific signals and messages.

Storytelling is highly recommended, also in the context of an authenticity-based strategy (Gilmore and Pine 2007, p. 62). The condition is that true stories are told which may be received as emotionally charged and mystical. Stories are featured which present the values of the organization and hence the convictions, principles, actions and activities to the outside world, stories about the origins, the region of the enterprise, about its history. Stories about the products and their natural ingredients, about the traditional manufacturing methods and their craftsmen. Stories about the founder personalities and the people who are crucial for success.

Print advertising appears exclusively in high quality magazines. Often the viewer is granted more freedom than the product advertised in the advertisement, because advertising in the luxury segment should primarily be used to allow people to dream and not to sell. This attitude agrees with the advertising aims of the authentic approach. Advertising should always be marginal. A higher aim of the organization should be in the foreground, not pure sales growth. The luxury watch manufacturer Patek Philippe, for example, advertises with the words: “You never actually own a Patek Philippe. You merely look after it for the next generation”. This campaign shows a father with his son or a mother with her daughter and suggests in this way that they have already bought the watch for their children.

Another authentic communication instrument is events (Tungate 2009, pp. 176–179). Events not only offer the clientele the chance to experience the brand intensively, but are also suitable for safeguarding the exclusivity for which the luxury industry stands. The invitation to an event decides on who belongs to the “club” of true luxury clientele and who does not. Small, exclusive events offer the luxury organizations the chance to look after their clientele in a special way, to present themselves to them authentically, face-to-face and to promote strong customer links. Moreover, it is popular to report on events, whether by the participants' word-of-mouth or by journalists reporting.

It is also recommended to sponsor major events (Chevalier and Mazzalovo 2012, pp. 228–229). These should be well considered, limited to one or two events and superficially coincide with the identity of the brand, as in the example of Hermès: The roots of the luxury enterprise lie in saddlery, and by sponsoring the horse race "Grand Prix de Diane" in Paris, a link is created to horses, tradition and aristocracy (Chevalier and Mazzalovo 2012, pp. 228–229).

Besides sponsoring, patronage of art and culture is highly suited for the luxury segment (Tungate 2009, pp. 176–179). As described previously, associating with art enhances the perception of authenticity. That is why luxury fashion companies like Salvatore Ferragamo or Brunello Cucinelli established foundations that promote art and culture in their country.

In conjunction with an authenticity-driven strategy, this kind of events and commitment makes a high contribution to the recommended pursuit of superordinate goals. Furthermore, they guarantee a high level of interest shown by the media. Because those who are not represented in today's leading media almost fail to exist. All the brand's actions should therefore be used for PR purposes, which again assumes authenticity-promoting behavior from the organization.

3.2.5 Distribution

The distribution strategy allows a luxury product organization to safeguard its originality and exclusivity. Here the scarcity of luxury goods is regulated, which, in turn, protects the exclusivity of luxury (Kapferer and Bastien 2012, pp. 99–103). Consequently, managing inaccessible products is a main component of a luxury-specific strategy, whether created physically due to scarce resources or processes, or artificially through communication. Scarcity should be perceivable at all stages of distribution: Whether on the basis of a limited number of points of sale, in selected metropolitan areas or based on the presentation and staging of the products or choice of personnel. Consumers needs to feel the exclusive status of the luxury. They may have to take time to be able to experience their favorite brand. In the case of Hermès, the clientele have to travel to a certain city to reach one of the shops, which are mostly run by the company itself. And should they want to purchase one of the classic bags, they have to accept a waiting time of at least two years. The company that is known to be one of the last true luxury suppliers could simply increase the number or craftsmen to meet the growth in demand. But this approach would contradict the company's philosophy (Thomas 2007, p. 174). Instead, as a result of this practice, it secures its authentic position. Profit motives thus take a back seat.

Generally, scarcity can be combined with rare resources, with social commitment and with sustainability and therefore strengthen the perceived authenticity of the organization. As a case in point, the cashmere producer Brunello Cucinelli saved an entire Italian village from unemployment and decline by starting wool processing for his cashmere products locally. Here too, this shows a higher aim than the simple concept of profit.

In addition, the company's own distribution channels offer the unique opportunity to present the brand and its identity, as desired. In its own stores, all the values for which a brand stands can be presented and authentically implemented. This way, the luxury store can act as an ambassador (Müller-Stewens and Berghaus 2014, p. 24). Here the customer has the chance to experience the brand as a whole in an interactive and multi-sensory sense and to perceive it as authentic. The origin, the history and the craftsmanship can all be documented in the store by (temporarily) displaying historical pictures of designers and craftsmen, their outstanding work and well-known classics (Beverland 2006, p. 258).

Additionally, the store succeeds in establishing and expanding long-term relationships with the customer. Customer loyalty and retention can be essentially influenced through authenticity and preserved with the "person-to-person" feature of authenticity. The sales staff is therefore of key importance for the luxury segment. They should be trained not to put their clientele under pressure, but rather to communicate the myth of the brand and to refer customers to the superb qualitative details of the products, their natural materials and their origin in skilled craftsmanship. This approach indicates the organization's authentic behavior. The customer is suggested that the focus is not on turnover and so the commercial motive does not take center stage. Furthermore, this leads to trust and therefore a long-term relationship can be built up with the customer.

3.2.6 Employees

The employees are the ones who confer authenticity to the organization and its luxury brand (Maloney 2012, 139–143). Whether it is the designer who creates the product and bestows their "signature", the craftman "brings it to life" and therefore "confers a soul" to it or the sales personnel that looks after all the customers' concerns. All employees of a luxury organization make a great contribution to its value creation and, in addition, to an authentic perception of their employer. The future of an enterprise in the luxury segment lies in personal, individual treatment and strong relationships between the organization and customer. Human contact should therefore be maximized, despite high costs. Because this raises authenticity.

The prerequisite for this is that the employee has a high "brand commitment" and fits with the identity of the brand and its company. Consistency in the brand presentation can be ensured and the identity of the organization confirmed and it can then be perceived as authentic (Eggers et al. 2013, p. 346). A natural and non-artificial, friendly manner of the sales advisers is just as necessary as their professional, well-considered use of language and an appealing appearance. Here too, it is important to protect the exclusive status of luxury through excellent service from the employees. The service offered by the Ritz Carlton hotel, which is known for its authentic and individual approach, serves as a good example. The hotel personnel succeeds in establishing trust by anticipating wishes and problems, consistently care, especially in difficult situations and always encounter the clientele naturally and, above all, open-heartedly and sincerely.

In relation to the authenticity concept, it is advisable, e.g. in production, to form cross-generational teams that are familiar with the tradition of the enterprise location and its region (Beverland 2005, p. 1020). These teams not only strengthen the link to traditional manufacturing methods presented outwardly, but innately introduce the necessary properties and knowledge. For the production of luxury wines, for instance, it is desirable to have employees who have known the vineyards since their childhood and are familiar with the quality requirements through their parents or friends from the region. In the case of craftsmen who produce luxury bags, it is advisable to have employees for whom small details matter and who place great attention and importance on the quality standards aimed for.

The aim of the luxury industry here too, as in dealing with the customer, should be to build up long-term bonds between the employees and the organization. The success of luxury producing organizations is based on their workforce and the stability of their teams. Furthermore, authenticity can be consolidated by employees who work a lifelong for their organization and have a wealth of experience.

On reflection of the individual main activities in the value creation chain of a luxury goods enterprise, it has to be noted that organizations from the luxury industry already act in an increasingly authentic way. The priorities are largely limited to marketing and communication, however. Fundamental authenticity criteria, such as transparency, honesty and sincerity, still appear to be little addressed, e.g. in conjunction with production and procurement. If, in the same breath, luxury goods manufacturers want to call themselves authentic, this should be anchored in all activities. Authentic behavior only in representative areas, combined with a lack of transparency, could awaken the impression that authenticity is only stage-managed. This approach may be legitimate, but is not expected from the European luxury industry. Its long history, its European roots, imply a real developed corporate identity and therefore non-staged authenticity.

Moreover, a luxury goods company, as part of pursuing its authenticity-based strategy, must consider that repeatedly over the course of time, the self-image of the organization has to confirm the image arising with the stakeholders. Both authenticity, as well as the enterprise's action and activities and the requirements of the luxury clientele associated with them, are of a dynamic nature and can therefore change over time.

Finally, we now want to round off our insights by dealing with the existing literature with selected, in-depth expert experience.

4 Authenticity from an Expert Point of View

In interviews with six experts from the luxury segment, the most important drivers and features of authenticity could be discussed in depth. An interview guide (see appendix) was created for these interviews on the basis of the theoretical findings. The guide consists of a variety of open and semi-structured questions. The interview partners come from various sectors of the Western European luxury industry and have a wealth of experience in strategy. These are:

- The CFO and member of the directorate of Buchinger Wilhelmi, a worldwide leading clinic for therapeutic fasting on a 5-star level with locations in Marbella and on Lake Constance. The main owner, Raimund Wilhelmi, is the grandson of Otto Buchinger, the founder of the therapeutic fasting method. The patients include VIPs from the global economy and politics, as well as the media world.
- A 2-star cook, owner and managing director of the Sylt Salt Manufactory. The manufactory extracts salt from the North Sea on the island of Sylt.
- The senior partner of Roland Berger Strategy Consultants. He is co-manager of the consumer goods and retail division, supervises projects including those in luxury brands and has published several papers on luxury.
- The CEO of the Steiff Group. Steiff is the world's best-known soft toy manufacturer and inventor of the "teddy bear", a cuddly bear with moving arms and feet, as well as a popular provider of limited collector's items for adults.
- The CEO of the designer furniture manufacturer Thonet. The company looks back over an almost 200 year-old history and is considered a pioneer of industrial design. Thonet became world-famous with the invention of the coffee house classic chair. Other innovations like the cantilever chair followed.
- The managing partner of Wempe KG who heads the traditional manufacturer of exquisite timepieces and jewelry in the fourth generation. Besides globally managed stores in the major cities New York, London, Paris and Peking among others, the company runs its own jewelry manufactory, as well as the watch manufactory, the observatory in Glashütte in Saxony.

In discussions, all the interviewees consider the necessity of authentic behavior of the organizations in the luxury segment as particularly important. They explain authenticity in enterprise management in the context of truth and genuineness, a long history, consistency in entrepreneurial actions and strong corporate values that are put into practice internally and externally. It is stressed that the luxury clientele wants products shaped by a "*genuine stance and attitude*" and "*where people have really thought things out*". Because also for the experts, the person, the employee of an organization represents the key driver of authenticity: "*Our company consists of people and not of machines and the factories in which it all takes place*". "*The employee is the key*". They ensure customer retention. They ensure "*spiritual kinship in the company*". They communicate. They convey "*passion*", "*enthusiasm*", "*pride*" and are understood to be "*part of the whole*". At this point, the interviewees speak of "*so-called spirit*". All these positive emotions that the employees radiate and emit form a pleasant and inspiring outlook that should prevail in an authentically operating company.

A precondition for this is identification of the employees with the enterprise. Ensuring this identification in the long term requires authentic behavior internally from the company: "*Employees find it easier to identify with the enterprise if they have the feeling: 'What is happening here is genuine*'" and "*if the enterprise is managed authentically and it really has the right values and aims and can involve the employees, then the energy can be steered in the right direction*".

This energy can then be realized in dialog and in contact with the customer. Customer relationships and customer retention can be established and expanded. Or it is apparent from the craftsmen in dealing with their products, as can be observed at Thonet: "*The [craftsmen] have missionary zeal. With each and every piece [manufactured] there is the feeling that they add part of their personality and their soul to the piece*". A luxury product should be able to mediate this soul. Luxury goods enterprises that still largely switch their manual production to automation and also shift their production facilities to low wage regions are "*exploiting their brand*" and "*robbing its soul*", it is argued. It should also be borne in mind that such companies accept the risk of quality lapses, loss of credibility and non-transparent through to dishonest behavior, which can lead to damage of the luxury brand. The original production location, its rootedness to the region and craftsmanship are other key criteria that allow the customers to perceive the luxury industry as authentic (Fig. 4).

Almost all experts questioned are agreed: Traditional craftsmanship is essential in the luxury segment. The weighting between manual and industrial manufacture should be undertaken by the respective sector and the individual goals of the respective enterprise. The emphasis should be on the best quality to be achieved in order to avoid compromising the authenticity of a luxury organization. Because "*if I automate a step, although it produces a far worse result, then I take away my authenticity. Everything has to stay within an acceptable framework and everything on a new level on which I don't pull the wool over the customer's eyes*".

It is similar with the views of the experts on the topic of "rootedness with the place of origin" and "shifting production". On the one hand, the experts speak of the "*holy place*" where luxury goods have been manufactured for decades or the "*credibility*" that a certain region conveys to their special field, such as the "*Swiss watch*" or the "*Italian suit*". And on the other hand, the continuously increasing personnel costs of craftsmen in the Western European countries is underlined, which present the luxury industry with a major challenge.

The representative from the service sector explains further that in the case of the Buchinger Wilhelmi Clinic, "*it would rather have been obstructive to have stayed in the original region. Buchinger Wilhelmi has been on the market for almost 100 years, originates from Central Germany. Nobody wants to go to Central Germany, I believe Central Germany is more harmful for our reputation. We have concentrated on locations like Marbella and Lake Constance, i.e. top locations*". This example implies that not all industries should express their authenticity by consistently expressing the preservation of the place of origin. For enterprises

Fig. 4 Thonet craftsmen in their element (*Source* Thonet)

working in the fields of luxury hotels, healthcare and leisure activities, locations are primarily of importance that offer something special, such as beautiful nature around Lake Constance or establish a so-called "hotspot for the wealthy". The location then forms the basis for differentiation. If this aspect is adapted to the entire service industry in the luxury segment, it appears crucial that the service providers are located where they meet with their customer group and not stay at a place of origin that does not attract this kind of clientele (Fig. 5).

The majority of the interviewees agree, however, that shifting production to low wage countries, in the worse case, in production facilities that also produce mass products, is not compatible with an authentically operating luxury industry. In addition, a place of origin should be preserved, whether it be only in the form of a central area, such as the main administration, or limited product for individual manufacture. Of course, differentiation must also be undertaken here by industry, corporation and its aims and standards in order to meet the highest quality standards.

Transparent and honest communication towards the luxury clientele with regard to deciding between craftsmanship and automation or shifting production facilities

Fig. 5 The Buchinger Wilhelmi locations Lake Constance and Marbella (*Source* Buchinger Wilhelmi)

is essential. The basic principle: "*Don't say anything that isn't true*". Because "*life is simplest if you are absolutely honest and your answers are quite clear*". By way of illustration, the design furniture manufacturer Thonet, alongside its handmade furniture, also offers a product line that is a mixture of automated and individually manufactured components. Thonet communicates this approach unashamedly: "*We say it is necessary to combine industrially produced parts with handcrafted ones, to keep the price more or less under control*". There is a common understanding that honesty and sincerity, as well as transparency, make it easier to do business in the long term.

This outlook is reflected with regard to stage-managed authenticity. Stage-managing authenticity takes too much energy to put into practice. In the interviews, reference is made to the "*roots*" to which a company should remain loyal. It is shown that an organization cannot simply "*impose*" something staged on its employees, because "*then something is lost*". And the call is made: "*We don't want to be at all different than we are*". According to the study results, nothing has to be invented. For instance, a fictive history may indeed be highlighted as a successfully staged and practiced example in the luxury segment, but it is stressed that other paths exist at the same time, e.g. an infective passion and the spirit to successfully build up a clearly positioned and meaningful enterprise in the luxury industry.

Meaningfulness and long-term orientation are guiding principles, which an organization that wishes to be perceived as authentic should pay special attention to. This may include motives, such as preserving jobs or fostering historical and cultural heritage. Striving for profit, the experts underscore, is, of course, vital for the survival of an organization. Nevertheless, the targeted realization of profits is not solely decisive for fulfilled and lasting satisfaction. One expert signalize: "*There must be something else apart from the world of figures*". This "something else" is the higher aim, a mission that should be part of the identity of a luxury organization, which, along with its values, gives the enterprise a framework in which it should operate.

In this respect, several of those interviewed addressed topics like ethics, sustainability and social commitment and point out that this should be understood as a "*basic attitude*", as a "*natural component*" of an organization in the luxury segment, which are "*anchored in the fundamental values of the company*" and do not require self-promoting communication with the outside world.

What needs to be communicated externally are the enterprise's products and services, which are a "*means to an end in order to safeguard economic success*", yet at the same time act as the "*actual ambassador*" in which the stance, the identity and authenticity of the organization are conveyed. The interview participants also prefer storytelling for authentic communication: "*Stories with true content are essential*", which are told in sales dialogs and similar settings or can be reinforced with a tour of the manufactory. Besides verbal communication, non-verbal communication should not be underestimated. Music played in the stores, the charisma of the employees or the interior or exterior architecture should reflect the identity of the enterprise and in such a way that "*everything is said without a single word spoken*".

The excellence and high standards, for example, should be apparent in the entire presentation of the organization in order that it can be perceived as authentic in the luxury industry. In specific terms, this means "*that one takes yet another extra step, that one is actually not yet satisfied*". To be able to act excellently and perfectly, the factor of time should be considered. Three experts recommend taking plenty of time in product development and in dealing with the luxury clientele. Moreover, an organization in the luxury segment should be aware that when it conveys perfection and excellence externally, it also has to keep this promise. In an enterprise that wishes to be perceived as authentic, primarily there should be "*no split between the inside and outside*".

Faithfulness to identity forms a key aspect in this regard. It is decisive with regard to all actions and activities of a luxury company. So each organization must decide for itself what makes up its identity and live accordingly. The experts accentuate that the employees, the products, innovations and new areas have to fit the enterprise and the brand. It is also recommended to select one's external partners according to a related identity: A comparable philosophy of the distributors, the suppliers and subcontractors confirm and strengthen the perception of authenticity.

One expert called for the luxury industry not to leave anything to chance. First and foremost, authenticity means "*that nothing is by chance, but everything is precisely considered*" without it being stage-managed.

5 Implications for Management

Summarizing, it must be emphasized that the organization's identity is at the core of a strategy of an enterprise in the luxury segment focused on authenticity. The values, the mission and vision, as well as the performance and competences they offer, but also their origin, their roots: All these features provide a framework within which the company should operate, principles according to which the company should live. Therefore, it is of key importance to firstly analyze what constitutes the enterprise: Where do we come from? Where do we want to go, which superordinate goals do we pursue? What performance and competences are our hallmarks?

If the term "live" is used for an organization, this mainly means the employees who adopt an essential role in the authenticity concept. Firstly, they convey authenticity inwardly and outwardly, in that they match the identity of the organization, they identify with it, and secondly they give the luxury goods a personality and a soul. Primarily it is the employees with their passion, enthusiasm, their pride and their conviction who shape the positive outlook, the spirit in the organization. They are also the ones who build up the long-term connection to the clientele strived for. It is therefore of special relevance to select the right employees and to train and educate them appropriately. In addition, it should be considered that the identification of a employee with their employer is the prerequisite for these positive influences and implications and, in turn, this can only be achieved if the organization lives and promotes authenticity.

Certain characteristics are ascribed to a luxury organization that have to be fulfilled in order to be perceived as authentic. These are the top quality secured by craftsmanship, scarcity of supply, unique esthetics, the link to the origin, non-necessity, the high price, but also the excellence, perfection that is to be expressed in all areas of the organization. This perfection requires an increased investment in time and thus a higher personnel costs, which should be reflected in the number of employees.

Besides these main characteristics that define a luxury product, a whole series of further drivers emerged that favor an entrepreneurial approach focusing on authenticity. Figure 6 provides a summary overview of all the factors derived for promoting authenticity, integrated in the relevant value chain activities for the luxury industry. This representation may be taken as an orientation aid for implementation of a strategy promoting authenticity. The lower area of the graphic represents the foundation of a strategy based on authenticity. The criteria compiled should be considered in all the enterprise's activities and actions. In the best case,

they should be anchored in the values and the culture of an organization in the luxury segment. The upper area outlines the identity characteristics to be pursued,

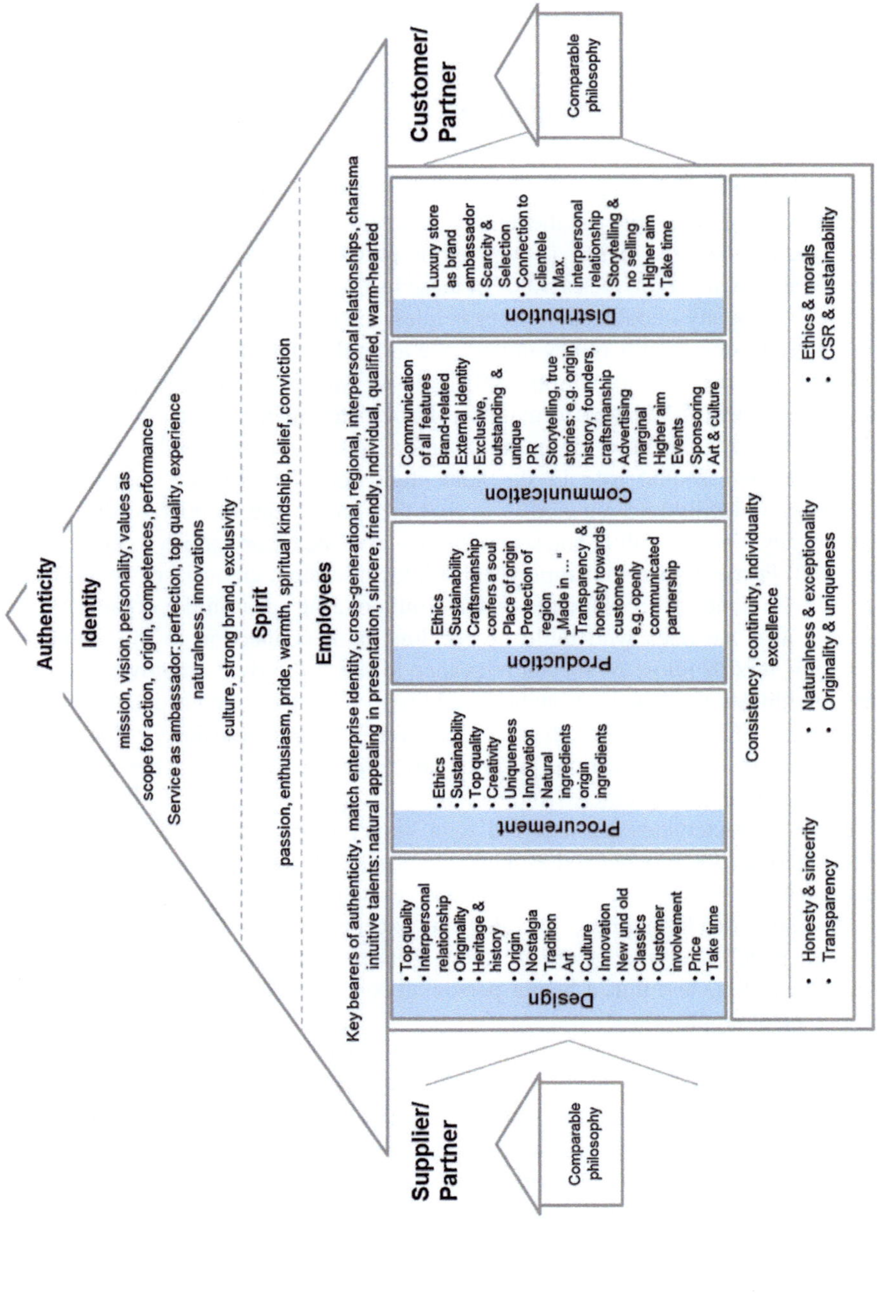

Fig. 6 Authenticity as the key factor of an enterprise strategy in the luxury segment (*Source* Authors, own illustration)

as well as the spirit shaping the identity. In the middle, there are the main activities of a value chain with a large number of voluntary and mandatory criteria. These can be selected variably depending on the enterprise and industry.

For successful differentiation and strategic positioning, it is recommended to filter out and mainly concentrate on six or seven core criteria that also correspond to the essential competences and the identity of the company. This focus allows an organization in the luxury segment to expand its unique selling point and to position itself authentically in the luxury market, to protect itself from imitators who exclusively count on high quality and authentically staged communication and to avoid dilution with imitators of the luxury strategy over the long term.

As presented in Fig. 7, this concentration of drivers can be seen from the companies of the experts interviewed, e.g. Buchinger Wilhelmi and Thonet. As presented in Fig. 7, this concentration of drivers can be seen in the companies of the experts interviewed, specifically at Buchinger Wilhelmi and Thonet. These two organizations represent, on the one hand, the differences in the luxury sectors: the dominance of services versus the dominance of the production and, on the other hand, they ideally show the bandwidth of the possible positioning options of the authenticity concept. For example, it becomes apparent, that Buchinger Wilhelmi (blue) concentrates on pursuing six core criteria. One of these is anchored in a strong history and the legacy of the founder. In addition, the passion of the employees, their above-average interpersonal competence, combined with very high personnel intensity (1.5 employees per patient) ensure a special spirit in the clinics, which the patients rave about. Despite the great demand, they take extreme care not to expand excessively and hence to maintain a natural scarcity.

Furthermore, the range of care and events offered, as well as the furnishing of the clinics only using natural and original ingredients, reflect the corporate philosophy. The most popular event for the patients fasting at Buchinger Wilhelmi is the morning walk in nature—here the luxury clientele is again sensitized for the original. Finally, it has to be pointed out that Buchinger Wilhelmi does not place any advertising. They rely on true stories instead. Journalists regularly stay in the clinics and report on their experiences. The owners are also sought after for interviews in the media.

The design furniture manufacturer Thonet (red) concentrates on seven core criteria. It should be stressed that Thonet already practices a design charter that reflects the company's stance. This includes the voluntary and mandatory topics, which the designers consider in product developments to maintain design consistency. This charter is also use for Thonet’s exhibition booths. It also has to be ensured here that a new booth matches the design furniture manufacturer’s identity. The company can look back over a very long history and a valuable legacy. It has remained loyal to its roots and its place of origin and still largely produces in Germany. 80% of the value creation for classical furniture is produced traditionally by craftsmen in the company’s own manufactory. Product lines that are partly produced industrially are also communicated and offered outwardly in a transparent way. Visitors are also able to come and see the manufactory in Central Germany.

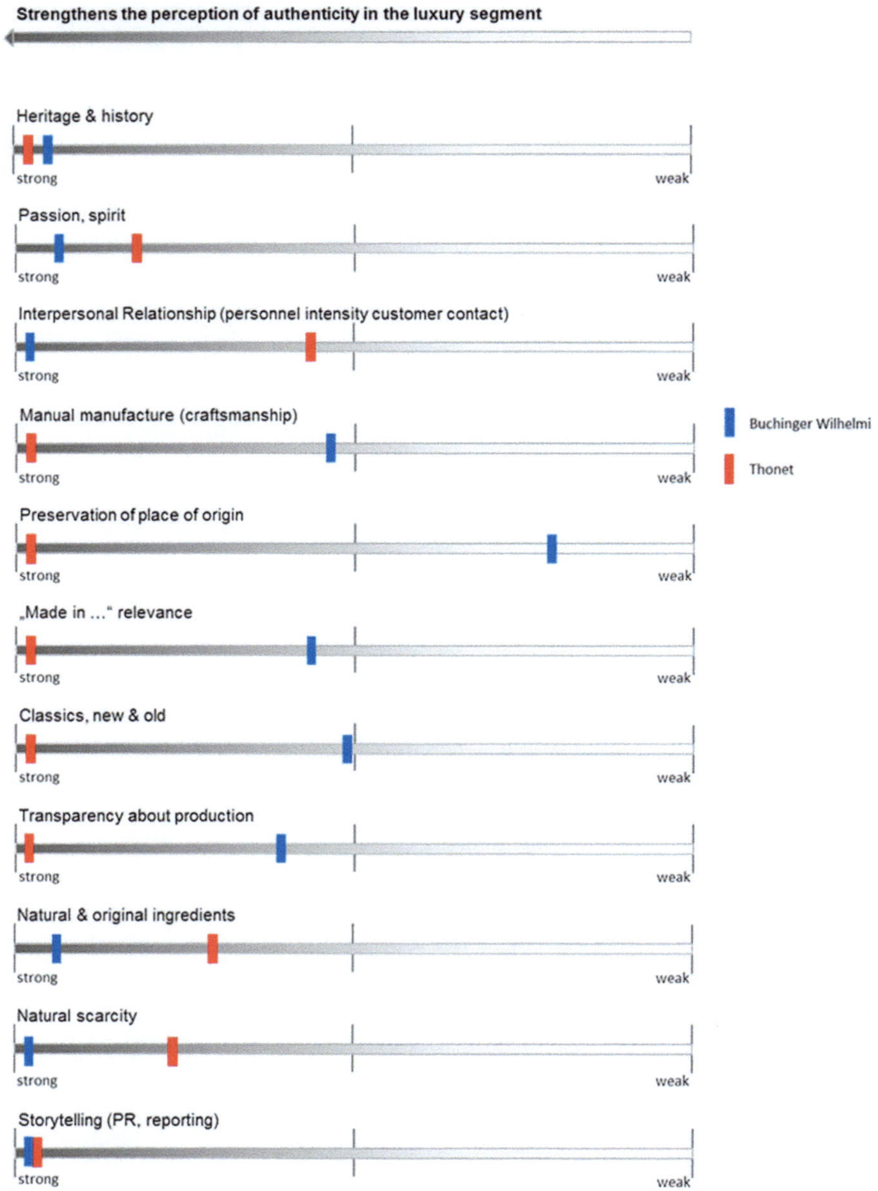

Fig. 7 Case study Buchinger Wilhelmi and Thonet and their core criteria (*Source* Authors, own illustration)

No advertising is placed. Only stories about the cultural and historical importance of Thonet, its well-known designers and customers, as well as the origin of the classics, are communicated externally by storytelling. The mission of safeguarding the continuation of a historically value cultural asset is strongly conveyed here.

The authenticity concept calls for integrated management of its drivers and characteristics. Only if all areas and functions of the enterprise act according to these guides can the organization in the luxury segment be perceived by the outside world as authentic and its further existence ensured in the longer term.

6 Future Research

The study's findings raise several questions for the future. Firstly, do we have to consider the various sectors of the luxury market separately? Does authenticity means something different for a watch brand and for a fashion brand? Such an approach would allow a more detailed classification of features promoting authenticity and an explicit definition of optional and mandatory criteria, depending on the luxury goods or services sector.

Further, it is necessary to learn more about how the luxury clientele defines a company that is operating authentically. What are their wishes, what are their suggestions and criteria? A study in the field of consumer research should be seen as particularly important. Today it is very difficult for the consumer, due to the professional and high-quality development and implementation of communication strategies, to differentiate the truly authentic organizations from the purely communicative authentic performer. Therefore, future research should look for concrete answers regarding this very significant distinction to secure a luxury segment that operates authentically in the long term.

Appendix: Interview Guide

Introductory question (open)

- Authenticity is one of the major challenges of the 21st century. Small, authentic craft businesses and niche suppliers thrive in the luxury segment.

  How do you see the need for authentic enterprise management in the luxury goods industry?

Detailed questions (semi-structured)

- Often it is the employee that gives a luxury brand a kind of "soul". Authentic enterprise management requires increased interpersonal contact. How do you see that?
- To judge by a study carried out by Newman and Dhar (2014), the original place of origin of a luxury goods supplier is decisive for its authenticity. Often, the respective regions are seen (e.g. on the basis of their history) as a seal of quality of their trade.

In your opinion, is it necessary to continue production where the company has been rooted since its founding or is it justifiable to relocate to lower-cost regions?

- Craftsmanship is the hallmark of the luxury industry and a determining factor for authenticity. However, most luxury companies manufacture just a small fraction of their range or only a small component of their product by hand. The majority is produced industrially.

To what extent is automatization in the luxury sector acceptable? How would you solve the dilemma between automatization and craftsmanship?

- Companies that operate authentically in the luxury segment are advised to pursue a higher goal than a purely profit-driven philosophy. How do you see that?
- Some scientists are of the opinion that authenticity can be staged in a company. In your opinion, is staging in the luxury industry justifiable, or may authenticity only be achieved in a genuine way, e.g., based a company's history, its roots and the associated natural development?
- Grayson and Martinec (2004), Eggers et al. (2013) and other scientists stress that transparency, honesty and integrity represent the foundation of authentic behavior. The luxury sector seems to pursue activities that are increasingly communicated (e.g. craftsmanship) and those that are practiced in the background (e.g. full automatization, relocation of production).

How do you see the need for transparency and honest communications?

Final question (open)

- In your opinion, how must a company act to be authentic and to be seen to be so? What other factors should be considered?

References

Albers M (2014) Das Luxus—Problem. Brand Eins 16(12):116–119

Anders T (2014) Increasing education brings more customers. In: Wittig MC, Sommerrock F, Beil P, Albers M (eds) Rethinking luxury. How to market exclusive products and services in an ever-changing environment. LID Publishing Ltd, London, pp 103–108

Arthur W. Page Society (2007) The authentic enterprise—an Arthur W. Page Society Report. http://www.awpagesociety.com/images/uploads/2007AuthenticEnterprise.pdf. Accessed 28 Mar 2015

Bain & Company (2014) Bain-Studie zum Luxusartikelmarkt: Die sieben Gesichter der Luxuskäufer. http://www.bain.de/press/press-archive/die-sieben-gesichter-der-luxuskaeufer.aspx. Accessed 16 Jun 2015

Bellaiche M-J, Mei-Pochtler A, Hansich D (2010) The new world of luxury. Caught between growing momentum and lasting change. The Boston Consulting Group (ed). https://www.bcg.com/documents/file67444.pdf. Accessed 16 Jun 2015

Beverland MB (2005) Crafting brand authenticity: the case of luxury wines. J Manage Stud 42(5):1003–1029
Beverland MB (2006) The 'real thing': branding authenticity in the luxury wine trade. J Bus Res 59(2):251–258
Beverland MB, Farrelly FJ (2010) The quest for authenticity in consumption: consumers' purposive choice of authentic cues to shape experienced outcomes. J Cons Res 5:838–856
Boyle D (2004) Authenticity. Brands, fakes, spin and the lust for real life. Harper Perennial, London
Brown S, Kozinets RV, Sherry JF Jr (2003) Teaching old brands new tricks: Retro branding and the revival of brand meaning. JM 67 (3):19–33
Burmann C, Blinda L, Nitschke A (2003) Konzeptionelle Grundlagen des identitätsbasierten Markenmanagements. Universität Bremen und Fachbereich Wirtschaftswissenschaft (ed). Lehrstuhl für innovatives Markenmanagement. Bremen (Arbeitspapier, 1). http://www.lim.uni-bremen.de/files/burmann/publikationen/LiM-AP-01-Identitaetsbasiertes-Markenmanagement.pdf. Accessed 15 Nov 2015
Burmann C, Schallehn M (2008) Die Bedeutung der Marken-Authentizität für die Markenprofilierung. Universität Bremen und Fachbereich Wirtschaftswissenschaft (ed). Lehrstuhl für innovatives Markenmanagement. Bremen (Arbeitspapier, 31). http://www.lim.uni-bremen.de/files/burmann/publikationen/LiM-AP-31-Marken-Authentizitaet.pdf. Accessed 13 May 2015
Coignet B (2014) Explain what you do and explain why you do it. In: Wittig MC, Sommerrock F, Beil P, Albers M (eds) Rethinking luxury. How to market exclusive products and services in an ever-changing environment. LID Publishing Ltd, London, pp 191–196
Carroll GR, Swaminathan A (2000) Why the microbrewery movement? Organizational dynamics of resource partitioning in the U.S. Brewing Industry. Am J Sociol 106(3):715–762
Chevalier M, Mazzalovo G (2012) Luxury brand management. A world of privilege, 2nd edn. Wiley, Singapore
Culler J (1981) The semiotics of tourism. Am J Semiot 1981(1):127–140
Eggers F, O'Dwyer M, Kraus S, Vallaster C, Güldenber S (2013) The impact of brand authenticity on brand trust and SME growth: a CEO perspective. J World Bus 48:340–348
ECCIA (2014) The contribution of the high-end cultural and creative industries to the European economy. Frontier Economics report prepared for ECCIA. European Cultural and Creative Industries Alliance (ed). http://www.eccia.eu/uploads/media/FINAL_Frontiers_Economics_report_prepared_for_ECCIA_01.pdf. Accessed 17 Jun 2015
Fine GA (2003) Crafting authenticity: the validation of identity in self-taught art. Theory Soc 32:153–180
Gardner W, Cogliser C, Davis K, Dickens M (2011) Authentic leadership: a review of the literature and research agenda. Leadersh Q 22:1120–1145
Gilmore JH, Pine BJ (2007) Authenticity. What consumers really want. Harvard Business School Press, Boston
Grayson K, Martinec R (2004) Consumer perceptions of iconicity and indexicality and their influence on assessments of authentic market offerings. J Cons Res 31(2):296–312
Harter S (2002) Authenticity. In: Snyder CR, Lopez SJ (eds) Handbook of positive psychology. Oxford University Press, London, pp 382–394
Holt DB (2002) Why do brands cause trouble? A dialectical theory of consumer culture and branding. J Cons Res 29(1):70–90
Huyssen A (2006) Zur Authentizität von Ruinen: Zerfallsprodukte der Moderne. In: Knaller S, Müller H (eds) Authentizität. Diskussion eines ästhetischen Begriffs. Wilhelm Fink, München, pp 232–236
Jones C, Anand N, Alvarez JL (2005) Manufactured authenticity and creative voice in cultural industries. J Manage Stud 42(5):893–899
Kapferer J-N, Bastien V (2012) The luxury strategy. Break the rules of marketing to build luxury brands. 2nd edn. Kogan Page, London
Kapferer J-N (2014) The artification of luxury: from artisans to artists. Bus Horiz 57(3):371–380

Kennick WE (1985) Art and inauthenticity. J Aesthet Art Crit 44(1):3–12
Kimberley Process (2015) KP Basics. http://www.kimberleyprocess.com/en/about. Accessed 14 Oct 2015
Knaller S (2006) Genealogie des ästhetischen Authentizitätsbegriffs. In: Knaller S, Müller H (eds) Authentizität. Diskussion eines ästhetischen Begriffs. Wilhelm Fink, München, pp 17–35
Leigh TW (2006) The consumer quest for authenticity: the multiplicity of meanings within the MG subculture of consumption. J Acad Market Sci 34(4):481–493
Maloney P (2012) Vertrieb von Luxusmarken. Von Absatzmittlern zu Markenbotschaftern. In: Burmann C (ed) Identitätsbasierte Luxusmarkenführung. Grundlagen - Strategien - Controlling. Springer Gabler, Wiesbaden, pp 139–143
Molleda JC (2010) Authenticity and the construct's dimensions in public relations and communication research. JCM 14(3):223–236
Müller-Stewens G, Berghaus B (2014) The market and business of luxury: an introduction. In: Berghaus B, Müller-Stewens G, Reinecke S (eds) The management of luxury. A practitioner's handbook. Kogan Page, London, pp 3–25
Napoli J, Dickinson SJ, Beverland MB, Farrelly F (2014) Measuring consumer-based brand authenticity. J Bus Res 67(6):1090–1098
Newman GE, Dhar R (2014) Authenticity is contagious: brand essence and the original source of production. JMKR 51(3):371–386
Pandel H-J (2009) Authentizität. In: Mayer U (ed) Wörterbuch Geschichtsdidaktik. 2nd edn. Wochenschau-Verl., Schwalbach/Ts., p 30
Pedraza M (2014) Algorithms increase the probability of making sale. In: Wittig MC, Sommerrock F, Beil P, Albers M (eds) Rethinking luxury. How to market exclusive products and services in an ever-changing environment. LID Publishing Ltd, London, pp 127–132
Peterson RA (2005) In search of authenticity. J Manag Stud 42(5):1083–1098
Rose RL, Wood SL (2005) Paradox and the consumption of authenticity through reality television. J Cons Res 32(2):284–296
Schallehn M (2012a) Marken-Authentizität. Konstrukt, Determinanten und Wirkungen aus Sicht der identitätsbasierten Markenführung. Gabler Verlag, Wiesbaden
Schallehn M (2012b) Identitätsbasierte Führung von Luxusmarken unter besonderer Berücksichtigung der Marken-Authentizität am Beispiel von Bugatti und Maybach. In: Burmann C (ed) Identitätsbasierte Luxusmarkenführung. Grundlagen - Strategien - Controlling. Springer Gabler, Wiesbaden, pp 53–67
The Economist (2014) Marques for millennials. Demographic trends. In: The Economist online, 14 Dec 2014. http://www.economist.com/news/special-report/21635762-young-people-choose-and-buy-differently-marques-millennials. Accessed 03 Mar 2015
Thomas D (2007) Deluxe. How luxury lost its luster. Penguin Press, New York
Tungate M (2009) Luxury world. The past, present and future of luxury brands. Kogan Page, London
Vigneron F, Johnson LW (1999) A review and a conceptual framework of prestige-seeking consumer behavior. AMS Rev 1999(1)
Walumbwa FO, Avolio BJ, Gardner WL, Wernsing TS, Peterson SJ (2008) Authentic leadership: development and validation of a theory-based measure. J Manag 34:89–126
Wang N (1999) Rethinking authenticity in tourism experience. Ann Tourism Res 26(2):349–370
Ward MK, Dahl DW (2014) Should the devil sell prada? Retail rejection increases aspiring consumers' desire for the brand. J Cons Res 41(3):590–609
Wittig MC, Sommerrock F, Beil P, Albers M (eds) (2014) Rethinking luxury. How to market exclusive products and services in an ever-changing environment. LID Publishing Ltd, London
Young K (2014) The young and the excess. JCK Mag. http://www.jckonline.com/2014/10/05/young-and-excess-what-jewelers-need-to-know-about-millennials. Accessed 22 Jun 2015

Creating a Culture of Shared Value Through Luxury Branding

Misha Pinkhasov and Rachna Joshi Nair

Abstract The pursuit of shared value relies on the individual as the common element between business and society to reconcile these two institutions' often competing goals. Brands are composed of people, so they must align their values with those of society, define their purpose as citizens, and shape the behaviour of their people to suit. As a subcategory of brands, luxury brands are not just citizens but leaders, which gives them the ability to raise their stature by influencing the wider community.

Keyword Shared value · Social responsibility · CSR · Sustainability · Values · Corporate culture · Responsible business · Luxury

Today there is pressure on companies to do more than make money. Customers still want the best product at the lowest cost, but "best" now means that the product, and thus the brand, must reflect their way of thinking and living. And "cost" now encompasses the holistic costs—psychological, environmental, and so on—of obtaining the product. In a similar way, employees want more than just a job and a salary. Modern workers are attuned to the opportunity for their work to be personally fulfilling as well as financially profitable. And company owners increasingly want more than just the biggest possible return on their investment. Shareholders are waking up to the knowledge that a sound investment requires a

---

Excerpted from Pinkhasov, M. and Joshi Nair, R., *Real Luxury: How Luxury Brands Can Create Value for the Long Term*, 2014, Palgrave Macmillan. Reproduced with permission of Palgrave Macmillan.

M. Pinkhasov (✉) · R. Joshi Nair
NAIR-SAFIR Strategy Consultants, Paris, France
e-mail: mp@nair-safir.com

R. Joshi Nair
e-mail: rjn@nair-safir.com

M. Pinkhasov · R. Joshi Nair
NAIR-SAFIR Strategy Consultants, Lisbon, Portugal

M.A. Gardetti (ed.), *Sustainable Management of Luxury*,
Environmental Footprints and Eco-design of Products and Processes,
DOI 10.1007/978-981-10-2917-2_3

company that is responsibly run, legally compliant and responsible in managing risk. Finally, indirect stakeholders, such as the people living near a factory or the fans who follow a brand on social media, want more than just a name they recognize. The public now wants to be assured that the company's presence in their lives and communities is a positive influence.

These developments show a convergence in the attitudes of all the stakeholders of a brand toward qualitative criteria in addition to quantitative ones. In an industrial, but pre-information economy world, quantity versus quality was a trade-off. Today, people want both.

This is especially true of luxury. The luxury customer wants the best of something, they are not willing to accept compromise in terms of design, manufacturing, or functionality. At the same time, the luxury customer wants the product they want, how they want it, when they want it, where they want it. This responsiveness is part of what they are paying for.

This mind-set is not limited to luxury. Over the last half-century, there has been a global increase in economic prosperity, turning what was once considered luxury consumption into a mass phenomenon. At the same time, technology has diluted institutions' once total control over brand image and messaging. Further, technology has shifted the definition of value from material wealth to information and knowledge, pushing everything to address more than our physical and security needs.

What does all this mean? It means that, since industrialization and the early 20th-century trials of socioeconomic dislocation and war, society as a whole has moved up the rungs of Maslow's Hierarchy. The post-war period, dominated by the influence of mass media and globalization, was marked by homogenization and the need to belong. And at the turn of the 21st century, technology allowed us to assert ourselves as individuals.

So now the focus is on developing our esteem and self-actualization. Up until this point in the sociological evolution, esteem was derived from the esteem shown to us by others: our status in the community. It required external validation. Society's propensity toward the lowest common denominator made this more about fitting in with mass opinion than challenging it. Standing out required an enormous tolerance for resistance and criticism from others. But the possibilities for individual expression now open to us via digital technology—social media and so on—allow and even encourage us to challenge the mass approval of others. Seeking esteem today means standing out from the crowd: showing our uniqueness and originality.

This was once the exclusive domain of artists, but is now accessible to the rest of us. It leads us to self-actualization, the highest level on Maslow's hierarchy, at which individuals live according to their values and pursue their personal sense of fulfilment.

Businesses must figure out how to integrate this new, individualistic focus into processes that have until now been designed to focus on growth by commodification. To do this, they must reconcile the competing value demands of different stakeholder groups into one value proposition that is shared among those groups. The challenge for business is how to move from thinking about economic value to shared value, and how to implement value systems, a corporate culture and behaviour to turn this ambition into reality.

The shared value model proposed by Michael Porter and Mark Kramer of Harvard Business School, also the founders in 2000 of non-profit strategy consultancy FSG, identifies two spheres: (1) creating business value, defined as investments in long-term competitiveness, and (2) creating social value, defined as investments that address social and environmental objectives (Porter and Kramer 2011). Porter and Kramer locate shared value in the area where these overlap (Fig. 1). Their definition of shared value is a straightforward combination of the two: investments in long-term business competitiveness that simultaneously address social and environmental objectives. On the surface, this is a simple repackaging of the three sustainable development pillars—economic, social, and environmental—into something similar to the holistic approach of social responsibility. (Bear in mind that "social" responsibility does not exclude the environment, but integrates it completely into a broader responsibility to society.)

But Porter and Kramer also go beyond sustainability or social responsibility in their model. By focusing on creating value shared between business and society, business profits by serving society. So rather than seeking to capture a bigger share of the economic pie, businesses should seek to make that pie bigger by solving the problems facing society.

In this model, business growth imperative is satisfied simply by the exponential growth of the human population, the consequent explosion in the scale and

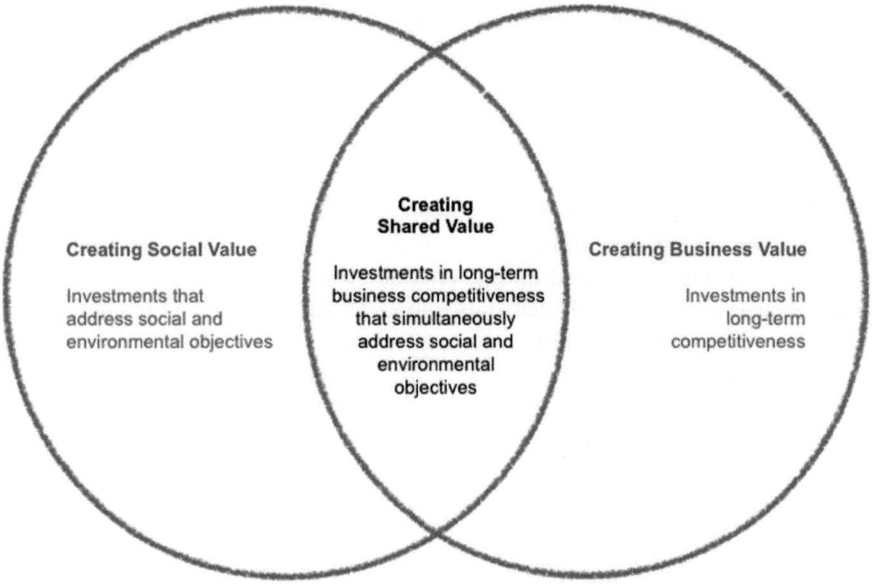

Fig. 1 FSG model of shared value (Excerpted from Pinkhasov, Misha and Joshi Nair, Rachna, *Real Luxury: How Luxury Brands Can Create Value for the Long Term*, 2014, Palgrave Macmillan. Reproduced with permission of Palgrave Macmillan.)

complexity of human need, and helping the world manage an increasingly complicated, fragmented, interconnected, accelerating and growing modern society.

But this model is of only limited relevance for luxury brands because luxury addresses our desires rather than our needs. If anything, luxury addresses higher level, metaphysical needs towards the top of Maslow's Hierarchy rather than the physical needs at the bottom. Even at a physical level, the luxury product does not exist to satisfy a person's basic requirements, but to do so in a way that connects it with their emotional experience. So luxury needs a more sophisticated model for creating shared value; one which accounts for the individual experience.

1 The Three Rings

There is an innate tension between business and society. First, business, by its nature, is competitive while society must be cooperative. Businesses look to capture market share and attract and keep the best resources at the best conditions for themselves. The success of a community hinges on its ability to share resources and meet the needs of all its members. Second, business thrives on dominance, while society aims for an equitable distribution of opportunities. Businesses look for fluidity, as in unregulated markets, in which to do whatever they must as they jostle for a better position. Society, on the other hand, makes rules to promote stability and security, and protect against injustice.

Even with the best intentions, there will inevitably be tension between the two. For example, why would a business pay higher wages if the market accepts lower salaries, particularly in times of economic crisis when unemployment is high and company revenues are low? The US debate about raising the minimum wage in the context of high unemployment following the 2008 crisis was not just about raising the minimum level of pay, but about providing a living wage to even the lowest level workers. Its opponents on the business side argue that keeping the minimum wage low would encourage businesses to hire more workers and allow them to keep prices low for struggling consumers. Its proponents on the social side, meanwhile, argue that employment that does not provide workers with sufficient means for subsistence is a false benefit that only serves to keep consumers struggling and exacerbates economic inequality.

It has become accepted that in order to make money, business must be selfish, merciless, even unscrupulous. It is only after having accumulated wealth that a business can afford to indulge more socially minded impulses by engaging in philanthropy to support charitable works and do good for society as a whole.

But what does it really mean when we say "business does this" or "society wants that"?

The *World English Dictionary* defines a company as both "a business enterprise" and "a number of people gathered together." So, the firm is not one monolithic entity, but is assembled from a group of people joined in the pursuit of its mission. The action of a company is then the sum of the actions of all the individuals, from the decisions made by management to the activities undertaken by labor in implementing those decisions. Those activities are the product of the decisions made by individuals about whether and how to execute the work requested by their bosses. The individual is absolved of personal responsibility for the consequences of their actions outside the firm (within the limits of legal compliance), because these consequences result from the cumulative action of the company.

Similarly, the *World English Dictionary* defines society as both “a system of human organizations” and “an organized group of people associated for some specific purpose.” So neither is society one monolithic entity, but rather assembled from a group of people joined for their shared benefit. The parallel between the two is even more obvious in French, where *une société* can refer either to a specific society or to any firm and *la société* can refer either to a specific firm or to society in general.

The common phenomenon in business and society is the gathering together of individuals. The difference is in these individuals’ reasons for coming together. Tension arises when the reasons diverge, forcing individuals to act in one sphere in a way that is counterproductive to the other. Having to exist simultaneously in both spheres under these conditions pushes individuals into conflict with each other and even with themselves. At best, this causes complex and competing requirements among different stakeholders, which companies have to balance. At worst, it spills over into society causing polarization, class conflict, political gridlock, even civil war.

Porter and Kramer’s model falls short by missing this individual dimension. It misses our need to fit within the values of our social groups as well as the need to fit within the values of the corporate culture for which we work. This sense of belonging furthers our self-esteem, whether by allowing us to rise in social standing or to move up the corporate structure.

But people can behave one way in their personal lives and a completely different, sometimes antagonistic way in their professional lives. This difference is at the heart of the challenge of social responsibility. As citizens, we may complain about businesses polluting the environment, avoiding taxes, or corrupting the political process. We may debate with friends about how money and marketing drive us toward an ever more meaningless, materialistic existence. We may worry that our children spend too much time obsessing over celebrities and fashion trends.

Then, when we go to work, we often check those values at the door. We look for ways to help our companies cut costs by finding shortcuts and loopholes, and by influencing governments to loosen regulations that stifle the company’s profit margins. We look for ways to help our companies grow revenues by coming up with new and more tempting product offers and marketing campaigns. We look for ways to engage consumers by playing into the popular culture and getting the media to cover an accelerating cycle of “must haves.”

Finally, as consumers, we reward ourselves for our hard work by buying things that give us pleasure before going home and worrying about how society is going down the drain.

Brenda Romero is a game designer famous for the successful *Wizardry* video games franchise. In 2009, she developed a board game called *Train*, which was not meant to be commercialized, but which caused a stir in the games industry because of its ability to make players confront the difficult emotion of taking responsibility for their actions. Players follow instructions to efficiently load little, yellow pegs, representing people, onto trains. The winner is the one who gets the most people to their destination. That is a good thing, right? But there is a trap in intentionally designed procedural gaps that oblige the players to agree on the rules going forward.

The twist comes when the destination is revealed to be Auschwitz, with the consequent realization that the players have been competing to send a maximum number of people to their deaths. According to Dean Takahashi: "The key emotion that Romero said she wanted the player to feel was 'complicity'." (Takahashi 2013). This complicity comes from blithely following instructions and even contributing to them by coming to a consensus with other players about how to fill the procedural gaps. Once acknowledged, this sense of complicity causes people to react with shame and even try and subvert the game, as by hiding the pegs.

Train demonstrates how easily individuals can and do subordinate their own values to the larger value system imposed by the context—be it by their community or their work, how easily people are "trained"—particularly when the full consequences of their actions are unclear.

The difference between real life and a game is that once the game player understands the point of the game, they can opt out. Things are not so simple in real life. Leaving the "game" has financial, personal, and professional costs. Few people can easily walk away from a job when they have obligations to their family and creditors, and aspirations for their own careers. Even as consumers, even as citizens, people look away from inconvenient truths. They may choose to wilfully not know the consequences of their actions because their obligations within the rules are enough of a burden. Even if companies can form a consensus among their people to change the system, it is still difficult to abandon existing practices—difficult and often pointless.

In real life, if a "player" leaves the "game," this does not mean that the game ends. Until the whole of society has agreed that the goal of the game is inconsistent with the goals of society, and implements the political and legal reforms to change the game, walking away from the table only frees a seat for another player. Boycotting the game serves one's own sense of integrity. But in the bigger picture, it accomplishes nothing until enough individuals opt out so that their seats are left vacant and the game ceases to exist.

We rarely openly articulate our deepest values and ideals in daily life, but they are visible through our actions. However, it can happen that we feel obliged to act in ways that differ from what we hold in our hearts. To be in harmony with oneself, there must be conformity between one's values and how one actually lives. But the truth is that we often live in a state of personal–professional conflict because we have not determined which values are responsible for our choices, and our actions and choices do not stem from our most firmly held beliefs.

Only at a turning point of crisis do we stop to reassess. This is why people will often change career paths after a life-changing event or a near-death experience, such as an accident or a serious illness, or a strong emotional experience such as the loss of a job or a loved one, or even in response to external events such as a natural disaster or a catastrophe like 9/11. These existential moments cause us to reflect on our lives up until that point and they confront us with the combination of our deepest values and our current priorities and behaviors.

It does not even have to be that traumatic. The roots of the word "crisis" come from the Greek *krísi*, which connotes decision or judgment. So, any turning point will do. The almost universal experience of the midlife crisis comes when people have reached a point in their lives, often their careers, where they get a sense of the time they have left to live and ask themselves how they want to spend it and the legacy they want to leave.

According to Sylvie Bénard, environment director at LVMH: "I often get approached by people looking to change jobs. They have reached a certain level of career success, but having achieved that, they realize that they also want to do good. I tell them to stay where they are. The environment needs allies throughout an organization, and I tell them that they can do even more good by changing things in their current role."

So, we can bring our actions in line with our values without abandoning the game, but by changing how it is played. Classical Greek philosopher Socrates advised: "it is the greatest good for a man to discuss virtue every day." Socrates was one of the earliest in a long line of thinkers who believed that the more dialogue we create around our deepest values and ideals, the more we reinforce them, and the more likely we are to act accordingly. Even if—perhaps, especially because—that dialogue leaves people with a keener awareness of how much we do not know about our values and our willingness and ability to enact them.

Research shows that critical self-examination of values affects subsequent behavior. In psychological studies, subjects were asked to participate in an activity that measured how much they discriminated against a randomly created group of people in the lab. One group first spent ten minutes expressing and defending their attitude towards equality. A second group was simply reminded of the value of equality but not asked to defend their views. In the subsequent activity, the first group demonstrated less discrimination than the second (Maio 2010). Those subjects who articulated their values were more likely to apply those values to their behavior, supporting Socrates' use of critical self-examination as a bridge between values and action.

Recognizing the role of individual values in collective outcomes, the shared value model in this book has three spheres: business, social, and individual. Taking Porter's and Kramer's definitions as a starting point, our definition for the individual sphere can be seen as:

> Investments that encourage leadership and build relationships between individual stakeholders that help align social and business interests.

This opens up critical self-examination within people to reconcile their conflicting personal and professional expectations. It also opens up dialogue between individuals, based on this reconciliation, which makes consensus with others and solution-seeking possible.

Rather than pitting stakeholder interests against one another, it opens up channels for mutual understanding, aligning objectives, and sharing the ideas and resources necessary to change the game. By being the common element of each, the individual binds the molecule of business and society together.

We then take it a step further by removing the associations imposed by the word "investment" and focusing directly on the value delivered in each sphere. This shifts the emphasis from the costs to the benefits. While focusing on the costs keeps people thinking in a mind-set of less, focusing on the benefits opens up the possibility of more. This is also how luxury approaches product value. Individual value means providing opportunities for self-actualization, accomplishment, and validation. These connect to the business values of product offerings for consumers, meaningful work for employees, and profits for investors. They connect to social value by providing responsibility and leadership that lead to community enhancement.

These additional dynamics make three-dimensional shared value possible. Porter's and Kramer's approach is strengthened by the addition of two new overlaps. By linking the individual and social spheres, we unleash the dynamic of humanity and solidarity; the compassion and cooperation that are at the root of what differentiates humans from animals in being able to form a society and build a civilization. By linking the individual and business spheres, we formalize relationships and objective-setting to create a dynamic of consensus and commitment. At the core, we find three-dimensional shared value, which engages the individual to solve business and social problems. The original shared value model is now expanded from sustainability to include engagement, through the ability to align objectives and share resources, and establishes companies as citizens of the community (Fig. 2).

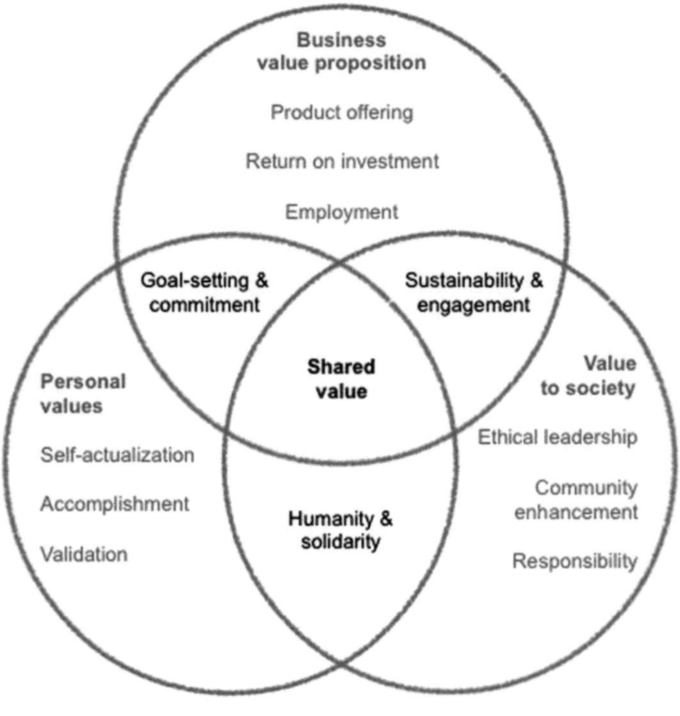

Fig. 2 Three-dimensional model of shared value

2 The "Give" Model

The question then, is how to reconcile the vast diversity of individual perspectives and point them all in the same direction to achieve this goal. When speaking about shared value, giving financial incentives alone is not enough, as that would run counter to the entire model. People are waking up to the power of qualitative motivations and of having a sense of purpose: that one's actions have impact and change something. More and more, this is about effecting change rather than just filling personal coffers. So companies can provide tremendous shared value by helping individuals achieve their qualitative goals. Their purpose is bound up in what it is about the world they want to change.

This is especially true of luxury, which addresses higher level needs of belonging, esteem, and self-actualization. Luxury is not in the security business, so it is free from defending the status quo. Like artists and other creators of social value, luxury firms envision an ideal world, one that is essential to creating the appetite for luxury. Once they themselves are in harmony with that vision, they can go about changing the game. This means that they have to realign the relationships between individuals through which the brand functions, both internally and externally—the brand's owners, its employees, its customers, and the broader community.

Today, a brand's relationship with its stakeholders functions according to what we call a "take" model (Fig. 3). This is not an impeachment of the company as greedy, taking from its stakeholders. On the contrary, the company is constantly being bled of resources. In this system, all the actors are perpetually "on the take" because the "take" model is based on transactional exchanges rather than relationships. The brand exchanges money and benefits for the labour of employees. It exchanges products for customers' money. It exchanges marketing investment for the expectations of the community, some of which converts into customer demand. All this in order to exchange profits for the continued support of investors. In a transactional view of work, the parties meet to take from one another. Money and utility, which are easily sourced elsewhere, are the basis of this model. As a result, a company's resources are taxed in multiple ways just to feed a single stream of returns.

Fig. 3 The "take" model of stakeholder relationships

Businesses may think the "take" model works in their favour because it is transactional and their investment is measurable against the returns they gain. For example, how many hours of work or how much output they can expect in return for the salary paid to employees, or how many products will sell at a particular price. The "take" model works on the assumption that each of the exchanges is hermetically sealed, and that the value chain from a company's owners through its employees to its customers and community is a simple linear construct. But, in a transactional relationship, each party gets only what they pay for and will therefore give only as much as they feel is justified by the price. When they feel they can get a better deal elsewhere, they will move.

To use a concept articulated by James Moore, companies exist in a business ecosystem where they are in simultaneous competition and cooperation with other players (Moore 1993). Like animals in the wild, they both need each other to survive and yet compete for the same resources. Example: a lion predator needs his antelope prey to survive, but he must share the watering hole with the antelope if he is to be nourished by its meat. Alternately, the antelope needs his herd for the safety it provides, but competes with them for grazing land. When one species disappears from an ecosystem, it can take a range of other species with it. But at the same time, that creates opportunities for other species that may have been marginal to the ecosystem to suddenly flourish and become dominant.

However, a company is not just an actor in the larger business ecosystem. It is also an ecosystem itself, one with porous inner and outer boundaries. Thanks to technology and rapidly evolving attitudes to the work–life divide, individuals are no longer confined to just one role in the organization. Increasingly, an employee can simultaneously also be a client, a shareholder, an external stakeholder, or any combination of these. So the organization is part of an ecosystem where the different players are linked, internally and externally, not in a linear or one-to-one relationship but in a complex web of multifaceted relationships. Twitch one strand of this web and it sets several, perhaps even all the strands vibrating, creating a resonance across the entire system.

In contrast to the "take" model, the "give" model gives every actor more than they pay for (Fig. 4). The company gives its employees not just a job, but a purpose. The sense of meaning generated by having that purpose rather than just a

Fig. 4 The "give" model of stakeholder relationships

set of tasks and responsibilities contributes to employee motivation and increases their productivity. Employees, who are the interface of the brand with the customer, now carry out more than the simple sales transaction of products for money. They provide solutions rooted in the company's purpose, recognizing that customers are not just consumers looking for products, but part of a much bigger community where complex issues and decisions intersect with their consumer behavior.

The move from products to solutions validates the consumer as a whole person. People naturally reward validation with loyalty. This is a heartfelt and intimate kind of loyalty, which goes beyond the loyalty created by loyalty and customer engagement programs such as private sales, frequent flyer miles, and social media campaigns designed to keep customers consuming. Those programs are just another form of transaction. More important, however, is the intent and substance of the relationship. By providing people with solutions that go above and beyond the product, the brand is making a contribution to the community. This raises the brand's standing by increasing the amount of goodwill felt toward the brand by the larger public. Goodwill is the key ingredient of brand love and reputation, and increases the brand's value as an asset. The brand comes to be seen as a leader, and its owners—who direct it toward a purpose—gain not only financial returns, but also prestige, with the trust and standing to exert even more leadership.

The "take" model drains the company of various resources in order to return revenues, keeping the company focused on marginal benefits and cost mitigation. It is powered by thinking small, which results in a race to the bottom. On the other hand, the "give" model creates a virtuous cycle in the behavior of the company by reinforcing its output of purpose with multiple streams of support, allowing the company to reinvest these in continuing its work. It aims for bigger. It is a race to the top.

The "give" model does not replace the "take" model but adds another dimension to it, giving us the combined model shown in Fig. 5. The relationships inherent in a

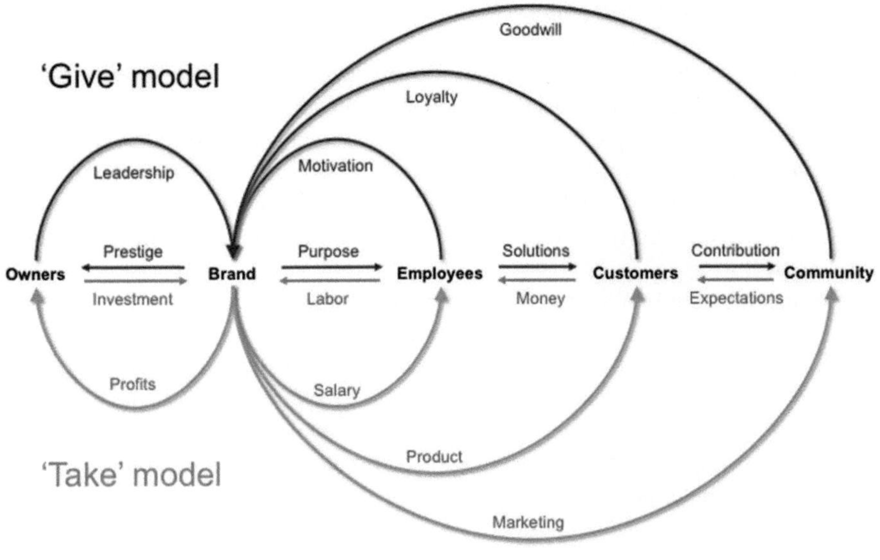

Fig. 5 The "give" and "take" models together

shared value model do not replace the transactions, but augment them by giving meaning to the whole process. This contributes to individuals' sense of fulfilment. Stakeholders are no longer beholden to one another in the moment. The brand no longer has to buy their loyalty. Instead, they voluntarily and continually give back to it in the form of the strength of its position in the community, the trust it enjoys, and its consequent influence. Self-interest still drives this approach, but seeks to increase the total amount of value in the system. By contributing to the company's internal and external ecosystems, it recognizes that individual interests thrive best when the whole environment thrives around them.

Take Apple, Inc. as an example. Apple is an innovator, but contrary to popular belief, Apple is not an inventor. Personal computers existed long before Apple came on the scene. Even the smartphones for which Apple is now celebrated were first developed more than a decade before the iPhone, by Ericsson, BlackBerry and others who combined the functionalities of mobile phones, digital organizers, home computers, cameras, and so on.

Apple's purpose is not even to be a technological innovator. Rather, the purpose behind their innovation is to humanize the technological experience. Apple realized this vision through "skeuomorphism," in which the technological user interface emulates objects in real life. On an Apple "desktop," for example, you move and layer documents the way you would papers on a desk. This was first conceived by Apple before flowing back into Microsoft's and other companies' products. An Apple product is designed to be intuitive: switch it on and start to use it. Most Apple products do not even come with a user guide.

By humanizing the use of technology, Apple has built an almost evangelical loyalty to the brand. And this is not just among the customers who wait in line for days before a new release. For example, Apple is not known for paying particularly high salaries to its store staff. But the personnel at Apple stores are proactively helpful and highly informed because they are driven by a passion that is converted into caring for the customer. This is a manifestation Apple's purpose shared right up the value chain, from the programmers who develop the software, to the designers of the hardware, to the standards pursued by managers. The sense of purpose has resulted in Apple's commercial success and the stratospheric rise in their share price.

That is not to say Apple is a gentle saint. Steve Jobs was notoriously difficult to work with. The company has been criticized for working with manufacturers in China that employ abusive labor practices. Their environmental track record needs work. And its community of followers constantly pushes them to improve in those areas as well.

The specific nature of a particular company's purpose depends on its own story. Not every company is in the business of technological innovation. A brand derives its purpose from its backstory of values and origin; the founding spark that motivated an entrepreneur to change something in the world. This is ultimately the road to citizenship. It uses the company's competitive nature to meet the cooperation needs of a society. Across all examples of corporate purpose, the common ingredient is the human dimension, which in one way or another is about caring for other people.

Gary Vaynerchuk, author of *The Thank You Economy*, says that: "Especially in the luxury category … we are living through the humanization of business … We have now conditioned ourselves as humans to actually have relationships with logos … Now it's going to come down to: Does business have the ability to be human? Do we strategize for sales or do we actually give a fuck?" (Vaynerchuk 2011)

A tremendous body of knowledge has emerged on corporate purpose. The *Good Purpose Study* by Edelman, the world's largest independent PR firm, found in 2012 and confirmed in following years that 87% of consumers believe business should place at least equal weight on societal and business issues, but only 28% of consumers believe business is performing well in this regard (Edelman 2012). It also found that consumers are more willing to name and publicly shame companies they feel do not live up to their social obligations. Beyond social media activism, consumers say they will avoid buying a company's products and even refuse to invest in it if they do not feel that the company contributes to a good cause. (In practice, companies have found this to be not quite so clear-cut: actual consumer and investor behavior has yet to fully reflect the survey responses.) Consumers' expectations for a company's contribution take familiar form: philanthropy and process or partner innovation. But the stakes here are rising as well, with consumers increasingly asking companies to provide education, information and dialogue opportunities, and to work with government, with the burden being on CEOs to proactively address societal issues.

Adam Grant, management professor at the University of Pennsylvania's Wharton School, identified three types of corporate culture: takers, givers, and matchers (Grant 2013). *Taker cultures* have high levels of internal competition, which causes colleagues to withhold assistance and information from one another, and undermine each other's work in order to get ahead individually. In *matcher cultures*, the category into which most organizations fall, colleagues help one another, but in quid pro quo exchanges of traded favors that occur within closed loops of relationships that do not include outsiders. In *giver cultures*, colleagues help one another without expecting anything in return, by sharing information, mentoring, and making connections. While they have to contend with issues such as distraction and inefficiency, giver cultures achieve the highest levels of performance.

Grant found that "The frequency with which employees help one another predicts sales revenues in pharmaceutical units and retail stores; profits, costs, and customer service in banks; creativity in consulting and engineering firms; productivity in paper mills; and revenues, operating efficiency, customer satisfaction, and performance quality in restaurants."

He cites research by the University of Washington, which found that competitive teams finished tasks more quickly than cooperative teams, but did so at lower levels of accuracy because members would not share crucial information with one another. Furthermore, the study found that teams that were incentivized to transition from competition to cooperation had trouble doing so. People had lost trust in colleagues who had previously been seen as competitors. Speed decreased, but quality did not improve.

This is clear evidence that the collaborative culture of communities—with aspects such as sharing, trust, and even altruism—transposed to companies through the "give" model can deliver measurable performance benefits beyond those offered by traditional competitive approaches. The returns on investing in being a good corporate citizen are not only a matter of faith. Furthermore, the "give" model creates fertile ground for cultivating leadership within and by the company, which generates prestige in return for the leader's contribution to individual as well as collective well-being. It is a much stronger leadership status, given by the support of the constituency rather than taken in the pursuit of power.

3 Connecting the Brand to the Individual

Passionate support for a brand cannot be demanded. A company must inspire it. Once a company has aligned its business, social, and individual value propositions, and identified its purpose, the question is: How can the company orient its stakeholders to support this purpose? The answer is to empower individuals to also align their own values across the three spheres in order to exercise their autonomy of judgment. This kind of individual empowerment can be seen as dangerous by organizations because it gives them less control over employee behaviour.

Most brands are very self-protective and closely control internal and external information and decision-making processes. Luxury brands can be particularly ferocious in this regard because so much of their cachet is bound up in the brand's mystique. But the transparency inherent in combining the business, social, and individual spheres of shared value builds trust internally and externally. The rules that determine the level of personal autonomy are dictated by a company's corporate governance structures, so what are these structures responding to? Is corporate governance a simple act of legal compliance, or does it support shared value creation? In a similar way, are companies encouraging their people to follow rules or fulfill its purpose?

LRN is a consultancy that inspires companies to "principled performance" and helps them build resilient and innovative cultures. LRN's work identifies three types of corporate culture: blind obedience, informed acquiescence, and self-governance (LRN 2012). *Blind obedience* organizations, like military organizations, show strong top-down leadership, and even coercion through rules and policing. *Informed acquiescence* organizations, like religious institutions, rely on performance-based rewards and punishments to motivate people to follow the rules, policies, and procedures established by what they believe to be a skilled management team. *Self-governance* organizations, like neighborhood cooperatives, are primarily values based, where purpose informs decision making and guides all employee and company behaviour.

These cultures have consequences for a company's ability to balance long- and short-term thinking. While obedience and acquiescence cultures place importance on the immediate benefits of following rules, self-governance encourages all employees to be leaders with a focus on long-term legacy. This gives their work meaning, which contributes to their sense of fulfilment.

Dov Seidman is the founder of LRN and the author of *How: Why How We Do Anything Means Everything*. His work with the Boston Research Group found that employee engagement has been falling across the board. A 2011 study by the Boston Research Group found that 27% of corporate bosses and only 4% of employees believe their firms are inspiring places to work (Seidman 2012). The lower figure is perhaps less surprising, because if those responsible for a firm's work do not find it meaningful, how can those who work for them care about it at all? Seidman calls current employee engagement cultures "out to lunch"—literally, companies feel they can improve their employee relationships with a proliferation of social events and team-building exercises. In truth, these often backfire by encroaching on an employee's personal life and their productivity. What is missing is the trust that allows organizations to empower their people.

LRN's *The How Report* found in 2012 and confirmed in subsequent research that there is a large gap in perceptions of self-governance between bosses and employees (LRN 2012). Senior management consistently felt that their organizations encourage more self-governance and less coercion than the overall employee population. To some extent, this is natural, as senior managers' responsibilities involve them in more lateral thinking and strategic decisions, which innately imbue their roles with more creativity than those charged with implementation. But this also concentrates a feeling of autonomy at the top, which does not disseminate trust throughout the organization, undermining the reciprocal engagement that leads to a sense of meaning and fulfilment for employees.

According to LRN's findings from observations of 36,000 employees in 18 countries, "Employees who experience a high-trust environment are 22 times more likely to be willing to take risks that could benefit the company. Employees functioning in an organization of high trust are eight times more likely to report higher levels of innovation relative to their competition. And finally, employees functioning in a culture of high trust, risk-taking, and innovation are six times more likely to report elevated levels of financial performance compared to the competition."

To build trust, companies need to extend the sense of ownership that senior managers feel to all levels of the organization. Only then will the corporate culture and the benefits of its purpose flow out into the community. If you are going to instil employees with a sense of purpose, they also need the liberty to pursue that purpose according to their own methods. Otherwise, it is not a purpose, it is an obligation.

To allow workers a feeling of purpose, they need the autonomy to express their values through the company's actions in a way that is coherent with the company's values. A company should not police employees' actions, but build a culture where employees know what is expected of them and can fulfil those expectations according to their own values, knowing that a certain core set is aligned with the company's. Further, because actions can transform values, the company has a pedagogical role in transforming the person, liberating them from blindly following instructions to the legal minimum.

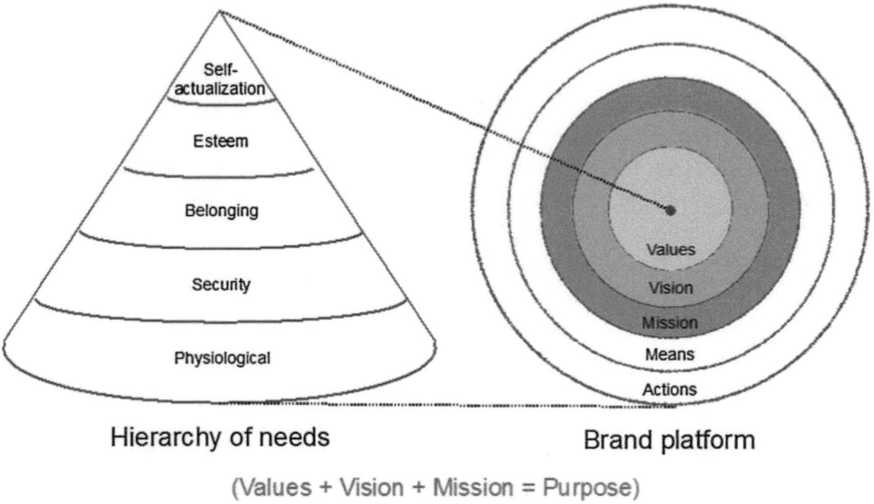

Fig. 6 Connecting the brand to the individual

The brand can be a tool for building a culture of self-governance. Individuals share an essential structure of needs based on our biological origins and the evolution of the human psyche. Abraham Maslow's "theory of human motivation" presents these in a concise and useful form, which is a familiar business tool (Maslow 1943). Viewed from above, Maslow's hierarchy of needs can be seen as a set of concentric circles, providing an opportunity to establish parallels between the layers of human needs and the common elements of a brand platform. Translating the brand to the individual experience shows how the brand can be the basis for need fulfilment on many levels (Fig. 6).

Businesses are often conceived in terms of their utilitarian actions: the manifestation of their value proposition to customers. More sophisticated models seek to frame these in terms of an overarching mission, with vision and values making the transition to the execution of actual work. Shared value is partly an emotional intelligence appeal, and luxury brands operate at the emotional and business nexus. A brand platform centred on the individual puts values at the core. Defining the self first gives meaning and direction to all the other elements as a source of motivation.

In the same way that a company can either serve compliance or create shared value, a company can also either let its actions determine its mission, vision, and values, or use its values to inform its vision, formulate its mission in service to that vision, and align its actions to suit. This parallels how people either can move up Maslow's hierarchy, pursuing higher level needs only once their physical and security needs are met, or use their need for self-actualization to meet their lower level needs by achieving success.

The vast majority of people work to move up Maslow's hierarchy. The fear of not meeting their basic and security needs keeps them from taking the risks that lead to self-actualization. But a special category of people, those driven by belief and purpose—artists like Vincent van Gogh and Leonardo da Vinci, or social activists like Nelson Mandela and Mahatma Gandhi—actually prioritize their need for fulfilment over their needs for security, money, and even sex.

Luxury brands function like artists. They are not in the business of creating just any product to sell and make profit. They are capable of sacrificing boundless growth in order to stay true to their vision. They are in the business of ennobling the human experience. This obviously happens at the physical and emotional levels, but they can influence the purpose-driven level as well.

When PPR changed its name to Kering in 2013, it was intended to signal a new strategy for the company. Under the leadership of François-Henri Pinault, Kering was evolving from a diversified holding company into an apparel and accessories group focused on luxury and the sport and lifestyle sector. In addition, Kering was making sustainability a core pillar of its corporate strategy. The name Kering resounds with the pronunciation of "caring" and takes its root from "ker," which means "home" in Breton, Brittany being the native region of the Pinault family. The new name also traces its roots to *Home*, the nature documentary made by photographer Yann Arthus-Bertrand, the production of which had been supported by PPR. The original name of PPR's sustainability department was PPR Home, showing how the company's commitment to sustainability came to define its corporate identity.

In implementing this commitment, Kering relies on an operational style called "freedom within a framework," by which its brands and its people are empowered to find their own ways to accomplish the group's goals in accordance with the group's values. Kering states its mission as allowing its customers to express, fulfil, and enjoy themselves through their products. This is all summed up in the Kering signature statement: "Empowering Imagination," which we conclude to be their purpose. Empowering the imagination of people within and outside the organization is the contribution that fuels their mission.

If companies do not manage their corporate culture, a culture will emerge anyway, like weeds in a garden. Rather than policing their employees' behaviour, companies must establish a culture where the rules of behaviour are not just explicitly written out, but where the expectations for what is appropriate are implicitly understood by all.

4 One Ring to Rule Them All

Luxury is the marriage between art and business, which means that it does not have to choose between working up or down Maslow's hierarchy, but can reconcile the two. What differentiates the starving artist from the successful artist, and the angry street protester from the leader of social change is their ability to merge their values with those of the market and community.

Luxury firms have already mastered many of the challenges facing other companies, such as providing personally attentive customer service, superior working conditions, creative freedom, and respect for raw materials. They are also good at engaging social causes, such as women's rights, biodiversity, cultural preservation, and education. But in an ever more competitive world, do they give up this lead to competitive pressure or can they exert outward influence so that the values that make luxury a leader can be shared more broadly throughout the business and social cultures?

Many non-luxury companies have picked up on the verbal and visual vocabulary of luxury. So, luxury does have an influence. How can it be more than stylistic and actually contribute to the creation of shared value? If luxury spreads its culture throughout business and society, will it not surrender the very thing that sets it apart?

No. Because luxury's advantage is not just its ideals, it is also the execution of luxury, and not every company is able or even willing to invest the time and talent needed for the dazzling level of artistry and quality that define luxury. And not every customer is able or even willing to appreciate them. Even in an ideal world, there will always be a lag. So, because luxury's culture will spread from luxury, it will always belong to luxury. Furthermore, as luxury's culture spreads to the rest of the market, it will push luxury ahead to open up perspectives that have not yet been experienced. That is, after all, how luxury remains luxurious.

Monet's *Water Lilies* is one of the most familiar images in art. Countless reproductions of the iconic work hang as posters in doctors' waiting rooms, adorn desks as calendars, liven up sofas as throw pillows, and protect people from the weather as scarves and umbrellas. This has not diminished the number of people wanting to see the original work. On the contrary, the painting's fame has increased the number of people wanting to come to Paris precisely to see the massive original adorning the curved walls of the Musée de l'Orangerie for themselves. And to Giverny to see the real gardens that inspired Monet, and the universe within which he lived and worked. Of course, *Water Lilies* is one of a kind. Forget about aspiring to one day be rich enough to afford the painting. You cannot even aspire to own it. Everyone knows that the version you have is an imitation. But the imitation has only driven the appetite for the real thing.

Luxury goods are different from *Water Lilies*. They exist in multiple examples. We can aspire to own them, so it is easier to pass an imitation off as the real thing. This makes them vulnerable to the effect of forgeries, the bête noire of the most popular luxury brands. But this popularity has also caused the legitimate stylistic imitation that has been hailed as the democratization of design. It has raised the bar for consumer tastes by opening our eyes to the possibilities luxury has discovered. In the same way that the impressionists broke barriers about the way artists represent the world, luxury breaks barriers about the way we experience an object through the senses and through the emotions the product can evoke.

The posters and umbrellas are pretty, but they have nowhere near the impact and emotional satisfaction of the original work of art. So, luxury can drive the appetite for the real thing, even in a world where imitations are widely available. By increasing the appeal of luxury and heightening our appreciation for what luxury really represents, luxury can drive consumers to desire better things. They may not all be able to afford "luxury" goods, but they can be pushed to look for products that are authentically better rather than for a proliferation of cheap imposters. When you increase the understanding and appreciation for the real thing, you are also increasing the pool of potential luxury customers.

As the most emotional of business sectors, luxury establishes a dialog with individuals paralleling the dialogue that art has with individuals. For this dialogue to work, it is crucial to get past the urge to market. Just chatting with fans on social media is not enough. The quality of that conversation is defined by its substance, and the substance is defined by the character of the interlocutors.

A self-absorbed person at a dinner party will have an equally vapid Facebook page, likely an endless stream of duck-faced selfies; whereas an interesting, intelligent person is likely to use Facebook to share their insights and observations about the world, articles and interesting ideas from others, as well as support causes and call attention to important issues. They enlarge your horizon by being a window onto the wider world.

When brands see social media in particular, and their stakeholder engagement more generally, as a platform solely for self-promotion, they effectively become a billboard that demands our attention but obstructs our view. We may look at it for a bit, but the instinct is to get past it and see the world beyond.

Our values shape our behaviours, but the reverse is also true: by acting in a certain way, we mould our thoughts to justify our actions, and gradually adapt our values to support them. This is the key to enculturation, which may happen unconsciously, as in parenting or peer pressure in schools, or consciously, as in religious education or indoctrination (Kottak 2011).

Because companies exist in ecosystems, they are subject to coevolution. They influence their environment to evolve. In turn, they evolve in response to environmental influences. Just as organizations coevolve in their ecosystems along with their suppliers and competitors, employees internally coevolve with the company they work for and thus shape the company itself. The employee is a real connection between the inside and outside worlds of the company.

Today more than ever, with a single individual coexisting in multiple stakeholder categories, companies are less hermetically sealed entities than they are porous vehicles. This dynamic can put the company either in harmony or in dissonance with the objectives of society. By creating a culture that causes people to act in a particular way, they can have a viral impact on the community. They can also learn from their environment by listening to the behaviours and expectations their own people bring in from outside.

The overarching debate about the role of companies in society boils down to corporate citizenship. If a corporation is legally considered to be a person, it is subject to the same responsibilities as all other members of society and is to be judged based on its contribution and moral character. So, the behaviour of a brand cannot be separated from the behaviour of the individuals that comprise that brand.

Luxury is exemplary. This makes it more than a citizen. It makes a luxury brand a leader. By establishing the brand as a culture that drives individual behaviour, a company can spread its leadership culture to the entire community. By basing its culture on a purpose that improves the ecosystem, a company can raise its stature and enhance its ability to lead. By using the values widely shared by individuals, it can align its values as a business with the values held by society.

References

Edelman (2012) Goodpurpose 2012: global consumer survey. Edelman, New York

Grant A (2013) Givers take all: the hidden dimension of corporate culture. McKinsey Quarterly. http://www.mckinsey.com/business-functions/organization/our-insights/givers-take-all-the-hidden-dimension-of-corporate-culture. Accessed 24 June 2016

Kottak C (2011) Window on humanity: a concise introduction to anthropology, 5th edn. McGraw-Hill, New York

LRN (2012) The how report: new metrics for a new reality: rethinking the source of resiliency, innovation, and growth. LRN Corporation, New York

Maio G (2010) Psychology of human values. European monographs in social psychology. Psychology Press, London

Maslow AH (1943) A theory of human motivation. Psychol Rev 50:370–396

Moore J (1993) Predators and prey: a new ecology of competition. Harvard Bus Rev 71(3):75–86

Porter M, Kramer M (2011) Creating shared value: how to reinvent capitalism—and unleash a wave of innovation and growth. Harvard Bus Rev 89(1/2):62–77

Seidman D (2012) (Almost) everything we think about employee engagement is wrong. Forbes. http://www.forbes.com/sites/dovseidman/2012/09/20/everything-we-think-about-employee-engagement-is-wrong/#5bb0525d1b67. Accessed 24 June 2016

Takahashi D (2013) Brenda Romero's train board game will make you ponder. VentureBeat. http://venturebeat.com/2013/05/11/brenda-romero-train-board-game-holocaust/. Accessed 24 June 2016

Vaynerchuk G (2011) The thank you economy: how business must adapt to social media. HarperBusiness, New York

Sustainable Performance Management in High-End Luxury Goods Firms: The Use of the "Reputation-Clock" Model

France Riguelle and Didier Van Caillie

Abstract High-end luxury firms have to elaborate even more precise and complex strategies depending on a large variety of key success factors in order to manage their value chain. Meanwhile, they have to incorporate the ethical and societal strong requirements imposed by sustainability. Therefore the performance management tools that they use have to go far beyond the traditional models. In particular, these have to take into account their potential huge impact on reputation if some related risks occur, as it is considered as one of the most valuable assets a firm can own, in particular for small organizations in this industry. However, this is rarely considered as an integrated part of the performance management process of the firm, disregarding the intrinsic complex nature of the relationships that govern the process of accumulation and dilution of reputational capital in particular, and the key success factors that can be mobilized by the organization in order to engage and feed it. Building on the specific case of a small Belgian firm of the luxury industry, this paper describes the motivations and the content of a reputational performance management tool, the "Reputation-Clock" Model, which incorporates the different key success factors that a firm from the high-luxury goods industry has to leverage in order to master its strategic performance from a reputation point of view. Our findings suggest that accumulating reputational capital is influenced by the organization's capacity to manage the related reputation risk. In particular, we distinguish the co-existence of a virtuous and a defective triangle of reputation, which mobilizes a short-list of key success factors in order to enhance the performance of the organization under study. Finally, we observe that the development of internal reputational capital measurement tools, or else the

F. Riguelle (✉) · D. Van Caillie
HEC Liege, CEPE (Center for the Study of Corporate Performance),
University of Liege (ULg), Rue Louvrex 14, 4000 Liege, Belgium
e-mail: France.Riguelle@ulg.ac.be

D. Van Caillie
e-mail: d.vancaillie@ulg.ac.be

M.A. Gardetti (ed.), *Sustainable Management of Luxury*,
Environmental Footprints and Eco-design of Products and Processes,
DOI 10.1007/978-981-10-2917-2_4

definition of a systematic procedure in order to react a reputation crisis would require much resources for this kind of firms, while flexibility and education can be key for both accumulating and protecting reputational capital.

Keywords Luxury · Luxury firms · Reputation management · Reputation clock · Key success factors sustainable luxury · SMEs

1 Introduction

Value chain management in high-end luxury firms becomes even more challenging, especially for small local ones that remain independent from luxury groups (Gardetti and Torres 2014). This requires from them to elaborate a precise and complex strategy depending on a large variety of key success factors (Riguelle and Van Caillie 2012b). When these strategies have to take into account the ethical and societal strong requirements imposed by sustainability, they need to implement management tools that go far beyond the traditional Sustainable Balanced Scorecard (Figge et al. 2002), notably when they have to prevent the potential huge impact on their reputation (Fombrun 2007) of risks arising when some elements are deriving wrong.

Corporate reputation can be seen as a rare and difficult to imitate intangible asset a company can own (Caves and Porter 1977; Mahon 2002), and can be measured by its reputational capital (Fombrun 1996; Fombrun and van Riel 1997). Therefore the capacity of a small high-end luxury firm to accumulate reputational capital, but also to prevent its dilution due to the arising of reputation risks may become critical for leveraging key competitive advantages (Riguelle and Van Caillie 2014). As a consequence, it requires understanding the dynamics that underline the process of accumulation and dilution of reputational capital.

Corporate reputation emanates from the aggregation of all stakeholders' perceptions of an organization's past actions and future prospects, i.e. in its ability to meet their demands and expectations compared against some standards (Chun 2005; Fombrun 1996; Rindova and Fombrun 1999; Walker 2010; Wartick 1992, 2002). Just like in any other firm, luxury companies' stakeholders can be classified into two categories: the external stakeholders which include the firm's customers, suppliers, shareholders, the government, the society, or else the media, and the internal stakeholders, namely the organization's employees, managers and owners. Corporate reputation is then "at risk" in every action with them, while standards stand for their reference framework (Carmeli and Tishler 2005). Standards emerge from social norms and aspirations at both the local and the international levels, while luxury precisely reflects these social norms and aspirations (Berry 1994). Reputation of a luxury firm is then totally embedded into its own environment and context. Meanwhile, sustainability now manifests itself in commonly accepted standards of corporate social responsibility (CSR) (Davies et al. 2012). But consumer responses to CSR are more nuanced than an explicit endorsement of corporate CSR policies (Smith et al. 2010). The expected quality of luxury originally embedded into its capacity to transmit ancestral skills and to preserve raw materials and local activities now includes the social and environmental dimensions (Lochard and Murat 2011). Sustainable development might thus be considered as a catalyst for brand differentiation and corporate image among customers and other stakeholders that are concerned by CSR issues (Kim et al. 2012). Consequently, sustainable luxury as a synonym of environmental and social development, but also of local culture, art, innovation, and craftsmanship preservation may then clearly become a powerful engine for the management of luxury firms' reputation risks in order to sustain their reputational capital.

The existing literature on reputation risk management has several limits for this purpose however. First, as it is based on various approaches to reputation risk conceptualization, which refers themselves to different concepts of reputation (Fombrun and van Riel 1997), reputation risk management still appears as a segmented management discipline that remains difficult to put into practice (Chun 2005). This is particularly the case for small firms that often lack attention in this field of research. Second, the idea that reputation risks are embedded in social constructions based on the perceptions of organizations' stakeholders (Rao 1994) as it is clearly the case for luxury firms is rarely considered (e.g. Kartalia 2000; Scandizzo 2011), thereby failing to question the existence of complex systems fuelled by social constructions at each level of assessment of reputation risk (Scott and Walsham 2005). Relatively recent contributions associate reputation risk—or assume that it is strongly correlated—with corporate social responsibility (Bebbington et al. 2008; Gaultier-Gaillard and Louisot 2006; Hogan and Lodhia 2011), but continue to bound the reputation risk construct into a circumscribed management discipline disregarding its complex status and nature (Suomi and Järvinen 2013). Although reputation risk management can only be effective if it operates as a continuous preventive and reactive influence on the organization's actions, behaviour and standards (Davies 2002), these approaches to reputation risks engender mainly static, asset-oriented, and reactive risk management systems that tends not to consider the dynamic aspects of the accumulation and dilution of reputational capital (Scott and Walsham 2005). However, as the business model of a luxury firm is embedded in societal values, failing to recognise changes in these later can result in substantial risk to the business (Berghaus et al. 2015). Therefore in the case of reputation risk management, preventing courses of actions rather than reacting to crises remains critical for a luxury firm to manage its reputation.

In this context, this paper describes the motivations and the content of a performance management tool, the "Reputation-Clock" Model, applied to the specific case of Raidillon, a small Belgian luxury firm active in the sector of high-end watches and leather goods. Based on the assumption that reputational capital accumulation and dilution is a complex adaptive system embedded in its own environment and context, the "Reputation-Clock" integrates the major key success factors that a small firm from the high-end luxury goods industry has to leverage in order to manage reputation risks within its conception/production/distribution processes from a sustainable point of view.

In the first part of this chapter, we expose the conceptual background of our contribution. We first discuss the construct of reputation, its signification and dynamics for a small firm of the high-end luxury industry. Then, we define how its key success factors leveraged under the overwhelming of a sustainable strategy may enhance reputation and organizational performance. This sets out the bases for the application of the "Reputation Clock" Model (Riguelle 2016) to the case of Raidillon. In the second section of the chapter, we describe the model briefly, and we describe how the model can be applied to the case of a given small luxury firm active in a determined sector. It illustrates how this can be integrated into the whole managerial framework of this organization, despite its limited available resources. Finally, we derive lessons from this in-the-field analysis and provide tips to managers and CEOs of comparable firms in order to help them accumulating reputational capital and thereby enhancing the organizational performance.

2 Reputation of the Small Firm of the High-End Luxury Industry: A Complex System

Luxury usually encompasses a set of common elements that characterize it, namely the scarcity or exclusivity of the product or service (e.g. Berry 1994; Vigneron and Johnson 2004; Mortelmans 2005; Atwal and Williams 2009; Kapferer and Bastien 2009), a high value associated by the customer to this later (Mortelmans 2005), its high quality (Berry 1994; Roux and Floch 1996), a related high price (Mortelmans 2005), a very pronounced hedonistic component (Vigneron and Johnson 2004; Kapferer and Bastien 2009) that allows customers to find an extension of their own personality (Vigneron and Johnson 2004), and finally, a very strong and noticeable human component in the final offer (Chevalier and Mazzalovo 2008; Kapferer and Bastien 2009; Lipovetsky and Roux 2003).

This definition has been developed through times (Vickers and Renand 2003), following the relative evolution of the luxury industry that has undergone a deep reversal of the economic fundamentals that drive it, as well as a profound modification of the social role of luxury for about four decades. At the origin, the luxury industry was predominantly composed of small to medium organizations characterized by labor intensive, craftlike processes (Djelic and Ainamo 1999) with a strong entrepreneurial orientation (Bertoldi et al. 2013). But the craftsmanship logic has progressively turned into an industrial and a financial logic dominated by the increasing importance of groups and of private investment funds specialized in the luxury sector (Roux and Floch 1996).

Compared to these big international actors whose large and diversified high-prestige brand portfolios yield huge and still increasing turnovers, smaller independent and often family-owned firms benefit from far less resources to sustain competitive advantages. Next to a narrower international visibility compared to bigger actors, some of them also suffer from an additional geographic identity problem.

Belgium, in particular, is not naturally considered a natural home of luxury, unlike France or Italy, though Belgian creativity is paradoxically recognized worldwide. Moreover, the domestic market remains small in volume while even more affected by the decline in the purchasing power of the middle class. Meanwhile, the emotional meaning of luxury goods has become stronger and the price-quality ratio even more important for buyers. And because of an upper profitability compared to other sectors in the current globalized crisis context, competition has increased dramatically.

While luxury might be divided into three distinct categories, namely the inaccessible (unique and made to measure products such as *haute couture*, high-end jewelry, crystal, etc.), intermediate (the traditional luxury sector with *prêt-à-porter*, leather goods, etc.) and accessible luxury (sensations and pleasures luxury such as gastronomy, perfumes, etc.) (Allérès 2006; Castarède 2014), firms that are part of the first two categories are clearly the most possibly affected by the threats identified. However, though mentioned as marginal and isolated in some countries (Djelic and Ainamo 1999), these original structures yet survive and continue to thrive against the increasing threat that these giants represent, so that we now attend the coexistence of multiple luxury firm structures sharing the common universe of the luxury industry.

In order to protect themselves from buyouts or even bankruptcy, high-end small luxury firms have to leverage key success factors based on specific assets that are rare and difficult to replicate (Riguelle and Van Caillie 2014). According to the resource-based view of the firm (resource based view), businesses with valuable assets that are rare, and difficult to imitate possess a competitive advantage and may expect to achieve sustained superior [financial] performance (Amit and Schoemaker 1993; Barney 1991; Grant 1991; Hall 1993; Peteraf 1993; Roberts and Dowling 2002). The limited resources small firms of the high-end luxury industry can benefit from make thereby intangible assets such as know-how, corporate culture and reputation the best candidates for this purpose (Branco and Rodrigues 2006; Hart 1995; Russo and Fouts 1997). This is particularly the case for reputation whose intangibility renders its replication difficult for competing firms (Deephouse 2000). Corporate reputation emanates from the intangible properties that customers constantly associate with the brand, and that stem from their psychological and social needs (Keller 2009). Because it requires time and continuous effort to build internally and externally, reputation is a stable asset (e.g. Balmer and Greyser 2006; Chun 2005; Walker 2010). For the purpose of sustaining the performance of these firms in the long run, it then reveals as a strategic asset (Deephouse 2000; Fombrun 1996; Fombrun and Shanley 1990; Fombrun et al. 2000; Forstmoser and Herger 2006; Gaultier-Gaillard and Louisot 2006; Hall 1993; O'Callaghan 2007; Roberts and Dowling 2002; Sheehan and Stabell 2010).

However, dealing with corporate reputation means then to handle the set of stakeholders' perceptions, which requires managing its intangibility in its whole complexity. In order to provide the concept a somewhat materiality and thereby making it more applicable, there is a need to understand what the construct is composed of in the particular case of a small firm of the high-end luxury industry on the one hand, and to formalize what it results in for the organization on the other hand.

Corporate reputation results from the aggregation of all stakeholders' perceptions of an organization's past actions and future prospects, and is influenced by the firm's ability to meet its stakeholders' demands and expectations compared against some standards (Chun 2005; Fombrun 1996; Walker 2010; Wartick 1992, 2002). Stakeholders may be either internal (employees, managers, etc.) or external (customers, media, governments, etc.) to the company. While external stakeholders play a role in the two-way bilateral relationship that exists between corporate reputation and image (Balmer 1998; Balmer and Greyser 2002, 2006; Bromley 1993, 2002; Fombrun 1996; Fombrun and Shanley 1990; Gotsi and Wilson 2001; Gray and Balmer 1998; Rindova 1997), internal stakeholders rather live and feel their own personal connection with the firm's identity (Balmer and Greyser 2006). Consequently, both image and identity as perceived by external and internal stakeholders respectively generally characterize corporate reputation (Wartick 2002). This suggests that maintaining a premium image and a strong identity might help sustaining a good reputation. For the firm, this would result in the accumulation of reputational capital as a more tangible and measurable asset than reputation itself, which can be defined as the "*residual value of the company's intangible assets over and above its stock of patents and know-how*" (Fombrun 1996).

In the case of a luxury firm, however, the concept of image usually refers to the brand rather than to the firm itself. Indeed, brand image is a key constituent of the luxury brand (e.g. Berthon et al. 2009; Husic and Cicic 2009; Kapferer and Bastien 2009; Keller 2009; Vickers and Renand 2003) as external stakeholders consume, refer to, or simply see the luxury brand but usually ignore the owner's name. The brand image is built upon external stakeholders' non-personal-oriented perceptions of the brand conspicuousness, uniqueness, and quality on the one hand, and on their personal-oriented perceptions of hedonism and extended-self on the other hand (Vigneron and Johnson 2004), while these perceptions are moderated by their culture of origin (Christodoulides et al. 2009). Brand image taken independently is itself a complex construct based on stakeholders' perceptions of actions and signals that they associate with the brand.

In contrast, corporate identity from an internal stakeholder's point of view is supposed to be exclusively related to the firm, because it emanates from the collective feeling of employees regarding the features of the company that appear to be central and enduring to them (Balmer and Greyser 2006; Chun 2005; Fombrun 1996). However, the same people are brought to manage the brand image and to represent it, meaning that they also have to be aware of how they want external stakeholders to see the brand (Chun 2005). This gives them a strategic role in the interplay between the external brand image they manage and their own internal perceptions of the identity of the organization they work for (Chun 2001), adding subsequent complexity to the corollary reputation management system.

Regarding these definitions of image and identity in the specific case of luxury firms, we can assume that the reputation of a luxury firm lies at the edge of brand image and corporate identity, which in turn results in the accumulation of reputational capital as the related measurable and tangible asset for the luxury firm.

However, if reputational capital may be acquired, it can also be diluted reciprocally as corporate reputation is "at risk" in every action with its stakeholders. Therefore it requires constant protection (Branco and Rodrigues 2006; Fombrun 1996; O'Callaghan 2007; Walker 2010). In this context, sustainability not only generates reputational gains, but also moderates the risk of reputational losses (Fombrun et al. 2000). However, it may also be the source of reputation crises if the firm deviates from standards it stands for. The key point is to know the "how to" for a small luxury firm in order to manage this delicate equilibrium.

Despite the link between risk management and reputation has been highlighted almost thirty years ago (March and Shapira 1987), reputation risk management (reputation risk management) has come under interest quite recently in the scientific literature (Scott and Walsham 2005). On the one hand, specific industries have been scrutinised at the organizational level, but no attention was paid to the type or size of the firm (Carroll 2009; Power et al. 2009; Sullivan 2000; Scandizzo 2011; Xifra and Ordeix 2009; Csiszar and Heidrich 2006; Gaultier-Gaillard and Louisot 2006; Schanz 2006), though this can have a huge impact on its strategic performance as discussed above. On the other hand, the global reputation management concern has generally been split up for analytical reasons. Even if this breakdown of the process is necessary for a finer comprehension of its dynamics, it lacks operationalization for professionals as it fails to consider that reputation risk management is only one part of the whole corporate reputation management process for a firm, and thereby does not illustrate how this can be integrated within the whole firm's strategic management process.

Enhancing reputation then requires understanding all of these interdependencies and complex relationships (Boyd et al. 2010), and existing studies that lie on the principles of resource based view still lack operational prescriptions for managers facing complex reputation issues that wish to accumulate key valuable assets such as reputational capital (Walker 2010).

As reputation is made up of a large number of parts that constantly interact between each other based on the interaction of individuals directly or indirectly connected together, it can be seen as a *complex adaptive systems* (Anderson 1999; Capra 1996; Simon 1996). These are characterized by a large number of agents that behave according to their own principles, and whose collective behaviour results in the appearance of stable, observable properties or patterns of the system (Morel and Ramanujam 1999; Stacey et al. 2000), such as corporate reputation. With respect to organizations, complexity refers to a number of activities or subsystems within the organization (Daft 2001), which together make up a whole that is interdependent with some larger environment (Thompson 1967), and with which the organization must deal (Scott 2013). In this framework, the most effective organizations are those that set up strategies that lie at the edge of chaos (Brown and Eisenhardt 1998), maintaining a balance between flexibility and stability (Anderson 1999). Therefore complex adaptive systems need to be constantly creative in order to adapt to emerging changes (Goldstein 1999). This process of adaptation builds complexity, from which in turn emerges perpetual novelty (Holland 2006). If complex adaptive systems may help explaining the dynamics of the system it self, it does not help understanding how to accumulate an asset. However, if we combine the principles of resource based view and complex adaptive systems, we can focus on the complex processes that build system-level resources over time (Colbert 2004). We know that the value of rare organizational resources, such as reputation, is grounded in their connection with the firm's operating context, while their inimitability is embedded in the complex interactions within the organization. This resonates with the resource based view central principles as far as these organizational resources, and in particular corporate reputation, will be positioned to enhance corporate performance when considering system's boundaries extended to the intent, choices, and actions of all the organization's stakeholders (Colbert 2004). We will then build on this theory, namely the complex resource based view, in order to present the "Reputation Clock" Model as a management tool that helps taking into account key inducers of this whole complexity.

3 Key Success Factors and Luxury Sustainability as Performance Drivers for the Small Luxury Firm

Whatever their fundaments, however, brand image and corporate identity may be conveyed towards stakeholders much faster than reputation itself through the use of short messages and pieces of information. Indeed, even if it acts itself as a latent and stable idea of a whole sight of a firm, corporate reputation may be quick to change by means of external communication (Lange et al. 2011; Thevissen 2002; Williams and Moffitt 1997). This suggests that the more flexible the management process, the quicker the launch of information and the stakeholders' reaction.

The small size of luxury SMEs makes them intrinsically flexible, which enables them to produce and deliver a very personalized and integrated offer, but also more accurate converging arguments in a smaller amount of time. They can then meet customers' expectations better and faster than bigger firms, and therefore contribute to the construction of their reputation in a more proactive way. This is in turn reflected in the accumulation of their reputational capital, which is a rare and difficult to imitate asset, thereby creating competitive advantages in the sense of resource based view (e.g. Barney 1991; Grant 1991).

However, small firms benefit from far less resources than bigger ones (Crutzen and Van Caillie 2008), while reputation construction requires constant interaction with stakeholders. For small firms of the high-end luxury industry, this means having the capacity to enhance the quality of marketing resources that are essential for the improvement of business performance (Hooley et al. 2005), rather than their quantity.

For this purpose, these firms do not deliver only products but dispose of diverse sets of aesthetic values (e.g. Berry 1994; Vigneron and Johnson 2004; Mortelmans 2005; Atwal and Williams 2009), such as a specific know-how that slow and steady became part of the offer itself as the outcome of the firm's heritage and history (e.g. Allérès 2006; Dion and Borraz 2015; Fionda and Moore 2009; Kapferer and Bastien 2009), the local culture, art and innovation of their country or region of origin that they represent (e.g. Beverland 2005; Djelic and Ainamo 1999; Gardetti 2016), but also their ability to meet environmental and social standards (Carcano 2013; Moraes et al. 2015). In particular, this is all what makes the very essence of the idea of (authentic) luxury delivered by small luxury firms as they are often much more embodied into their local context because much less internationalized than bigger companies, and as product and service quality, craftsmanship, exclusivity, and values remain the fundamentals of their business strategy (Riguelle and Van Caillie 2012a).

In the current trend to return to the true well-springs of ones roots, luxury has progressively become a way for people to express their own personal values, themselves embodied into current living standards that encompass much more than the only social and environmental considerations, but well all of these elements that make luxury authentic. As co-creators of the brand-experience (e.g. Roper et al. 2013; Tynan et al. 2010), stakeholders made purely conspicuous luxury progressively give away to sustainable luxury, far away from the ambiguous association between luxury and sustainable consumption (Davies et al. 2012; Janssen et al. 2014; Nelissen and Meijers 2011). Sustainable luxury appears now as a distinct paradigm for luxury that can sustain the constitution of corporate reputation, and thereby the accumulation of reputational capital. This return to the sum and substance resonates with a return to the idea of authentic luxury. Sustainable luxury refers to sustainability in a broader sense, dealing with many other matters than the only social and environmental ones. This brings all aesthetic values of the small firm of the high-end luxury industry as described above at the core of the management of its reputation, as they become key success factors when leveraged for strategic purposes. In this system, sustainability becomes the engine that activates the machinery in order to enhance the global luxury firm's performance.

4 The "Reputation Clock" Model: A Brief Presentation

The "Reputation Clock" model (Riguelle 2016) provides a global and dynamic view of the reputational capital accumulation and dilution process. In particular, it highlights the specificities of this later within small firms of the high-end luxury industry compared to any other luxury company, through the integration of the most relevant key success factors for these companies to success in sustaining performance in the current global context of the industry. It demonstrates further that the capacity of an organization to accumulate reputational capital is intrinsically influenced by its capacity to manage the related reputation risk.

The model breaks down into 3-elements gears, each one breeding the other: key success factors lead to the accumulation or dilution of reputational capital which in turn results in performance improvement, continuously adapting and enhancing the resulting benefits of the system. Within the reputational capital accumulation and dilution process, two complementary triangles coexist: the virtuous and the defective triangle. Their completion directly influences the firm's strategic performance.

The virtuous triangle explains how a firm can improve its reputation, and starts with arrow n°1. Managers constantly mobilize the set of key success factors that the firm disposes of at a given time in order to initiate signals that are transmitted towards internal and external stakeholders (blue circle on the arrow), thereby influencing the brand image and firm's identity. These signals reach not only organizational stakeholders but are broadcasted and transit through the whole organization and stakeholders' environment, which both play a role in the stakeholders' interpretation of the initial signals. As a consequence, the same signal at the origin is susceptible to be perceived differently by each stakeholder in the end as it has been filtered through the prism of his or her global environment.

These different perceptions of signals by each stakeholder might be grasped by the organization (arrow n°2). This results in the "perceived reputation", which refers to the managers' perception of what the firm's reputation stands for in the eyes of stakeholders. Indeed, the understanding of the luxury firm's reputation is itself submitted to the managers' own interpretation of information that is sent back to the firm.

The virtuous triangle ends with arrow n°3, when the aggregation of the perceptions of all stakeholders as perceived and measured by managers then finally contributes to the accumulation of reputation capital.

However, risks exist that can tarnish the reputation and thereby dilute reputational capital. The defective triangle shows the mechanism behind this. In case of the emergence of a reputation crisis as perceived by managers (arrow n°4), the organization has to react in order to limit damages to its reputation, or even to leverage reputational opportunities that can step out an appropriate reaction (Gaultier-Gaillard and Louisot 2006).

Following arrow n°5, managers transmit remedial signals (blue circles) towards the firm's internal and external stakeholders. However, while the objective here is to limit damages to reputation and if possible, to transform the crisis into a strategic opportunity, the set of key success factors mobilized might be different than in step 1. From this point, signals that are sent to stakeholders follow the same process of transformation and interpretation than previously described in the virtuous triangle.

As soon as the reputation crisis ends up, the impact on the organization's reputational capital can be assessed in order to observe if the adopted reputation risk management strategy before and during the crisis will result in either the accumulation or the loss of reputational capital (arrow n°6) (Fig. 1).

As far as the system is intrinsically governed by the principles of the complex resource based view, it is thereby embedded within its own context and determinants. All firms do not unfold exactly the same process, which is influenced by the organization's age, size, sector, and local context of every small firm of the high-end luxury industry (Riguelle 2016).

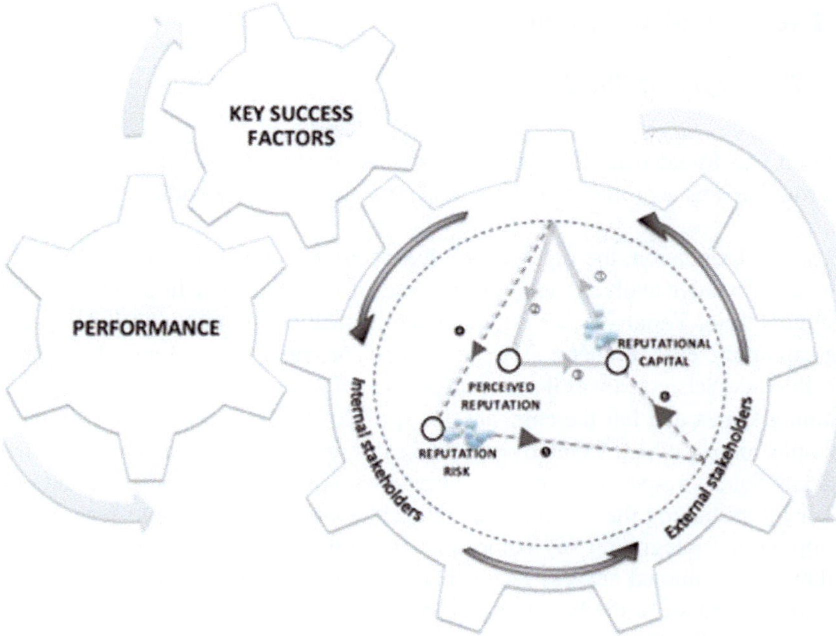

Fig. 1 The "Reputation Clock" model (Riguelle 2016)

Similarly, all the steps of the "Reputation Clock" do not necessarily occur at each completion of the clock rotation. The model applied to a specific small luxury firm in a given environment can demonstrate different characteristics than comparable organizations active in relative though different luxury sectors.

Therefore there is a need to validate this model in order to understand empirically how reputation can be managed in a proactive and dynamic way for small luxury firms that take into account the intrinsic complexity of its accumulation and dilution.

We choose to apply it to the specific case of Raidillon, a Belgian small firm active in the high-end watches sector, that extend its brand in leather goods recently. We give priority to the case study methodology for two reasons. First, it constitutes the main methodology used by scholars to assess reputation risk issues (Carroll 2009; Sullivan 2000; Xifra and Ordeix 2009). It can therefore enable us to compare obtained results if applicable. Second, it enables to capture the whole complexity of the results obtained (Yin 1993). In particular, as we try to analyse or explain why and how reputational capital is accumulated or diluted, we use explanatory research in which we include several units of analysis: 2 interviews of the CEO that both lasted more than 1 hour, data collected on the official website, on different websites that refer to the brand or firm identified via Google and in paper magazines, and financial data from 2009 until 2014 available on the public website of the National Bank of Belgium.

5 The Case of Raidillon

5.1 The Story of Raidillon

Raidillon was founded in 2001. Driven by a passion for car racing, the experience of its founder was acquired in international companies of the jewellery and watches sector.

From its foundation, the organization has been developing. In the first instance, the creator-founder evolved alone at the head of the company. In 2009, a private investment fund, Yamm SA, invested in the organization. It bought back 72% stake via a capital increase. Frédéric Dawans became chairman of the board of directors, and Fabien de Schaetzen was designated as CEO. The initial founder finally sold its remaining shares and left the company in April 2014. The company now employs 12 people, among which 9 employees and 3 sales representatives, and has just hired Christian Salez, a specialist in the management of luxury firms as an internal advisor for sustaining the organization's development. The continuous research for developing and integrating key competencies whithin the firm's value chain is the first key organizational success factor that the management team has mobilized in order to succeed since the beginning of the story.

Meanwhile, Raidillon aims at professionnalize its processes. So far, Fabien de Schaetzen is progressively developing the use of formal budgets and cash plans. Raidillon also works with a recruitment agency for hiring new collaborators, which helps formalizing their recruitment system, and the Board meets regularly. Rapidly, Raidillon has opened to its external environment, and seized opportunities to grow and to professionalize in order to enhance performance.

From an organizational point of view, the firm has shifted from a privately-owned structure driven by its founder to an organization held in majority by a private investment fund, though represented by one single person. According to the perceptions of several CEOs of small Belgian firms of the luxury industry, the input of an external investor into the capital of such a business could damage the brand reputation, except if this information remains unknown from the stakeholders (Riguelle and Van Caillie 2012a). Although it is not shown in official releases of the brand in the case of Raidillon, the information is available on public websites such as Wikipedia however, and has demonstrated no impact on the brand reputation. On the contrary, the provision of funds has been enabling investments in the brand and product development, and to seize strategic opportunities such as the hiring of people with key competencies that are required for the organization to develop. Nevertheless, entrepreneurial characteristics of small firms remain rooted in the owners' mentalities, namely the willing to finance everything on equity, the innovativeness in products and in services, and the taste for moderate risk in strategy deployment. As a result, Raidillon opened its own boutique in the Galeries de la Reine in the heart of Brussels (Belgium) where it located its headquaters in 2009, and is about to open a second one in Antwerp (Belgium) in 2016.

Besides, the different watches and leather goods collections are sold by 36 retailers in 12 countries all around the world. The brand and the product range have been developing as well. Watches evolved from 25 models in 2009 to more than 60 now, and a leather goods line has been developed since a couple of years, which now represents about 40% of the total turnover. This made the brand recognition transcend Belgian borders, although Raidillon is still young compared to other actors in its sector.

This suggests that developing brand and product, while still cultivating an entrepreneurial and flexible structure, and constantly sending related signals toward stakeholders enables a small luxury firm to constitute a cushion that can prevent reputational capital dilution. This also reveals that professionalization may be a vector of innovation and global performance that does not impact its reputation. On the contrary, this enables to develop structured processes that sustain the development of the brand and of the firm's identity in the eyes of stakeholders throughout the leveraging of various key success factors. All together, they work at the creation of a sustainable luxury strategy for the small firm that enhances its reputation and in turn its whole performance.

5.2 Key Success Factors of Raidillon

The name Raidillon is well known worldwide as the famous Raidillon de l'Eau Rouge slope on the legendary Spa-Francorchamps circuit in Belgium. Built in 1939, this corner has progressively become the cynosure of every racing driver, while its name comes from the related term, which signifies a very sloping alley or footpath in French.

From the very outset, key success factors were identified and mobilized by the founder in order to build the brand, develop difficult-to-replicate key competitive advantages compared to competitors, and thereby enhance performance.

From an external stakeholders' point of view, principally, exclusiveness is the first one. The number 55 is considered one the major symbols of the brand referring to the number of old cars allowed to start some races in the 50s. Therefore all models are limited to 55 exemplars. On the left side of each case is engraved the watch's serial number from 0 to 55, never having the number 13, just as in car racing where it is considered unlucky. The number 55 is also put forward to 11 on the dial of the watch. Next to the explanation of these characteristics on the website, they are reminded regularly on the social network page. One single poster is dedicated to each numbered watch, with picture and a short history written by the photograph on where the picture was taken. The background of the photography itself is always different and studied in order to characterize each piece with a distinct personality.

Based on the same elements that justify brand and product exclusiveness, the brand heritage is the second key success factor which is mobilized for the construction of the brand image. The brand is built around the legendary Spa-Francorchamps circuit, making its history the brand's own in some way. Raidillon goes even further, building on the image of famous racing drivers that are reputed for being equally sensitive to engines and the beauty of bodywork, than to fine watches. This is clearly what helps them to fill in the lack of heritage of tens of years of existence from which other older brands could benefit. In this vein, the Belgian circuit itself was the scene of the presentation of the first Raidillon watches in June 2001. Watches have also been created as the effigy of car racers. Partnerships with car makers were also explored by the brand, with the development of the RCZ model per se. This kind of endorsement strategy was then perfectly aligned with the brand image, and reminded the brand's country of origin. This helped matching stakeholders' expectations regarding their own sensitiveness towards car racing or the Spa-Francorchamps circuit, and the values and codes the brand proposes to identify to in a sustainable way. This started to prevent disruption between stakeholders' expectations and the brand image as they can identify themselves to the brand values. Wearing Raidillon watches entrench them into their own heritage, identity, tastes, and values, and thereby enables them to show it to others. Sustainable luxury characteristics such as the reference to a region, its heritage and its culture are then clearly at the basis of the creation of the Raidillon's reputation, making the "Reputation Clock" turn around.

Besides, other characteristics have been mobilized in order to build the brand: product and service quality supported by a specific product warranty, innovative designs, and brand experience.

What product quality concerns, first, technical information appear on the brand website that put forward the Swiss movements. What the leather goods concerns, pieces are made by specialized manufactures that work for big luxury brands in Europe. All of these characteristics that participate to the product quality are mentioned by each team member as one gets into contact with any stakeholder in order to educate and make him or her feel and understand the quality of the product. Whatever the number of repeats required, each team member should demonstrate availability in order to explain patiently to anyone interested the fundamentals that constitute the product quality. Therefore all collaborators are coached by the CEO himself in order to get a "luxury attitude". Buildings by professionals around this theme are also organized in order to learn how to act with a "luxury customer". This in turn improves service quality.

Reciprocally, in order to prevent reputation risks regarding what has been demonstrated to stakeholders, Raidillon also pays attention to the attitude of its collaborators on the internet, and especially on social networks. As representatives of Raidillon, they are on the front line to convey the brand signals and investments that are made towards each stakeholder. They truly embody the brand and the firm, its values, and thereby contribute to its reputation construction on the one hand, but can also lead it to its destruction on the other hand.

The brand has also continued to build from its solid foundations with an array of a successful brand extension in leather goods, and watches variants, with foremost concern the innovation in products designs. Bracelets are specifically designed in order to refer to racing cars steering wheels, and this element is declined throughout the whole product lines as a discrete but strong reference to its origins.

Brand experience for any stakeholder emanates from this set of measures, but not only. The location of the boutique in Brussels is also chosen for its reminder of the brand values, its encourage within the ancient history of Belgian luxury that the Galeries de la Reine represent, and its alignment with the product quality, exclusivity, and timelessness.

Raidillon also uses to invest for charity and social concerns. A rally of old racing cars is organized by Raidillon each year in favour of a Belgian organism that aims at helping handicapped children. Next to the demonstration of its concern for social issues, this is another occasion for Raidillon to reinforce its exclusiveness and brand codes. Only 55 cars are allowed to take part to the exclusive 2-days race, and accommodations for participants are also carefully selected in order to be perfectly aligned with the brand image. The event is communicated to a mailing list, reaching potential participants directly. This gives stakeholders the feeling to benefit from an exclusive treatment and service.

But above all, communication around these attributes has been at the center of the strategy in order to enhance performance. For years, Raidillon has cultivated the combination of these elements throughout the different signals and investments it has been launching continuously towards its external stakeholders in order to build the brand. Sustainability in a broad sense has also been directly put at the centre of its development in terms of anchorage in Belgian and car racing heritage in particular, respect and admiration for expertise and craftsmanship of Swiss watchmakers and Italian leather goods manufacturers, and investment in social concerns.

From an internal stakeholders' point of view, attention is given to the attachment and self-identification of the team members to Raidillon and its values. When selecting collaborators and even trainees, whatever the occupation, the person should be passionate about the product first and foremost. This is the condition for signals to be properly communicated among external stakeholders. If the presenter is passionate, then the product and the brand will be enhanced and put forward.

Moreover, opportunities are given to people to take initiatives while flexibility and versatility are required as they still evolve in a small entrepreneurial structure. This gives them the opportunity to learn different though complementary things that they can put into practice rapidly while possibilities exist to evolve as the firm grows. People are aware of that, and it participates to their own perception of the Raidillon's identity as a firm next to the brand.

Besides, the CEO deploys a strategy of transparency and trust among the teams. This in turn requires from them loyalty and confidentiality.

This all makes them feel involved personally in the development of the organization and to the improvement of its reputation and performance. They feel proud of their work, but above all, of the history they are constructing together.

5.3 The "Perceived" Reputation by the Raidillon's CEO

Raidillon regularly sends signals throughout the set of stakeholders and the firm's global environment. In the "Reputation Clock" model, we have completed arrow number 1, describing the set of key success factors that are mobilized in every message sent. At this stage, the ball is in the stakeholders' court. This is where their own environment and cognitive analysis interact in order to interpret the signal they received, and to give a positive feedback to the firm. This is also where the risk can emerge and tarnish reputation, thereby diluting reputational capital.

But measuring and assessing reputation remains difficult and intangible though. The CEO and his team members regularly try to catch the successive aggregation of perceptions of these signals in order to measure the "perceived reputation", but this where the models has its limits for small firms.

Actually, they can gather data either at the beginning or at the end of the chain, but seldom in-between. They use small questionnaires that they submit to people who enter the boutique in order to get information about brand recognition; they compare buying intentions and effective purchases depending on the previous measures. They could try to interview existing customers, but they consider that they are already convinced about the brand, and that has less added value than understanding what could make them known by others. However, this would required more resources, and priorities should be made as resources are limited due to the structure size, while they wish to keep the firm out of debt.

Any other information is caught informally during conversations, readings of magazines, and above all, surveillance of social networks and blogs. Media, according to the CEO, usually do not denigrate a brand for a couple of reasons, especially when they are not largely known. Actually, they usually do not tackle small brands because this is not as interesting for sales as when this is a famous brand that is concerned. Moreover, PR relations aim at managing the win-win collaboration between the brand and media. However, blogs and social networks do not pursue comparable issues. Bloggers and people on social media rather seek to communicate their own perception about a brand, a product, an event linked with a firm or with a sector in overall toward their environment. This is what makes the management of the reputation and the prevention of reputation risks difficult, because the more communication channels, the more risks due to misinterpretation of any signal about the brand or the firm.

Assess the "perceived reputation" of Raidillon then requires constant attention from managers, but remains difficult for them to gather elsewhere than in direct contact with stakeholders. This is why this is important to listen to them and identify their expectations.

5.4 When the Defective Triangle Moves Off

Although Raidillon does not measure customers' satisfaction properly, evidence showed that potential reputation crises came only from unsatisfied customers in the case of Raidillon, and were conveyed via blogs or social networks.

The first example is the case of a customer whose watch started to loose time. He wrote messages on a blog that discredited the quality of all the products of Raidillon. Thanks to the team's careful attention, Raidillon identified the problem rapidly before it propagates, and the CEO offered to contact the customer personally in order to fix it. The guy just ignored that a watch should be maintained, just as a car.

The second example is quite similar. A woman let a watch lie on a desk for months and was astonished that it did not work anymore when she came to get it back. Her first reaction was to propagate the information on social networks. But the mechanism of the Raidillon watches makes them wind themselves with the natural movements of the wrist. Again, she had not been advised when she bought the product. The same approach was adopted by the CEO himself in order to transform these little crises into opportunities to enhance the brand reputation toward customers.

Both of these short stories bring to light that a first category of reputation risk for small high-end luxury firms is the communication risk that occurs when an inappropriate communication is launched and propagates about the brand, especially due to ignorance or misunderstanding. This confirms that previous education of stakeholders helps preventing this kind of deviance, and enhances reputation as a cushion that can prevent damages. It also illustrates that the similar key success factors were declined in the messages delivered by Fabien de Schaetzen toward unsatisfied customers: product quality, timelessness, exclusivity requiring greater attention to the product, and above all, the insurance to get a personal response to any problem, and a close relationship with the brand throughout an exclusive customer service. This resulted in a stronger relationship with existing customers, and a better reputation for Raidillon as perceived by its CEO when hanging up the phone.

A slightly different example of reputation risk for Raidillon appeared when it started to extend the brand. During the presentation of the leather goods line, an observer told to the CEO that the collection "did not correspond" to the related watches. Further investigations revealed that some adjustments in product design were necessary in order to perfectly align with the quality, design and refinement of watches. Though this did not result in any crisis, and therefore had no direct impact on reputation, efforts made by the team in order to realign product lines contributed to show to stakeholders that comments are welcome and that their opinion is important for Raidillon. This contributes to the brand development as well. However, the reputation risk linked with the misalignment between the product quality and the brand image as perceived by stakeholders might damage the brand's reputation as well, and constitutes the second category of major reputation risks that we observe.

Although these events could be considered as anecdotal, evidence shows that reputation is so important for luxury firms that even big groups of the luxury industry are the first ones to consider that each complaint and each single signals from its environment that aims at discrediting the brand or the firm's identity towards its stakeholders or its environment has to be investigated as quickly as possible, and that a solution has to be found for each problem identified. Whatever the strength of the brand, one could easily imagine that a similar risk occurring for a smaller actor can rapidly become a devastating quake for its reputation.

5.5 *The Raidillon's Reputation Clock*

Building on the elements highlighted above, Fig. 2 illustrates the "Reputation Clock" model applied to the specific case of Raidillon. In particular, we can notice that Raidillon mobilizes a specific and limited set of key success factors that are at the basis of its brand image and firm's identity, thereby contributing to accumulate reputational capital. Depending on the situation (virtuous or defective triangle), selected key success factors might be different, principally in order to prevent or to

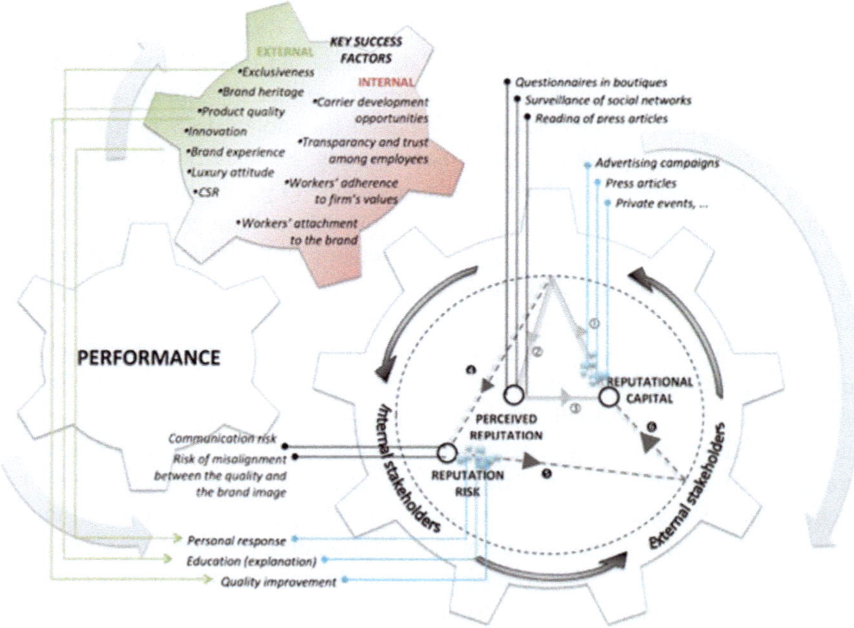

Fig. 2 "Reputation Clock" model applied to the specific case of Raidillon

tackle reputation crises. In this case, education of the stakeholders is privileged, principally emphasizing the quality of the product and its exclusiveness, while contributing to the brand experience by extension.

6 Conclusion: The "How to"

This chapter aims at discussing a performance management tool, the "Reputation-Clock" Model based on its application to a specific case study. The model highlights the fact that accumulating reputational capital is intrinsically influenced by the organization's capacity to manage the related reputation risk. Based on the assumption that reputational capital accumulation and dilution is a complex adaptive system embedded in its own environment and context, the "Reputation-Clock" integrates the major key success factors that a small firm from the high-end luxury goods industry has to leverage in order to manage reputation risks within its conception/production/distribution processes from a sustainable point of view.

Two integrated though different sub-processes are distinguished: the virtuous and the defective triangles. Both are integrated into the "Reputation Clock", and are operated by the force of sustainable luxury that underlines the whole system. Gears then result in the enhancement of the global corporate performance.

The model proposed is a first step toward the consideration of reputational capital accumulation and dilution as a complex adaptive system embedded in its global environment and context. An application of this model to the case of Raidillon, a small Belgian luxury firm active in the sector of high-end watches and leather goods illustrates one of its possible applications. This made key elements emerge from the field that can explain how the "Reputation Clock" can work for a small firm of the high-end luxury industry.

First, capitalizing on key characteristics that define the brand such as exclusiveness, refinement, product and service quality, brand experience, and finally the entrenchment into an history and an heritage from the very beginning of the process is the first best practice for a small firm of the high-end luxury industry that aims at accumulation reputational capital as a key competitive and difficult-to-replicate asset.

Second, we discovered that the modification of the firm's organizational structure did not necessarily result in reputational capital dilution for Raidillon, as far as it was accompanied by a strong brand development, thereby demonstrating to stakeholders that the brand offer and the firm's identity did not change with organizational characteristics. This also suggests that the nervousness of CEOs of certain Belgian small luxury firms to open their capital to solid partners prevent them to seize opportunities and to accumulate reputational capital, whatever the age of the firm.

Third, enhancing the firm's identity through the motivation of employees, the reward of their work, and the management of their image as they convey the brand's one also important for the reputation to improve.

Fourth, developing internal reputational capital measurement tools seems to be time-consuming and difficult because of complexity of the process. Similarly, defining a systematic procedure in order to react a reputation crisis would be too difficult for a small firm to afford in terms of investments in resources, however. In this case, evidence at Raidillon demonstrates that flexibility, which is an intrinsic characteristic of small firms compared to bigger ones, is key for both preventing and reacting to risks. It plays the role of more standardized risk prevention and reaction strategies, but is less static, and enables to be more creative in the response given, and to transform the risk into an opportunity for enhancing reputation more easily. Quick adaptations and communication of positive signals toward stakeholders at the source of risk emergence appears as one of the best ways for small firms of the high-end luxury industry to prevent potentially huge damages to their reputation as soon as a risk occurs. But above all, attention and investments should be put on prevention through education and constant interaction with stakeholders aligned with the brand image and the firm's identity.

Finally, we did not observe in the case study that complexity plays an important role in the operating of the model. However, this does not mean that complexity does not exist in this case, but rather than complexity is subtler and requires analysing the underlying forces that influence the dynamics of the "Reputation Clock". Therefore it would be interesting to investigate this in further researches.

References

Allérès D (2006) Luxe, métiers et management atypiques. Economica, Paris

Amit R, Schoemaker P (1993) Strategic assets and organizational rent. Strateg Magag J 14:33–46. doi:10.1002/smj.4250140105

Atwal G, Williams A (2009) Luxury brand marketing—the experience is everything! J Brand Manag 16:338–346. doi:10.1057/bm.2008.48

Anderson P (1999) Perspective: complexity theory and organization science. Organ Sci 10:216–232. doi:10.1287/orsc.10.3.216

Balmer J (1998) Corporate identity and the advent of corporate marketing. J Mark Manag 14:963–996. doi:10.1362/026725798784867536

Balmer J, Greyser S (2002) Managing the multiple identities of the corporation. Calif Manag Rev 44:72–86. doi:10.2307/41166133

Balmer J, Greyser S (2006) Integrating corporate identity, corporate branding, corporate communications, corporate image and corporate reputation. Eur J Mark 40:730–741

Barney J (1991) Firm resources and sustained competitive advantage. J Manag 17:99–120. doi:10.1177/014920639101700108

Bebbington J, Larrinaga C, Moneva J (2008) Corporate social reporting and reputation risk management. Account Audit Account J 21:337–361. doi:10.1108/09513570810863932

Berghaus B, Bossard C, Baehni L (2015) Strategic foresight in the luxury industry: managerial perspectives. Lux Res J 1:76. doi:10.1504/lrj.2015.069827

Berry C (1994) The idea of luxury. Cambridge University Press, Cambridge [England]

Berthon P, Pitt L, Parent M, Berthon J (2009) Aesthetics and ephemerality: observing and preserving the luxury brand. Calif Manag Rev 52:45–66. doi:10.1525/cmr.2009.52.1.45
Bertoldi B, Giachino C, Prudenza V, Zalica M (2013) How "familiness" influences the business: cases from the luxury industry. J Bus Rev 3:74–84. doi:10.5176/2010-4804_3.1.289
Beverland M (2005) Crafting brand authenticity: the case of luxury wines*. J Manag Stud 42:1003–1029. doi:10.1111/j.1467-6486.2005.00530.x
Boyd B, Bergh D, Ketchen D (2010) Reconsidering the reputation-performance relationship: a resource-based view. J Manag 36:588–609. doi:10.1177/0149206308328507
Branco M, Rodrigues L (2006) Corporate social responsibility and resource-based perspectives. J Bus Ethics 69:111–132. doi:10.1007/s10551-006-9071-z
Bromley D (1993) Reputation, image, and impression management. Wiley, Chichester, England
Bromley D (2002) Comparing corporate reputations: league tables, quotients, benchmarks, or case studies? Corp Reput Rev 5:35–50. doi:10.1057/palgrave.crr.1540163
Brown S, Eisenhardt K (1998) Competing on the edge. Harvard Business School Press, Boston, Mass
Capra F (1996) The web of life: a new scientific understanding of living systems. Colonial Waterbirds 20:152. doi:10.2307/1521798
Carcano L (2013) Strategic management and sustainability in luxury companies: the IWC case. J Corp Citizensh 2013:36–54. doi:10.9774/gleaf.4700.2013.de.00006
Carmeli A, Tishler A (2005) Perceived organizational reputation and organizational performance: an empirical investigation of industrial enterprises. Corp Reput Rev 8:13–30. doi:10.1057/palgrave.crr.1540236
Carroll C (2009) Defying a reputational crisis—cadbury's salmonella scare: why are customers willing to forgive and forget? Corp Reput Rev 12:64–82. doi:10.1057/crr.2008.34
Castarède J (2014) Le luxe. Presses universitaires de France, Paris, France
Caves R, Porter M (1977) From entry barriers to mobility barriers: conjectural decisions and contrived deterrence to new competition*. Q J Econ 91:241. doi:10.2307/1885416
Chevalier M, Mazzalovo G (2008) Luxury brand management. Wiley, Singapore
Chun R (2001) The strategic management of corporate reputation: aligning image and identity. University of Manchester, Ph.D
Chun R (2005) Corporate reputation: meaning and measurement. Int J Manag Rev 7:91–109. doi:10.1111/j.1468-2370.2005.00109.x
Christodoulides G, Michaelidou N, Li C (2009) Measuring perceived brand luxury: an evaluation of the BLI scale. J Brand Manag 16:395–405. doi:10.1057/bm.2008.49
Colbert B (2004) The complex resource-based view: implications for theory and practice in strategic human resource management. Acad Manag Rev 29:341. doi:10.2307/20159047
Crutzen N, Van Caillie D (2008) The business failure process: an integrative model of the literature. Rev Bus Econ 53:288–316
Csiszar E, Heidrich G (2006) The question of reputational risk: perspectives from an industry. Geneva Pap Risk Insur Issues Pract 31:382–394. doi:10.1057/palgrave.gpp.2510096
Daft R (2001) Organization theory and design. South-Western College Publishing, Cincinnati, Ohio
Davies D (2002) Risk management—protecting reputation. Comput Law Secur Rev 18:414–420. doi:10.1016/s0267-3649(02)01108-1
Davies I, Lee Z, Ahonkhai I (2012) Do consumers care about ethical-luxury? J Bus Ethics 106:37–51. doi:10.1007/s10551-011-1071-y
Deephouse D (2000) Media reputation as a strategic resource: an integration of mass communication and resource-based theories. J Manag 26:1091–1112. doi:10.1177/014920630002600602
Dion D, Borraz S (2015) Managing heritage brands: a study of the sacralization of heritage stores in the luxury industry. J Retail Consum Serv 22:77–84. doi:10.1016/j.jretconser.2014.09.005
Djelic M, Ainamo A (1999) The coevolution of new organizational forms in the fashion industry: a historical and comparative study of France, Italy, and the United States. Organ Sci 10:622–637. doi:10.1287/orsc.10.5.622

Figge F, Hahn T, Schaltegger S, Wagner M (2002) The sustainability balanced scorecard—linking sustainability management to business strategy. Bus Strateg Environ 11:269–284. doi:10.1002/bse.339

Fionda A, Moore C (2009) The anatomy of the luxury fashion brand. J Brand Manag 16:347–363. doi:10.1057/bm.2008.45

Fombrun C (1996) Reputation. Harvard Business School Press, Boston, Mass

Fombrun C (2007) List of lists: a compilation of international corporate reputation ratings. Corp Reput Rev 10:144–153. doi:10.1057/palgrave.crr.1550047

Fombrun C, Gardberg N, Sever J (2000) The reputation quotient: a multi-stakeholder measure of corporate reputation. J Brand Manag 7:241–255. doi:10.1057/bm.2000.10

Fombrun C, Shanley M (1990) What's in a name? Reputation building and corporate strategy. Acad Manag J 33:233–258. doi:10.2307/256324

Fombrun C, van Riel C (1997) The reputational landscape. Corp Reput Rev 1:5–13

Forstmoser P, Herger N (2006) Managing reputational risk: a Reinsurer's view. Geneva Pap Risk Insur Issues Pract 31:409–424. doi:10.1057/palgrave.gpp.2510085

Gaultier-Gaillard S, Louisot J (2006) Risks to reputation: a global approach. Geneva Pap Risk Insur Issues Pract 31:425–445. doi:10.1057/palgrave.gpp.2510090

Gardetti M (2016) Loewe: luxury and sustainable management. In: Gardetti M, Muthu S (eds) Handbook of sustainable luxury textiles and fashion, 2nd edn. Springer, Singapore, pp 1–16

Gardetti M, Torres A (2014) Sustainable luxury. Greenleaf Publishing, Sheffield, UK

Goldstein J (1999) Emergence as a construct: history and issues. Emergence 1:49–72. doi:10.1207/s15327000em0101_4

Gotsi M, Wilson A (2001) Corporate reputation: seeking a definition. Corp Commun Int J 6:24–30. doi:10.1108/13563280110381189

Grant R (1991) The resource-based theory of competitive advantage: implications for strategy formulation. Calif Manag Rev 33:114–135. doi:10.2307/41166664

Gray E, Balmer J (1998) Managing corporate image and corporate reputation. Long Range Plan 31:695–702

Hall R (1993) A framework linking intangible resources and capabiliites to sustainable competitive advantage. Strateg Magag J 14:607–618. doi:10.1002/smj.4250140804

Hart S (1995) A natural-resource-based view of the firm. Acad Manag Rev 20:986. doi:10.2307/258963

Hogan J, Lodhia S (2011) Sustainability reporting and reputation risk management: an Australian case study. Int J Account Inf Manag 19:267–287. doi:10.1108/18347641111169269

Holland J (2006) Studying complex adaptive systems. J Syst Sci Complex 19:1–8. doi:10.1007/s11424-006-0001-z

Hooley G, Greenley G, Cadogan J, Fahy J (2005) The performance impact of marketing resources. J Bus Res 58:18–27. doi:10.1016/s0148-2963(03)00109-7

Husic M, Cicic M (2009) Luxury consumption factors. J Fash Market Manag Int J 13:231–245

Janssen C, Vanhamme J, Lindgreen A, Lefebvre C (2014) The catch-22 of responsible luxury: effects of luxury product characteristics on consumers' perception of fit with corporate social responsibility. J Bus Ethics 119:45–57. doi:10.1007/s10551-013-1621-6

Kapferer J, Bastien V (2009) The specificity of luxury management: turning marketing upside down. J Brand Manag 16:311–322. doi:10.1057/bm.2008.51

Kartalia J (2000) The four building blocks of a reputation management program. Strateg Leadersh. doi:10.1108/sl.2000.26128cab.002

Keller K (2009) Managing the growth tradeoff: challenges and opportunities in luxury branding. J Brand Manag 16:290–301. doi:10.1057/bm.2008.47

Kim K, Ko E, Xu B, Han Y (2012) Increasing customer equity of luxury fashion brands through nurturing consumer attitude. J Bus Res 65:1495–1499

Lange D, Lee P, Dai Y (2011) Organizational reputation: a review. J Manag 37:153–184

Lipovetsky G, Roux E (2003) Le luxe éternel. Gallimard, Paris, France

Lochard C, Murat A (2011) Luxe et développement durable: La nouvelle alliance. Eyrolles, Paris, France
Mahon J (2002) Corporate reputation: research agenda using strategy and stakeholder literature. Bus Soc 41:415–445. doi:10.1177/0007650302238776
March J, Shapira Z (1987) Managerial perspectives on risk and risk taking. Manag Sci 33:1404–1418. doi:10.1287/mnsc.33.11.1404
Moraes C, Carrigan M, Bosangit C, Ferreira C, McGrath M (2015) Understanding ethical luxury consumption through practice theories: a study of fine jewellery purchases. J Bus Ethics. doi:10.1007/s10551-015-2893-9
Morel B, Ramanujam R (1999) Through the looking glass of complexity: the dynamics of organizations as adaptive and evolving systems. Organ Sci 10:278–293. doi:10.1287/orsc.10.3.278
Mortelmans D (2005) Sign values in processes of distinction: the concept of luxury. Semiotica. doi:10.1515/semi.2005.2005.157.1-4.497
Nelissen R, Meijers M (2011) Social benefits of luxury brands as costly signals of wealth and status. Evol Hum Behav 32:343–355. doi:10.1016/j.evolhumbehav.2010.12.002
O'Callaghan T (2007) Disciplining multinational enterprises: the regulatory power of reputation risk. Glob Soc 21:95–117. doi:10.1080/13600820601116583
Peteraf M (1993) The cornerstones of competitive advantage: a resource-based view. Strateg Magag J 14:179–191. doi:10.1002/smj.4250140303
Power M, Scheytt T, Soin K, Sahlin K (2009) Reputational risk as a logic of organizing in late modernity. Organ Stud 30:301–324. doi:10.1177/0170840608101482
Rao H (1994) The social construction of reputation: certification contests, legitimation, and the survival of organizations in the American Automobile Industry: 1895, Äì1912. Strateg Magag J 15:29–44. doi:10.1002/smj.4250150904
Riguelle F (2016) The "reputation clock" model: a sustainable performance management tool for high-end luxury goods firms. Working paper. HEC Liege, Business School of the University of Liege, Liege
Riguelle F, Van Caillie D (2012a) La perception de la présence des investisseurs institutionnels publics dans la structure de capital des "Maisons de luxe" belges: le point de vue des entrepreneurs dirigeants. Les Investisseurs Institutionnels: diabolisation ou sacralisation?
Riguelle F, Van Caillie D (2012b) Toward a taxonomy of key success factors for SMEs in a changing environment: the case of the luxury industry. In: The 12th IAMB conference
Riguelle F, Van Caillie D (2014) Competing as a luxury SME. In: Berghaus B, Müller-Stewens G, Reinecke S (eds) The management of luxury, 1st edn. Kogan Page, London, pp 275–286
Rindova V (1997) Part VII: Managing reputation: pursuing everyday excellence: the image cascade and the formation of corporate reputations. Corp Reput Rev 1:188–194. doi:10.1057/palgrave.crr.1540042
Rindova V, Fombrun C (1999) Constructing competitive advantage: the role of firm, Äìconstituent interactions. Strateg Magag J 20:691–710. doi:10.1002/(sici)1097-0266(199908)20:8<691:aid-smj48>3.0.co;2-1
Roberts P, Dowling G (2002) Corporate reputation and sustained superior financial performance. Strateg Magag J 23:1077–1093. doi:10.1002/smj.274
Roper S, Caruana R, Medway D, Murphy P (2013) Constructing luxury brands: exploring the role of consumer discourse. Eur J Mark 47:375–400. doi:10.1108/03090561311297382
Roux E, Floch J (1996) GÉRER L'INGÉRABLE: La contradiction interne de toute maison de luxe. Décis Market 9:15–23
Russo M, Fouts P (1997) A resource-based perspective on corporate environmental performance and profitability. Acad Manag J 40:534–559. doi:10.2307/257052
Scandizzo S (2011) A framework for the analysis of reputational risk. J Oper Risk 6:41,Äì63
Schanz K (2006) Reputation and reputational risk management. Geneva Pap Risk Insur Issues Pract 31:377–381. doi:10.1057/palgrave.gpp.2510092
Scott R (2013) Institutions and organizations. Sage, Thousand Oaks, CA

Scott S, Walsham G (2005) Reconceptualizing and managing reputation risk in the knowledge economy: toward reputable action. Organ Sci 16:308–322. doi:10.1287/orsc.1050.0127
Sheehan N, Stabell C (2010) Reputation as a driver in activity level analysis: reputation and competitive advantage in knowledge intensive firms. Corp Reput Rev 13:198–208. doi:10.1057/crr.2010.19
Simon H (1996) The sciences of the artificial. MIT Press, Cambridge, Mass
Smith C, Palazzo G, Bhattacharya C (2010) Marketing's consequences: stakeholder marketing and supply chain corporate social responsibility issues. Bus Ethics Q 20:617–641
Stacey R, Griffin D, Shaw P (2000) Complexity and management. Routledge, New York
Sullivan K (2000) Identity, conflict and reputation in the university setting: an illustrative case study. J High Educ Policy Manag 22:177–185. doi:10.1080/713678139
Suomi K, Järvinen R (2013) Tracing reputation risks in retailing and higher-education services. J Retail Consum Serv 20:207–217. doi:10.1016/j.jretconser.2012.12.003
Thevissen F (2002) Belgium: corporate reputation in the eye of the beholder. Corp Reput Rev 4:318–326. doi:10.1057/palgrave.crr.1540153
Thompson J (1967) Organizations in action. McGraw-Hill, New York
Tynan C, McKechnie S, Chhuon C (2010) Co-creating value for luxury brands. J Bus Res 63:1156–1163. doi:10.1016/j.jbusres.2009.10.012
Vickers J, Renand F (2003) The marketing of luxury goods: an exploratory study, Äì three conceptual dimensions. Market Rev 3:459–478. doi:10.1362/146934703771910071
Vigneron F, Johnson L (2004) Measuring perceptions of brand luxury. J Brand Manag 11:484–506. doi:10.1057/palgrave.bm.2540194
Walker K (2010) A systematic review of the corporate reputation literature: definition, measurement, and theory. Corp Reput Rev 12:357–387. doi:10.1057/crr.2009.26
Wartick S (2002) Measuring corporate reputation: definition and data. Bus Soc 41:371–392. doi:10.1177/0007650302238774
Wartick S (1992) The relationship between intense media exposure and change in corporate reputation. Bus Soc 31:33–49. doi:10.1177/000765039203100104
Williams S, Moffitt M (1997) Corporate image as an impression formation process: prioritizing personal, organizational, and environmental audience factors. J Public Relations Res 9:237–258. doi:10.1207/s1532754xjprr0904_01
Xifra J, Ordeix E (2009) Managing reputational risk in an economic downturn: the case of Banco Santander. Public Relations Rev 35:353–360. doi:10.1016/j.pubrev.2009.08.004
Yin R (1993) Applications of case study research. SAGE Publications, Newbury Park, Calif

Part II
Sustainable Practices Within the Luxury Universe

Part II
Sustainable Practices Within the Luxury Universe

Sustainable Luxury and Fashion: From Global Standardisation to Critical Customisation

Patrizia Calefato

Abstract The concept of luxury is often linked to fashion. Even if those two terms have obvious different meanings, they both refer—in terms of goods and expenses—to the economic and cultural field of uselessness, excess, waste. Furthermore, in these years, a new element has been introduced to this relationship: distinction. Although we live in the age of global fast-fashion, distinction means nowadays creating a personal fashion, caring for the production processes, the workers' conditions, environmental issues and the material "history" of what one wears. In other words: sustainability. Therefore fashion becomes luxurious, not only economically, but also highlighting values such as uniqueness, customisation and wellness—the same values on which luxury is based. Fashion and luxury can thus create a new critical and ethical paradigm. This paper will focus on this new paradigm of luxury concerning fashion, the word "critical" referring above all to the concept of "choice". One can choose his/her clothes and can distinguish his/herself by taking care of the senses, of materials, of time. The first section of this paper will analyse how the concepts of fashion and luxury were defined and differentiated by Social Sciences at the beginning of the Twentieth Century (Sombart in Luxus und Kapitalismus. Duncker & Humblot, Berlin, 1913; Veblen in The theory of the leisure class. Macmillan, New York, 1899; Simmel in Psychologie der Mode. Die Zeit: Wiener Wochenschrift für Politik, Volkswirtschaft, Wissenschaft und Kunst 5 (54):22–24, 1895). The main aspect of this difference lies in the relationship with time of the two concepts: luxury is eternal and fashion is always changing. Then this paper will consider the critique of this "classical" differentiation operated by Fashion Theory—born in the last decades of the Twentieth Century as an autonomous field of research in Cultural Studies. The symbolic principles of modern luxury will be compared with the symbolic meaning of contemporary fast-fashion. The core of this paper is represented by the analysis on how fashion is rethinking its

English translation from Italian by Sveva Scaramuzzi

P. Calefato (✉)
Dipartimento di Scienze Politiche, Università degli studi di Bari Aldo Moro, Bari, Italy
e-mail: patriziaccia@gmail.com

M.A. Gardetti (ed.), *Sustainable Management of Luxury*,
Environmental Footprints and Eco-design of Products and Processes,
DOI 10.1007/978-981-10-2917-2_5

own patterns: from haute couture to the "fastest" brands, many fashion companies and designers are aware of ethical issues and sustainability, nowadays. Therefore sustainable luxury seems to be the only path for contemporary fashion culture and creativity. Finally, this paper will consider some examples as case studies, focusing especially on some new contemporary Italian fashion brands. One of these brands—Cangiari—will be analysed in detail as an ethical luxury brand, caring deeply for social commitment and sustainability.

Keywords Fashion · Luxury · Sustainability · Senses · Body · Reuse · Customisation · Ethics · Social commitment · Italian fashion

1 Fashion and Luxury: Shared Elements and Differences

The concept of luxury has often been linked to fashion, clothing and dressing. This link accompanied the history of the notion of luxury in the use of the word itself, in the laws that restricted it throughout the centuries (*leges suntuariae*), and in sociological, economic and cultural theories that defined its motivations and rules (Enzensberger 1997). Although luxury is not a synonym for fashion, as fashion is not luxury per se, these two areas have actually something in common. Luxury goods, expenses, luxury as display of conspicuous consumption—as Veblen (1899) defined it—all belong to the economic field of "uselessness" and excess. Food, clothing, jewellery, furnishings, and architecture are the branches which mostly fuel this field. Therefore fashion is entirely part of it (Calefato 2014).

Luxury has always been conceived as a fundamental element which favoured economically, socially and culturally the capitalistic mode of production since the beginning of its original accumulation phase. This hypothesis was supported by Social Sciences at the beginning of the Twentieth Century, during the crucial shift from the capitalistic mode of production of the Nineteenth Century to the accomplishment of mass production of goods, which occurred at the turn of the Century. Luxury symbolised then a specific case in the process of productive and communicative standardisation, and perhaps this was the reason of the amount of attention given by Social Scientists to this subject, in analysing its deep causes (Berry 1994; Borghero 1974).

Among those scientists, Sombart (1913) theorised the effective necessity of luxury. He underlined the wasting spirit of capitalism, as opposed to Weber (1904–1905), who highlighted the Calvinist strictness of this mode of production instead. According to Sombart's analysis, the function of luxury for capitalism manifested itself through the complex transformation of lifestyles of both aristocracy and middle class during the Seventeenth and Eighteenth Centuries. This transformation is due also to the fundamental contribution of women, both as a social role in everyday life, and as courtesans. Sombart maintains that upper middle class women fuelled the luxurious entries of consumption for the first time: textiles, furniture, clothing, food, toiletries, fashion accessories.

Sombart's analysis gives a global overlook of capitalism, and the fundamental role of fashion is highlighted through the leading role of textile industry in the European Industrial Revolution. Especially in Britain, the Industrial Revolution necessarily depended on the supply of raw materials from Asia. Sombart writes:

> It was toward the end of the Seventeenth and during the beginning of the Eighteenth Century that Indian cottons became the fashion among the rich and threatened home industries with serious competition (Sombart 1967: 123).

If it is historically true that the production of a system of social cohesion that we call "fashion" is intimately linked to "modernity" as the primal characteristic of European middle class societies (Paulicelli 2006), it is also true that this modernity, the birth of European middle classes and the constitution of Western capitalistic States and economies have fed themselves since their origins on raw materials, human beings and goods that other areas of the world supplied to Europe. The textile industry, for example—engine and basis for the very possibility to create a fashion—was born and strongly developed in Eighteenth century England in close relation with the rise of the British colonial empire. The "exotisms," "chinoiseries" and "japaniseries" have fed for centuries the refined tastes for clothes, porcelain, furnishing and tapestry in Europe and later in North America. In the same way, the slaves taken from Africa worked in American plantations to produce the cotton destined for the clothing industry.

In his *Theory of the Leisure Class* Thorstein Veblen introduces the concept of "conspicuous consumption". The notion of "leisure class" refers mostly to the upper middle class of his time, even if some characteristics, tastes and lifestyles plunged their roots in the previous Centuries. According to Veblen, a certain kind of consumption aimed to personal wellness gain a "honourable" function for those in society who can afford and show the possibility of enjoying it. The dress is one of the most prominent elements of this kind of consumption: "The need of dress is eminently a 'higher' or spiritual need" (Veblen 2007: 104). The function of fashion goes even further: the luxurious signs of clothing do not only show a conspicuous consumption, but also a conspicuous waste:

> The imperative requirement of dressing in the latest accredited manner, as well as the fact that this accredited fashion constantly changes from season to season, is sufficiently familiar to everyone, but the theory of this flux and change has not been worked out. We may of course say, with perfect consistency and truthfulness, that this principle of novelty is another corollary under the law of conspicuous waste (Veblen 2007: 106).

Also Veblen lingers on the relationship between fashion, luxury, and women, explaining how the conspicuous waste of fashion by leisure class women marks their social unproductiveness. The role of women as consumers, free from any kind of work form, is perfectly symbolized by her bulky dresses and by constrictive lingerie such as corsets:

> The corset is, in economic theory, substantially a mutilation, undergone for the purpose of lowering the subject's vitality and rendering her permanently and obviously unfit for work (Veblen 2007: 106).

2 Imitation and Distinction

In his famous 1895 essay—*Fashion*—Georg Simmel writes that fashion represents one of the best ways through which the fundamental tension fueling social life shows itself. In his opinion, the motivation of fashion resides in two typical characters of humanity: imitation and distinction. The former highlights the social dimension of culture, the latter, on the contrary, underlines the separation of the individual from the social group and remarks his/her individuality. According to these two polarities Simmel conceives fashion as a system for social cohesion which dialectically conciliates the individual's participation to a group, on one side, and his spiritual independence, on the other. Following fashion, substantially, does not undermine, for example, religious, spiritual or ethical beliefs of the individual, but controls and ratifies the bond between the individual and society through the adherence to an aesthetic, a taste, a common sense. According to this classic theory of fashion, which saw its accomplished manifestation during the first phase of mass society, imitation and distinction govern a top-down motion in the social pyramid. Distinction is a prerogative of upper classes, which are imitated by the masses in a trickle-down effect.

In the genesis of this phenomenon Simmel singles out a peculiar mechanism that regulates the relationship between his time's western civilisation and non-western ones. He writes that the more the origin of a fashion is exotic, the more the social group which adopts it will be cohesive:

> Because of their external origin, these imported fashions create a special and significant form of socialization, which arises through mutual relation to a point without the circle. It sometimes appears as though social elements, just like the axes of vision, converge best at a point that is not too near (Simmel 1904: 136).

In this sense Simmel introduces a particular idea of social distinction—actually related to fashion—as something around which the social cohesion of a group develops, but which cannot be received by the group itself. A fashion, writes Simmel, belongs only to a part of the group, while all the rest goes towards it; as soon as everybody does what was done by few at the beginning, the fashion stops being a fashion. Every increase drives the fashion to its death because it obliterates diversity; the charm for novelty is therefore also a charm for transitoriness (Simmel 1904: 139).

Simmel writes that fashion has a peculiar relationship with time: a relationship which has "the attraction of a simultaneous beginning and end, the charm of novelty coupled to that of transitoriness" (Simmel 1904: 139). According to Simmel, women find in fashion the possibility of combining imitation and distinction and personal significance, features that were denied to them in other matters:

> In a certain sense fashion gives woman a compensation for her lack of position in a class based on a calling or profession (Simmel 1904: 145).

In the end, according to Simmel:

> the peculiarly piquant and suggestive attraction of fashion lies in the contrast between its extensive, all-embracing distribution and its rapid and complete disintegration (Simmel 1904: 155).

This definition approaches two concepts that bind fashion and luxury, even contradictorily: "all-embracing distribution" and "disintegration". Luxury, as fashion, tends to being limitless trough time in its forms and symbols, but it is also built on the ideas of waste, squander, disintegration. Fashion and luxury are therefore akin and, at the same time, opposed. Akin because the laws of fashion are founded on the paradox of making mandatory the unnecessary, the groundless, that which appears as waste and unreasonable consumption. Opposed because fashion is constantly changing, while luxury is basically eternal.

The notion of luxury is much older than the one of fashion (Calefato 2003). The former travels through the history of diverse ages and social formations, while the latter reaches its full appearance only in middle class society. In fact, only after the French Revolution the "citizen" could use fashion as a symbolic device to participate both to the rights of equality and the prerogatives of distinction. As an example, the middle class male dress, as Barthes (2006) observes, has fundamentally democratic origins. After the French Revolution, the idea of democracy gave birth to an idea of male dress—which lasted substantially unchanged until the Twentieth Century—originated from military uniforms and from the values of equality which mirrored the fracture with the *ancient régime*, even aesthetically. Barthes writes:

> the great movement within masculine dress towards standardization and democratization launched by the Revolution and inspired in form by reference to the austerity of Quaker dress, was bringing about a whole revision of vestimentary values; seemingly *déclassé*, clothing could signal social distinctions only via a new value, namely that of *distinction* (Barthes 2006: 23).

Distinction as a social phenomenon characterises the birth of fashion in modernity. It is indeed difficult to obtain into the communal and democratic dimension of the idea of taste on which the inner workings of fashion are founded. In the tension between distinction and taste as common sense, Barthes singles out the birth of the aesthetics of the detail:

> the knot on a cravat, the material of a shirt, the buttons on a waistcoat, the buckle on a shoe, were from then on enough to highlight the narrowest of social differences (Barthes 2006: 66).

The detail is a pivotal element of one of its most peculiar expressive forms: dandyism. Fashion, in its accomplished social meaning in mass society, stands opposite the dandy, because the distinction of the dandy is elitist, while in fashion distinction is in a dialectic tension with imitation. As Barthes writes:

> Fashion is the collective imitation of regular novelty; even when it has the alibi of individual expression, or, as we say today, of a 'personality', it is essentially a mass phenomenon in which sociologists are happy to be interested so long as they find in it the privileged example of a completely pure dialectic between the individual and society (Barthes 2006: 68).

3 Luxury According to Fashion Theory

Fashion Theory arises as an independent branch of research of social sciences and humanities in the last decades of the Twentieth Century. It conceives fashion as a cultural form, therefore—from the academic point of view—it is part of the diverse constellation of Cultural Studies. Fashion Theory analyses the characteristics of fashion in relation to aesthetics, semiotics and critical discourse, as tightly bound to everyday life and body representations. Luxury is also analysed through this cultural perspective, both by referring to the classic sociological theory as previously outlined, and to the analysis of the typical symbolical forms of luxury nowadays.

According to Fashion Theory luxury does not only concern wealth but the actual embedding of values such as quality, rarity, lasting strength, craftsmanship and customisation in the objects, especially nowadays. In order to better explain this concept we can refer both to the etymology and to the semiotic quality of the word "luxury". The words which are used in the most common Western European languages—luxury, luxe, Luxus, lujo, lusso—derive from the Latin word *luxuria, -ae*, which referred to exuberance, even in the field of lasciviousness, and excess (Bataille 1967): traces of this meaning can be found in the use of the adjective "luxuriant" (in reference to a garden). Tacitus uses the expression *eruditus luxus* meaning "fine taste in the expenses": the squandering of monetary wealth relates to the cultural quality of the expenses themselves. Another Latin word with the same spelling, *luxus*, is used as an adjective, which derives from the past participle of the verb *luxare* and shares its Indoeuropean root with the ancient Greek verb *λύω*, meaning "to melt". From the same root derives the Italian word *lussazione* (dislocation), an interruption, an interjection of something through something else, that sometimes occurs to articulations. Words acquire their meaning from their usage through time and memory: so, even if these concepts refer to two different semantic fields, luxury and *lussazione* seem much more related than ever nowadays. Luxury is indeed represented by objects, attitudes, images, styles that—sometimes symbolically—interrupt the serial progress of life and *lussano* (dislocate) the habits.

Contemporary luxury, bearing this cultural and qualitative dimension, differs from the traditional concept of opulence, conspicuous consumption, ostentation of wealth. Luxury still represents excess, overabundance, unrestrained flux, but the key concepts that best define and circumscribe it are: uniqueness, rarity, distinction, eternity, time, space, worth, wellness. These concepts establish the qualitative dimension of luxury, understating the ostentation, which gradually disappears through the subtraction of signs that can relate to the idea of *sprezzatura* so perfectly described in Castiglione's *Book of the Courtier* (1528). Contemporary luxury contains the motives which Georg Simmel related to distinction more than a Century ago: something appealing to the relationship between the individual and the hazard of exception, the challenge of waste and of everything that disregards the law of usefulness and functionality. On the one hand, luxury evokes the idea of eternity (Lipovetsky 2003), meaning defying the concept of death itself, because it remains and survives through time. On the other hand, luxury shows itself through images of foolish waste, of desire exceeding the needs, of "priceless" worth.

"Price" is based, indeed, on a common agreement, on a negotiation, on an "average" value of an object, set by a community (Benveniste 1969). "Worth", on the contrary, deals with an intimate dimension, referring to the qualitative aspect of an object or a good, far—as much as possible—from its own "consumer's goods" nature. This is why this word is often used also referring to people. Worth is not bound to practical use or functionality, to mere wealth and market value, but to charm, to myth (Barthes 1957), to the even fetishistic value of an object, to its "appreciation" in terms of additional values. Valuable goods are often appreciated because of their uselessness. They are cherished because of their incredibly high price—symbolically, not necessarily economically.

Space is a further key word in the mythologies of contemporary luxury that narrate just how little space there is in the world today. The space you can see from the top of a tower in the middle of an overcrowded city is equal to dominion over that world, in the happy individualism conjured up by the image of home, albeit a temporary one, as ephemeral as stocks and bonds for a junior broker. The typical medley of streets and houses in the Nineteenth century Paris inspired Benjamin's reflections on the collective dimension which the dialectic between street and home represented for mass society:

> Streets are the dwellings place of the collective. The collective is an eternally unquiet, eternally agitated being that - in the space between the building fronts -, experiences, learns, understands, and invents as much as individuals do within the privacy of their own four walls (Benjamin 1982: 423).

The opposite of this collective dimension, the value of space in exclusive luxury homes in the contemporary metropolis, is based instead on the happy isolation of the new elites (Frank 1999). In this sense luxury is indeed space, though it is not a deserted space, as we can see from the image of the apartment (with five bathrooms) on the 54th floor of the Trump Tower, from whose windows you can look far down on the bright lights and bustle of the world below. An ease that takes shape in the enormous interval luxury places between world and home.

In *The Philosophy of Money* (1900) Simmel notes how money allows men to create the distances, boundaries and privacy that allow them to survive in the traffic and reciprocal pressures of modernity. For Simmel this distance is enhanced by the fact that personal wealth consists of means of production and not consumer objects, as in primitive cultures, and so various intermediate stages separate the proprietor from the ultimate goal of possession (Simmel 1978: 76). Today personal wealth and life style are based entirely on these intermediate stages, which are of the nature of signs, and of which luxury represents the quintessence.

Contemporary luxury activates the awakening of the senses and of intellect, of pleasure and reason. This kind of luxury features two apparently synonymous concepts, which actually differ from each other: uniqueness and rarity. What is "unique"? Not even a product of art, nowadays, for ours is the age of mass production and reproduction of art, where even its "aura" is reproduced and sold. Nor a designer dress or accessory is "unique", since the designer label itself is a serially reproducible sign, globally recognised. Therefore the uniqueness of luxury cannot be separated from the idea of "rarity", a concept—in opposition to "unique", which always refers to artefacts—related to natural goods or an object created or modified by natural or temporal elements.

4 A New Idea of Distinction

During the Twentieth Century, fashion favoured the element of imitation, through which the image of the clothed body serialised, until it became standardised. At the same time, fashion changed the way with which it spread through social classes, overturning the trickle-down movement to its opposite—bubble-up movement—, from the bottom to the top of the social pyramid. This movement characterised the influence of different "styles" and "subcultures" on many fashions, starting from the Sixties. "Style" emerges as an element of rupture of Simmel's principle about mutual independence of fashion and "individual spirit". According to Dick Hebdige's essay *Subculture: The Meaning of Style* (1979)—on British subcultures, especially punk—style embodies the strongest form of aesthetic and ethic submission to a way of life, not only to a fashion. Since the end of the Seventies styles have influenced fashions; even those which Hebdige presents as independent subcultures—contrasting the "institutional" culture and fashion—have been incorporated. Prêt-á-porter, casual, "young" fashion, punk-inspired couture of the Eighties and Nineties (Gianni Versace above all, Versace and Calabrese 1993) are the aesthetic and industrial results of this influence.

At the turn of the millennium, distinction arouse again in fashion as a sort of reaction to standardisation (Bourdieu 1979). One of the first answers of the fashion industry to this demand of distinction was "fast-fashion", according to which its production does not follow the traditional, slow, seasonal rhythm of prêt-á-porter, but it is based on fast consumption needs, low-cost. The stores upgrade their stocks every ten days or every two weeks, choosing their new items according to the preferences of their buyers. This fast-fashion system, typical of the biggest retail multinational corporations, is nowadays the standard for the textile-clothing mode of production.

In order to keep a fast pace of production, low-cost brands (as most of the biggest fashion brands, anyway) exploit sweatshop labour from underdeveloped countries and areas in Europe, Asia and Africa. The 2013 tragedy at the Rana Plaza in Bangladesh—the collapse of a building harbouring many illegal textile plants—saw the death of more than one thousand workers, or the same year fire in a warehouse near Prato, Italy, where seven Chinese immigrant workers died, are gruesome examples of the unspoken true cost of fashion.

Many millions of people all over the world work in this branch, enduring terrible working conditions—very low income, non-existent security, incredibly long working days. *The True Cost*, a 2015 documentary by American director Andrew Morgan, explains who actually is paying for the low-cost of globalised fashion, and how this kind of production on these condition is already unsustainable.

The fast-fashion standard, ideally and culturally, does not represent at all the ethical dimension of fashion which the consumers are demanding. Furthermore, it is nowadays incorrect speaking of "consumers", especially in fashion: it should be better keeping in mind that, as time goes by, individuals emerge as crafters of their own fashion and taste, disregarding common sense, through a daily DIY routine where everyone defines and adapts his/her own style. Therefore, the need for a fashion which respects other values—such as politics, culture, environmental awareness—grows stronger and bigger. With these concepts as staples, a different quality of the concept of "distinction" is arising in the last few years: the search for adjusting clothes and objects to the personal and intimate idea of body and individual wellness, for a profound and synergetic relationship between the clothing, the senses and a "slow" time.

Fashion has always been related to time, it is time per se, indeed. The Italian word for fashion, *moda*, evokes in its own etymology the idea of "modernity"; fashion lives in an eternal present, foreshadowing an eternal future, always facing towards the newest change. However, Walter Benjamin writes that fashion is also the form through which the present manages to do a "tiger's leap" into the past (Benjamin 1968: 261). Contemporary forms of reproductions of fashion signs seem to confirm and reinforce this image: fashion nowadays uses the past as a huge supply of images and quotes which come back to the present with new meanings. The forms of this throwback are vintage, revival and reuse: through these, the system of fashion allows us to recover and rework the objects and the signs left behind. This is particularly relevant especially for the current age, in which the fragmentation of life seems to have erased the past as memory. Rearrangements, contaminations between past and present incorporated by fashion fill the emptiness of this fragmentation, and give opportunities to memory once more, under the form of an individual and unique style. The demand for an "interior" dimension of fashion grows increasingly, in order to regain an emotional and ethical value as opposed to one representing an economic or social status.

In this sense, fashion goes beyond itself, and luxury does the same. Luxury is based nowadays on wellness as "well-being", as opposed to "being wealthy". As Mathieu and Monneyron write:

> We should also add the environmental concerns which, due to their interest on sustainability, strongly pose to the living the existential questions proper to luxury. (Mathieu and Monneyron 2015: 68).

The relationship of luxury with time, the idea of eternity and lasting quality that characterise it, recall immediately the issue of cultural and actual sustainability of global social reproduction. There cannot be luxury without sustainability. There cannot be fashion outside its ethical dimension. Luxury and fashion will survive only if they will make critique and awareness their founding grounds. The word "critique" means "choice": the one embodying the production, exchange and consumption of "well-made" goods. Awareness means knowing the raw materials, knowing the people that worked in the production, knowing how much the environment and the workers have been respected in order to create that piece which is bought or admired in a beautiful shop window. Actually, it is life itself at stake.

5 Ethics, Critique, Sustainability

"Life" is one of those words which are imbued with positive shared values, not needing any explanation for that: its mere pronunciation discloses merry images in our minds, not always as authentic as we would like, such as ones of a healthy planet, a happy baby, an always shining hope. The Italian translation of this word—vita—inspired the Los Angeles based fashion designer Amanda Shi in the naming of her eco-friendly fashion brand—Avita—by juxtaposing an "a" to it. The brand is eco-friendly because it incorporates the philosophy of ethical and sustainable clothing: its core idea is producing nice pieces of clothing using only natural raw materials, in order to contribute to a long and happy life for the planet. Bamboo, recycled cashmere, organic cotton and soy are the main raw materials for Avita's collections, in order to produce "clean" textiles and pieces of clothing which are able to rouse, in fashion and luxury industry, the knowledge of "life" meaning making a better world.

Can a dress give happiness and favour life? An affirmative answer is difficult, if fashion is a synonym of ephemeral values—the fashion of the moment, the aim for seduction of the one for taking part to a social clique because of a particularly exclusive brand. No effort at all, the opposite of it, indeed: this connotation of fashion lets us drift in the flow of common sense, of imitation, it stimulates at most ostentatious competition. The target of sustainable fashion and luxury are, on the contrary, making taste, desire for ostentation and even narcissism ends in themselves, but elements of the ideal effort for a better world, from a militant point of view. Being aware of what we use as clothing, we could feel involved in the essential quest on how to preserve the environment and protect human rights.

A very large number of clothing brands take part to this ideal current, nowadays. In the branch of sport shoes, for example, the French brand Veja's products are made with organic cotton from Tauà—a small Brazilian village where eighty-nine families live with their organic agriculture. WornAgain, of the Terra Plana group, is inspired by recycle, in fact it recycles many raw materials for their shoes, never adding chemicals or GMOs. Or the old German brand Zeha—which made the shoes for East-Germany athletes before the fall of the Berlin Wall—relaunched its production in 2003, exploiting the vintage cult of *Ostalgie* for objects and symbols of the old DDR, paying attention to a sustainable production in its plants in Slovakia.

There is a search for luxury in terms of quality, as opposed to appearance. A further example fort this is the Italian brand Relight-up, founded in 2008 through a relighting project—as in its name—on power sources, materials and lifestyles of our world: an idea perfectly symbolised by the windmill blades of its logo. Organic cotton, seaweed, corn and soy are the raw materials of its eco-friendly and (by now) limited edition T-shirts and sweaters. The motif of limited edition comes from the philosophy of luxury, in casual fashion which does not allow a casual choice on clothing, but demands awareness.

The designer-artist Patrick Lafrontière is also part of this ethical fashion branch, trying to introduce new concepts and ideas in a world for too long ruled by the ephemeral. Born in French Guiana, Lafrontière's raw material is the bark of palm trees, inspired by the modalities of invention of the objects of his country's everyday life. His works are unique pieces that resemble sculptures, and his runways look like a gallery or a museum.

The quicker and more hectic the pace of fast-fashion, the more relaxed and responsible is the time of slow-fashion, as it is perfectly highlighted by the artisanal branches of luxury. One of the most important brands of Italian knitwear, Loro Piana—acquired in 2013 by LVMH—expresses its company philosophy by telling how its products are made and for what they are made. The manufacturing is artisanal, the raw materials are collected respecting populations and animals, the quality of their products is perfect, their lasting power is almost eternal, they are simple and understating, and this is the reason of their luxury value. Luxury and moderation, almost frugality, seem to touch each other in this philosophy involving fashion, consumption cultures and lifestyles.

The ethical, environmental and fair values are the core of the notion of luxury nowadays. The "humanistic entrepreneurships" are the leaders of this new idea, joining sustainability, knowledge of the materials and respect for the workers. Another famous Italian cashmere brand—Brunello Cucinelli—considers its workers, its locations and its quality the core of the company, giving one third of its profits to the workers—who already earn a higher salary than the average of the branch—, one third to the territory and one third to the entrepreneur.

The capital cities of world fashion have also been recycled towards sustainability, and they host events supporting ethical fashion, aside the official ones, such as the Milan based *So critical so fashion*, taking place since 2013 during the Milan Fashion Week. Eco-friendly and quality targeted independent brands, fashion designers, artisans, industrial designers, creative and valued tailoring studios can show their products in this event. And all over the world many cities become capital cities of fashion because they are the centre of this productive movement: Sao Paulo in Brazil and Dakar in Senegal are the meeting points where huge social and cultural energies headed towards sustainability grew, in two continents which have been exploited for a long time and widely, such as South America and Africa.

6 Sustainability as Culture

In 2012 the Italian *Class* magazine and its publisher Class Editori, in occasion of their 25th anniversary, organised an international exhibition at the Triennale of Milan, whose title was *C25 Options of Luxury*. The exhibition showed 250 select objects, signs and concepts which expressed the qualities and the complex features of contemporary luxury. Next to high-end jewellery and five stars hotel and resort franchises, the exhibition chose as symbols of luxury some natural goods, such as the Sicilian Bronte pistachio and Trapani sea salt. This example shows how the idea of luxury is nowadays a necessity, whose founding base is completely different from the past. If luxury is bound to the ideas of uniqueness, rarity, eternity, it has the duty of protecting unique, rare or endangered goods. Rarity, and therefore luxury, deals with life and death because often it is caused by depletion and the wearing effect of time. Therefore rarity is a challenge to nature and life itself and requires a consideration on time, which also becomes a supporting value of luxury. Luxury must have an allure of eternity and endless lasting of the objects and even of the people who embody it. And yet, right here its contradiction is unveiled through waste and excess, which also define luxury and corrode the signs in a moment or in a season (Curcio 2005).

In this sense, we are living nowadays neither the age of "the end of luxury" nor what trend forecaster Li Edelkoort calls "the end of fashion" (Edelkoort 2015). Fashion and luxury in the classical sense of the Twentieth Century have certainly come to an end, but both are enduring a rebirth in the terms of sustainability, customisation, distinction, culture. As Gardetti and Muthu write:

Sustainable luxury is the return to the ancestral essence of luxury, to the thoughtful purchase, to the artisan manufacture, to the beauty of materials in its broadest sense, and to the respect for social and environmental issues (Gardetti and Muthu 2015: IX).

Dana Thomas highlighted in her 2007 book how the most important brands "switched off the luster of luxury" in the last years of the Twentieth Century, on the contrary we can maintain that nowadays fashion is reevaluating luxury as a cultural value. As British journalist Lucy Siegle writes, the fashion brands have started the "second act of luxury", and they did it exactly in the name of sustainability. Siegle cites many cases: Tiffany and Co., for example, one of the founding members of I. R.M.A. (Initiative for Responsible Mining Assurance), takes its raw materials only from mines whose supply chains are traceable. Kering, François-Henri Pinault's luxury holding, imposed many sustainability targets to its brands, among them: the attention to the locating of raw materials, paper and packaging; the decrease of water and waste consumption; the decrease of the emissions of carbon monoxide, dangerous chemicals and other materials; control on suppliers; an aware development for new professionalisms. Siege cites also Livia Firth, creative director of London-based agency for environmental consulting Eco-Age, who works with some of the most renowned fashion brands and who founded in 2012—along with Siegle herself—the Green Carpet Challenge project, using her film industry connections to persuade Hollywood celebrities to use eco-friendly fashion. Firth worked also with Lanvin and Stella McCartney, the most prominent brands of the Kering group, in order to restructure the supply chains, from silk to leather.

Siegle writes:

> Sometimes it is worth convincing an entire generation of producers of the worth of joining sustainability, in order to access a profitable and growing market, as happened for Gucci zero-deforestation handbags. It is a matter of reconsidering the artisanal tradition, and leading it into the present. The attention is all on those luxury brands which are composing this scenario. Those are the forerunners of the famous "turning point" that we are impatiently waiting (Siegle 2015).

Certainly, Siegle—one of the executive producers of *The True Cost* documentary—emphasises the positive value of the prominent brands of sustainability in the luxury industry, while stigmatising low-cost fashion as founding on complete negligence, or better on the extreme exploit of raw materials and workers, in order to obtain a low-cost final product and fast production. It is also true that, after all those criticisms towards them, even some low-cost brands are now promoting the awareness for materials and production conditions in their systems. Among these brands, Swedish brand H&M has been promoting since 2011 the H&M Conscious project, based on values such as recycling, ethics, awareness the responsible exploit of natural resources. On these values H&M founds its production policy and its marketing, aiming to making even the customers more responsible.

Although their ethical and critical engagement, luxury goods are still exclusive. On the contrary, low-cost fashion increased enormously the number of fashion consumers: actually everybody can afford clothing, now (Giannone and Calefato 2007). Low-cost is therefore a form of sustainability of fashion, as it is for tourism, furnishing and healthcare. One exemplar case of low-cost in the sense of social sustainability is the 2013 declaration of Indian Supreme Court on the right of the local industry of producing generic drugs against cancer. These drugs cost one tenth of the "branded" one, produced by the pharmaceutical corporation who held the exclusive until then, and who wanted to keep it. Therefore an ethical revolution—necessarily producing added value—even in low-cost production is welcome, in order to create a sort of "cultural luxury", accessible only to aware customers and producers. Luxury can become "democratic" only as culture.

The resources whose endangerment is constantly increasing are not only natural resources: culture can be one of the most neglected branches of economy; often, when governments have to cut down public expenses, the first ones suffering these cuts are those destined to museums, libraries, monuments archaeological sites and cultural heritage in general. In the last years some fashion brands invested in culture differently than the traditional sponsorship of cultural enterprises. Luxury brands actually built up cultural enterprises: Foundations. Cartier was the forerunner of this trend, and its 1984 established Foundation, in Paris, supports art through acquisitions, exhibitions, publishing, sponsorship to young artists. After it Benetton in Treviso, Montblanc in Hamburg, Trussardi in Milan, Pinault in Venice are just some of the many examples. Famous for their bond with architecture are Louis

Vuitton—its Paris headquarters are designed by Frank Gehry—and Prada—whose foundation was established in 1993 and it actually built a cultural pole in Milan in 2015 by reconverting an old distillery. The Prada Foundation project is by Rem Koolhaas, an architect whose long term collaboration with Miuccia Prada is well known (Koolhaas 2001), and it represents one of the most interesting examples of luxury in the meanings of culture, spaces, and redevelopment of peripheral neighbourhoods. It is also very interesting to note how the café—a place usually considered as "trivial" as opposed to the "sacredness" of a museum—gained in this case a peculiar dignity: The Bar Luce of Prada Foundation was indeed designed by director Wes Anderson, who recreated there the atmosphere of a typical bar of Milan in the Sixties. In the case of Foundations luxury rediscovers its primordial elements of gift and culture—art, architecture, design, cinema, photography—as a social value. As journalist Renata Molho writes: "the new frontier of luxury cannot ignore responsibility, gratitude, and giving back" (Molho 2016: 52).

7 Cangiari

At the 2013 Paris Fashion Week a new Italian Calabria based brand, whose production is entirely based on social enterprises—Cangiari—debuted in the international fashion scene. It is a symbolic and particular example of sustainable luxury: in fact Cangiari is by now the first eco-ethical high-end fashion brand in Italy. It produces handcrafted textiles, most of them weaved on a manual loom, a technique increasingly exploited by the brand. The aim of the brand is creating high-qality clothing on the one hand, while re-updating the ancient tradition of Calabrian weaving, which roots into Grecanic and Byzantine cultures. Combining this traditional weaving with research and innovation, the brand creates unique products with a tailored finish.

Cangiari belongs to the 2003 founded GOEL Cooperative Group, which comprehends many social enterprises of the Locride and the Gioia Tauro Plain, in the Reggio Calabria province. Its base in Southern Italy is extremely relevant, even if Cangiari has a promotional branch in Milan and its supply network is expanding especially in Northern Europe. The mission of GOEL Group is in fact changing Calabria and give value to local communities. The actual name GOEL has a biblical etymology and its meaning in ancient Hebrew is "the liberator", "the redeemer". The ethical engagement for redemption and change are represented also by the name Cangiari itself—meaning "to change" in Calabrian dialect. Calabria is one of the many Southern Italy regions which endure a structural disparity in comparison to the Northern ones: this is due to the centuries-old *Questione Meridionale* (southern issue), which was deeply analysed by Gramsci (1975), among many others, during the years of his imprisonment as an antifascist. After the Second World War, the economic policy of the Italian institutions and, later, the precisely focused interventions of the European Union tried to solve this issue, but they never accomplished the task completely. Especially Calabria is still nowadays a land of poverty and emigration. Furthermore the activities of the local racquet, *'ndrangheta*, deeply control territory, economics and politics.

Cangiari supply chain and production, therefore, has a highly valued cultural, political and social meaning, because the enterprises of its network rebel to the criminal control of territory and economics. GOEL Group actively opposes to *'ndrangheta* by reporting to the authorities its activities and actually showing that the best answer to criminality is social engagement and work quality. The network of local artisans which are involved in the Group is also committed to the inclusion of impoverished people and lower social classes.

The GOEL network of enterprises safeguards its environmental sustainability by overseeing all its production chain, from hand-weaving to communication. The textiles and the organic dyes are accurately certified and controlled in order to ensure the best environmental care and the well-being of the customers. Thanks to the direct control of the production chain, its clothes can be highly customised, too.

Together with Cangiari, GOEL Cooperative Group oversees many other activities in the fields of responsible tourism, organic farming, local development, marketing and communication, social and healthcare services. The Group also founded a farming cooperative, GOEL Bio, which reunites many businesses who experienced attacks by *'ndrangheta.*

The GOEL strategy for change is based on a twofold action: giving actual and emancipating answers to the needs of people and offering cultural offers for change to the territory through its entrepreneurship. The Group's philosophy is founded on the idea that ethics cannot be only just anymore, but it has to be also effective.

In December 2015, the brand presented in Rome a runway show in collaboration with Action Aid: in this occasion the principles of sustainable fashion were exhibited to the public, supported by the participation of *The True Cost* director Andrew Morgan. This initiative represented an example for how fashion nowadays goes beyond fashion itself, since the exhibition of the dresses on the runway was only a part of a bigger cultural enterprise. In the same month Cangiari released also a bridal dresses collection named "The Ethical Bride", dedicated to those women who want to live their wedding day coherently with their aesthetical and civil convictions.

Vincenzo Linarello, president of the GOEL group, declared:

> You cannot ignore what is behind some glittering fashion products, you cannot, no more. It is unacceptable to consider "beautiful" a garment which enslaves workers, devastates the environment and damages the health of the wearer. If fashion is not ethical, it is horrible, monstrous, and who signs the garments coming from these circuits underwrites the shame behind them.

Nothing more conclusive could be said on the sustainable present and future of luxury and fashion.

References

Ágalma (2002) Il lusso oscuro oggetto del desiderio. 2, Meltemi, Roma
Barthes R (1957) Mythologies. Seuil, Paris. English edition: Barthes R (1972) Mythologies (trans: Lavers A). Paladin, London
Barthes R (2006) The language of fashion. In: Carter M (ed) (trans: Stafford A). Berg, Oxford
Bataille G (1967) La part maudite précédé de La notion de dépense. Minuit, Paris. English edition: Bataille G (1988) The Accursed Share: An Essay on General Economy. Volume I: Consumption (trans: Hurley R). Zone Books, Cambridge, Mass
Benjamin W (1968) Illuminations. In: Arendt H (ed) (trans: Zohn H). Schocken Books, New York
Benjamin W (1982) Das PassagenWerk. In: Tiedemann R (ed) Suhrkamp, Frankfurt am Main. English edition: Benjamin W (1999) The Arcades Project (trans: Eiland H, McLaughlin K). Harvard University Press, Cambridge, Mass
Benveniste E (1969) Le vocabulaire des institutions indo-européennes. I. Economie, parenté, societé. Minuit, Paris
Berry C (1994) The idea of luxury. Cambridge University Press, Cambridge
Borghero C (ed) (1974) La polemica sul lusso nel Settecento francese. Einaudi, Torino
Bourdieu P (1979) La distinction. Minuit, Paris. English edition, Bourdieu P (1984) Distinction: A Social Critique of the Judgement of Taste (trans: Nice R). Routledge, London
Calefato P (2003) Lusso. Meltemi, Roma
Calefato P (2014) Luxury. Fashion, lifestyle, and excess. Bloomsbury, New York-London
Castiglione B (1528) Il libro del Cortegiano, English edition: Castiglione B (1997) The Book of the Courtier (trans: Bull G A). Penguin, New York
Curcio AM (ed) (2005) Sociologia della moda e del lusso. Franco Angeli, Milano
Edelkoort L (2015) Anti-fashion manifesto. Available at http://www.edelkoort.com/2015/09/anti_fashion-manifesto. Accessed 29 Feb 2016
Enzensberger HM (1997) Zickzack. Suhrkamp, Frankfurt am Main. English edition: Enzensberger HM (1998) Zig Zag: The Politics of Culture and Vice Versa (trans: Rugg LH). New Press, New York
Frank RH (1999) Luxury fever. Money and happiness in an era of excess. Princeton University Press, Princeton-Oxford
Gardetti MA, Muthu SS (eds) (2015) Handbook of sustainable luxury textiles and fashion. Springer, Heidelberg
Giannone A, Calefato P (2007) Manuale di comunicazione, sociologia e cultura della moda, Vol 5: performance. Meltemi, Roma
Gramsci A (1975) Quaderni del carcere. Edizione critica a cura di V. Gerratana. Einaudi, Turin. English edition: Gramsci A (2011) Prison Notebooks. Columbia University Press, New York
Hebdige D (1979) Subculture. The meaning of style. Routledge, London
Koolhaas R (2001) Projects for prada part 1. Fondazione Prada, Milano
Lipovetsky G (avec Roux E) (2003) Le luxe éternel. Gallimard, Paris
Mathieu P, Monneyron F (2015) L'imaginaire du luxe. Imago, Paris
Molho R (2016) Due mondi, una sola vision, marie claire #likes 2, HMC Italia: 52–53
Paulicelli E (ed) (2006) Moda e moderno. Meltemi, Roma
Siegle L (2015) Eco-luxury. L'etica ha un valore. marie claire #likes 1, HMC Italia. http://mclikes.marieclaire.it/alta-moda-lusso-ecosostenibilita. Accessed 29 Feb 2016
Simmel G (1895) Zur Psychologie der Mode. Die Zeit: Wiener Wochenschrift für Politik, Volkswirtschaft, Wissenschaft und Kunst 5, Band 1895, Nr. 54 (12.10.1895): 22–24. English edition: Simmel G (1904) Fashion. Int Q 10: 130–155. http://www.modetheorie.de/fileadmin/Texte/s/Simmel-Fashion_1904.pdf. Accessed 29 Feb 2016
Simmel G (1900) Philosophie des Geldes. Duncker & Humblot, Berlin. English edition: Simmel G (1978) The Philosophy of Money. Routledge, London

Sombart W (1913) Luxus und Kapitalismus. Duncker & Humblot, Berlin. English edition: Sombart W (1967) Luxury and Capitalism (trans: Dittmar WR). University of Michigan Press, Ann Arbor

Thomas D (2007) Deluxe. How luxury lost its luster. Penguin, London

Veblen T (1899) The theory of the leisure class. Macmillan, New York. Modern edition: Veblen T (2007) The theory of the leisure class. The Echo Library, Teddington, Middlesex

Versace G, Calabrese O (1993) Signatures. Abbeville Press, New York

Weber M (1904–1905) Die protestantische Ethik und der Geist der Kapitalismus. Archiv für Sozialwissenschaft und Sozialpolitik, Tübingen. English edition: Weber M (2002) The Protestant Ethic and the "Spirit" of Capitalism and Other Writings. Penguin, London

Websites[1]

Avita: http://www.avitastyle.com

Brunello Cucinelli: http://www.brunellocucinelli.com/it

Cangiari Moda Etica Artigianale: http://www.cangiari.it

Loro Piana: https://www.loropiana.com/it

Patrick Lafrontière: http://www.une-saison-en-guyane.com/article/culture/patrick-lafrontiere-projet-filage-tissage

So Critical So Fashion http://www.criticalfashion.it

Veja: http://www.veja-store.com

Worn Again: http://wornagain.info

Zeha: https://www.zeha-berlin.de

[1]All accessed 30 March 2016.

Identifying the Luxury Sustainability Paradox: Three Steps Toward a Solution

Oliver P. Heil and Daniel André Langer

Abstract In this chapter, we argue about a paradox of sustainability in the context of luxury goods and brands: Intuitively, luxury brands should be more sustainable versus normal brands, since consumer's willingness to pay is high which should allow for highest standards in quality, including sustainability. However, many of the most expensive luxury products appear to exhibit limited sustainability. Examples include sports cars that typically are gas guzzlers, yachts that often carry only a few people but require a high amount of resources to be operated, or fur products that require animals to give their lives for. For example, the Hermes Birkin bag recently received a lot of negative media due to the alleged (mis-)treatment of crocodiles. Hence, albeit a bit counterintuitive, many luxurious products seem less sustainable than they could and probably should these days causing what we call the "Luxury Sustainability Paradox" to surface. In essence, luxury products' superiority appears to not span across all attributes and, paradoxically, to do so may be very challenging if not impossible. In short: being at the absolute top along certain luxury or performance attributes may, paradoxically, preclude top scores on sustainability.

Keywords Luxury · Sustainability paradox · Managerial implications · Luxury marketing · Luxury management · Luxurystrategy

In the chapter we aim to significantly broaden the scope of luxury research aspects by dissecting the following:

- Ambiguous Meaning of Luxury—Blurred Aspects of the "Luxury Sustainability Paradox".
- Behavioral aspects from a luxury consumer perspective.

O.P. Heil (✉)
Johannes Gutenberg University of Mainz, Mainz, Germany
e-mail: oliverheil2055@gmail.com

D.A. Langer
Johannes Gutenberg University of Mainz, Scottsdale, USA
e-mail: drdaniellanger@gmail.com

M.A. Gardetti (ed.), *Sustainable Management of Luxury*,
Environmental Footprints and Eco-design of Products and Processes,
DOI 10.1007/978-981-10-2917-2_6

- Is there a consumer paradigm shift and how can it be leveraged?
- How to create Added Luxury Value with Sustainable Luxury?
- Description of durability, ownership vs consumption effects.
- Competitive Advantage Through Sustainability.

Luxury goods are distinctive from "normal" goods. Through our findings we provide managers with comprehensive toolsets and strategies to manage luxury in a more competitive, profitable and sustainable way.

1 The Paradox Between Luxury and Sustainability

Luxury is an elusive concept that relatively few seem to truly grasp and understand. Why is this the case?

Unfortunately, the term "luxury" is typically used with a lack of precision and clear definition. After all, it seems clear to everybody what luxury really means. But is that really the case? We suggest just to do a Google search on keywords and images and you will find a whopping variety of things under the label "luxury": From very cheap to very expensive items across a variety of topics and categories. Equally, if you look up luxury brands you will find an astounding range of brands, some surprisingly positioned at entry price level just using the term "luxury" to signal something somewhat aspirational. In most newspaper and press articles about luxury, you will find no clear use of the term which blurs the line even further. Looking up the portfolio of most brands that position themselves within the luxury space and you will find a lack of coherence in terms of pricing. Many "luxury" brands offer a surprising range of products and price points, which are not always coherent.

In short, the first challenge is to be precise about what luxury really is. We will address this in the next chapter and provide sharp operational definitions that help to set a common base to discuss the implications for sustainability and managing luxury in a more precise and more successful way.

At this point we state that luxury goods (brands and services) are **distinctive from "normal" goods (brands and services) in a significant way**, which makes distinct managing tools and strategies not only optional but a necessity. **Luxuries are the (few) top end items along price ranges within categories and offer specific properties that make them unique.** As a result, consumers who buy luxuries opt to buy products that are disproportionately expensive, typically in a non-linear fashion within their categories. If this happens, those products must offer a significant perceived consumer value that entices the purchase.

With this in mind, let's dig a bit deeper into the topic of sustainability. There are many definitions of sustainability but operationally we define **sustainability as a significantly reduced negative consumption effect at a given value a product creates.**

Let's look at the implications. **If two products create the same perceived value, the more sustainable product will have less negative consumption effects**, such as resource utilization at production, better production conditions, reduced carbon footprint along its lifecycle, higher longevity and durability, etc. Switching perspective, this also means that **if all negative effects between products are the same, the one that creates higher perceived consumer value is the more sustainable.**

As a result, it should be fairly easy to distinguish among products and create more sustainable luxuries. This causes the emergence of some kind of dilemma: Since luxuries create a "non-linear" value within a given category, the measurement of relative sustainability between luxury brands is challenging. Figure 1 illustrates this.

Most products in a given category have comparable price points closely correlated with the features (e.g. size, other functional and brand-related values). Luxuries behave non-linear, hence although features do not increase in a non-linear way, their value does: it "jumps". Those value jumps or steps can be extreme, and as categories evolve, the jumps tend to increase and get more extreme. Now, if the value jumps, the relative footprint in relation to the value decreases significantly, which make the measure difficult (both for consumers who want to do a purchase decision based on sustainable aspects as well as for companies who want to create more sustainable products).

This is the first signal of a Luxury Sustainability Paradox. Luxuries should be more sustainable, but how to measure this in relation to the value they create? Notably, this conceptual starting point seems to be confirmed by direct observations: Many of the most expensive products are not very sustainable if the external effects are not taken in relation to the value they create for individual consumers but if compared on an absolute level to "normal" products. To illustrate, let's have a look at some examples: practically all high end sports cars are gas guzzlers. The Bugatti Veyron, one of the world's most expensive production cars to date

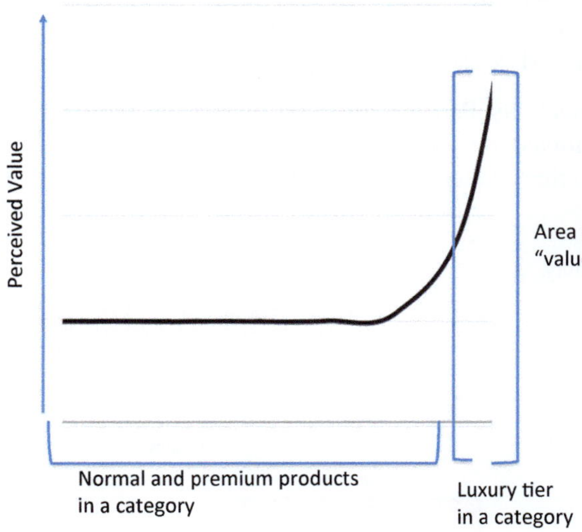

Fig. 1 Luxury value jumps or steps

consumes around 100 L per 100 km driven at top speed–even for standards of supercars an exceptional number as high gasoline consumption seems to amount to some kind of rule for high end sports cars.

Super yachts typically carry few people but require a high amount of resources to be operated. Starting from fuel to cleaning and maintenance requirements, energy and resources consumed for producing the boats, rare (and often hard-to-harvest) materials used, such as very robust woods, etc. Similar is the effect of using a private plane. The carbon footprint of flying 2–5 people in a small aircraft is similar to the one flying hundreds of people in a commercial airliner.

In Haute Couture and Luxury handbags it's not atypical that animals give their lives for the leather or fur. Since luxury brands will try to only use "highest quality" skins, those with even smallest defects will not be used. Negative production (= consumption) effects can be assumed (much) higher versus normal products. A point in case is the famous Hermès Birkin bag that received a lot of negative media coverage recently due to the alleged (mis)-treatment of crocodiles to a point that Mrs. Birkin was threatening not to be the "spokeswomen" for her iconic bag anymore. Those are only few examples, but they showcase that luxury brands and products are not necessarily as sustainable that one would expect at first glance from products that create the highest consumer value within their categories. As a result, we distill what we call a luxury sustainability paradox.

More precisely, the "Luxury Sustainability Paradox" appears to include the following aspects:

- Compared to non-luxuries, the negative consumption effect of luxuries seems to be higher.
- It is difficult to compare luxury offers on relative sustainability, since luxuries provide a "Value Jump" and make relative comparisons challenging.
- Taken the "Value Jump" into consideration, luxury brands can be perceived as (relatively) more sustainable in relation to the perceived value than the more linear "non-luxuries" (Fig. 2).

The Luxury Sustainability Paradox reiterates once more the elusive nature of luxury, which is not limited to its definition but impacts consumptions strategies by consumers and marketing strategies by brands in a significant way. Intuitively

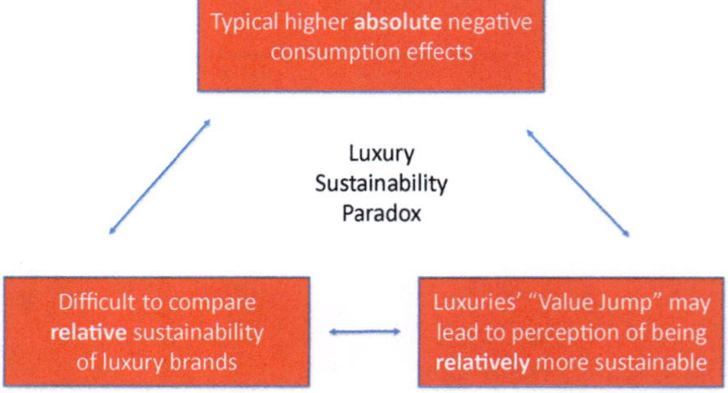

Fig. 2 Luxury sustainability paradox

luxury brands should be more sustainable versus normal brands, since consumer's willingness to pay is high which should allow for highest standards in sustainability. While certain absolute sustainability aspects (MPG/fuel consumptions for cars, absolute carbon footprint, etc.) can be compared in relative easy terms, the relative comparison can become very tricky, if not impossible due to the value part of the sustainability equation. Products that absolutely seem less "sustainable" if just the negative consumption effects are measures, may be perceived as more sustainable taking created value into consideration.

In the subsequent sections of this chapter we will elaborate implications and solutions to the paradox.

2 What Is Luxury and What Does It Imply?

In an early paper on luxury, Dubois and Duquesne (1993, p. 43) describe that "… many people buy such goods for what they *symbolize*. This is consistent with hedonic consumption and extended self-personality models, according to which purchasing luxury goods represents an extreme form of expressing one's values."

Their description includes important aspects of what luxuries are: Luxuries are (also) symbols that help people to express their personalities and their values. They are "extreme" in nature, which allows people to use them to clearly shape, define, enhance or underline certain aspects they want to be associated with. Langer and Heil (2013) showed that the effects of luxuries are as pronounced on the "self" (hence, the person that consumes a luxury) as on others (the observers), while the effects seem to a certain extent indifferent to beliefs. In other words, it does not matter if someone likes luxury goods or not and if he or she has a high or low perceived affinity towards luxury for a luxury item to have a perceptual effect on the self and on others.

This important finding explains why luxury brands and products are so important to people that they pay non-linear prices, far beyond "feature" value of items. This is also a very important aspect because it somewhat shows the danger for brands to offer "affordable luxury" or poorly defined luxury propositions. If luxury brands have too many affordable "entry products", get too "visible" by selling too many items, or get incongruent in their portfolios, the perception of "luxury" suffers, and with it its aspiration and willingness-to-pay.

Langer and Heil (2013, p. 90) define: "**Luxury is something rare and hedonic, difficult to acquire or use, that provides a perceived unique experience in combination with a perceived enhancement or reinforcement of the social position. It is an emotional social marker and differentiator**". This clearly propels luxury into a social context but also reinforces luxury as a tool to produce unique, memorable experiences. Both aspects are typical for sharply defined luxury brands and products, which as a result create an "**Added Luxury Value**" that drives the entire value perception.

In an extended luxury definition, Langer and Heil (2013, p. 91) incorporate the value aspect: "**A luxury is something rare that is able to signal status information. The signal can be received and processed by the owner and by others depending on signal clarity, strength and consistency. The signal has the effect to trigger the perception of an enhancement or reinforcement of the social status. It results in a perceived consumer value, the added luxury value ALV, which can become the major value component of a luxury product's total value for a consumer. ALV is composed of enhanced self-esteem, self-perception and attractiveness, attribution of financial liquidity, social power and expertise, protection in public and the perception of new, unique and ultimate experiences by a hedonic good that functions as ultimate treat. Additionally, a luxury is a social marker, classifying people as members of distinct social groups. As a reflection of added luxury value, luxuries tend to be overly expensive, they are difficult to purchase or use. Typically they belong to the top 1% cluster of the most expensive goods or services in a category. Luxuries tend to be compatible with social norms and values, a precondition for status reassurance.**"

Figure 3 summarizes the definition of luxury. Still, many "luxury" brands manage their equities in a rather lose manner while a precise application is crucial in order to keep the perception of a luxury consistent over time. Seemingly small (at a first glance) mistakes in positioning.

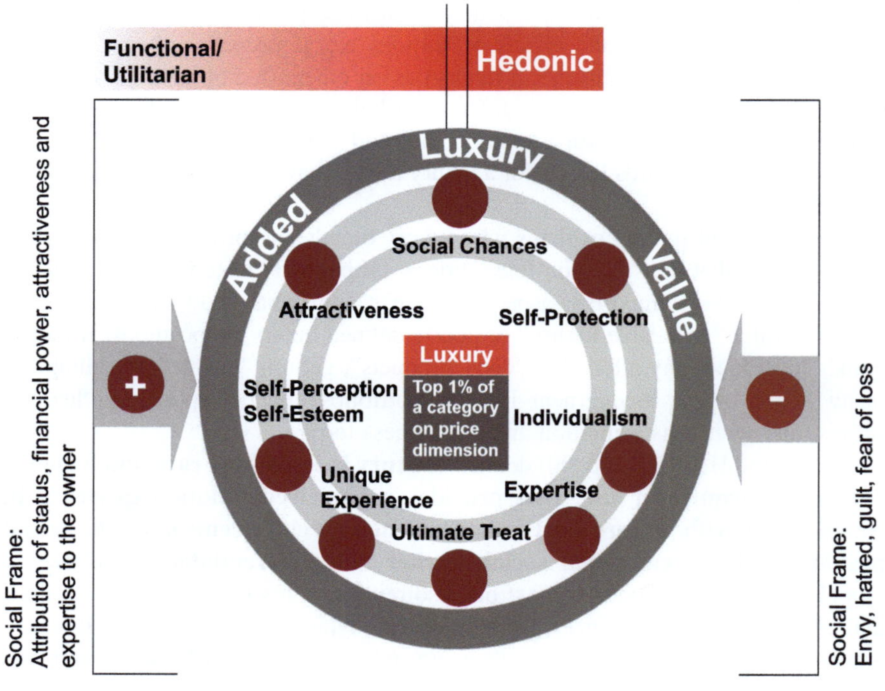

Fig. 3 Luxury definition (Langer and Heil 2013)

What are the implications of a sharp versus a somewhat conceptual or lose definition and execution of luxury for the topic of sustainability? Let's start with managing luxury in a rather lose way: brands and products that offer affordable luxury or are not consistent across all consumer touch-points will not be able to create significant non-linear added luxury value over time. Hence, those brands will be perceived in a very similar fashion to normal brands. Negative consumption effects will become more important as they can't be mentally offset by larger perceived value. For consumers who take negative consumption effects into consideration, those brands need to focus strongly on delivering a better footprint than competitors in order to to compete over time.

Brands that are more consistent in terms of their luxury positioning may have more degrees of freedom. The consistently high added consumer value they are able to create can potentially help them to offset some of the negative consumption disadvantages.

Another important aspect is the social frame within which a luxury is owned and consumed. Figure 3 shows that there are external drivers that influence the luxury perception significantly.

Among positive drivers are attribution of status and financial power by others, as well as attribution of attractiveness and expertise to the owner. The anticipation (conscious or unconscious) of those effects help to create Added Luxury Value and entice people to buy luxuries.

On the flipside, a negative social frame can significantly reduce ALV. Some important negative effects are envy, hatred, guilt and fear of loss. Envy is a feeling typically triggered by a perceived difference of the possession state between two people: The bigger house, car, TV of a neighbor, the fancy trip of an acquaintance, just to name a few.

In extreme cases this can trigger feelings resulting in hatred, especially when social differences are perceived to grow beyond an individual threshold that is seen as "fair". But hatred can also be triggered if someone perceives that others consume without caring about the negative consumption effects. Especially when the consumption is seen as non-essential, the acceptance by others of negative consumption effects may decrease significantly. Given the nature of luxury goods—highly relevant for the owner but potentially non-essential in the view of others—they are prone to non-acceptance by others. In the extreme, it can render a brand or product obsolete.

Luxury products also define the self to a certain extend. Belk (1988) revealed that the fundamental role that possessions play for people, functioning as an extension of the self. The more important a possession, the more important is its role in self-extension. But this has a risk: Any important possession that people lose or that becomes obsolete due to increasing negative frame effects makes people also lose part of their identity and memories.

If a person fears that a loss is imminent he or she may not buy a product that otherwise would be considered. Hence if people fear that negative consumption effects may increase, let's say because of increasing ecological consciousness of the society for example, they may shy away from buying a luxury brand they would otherwise purchase or aspire for.

But, also, external shocks can be vital for purchase and consumption decisions. China's fight against corruption has led to draconic fines if there is proof that government authorities accepted luxury products as bribes. This has not only reduced the number of luxury items bought as such gifts (as it would be expected) but also impacted the consumption pattern of remaining luxury consumers. Brands that became too flashy or ubiquitous and, most importantly, those that were easily associated with brand being typically used as "gifts" in the past, suddenly suffered while less visible luxury brands were less impacted.

One key takeaway is that luxury brands are part of an individual purchase and consumption portfolio decision which is embedded in the social frame of each person. External effects influence the frame and individual decisions. Luxury brands have to take this into consideration when defining and shaping their offers and portfolios.

3 A Closer Look to the "Luxury Sustainability Paradox"

So far, we distilled some important aspects: First, maybe counterintuitively, we argue that there is a "Luxury Sustainability Paradox" as many luxuries seem to have larger negative sustainability effects compared even to non-luxuries. Part of the paradox is the difficulty of comparing luxury brands easily in terms of sustainability due to the large impact of added luxury value. Even more so, brands with higher negative consumption effects may be perceived as more sustainable, since they create a higher value. Second, we defined what luxury is, linking it to unique perceived experiences and an improvement of the social positioning, which make those products so valuable for people while setting the bar extremely high for managing a luxury brand. Seemingly small mistakes can significantly decrease the perceived value of a luxury brand.

An conclusion at this point could be that the Luxury Sustainability Paradox can easily be solved by simply enhancing the added luxury value and hoping that its non-linear properties minimize the negative consumption effects in the total sustainability equation. But this would be incomplete and, to a degree, even myopic.

We have seen already that the luxury consumption is influenced by the social frame and individual decisions will take aspects of expected envy, hatred and fear into consideration. For sure a large, non-linear ALV may offset the negative but still consumption has to take place in a social context and people may simply fear to be perceived in a too negative way. While fur was well accepted for a long time, the category literally disappeared and today most luxury brands only offer fur in very few, very selective items, often just as a small feature. At a certain point in time many consumers simply did not want to be associated anymore with the category and its negative consumption effect and fur today is mostly seen among rap music stars as they protest everything.

Celebrities typically play a key role in terms of being role-models for "behavior", both positively and negatively. Some recent consumption trends were adapted and

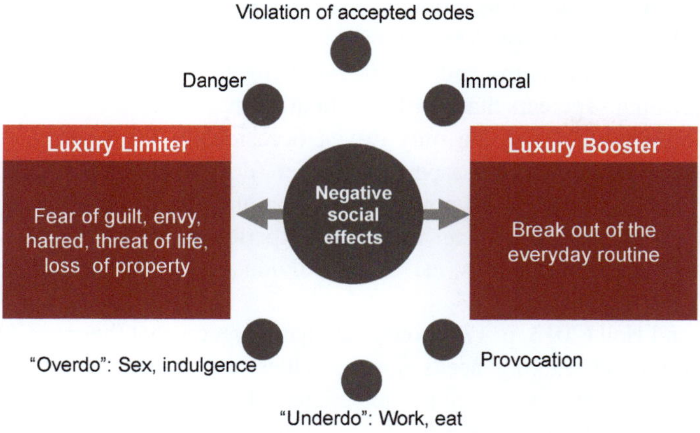

Fig. 4 Negative social effects as limiter and booster (Langer and Heil 2013)

endorsed early on by celebrities who made them part of their lifestyle. When electric cars started appear, many Hollywood celebrities were among the first adopters, thus signaling a behavioral change of people who many aspire to. Many celebrities publicly state their preference for organic and raw foods, endorsing a lifestyle choice that initially started as niche and is adapted only by a few. This signal then, over time, changes the behavior of a larger population.

Naturally, one might wonder why sustainability is not much stronger embedded in the nature of luxury? Langer and Heil (2013) provide an initial discussion of the phenomenon by suggesting that the negative social effects can have exactly the opposite effect of being a luxury limiter: They can become a luxury booster (Fig. 4).

The boost comes by providing an opportunity to break-out of the everyday routine almost all people have in one way or the other. Since luxuries provide unique experiences they are ideal in nature to provide an escape from normal. Its drivers appear to include:

- Association with danger
- Violation of accepted codes and norms
- Immoral consumption
- Provocation
- Being associated with an "excess" of sex and indulgence
- Being associated with an easy lifestyle ("underdo work") or extreme diets to appear more beautiful ("underdo eating")

This adds to the Luxury Sustainability Paradox: **While negative consumption effects may be unwanted in most situations, they can be wanted (or at least accepted) in other situations by the same consumer.** People may rationalize a consumption choice that has significant negative effects simply with the few usages that a "unique" consumption typically entices. A supercar with an excessive gas consumption is typically used only occasionally (e.g. on a racetrack on weekends) hence providing an excuse to rationalize an otherwise almost "unethical" or "immoral" level of consumption. The authors know people who drive electric cars that they power with solar energy and who actively engage in sustainable lifestyles but own gas-guzzling supercars that would not fit at all into the idea of reduced external consumption effects. Those are only driven occasionally for "special moments", which then could include the aspect of danger (driving the supercar at extreme speeds), violation of codes and norms (high wear and tear of tires and material on the race-track in exchange for an individual experience), excessive gasoline consumption and carbon emissions, etc. Even individual consumption patterns can be paradox.

Langer and Heil (2013, p. 172) conclude, that "it seems that there can't be luxury without the normal. Luxury needs norms, values and yardsticks, which are then exceeded, thus leading to a unique experience. However, as shown, exceeding norms might be a facilitator or a limiter to luxury depending on the amplitude."

Exceeding norms also means that production and supply chain aspects are typically significantly more challenging and complex versus normal goods. There are simply few products to be produced but with a high amount of craftsmanship and complexity as well as the utilization of rare and extreme materials which are often hard to find or harvest. Those aspects significantly complicate production of luxury goods. They also make a highly efficient supply chain in regards to minimization of negative footprint during production processes challenging if not impossible for many luxury products.

A luxury yacht will always require a very complex (even to the "extreme") sourcing, supply and production setup where almost every detail is highly individualized and many revisions and refinements are made during the production process. The same goes for many luxury items and is in principle not different for haute couture dresses (only the finest materials are used, they are difficult and timely to produce, require many fittings were either the tailor or the buyer have to fly in for), luxury mansions (utilization of a high amount of space, material, technology, often overcoming difficult architectural challenges) or supercars (very powerful engines and components that require extreme materials to withstand extreme forces, high individualization of colors, materials and features). The list can be extended to include jets, red wine, islands, watches, etc.

One can argue that luxury products may be used for a longer period of time giving them a higher longevity versus normal products and thus helping to offset negative consumption effects over time since less items need to be replaced. While this may sometimes be the case, it is not the rule. Given the unique and extreme experiences of luxury consumption, the extreme needs to be maintained from a consumer perspective to fulfill the task. And consumers who seek the extreme will easily switch brands and products whenever there is a new extreme, rendering the previous object of desire obsolete.

The authors recently interviewed a sales manager of one of the world's largest Porsche dealerships and distilled the insight that especially the top end customers who would not buy anything less than a fully loaded Porsche 911 Turbo S or GT3, often buy the car and drive it for just a couple of weeks (or days), and then sell it again once the excitement flattened. He called the phenomenon "Acute Automotive Purchase Disorder" and stated that it was somewhat typical for his (few) top customers who buy multiple cars a year.

Staying with cars, the depreciation of luxury cars is among the highest of all cars. Prices of used luxury cars tend to collapse whenever a new model is launched. Only very few brands with strict limitation and a large number of collectors (e.g. Ferrari, Bugatti, Lamborghini) manage to somewhat maintain or even increase the value over time. The high depreciation and value loss is another example indicating that longevity may be on many aspects rather an exception (or an excuse for the external consumption effects or the high prices paid) than the norm.

Even among watches only very few brands (e.g. Patek Philippe, Audemars Piguet) manage to maintain or even increase the prices of used watches. And even for the most valuable brands, maintaining the value typically only works on very few models that are highly sought after but were produced in very limited quantities. Most other watch brands also face a (steep) decrease in value when watches become older and are offered for resale.

We maintain: The Luxury Sustainability Paradox gets even more pronounced upon closer inspection. It is a complex construct consisting of forces that at the same time have the potential to boost and limit luxury consumption and perception.

4 The Consumer Perspective: Behavioral and Society Shifts

The world is changing and social media and better connectivity in real time is one of the forces behind this change. Social media allows people to connect, interact, exchange and comment without separation of space and time. When people share in social media they get rewarded by an increasing number of followers and likes. As a result, people communicate and share about all aspects of life including their experiments with brands.

For almost any product, even the rarest, there are Instagram posts, YouTube videos, Facebook discussion forums and reviews available. Never before has it been so easy to be informed about every detail of the product experience even without personally knowing people who own specific products. Everything is just a click or a tap away.

This changes the knowledge of people about products they would potentially buy. Even just a few years ago, the sales person was often the sole source of information and was able to emphasize or deemphasize product features or their shortcomings. Today, there is a fair amount of transparency available without the need of too crazy search efforts.

For luxury brands this amounts to a rather dramatic change and challenge. Luxury brands now need to withstand social discussion about any aspect. If the public (or the media or both) pick up issues, it can have severe effects on brand equity. Also, where an article may have remained local without being much noticed in the past, now almost anything is picked up fast in social media. And latest when consumers think of buying a rare piece and Google for reviews and information, they will easily find even a locally published article.

In 2015, the Mirror in UK published an article headlined "Hermes Birkin croc horror: Is this the true cost of the celebrities' must-have £25,000 handbag?" about an issue we have already mentioned earlier in this chapter. The article was drastic: "But tonight US animal rights activists claimed the thousands of alligators and crocodiles bred to provide the French firm's skins for the bags are subjected at times to dreadful mistreatment before seemingly being slaughtered inhumanely. Undercover filming by People for the Ethical Treatment of Animals (Peta) appears to show a world away from the glamour and luxury of A-lister fashion. The reptiles are seen in cramped concrete pits—in some cases surrounded by faeces in brown water—in footage filmed at alligator and crocodile farms in Texas and Zimbabwe. In one scene, allegedly filmed in Texas, workers stab knives into the backs of seemingly still conscious animals because bolt guns, which are usually used to stun them, cannot be found." (Mirror, Jun 24, 2016, Online).

While Hermès immediately reacted and announced to sanction all suppliers that do not respect their defined procedures, damage was done and may have led to some potential consumers not purchasing the bag or feeling embarrassed being seen with one in public.

Hummer was known in the early 2000s as rather expensive and martial over-the-top SUV and the go-to car for Hollywood celebrities and other affluent people in the US and elsewhere. The brand has literally disappeared after it was showcased and discussed across all media channels as the synonymous for a ridiculously oversized (even for US standards) gas guzzler that no-one really needed ever. While hyped when it was new and different, its lifespan ended the moment the hype was over, and negative consumption effects came into the spotlight.

Those two examples show that luxury brands cannot rely solely on extreme experiences and very high added luxury value as means to mitigate shortcomings on negative consumption effects. More transparency about product features, positive and negative, impose strong risks if negative consumption effects get too extreme in public perception and overshadow the positive value creation.

With the advent of social media, there is much more transparency not only on brands but also on consumers. Many of the top luxury consumers are publically known, have their own sites, feeds and channels with thousands or even millions of followers and like to broadcast about their lives and their favorite brands. The same Mirror article cites Victoria Beckham, Kim Kardashian and Tamara Ecclestone among Birkin bag consumers and cites that some of them have wardrobes filled with those handbags. Now, if customers that buy dozens or even hundreds of the most expensive luxury items in a category, it becomes obvious what may happen if negative aspects become so prevalent that they discourage from using the item or even publicly renouncing ownership.

But it is not only social media transparency that drives the need to rethink negative consumption effects. Most societies in the world evolved from being "consumption-only" focused towards more social responsible consumption. Climate change, differences between rich and poor, the more socially conscious Millennial generation, to name a few, are driving this change. In countries like Russia or China where luxury consumption was accepted to be rather conspicuous for a while, more conscious luxury consumption is evolving. Publications like the Financial Times Sustainability Index further pressure companies into more sustainable practices if they want to remain competitive.

What does this mean for a luxury brand? Both, the image of the brand as well as the image of the consumer has to be taken into consideration. The shift in consumer behavior and expectations is a threat for luxury brands given its paradoxical nature in regards to sustainability. However, it can also be an opportunity for luxury brands as we will show in the next part. The key question is: how can the paradigm shift we just described be leveraged for luxury brands to further enhance and strengthen brand equities and product experiences?

The easy answer would be to simply reduce the negative consumption effects by being more choiceful with materials and production setup. But we have seen that this is not how luxury works because negative effects also serve as luxury boosters. To illustrate, we take a look at the BMW i8. While groundbreaking in design and sustainable production methods (carbon-neutral production, recycled materials, electric engine to support a fuel-efficient gasoline engine, etc.), the car does not offer the same extreme driving experiences that many consumers seek in its segment. Hence, it falls behind competitive offers like the Porsche 911 Turbo or the Mercedes AMG GT, which do not put many sustainable materials, production methods and so-forth into the spotlight of their marketing campaigns. Hence, even with a general paradigm shift in consumer expectation towards sustainability, the extreme experience needs to be still provided to remain competitive and this may contradict certain sustainability aspects.

5 How to Boost Added Luxury Value with Sustainable Luxury?

To answer the question on the "how to" we take take a step back and look at how Added Luxury Value is created, the value component of a luxury product that is the key reason to buy it at non-linear price points within categories.

Added Luxury Value (ALV) is perceived as a result of a signal is emitted by a good that people identify as "luxury" (vs. non-luxury) based on certain tangible and intangible factors or "antecedents" of luxury (for a detailed description of the mechanism see Langer and Heil 2013). Figure 5 showcases this. The signal can be moderated (magnified or reduced) by influences of the sender as well as by the

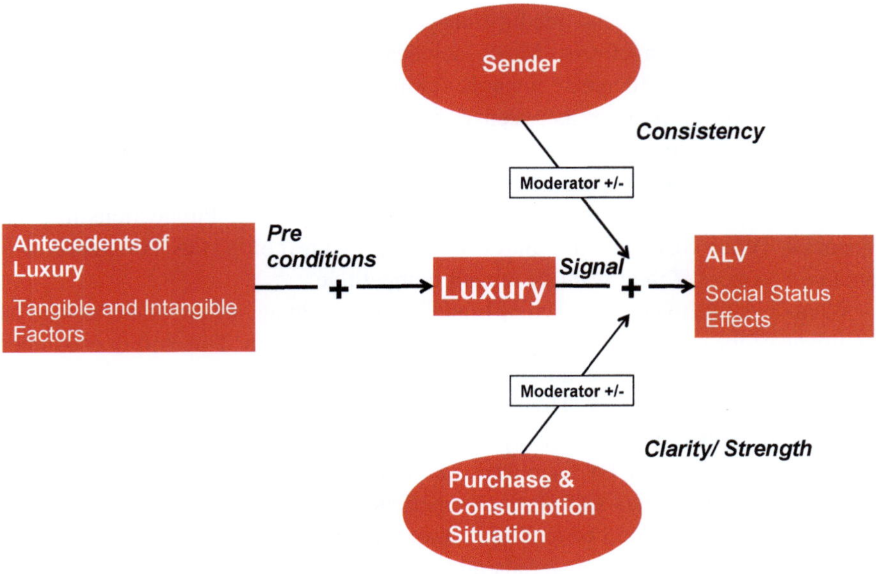

Fig. 5 General framework of luxury (Langer and Heil 2013)

purchase and consumption situation. ALV as result of the signal will be highest if the signal is consistent, clear and strong and ideally amplified by the moderating factors just mentioned.

This framework offers three key levers to solve the Luxury Sustainability Paradox:

- First, a product has to be clearly identified as luxury.
- Second, negative consumption effects should not moderate ALV in a negative way.
- Third, to amplify ALV, additional measures should be considered (see below).

Let's start with the first one. On a first glance this seems a no-brainer but it is surprising that its not always considered first. Going back to core definition, luxury has to provide unique (= extreme, once-in-a-lifetime) experiences in combination enhancement or reinforcement of the social status. It can be argued that this is the core proposition and that there can never be any compromise in terms of delivering it.

This means that the optimization of all tangible and intangible factors that identify a product as a luxury has to be the very first consideration: How can a super sports car become significantly faster, more exciting, more differentiating? How can a handbag become more spectacular? How can a dress flatter the person who wears it more? How can the experience of a high-end wine be amplified further? How can a vacation trip become even more unique and memorable?

Because this, in sense, is what creates a "luxury", which, as a result, emits its unique luxury signal: That is, the creation of luxury has to come first and needs to be elaborated down to its smallest detail to make the experience authentic, differentiated, surprising and with this create a unique, even once-in-a-lifetime experience.

Part of this effort requires to carefully select which negative consumption effects are needed–even wanted–and essential in an effort to create the experience. The strategy playbook needs to define all that's "in" but also all that's "out" so that the product or brand gets a sharp enough profile. (For completeness, we ad that in some cases certain attributes may be totally unwanted, in other cases they may be an important part of the experience—but, due to space constraints, we cannot elaborate on this contrast.) What is important is to have a clear strategy how the experience should be created and which attribute levels are important enough to allow them to challenge sustainability.

Once the playbook is defined, signal moderators have to be evaluated and as we pointed out, this has to be done twofold: how can negative moderation be minimized, how can positive moderation be amplified?

Langer and Heil (2015) point out that social values have to be taken into careful consideration even if provocation or other negative consumption effects are part of the playbook. Luxury managers have to carefully steer negative effects and make sure that they do not get too large, so that they impact the brand perception in a negative fashion. This means we need to differentiate between wanted and unwanted negative consumption effects.

The wanted negative effects are those that are part of the playbook and are used to drive the luxury perception. Hence they should always be consciously chosen as part of the experience creation, they can never be random elements as otherwise they would negatively moderate the luxury signal.

Any unwanted negative consumption effect that is not part of the playbook has to be minimized especially if those effects could influence the consumption situation significantly.

All this amounts to a rather new way of looking at value creation of sustainable luxury products: **Non-sustainable luxury can still "sell" but negative consumption effects are only permissible if-and-only-if they are key for the luxury experience, i.e. the experience would suffer significantly by omitting them.** This is a direct result from combining the General Framework of Luxury with the implications of the Luxury Sustainability Paradox.

One could argue that the BMW i8 may not have been consequently designed to provide the ultimate luxury experience due to its focus on minimizing negative consumption effects wrapped into a spectacular design. This approach could work for normal brands. For luxury brands with predominant focus to perform in an extreme way on certain aspects a too unilateral focus on minimizing negative effects can be dangerous. Especially if it impacts the overall aspiration and identification. **It's important to have a holistic playbook and manage all positive and negative aspects a luxury brand or product offers.**

As a consequence, despite all issues and risks discussed earlier, the strategy of Hermès to offer handbags made of crocodile skin seems to work better in regards to luxury. Why? Crocodile skins are very rare and provide unique experience in terms of feel, aesthetics, touch, etc. hence they make the product very special. As long as there is no (or not too much) cruelty done to animals some consumers may still go for a crocodile handbag to have a truly differentiating product. Hence one can argue that Hermès should consider to expressively built negative consumption effects into the playbook of their crocodile handbags. The emerging technology around the concept of NeuroChoice could be employed to identify what needs to be "in" and what has to be "out."

The key is to make sure that the strategy is executed with excellence on a consistent base and without any exception. Also, the playbook needs to be revisited regularly to make sure that changing social expectations and norms are constantly reflected. If there was ongoing public discussion on the use of crocodile skin on handbags, probably refraining from offering this type of leather and moving on to other leathers or fabrics could be advisable. If strategy execution is not excellent, negative press and outrage (thus damaging brand equity) can be the result as we could see in the case above.

Thus, a luxury manager has to make the right choices of which negative consumption elements are important and which are not. What should be in the playbook, what should be out?

If the playbook is well-defined and strategy execution flawless, then the third aspect of managing sustainable luxury comes into play: amplifying the signal. This can be seen as the reverse approach of minimizing the negatives. Indeed, this is about focusing on positive aspects of the luxury consumption in relation to social effects, social norms or actually improving the environmental status quo. This was part of the go-to-market model of the Tesla Model S. First of all, the Model S was designed to be not the best electric car (hence a compromise on the luxury perception) but the best car in the world, offering more space than other comparable cars, innovative design, features never before in a production car (touchscreen operation, autopilot function, seven seat option in a sedan) and super-car like acceleration.

But all those factors are in a sense evolutionary features with little effect on the negative consumption effect of driving a car per se. The key feature that Tesla pointed out in this respect is that the car is offering zero emissions. Hence it offers a suit of evolutionary facets with the disruptive feature of zero-emission driving. While many consumers still will have to use electricity made from fossil sources, at least the car offers the opportunity to be driven completely emission free if energy is created by solar or other renewable energy sources. This disruptive feature has the ability to even amplify the luxury signal because the offer now is not only better than other is many aspects, but has a feature that practically reverses negative consumption effects if a normal car is replaced by it. Importantly and remarkably, the fact the car's batteries, once at the end of the life cycle pose huge problems for generations to come. A fundamental threat to the car's positioning in terms of sustainability but that threat has yet to appear on the public's radar.

Finding signal amplification factors by turning negative consumption effects into positive ones is not easy and may not be possible for all product categories. But in case a brand can figure out how to turn any significant negative effect that is embedded into the category it competes into a positive effect it can significantly enhance the added luxury value via signal amplification.

This is extremely encouraging for managers of luxury brands. Consumers will be willing to pay a price that reflects the perceived value a product provides as long as it's relevant for them, there is no competitive offer that is seen as superior and the individual budget allows the purchase. Amplification of the luxury signal means that a value perception of a luxury that is already disproportionate can be further enhanced. This offers additional non-linear pricing opportunities for a "sustainable" luxury good vs a "usual" luxury good. We call this aspect the Additional Sustainability Value Jump or ASVJ, see Fig. 6 for a visualization.

In other words, in a category with significantly negative consumption effects (e.g. tailpipe emissions of cars and their use of fossil fuels) the replacement of a product with a luxury product that eliminates the negative consumption effects has the opportunity to significantly increase the total value proposition of the luxury offer with by amplification of the luxury signal. The value increase is what we call Added Luxury Value Jump and can be monetized by higher price points reflecting the additional value created.

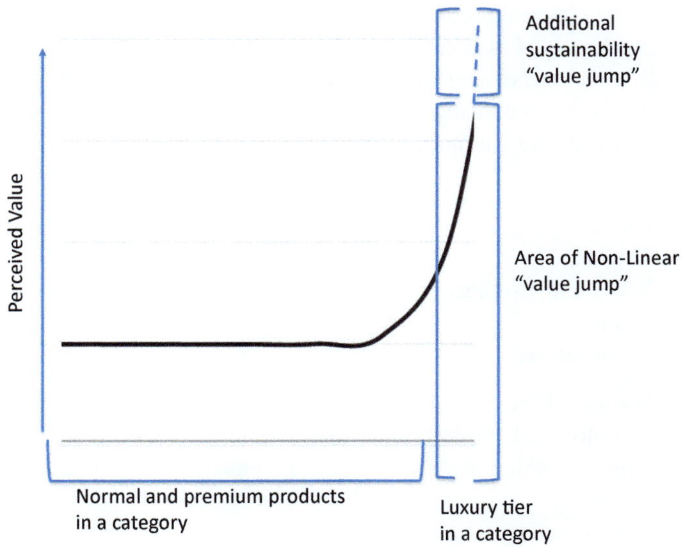

Fig. 6 Additional sustainability value jump

6 Moving Toward Solving the Luxury Sustainability Paradox

At the start of this chapter, we argued that luxuries may possess a kind of paradoxical relationship once sustainability is contrasted with luxuriousness—we call that the Luxury Sustainability Paradox. We consider this to be a complex contrast made up of forces that–at the same time have the potential to boost and limit luxury consumption and perception. Consequently, eliminating negative consumption effects may not be always advised as they can interfere with the value created by extreme experiences.

Our research provides a key to solve the luxury sustainability paradox for brands by following a three-step approach to sustainable luxury:

- Defining the luxury playbook: precisely describing how to create added luxury value, which negative factors to include, which to eliminate. This creates a consistent and clear luxury signal and ALV as a consequence.
- Analyzing significant negative consumption effect of the category and defining strategies to eliminate them to create Additional Sustainability Value Jumps to further boost ALV.
- Excellence in execution especially on all negative consumption effects that are built on purpose into the luxury proposition

With this in mind, our approach offers luxury brands the opportunity to become front-runners in sustainable luxury. First and foremost, a brand must provide added luxury value to be successful over time. Any compromise here will either lead to a brand not being perceived as luxury or lead to a short lifecycle where the brand gets irrelevant and loses cachet and appeal. Langer and Heil (2013) provide a comprehensive set of strategies to successfully manage luxury goods in a profitable and consistent way which are summarized as the 6 L of Luxury:

- Link to Consumer
- Luster Creation
- Leverage Pricing
- Link to Values and Cultures
- Legend Building
- Leverage Experiences

The combination of those strategies lead to brands that deliver continuously on creating highly individual products that offer consumers non-linear value propositions by building on unique experiences and the enhancement or reinforcement of the social position.

Those strategies need to be the starting point for creating the luxury offers because of a product is not perceived as luxury, there is no added luxury value. Based on successfully creating a luxury good, its luxury signal can be boosted by consequently eliminating significant negative consumption effects a category typically offers and turning the luxury good into the "better" offer that provides a positive consumption effect in relation to its category.

It is important to note: While the Luxury Sustainability Paradox will continue to exist as part of the nature of luxury, our approach offers the opportunity to create more sustainable luxury products that can outperform less sustainable alternatives on key luxury driving dimensions, while solving negative category effects.

Any remaining, carefully orchestrated "built-in" negative consumption effects need to be excellently executed and managed. Especially shifts in social norms need to be taken into consideration to ensure that the luxury offers are still in the realm of "acceptable consumption".

As one can see, we don't focus too much on classic arguments that luxury products may be more sustainable due to the opportunity of a more careful choice of materials, longer ownership and durability as well as the opportunity to have better work conditions as a result of higher margins. While all of those aspects may be valid for certain products, they do not guarantee success in the luxury world and they do not fully solve the luxury sustainability paradox.

We propose a solution of the paradox by "embracing" it, creating first what makes a luxury a luxury by applying a consequent luxury playbook that includes certain negative consumption effects (those that may boost luxury perception) while excluding others (those that may limit). This has to be followed by seeking category opportunities of eliminating negative category consumption effects, thus making the luxury the better alternative and creating a Sustainable Luxury Value Jump. In parallel executional excellence is a non-negotiable must for luxury managers.

This approach provides the opportunity to create continuous competitive advantage through sustainable luxury and a boost of profitability through the Sustainable Luxury Value Jump.

7 Key Implications for Managers and Future Research

We encourage managers to rethink their approach to sustainable luxury. It can be a tremendous tool to increase the perceived value of their luxury brands and boost profits.

A typical pitfall is putting the decrease of negative consumption effects above all as this may exclude important value drivers from the luxury playbook. Excluding important aspects will make the entire luxury offer obsolete in view of consumers. And an obsolete luxury offer will never be a sustainable alternative.

Value needs to be created first and then boosted by amplifying the signal. Luxuries are still luxuries and are bought in part to surprise people with unique experiences, provide a solution to their dreams. In short, the allure of luxury must not be lost by creating a sustainable luxury offer.

We encourage researchers to focus on developing concepts, methods and instruments to better measure and manage Added Sustainability Value Jumps ASVJ. This should also include market research tools specifically designed towards luxury brands and their inherently competing threats to sustainability. As a start, we provided a 3-step approach to manage luxury brands in a more sustainable fashion as a foundation to manage sustainable luxury brands in a more precise way.

Luxury products entail some of the largest managerial challenge since ALV and ASVJ are still to be classified as somewhat elusive, almost kind of a mirage. This is in addition to more classic notions of luxury entailing that luxury is rare by nature, highly aspirational and therefore highly valuable for consumers. There are few brands whose management entails to truly understand luxury and consistently perform over time in creating and maintaining a high aspirational value. Remarkably, the exception still amounts to the rule!

In times where consumers expect more from brands overall, sustainability becomes more important. It should be rather obvious at this point that management needs to continuously watch out to not fall prey to the Luxury Sustainability Paradox, as it needs to be taken into consideration when creating and updating—i.e. managing–luxury offers.

When the playbook is fully includes a conscious incorporation (or elimination) of "negative consumption effects" a luxury brand can dramatically increase its value proposition.

References

Belk R (1988) Possessions and the extended self. J Consum Res 15(2):139–168

Dubois B, Duquesne P (1993) The market for luxury goods: income versus culture, Eur J Mark 27(1):35–44

Langer D, Heil O (2013) Luxury marketing & management. Center for Luxury Research, Scottsdale, Mainz

Langer D, Heil O (2015) Luxury essentials. Center for Luxury Research, Scottsdale, Mainz

Mirror (2015) Hermes Birkin croc horror: is this the true cost of the celebrities' must-have £25,000 handbag? http://www.mirror.co.uk/news/world-news/hermes-birkin-croc-horror-true-5943831. Accessed 10 May 2016

Strike a Pose: Luxury for Sustainability

Hakan Karaosman, Gustavo Morales-Alonso and Alessandro Brun

Abstract Sustainability cannot be secured without a joint effort. From cleaner production to responsible sourcing and conscious consumption, a long-term perspective should be embarked. Accordingly, it is timely to interrogate to what extent luxury fashion could inquire more sustainable solutions and embrace sustainability. Luxury fashion could inspire and pursue quality, longevity and exclusivity in order to go confidentially towards achieving sustainability. In this vein, this chapter examines how sustainable luxury could lead to a reduction in resource consumption while authentically improving quality, and durability. What a luxury brand does today will be part of its heritage within fifty years, fostering sustainability can therefore become as equally important as the heritage. To this end, this chapter introduces sustainability and fashion paradox. Thus, a novel perspective is brought with an explanation of the degree to which luxury fashion could indeed have a positive impact on sustainability through design, local resource utilization, innovation, and longevity.

Keywords Luxury fashion · Luxury goods industry · Craftsmanship · Sustainability · Quality · Innovation · Sustainable design · Fashion supply chains

H. Karaosman (✉) · A. Brun
Department of Management, Economics and Industrial Engineering,
Politecnico di Milano, 20156 Milan, Italy
e-mail: hakan.karaosman@polimi.it

A. Brun
e-mail: alessandro.brun@polimi.it

H. Karaosman · G. Morales-Alonso
Department of Industrial Management, Business Administration and Statistics,
Universidad Politecnica de Madrid, 28006 Madrid, Spain
e-mail: gustavo.morales@upm.es

M.A. Gardetti (ed.), *Sustainable Management of Luxury*,
Environmental Footprints and Eco-design of Products and Processes,
DOI 10.1007/978-981-10-2917-2_7

1 Foundations of Sustainability

Sustainability was defined by the United Nation's World Commission on Environment and Development (WCED 1987) as the means of satisfying current needs without compromising the possibility for future generations to satisfy their own needs. In this vein, the term 'Triple Bottom Line' (3BL) dates back to the mid-1990s, when AccountAbility began using the term in its context (Norman and Macdonald 2004). In particular, 3BL was revealed within the 1997 publication of the British edition of John Elkington's '*Cannibals With Works: The Triple Bottom Line of 21st Century Business*'. In this regard, not only does a company's ultimate success constitute traditional financial bottom line, but it also encompasses social and environmental considerations. Companies might experience scarcity of time and capital to more effectively manage environmental and social issues; yet, the integration of social and environmental considerations into decision-making structures is highly required to meet global standards alongside societal needs.

The International Labour Office recognizes fashion as one of the most labour-intensive industries (ILO 2016); and companies in labour intensive industries are urgently required to revisit their business models and restructure their operations as a consequence of the need for environmental protection and growing natural resource demand (Wu and Pagell 2011). Nevertheless, due to the ambiguities surrounding the definition of sustainability, complications often occur while applying the principles into the practice. Sustainability in fashion struggles to balance different and often competing needs against an awareness of ethical and environmental issues of today. Globally dispersed, and highly complex fashion supply chains (SCs) comprise various processes through the stages of fibre production, yarn spinning, fabric production, apparel manufacture, retail and consumption. However, fashion products are manufactured by ever changing suppliers in relatively small production cycles, and the difficulty of embarking sustainability into fashion occurs due to highly fragmented and complex SCs (Joy et al. 2012).

Due to the cost restrictions, the complexity and the lengths of fashion SCs, tracing garments' ethical and environmental performance, and monitoring provenance of materials are not easy to proceed. In this sense, the biggest challenge of fashion SCs is linked with direct interactions with sub-contractors. Environmental and social impact the fashion industry generates, not only in developing but also in developed countries, presents increasing concerns. International unions and NGOs have been working to improve the conditions. Conversely, it is still questionable to what extent environmental, human and safety measures really exist in fashion SCs. The lack of an integrated approach toward sustainability hinders the impact of a fashion company's sustainability efforts, its ability to robustly thrive in the future and its opportunities to address today's global issues (Mosher and Smith 2015). Henceforth, it is essential to focus on the entirety of the SC from both internal processes to tackle sustainability challenges. In such larger systems companies often feel discouraged to embark sustainability. Nevertheless, once sustainability is truly integrated into the larger global fashion systems, the meeting point of commerciality with sustainability could be reached.

2 Sustainability in Fashion Myth

'Fashion' and 'Clothing' represent different aspects of the relationship with what is worn. Fashion is a way to reflect and communicate an individual's vision within society. Clothing, on the other hand, is the material that gives fashion a contextual vision. The fashion and clothing industry is one of the few remaining craft-based industries and manually operated sewing machines are still the principal means by which garments are made (Black 2012). During the pre-industrial era, the main attributes of textile and clothing were associated with limited production, limited division of labour and limited number of specialized crafts. Nonetheless, with the advent of sewing machine on 1790, and with the industrial revolution in the period from 1760 to about 1840, the connection between production and consumption were broken down into numerous processes. In the following centuries, the concept and the philosophy of fashion and textile were further transformed. By the mid 1970's mass production took over the industry to make customization technically feasible; subsequent to the SC revolution of the 1990's, globalization, and the phase-out of the Multifibre Arrangement in 2005, fashion has become a massive global business with constant production that has been moved to countries where lower wages are offered and less protection for workers and the environment is provided.

In this setting, the global textile and apparel industry reached a total revenue of €1.17 trillion in 2014 with a significant annual growth rate of 3.8% between 2010 and 2014 (Research and Markets 2015). The performance of the industry is forecasted to accelerate, which is expected to drive the industry to a value of €1465 billion by the end of 2019. According to the European Apparel and Textile Confederation, the overall size of the industry in the EU-28 represented a turnover of €165 billion in 2014 and with investments of €4 billion (EURATEX 2015). Yet, despite the growth, the global fashion industry is affected by economic and socio-political events. Demand unpredictability, rising labour costs in emerging countries, supply network complexity, production fragmentation, geographical dispersion, economic crisis, sustainability issues, time-based competition and the mounting sensibility to environmental issues challenge the industry.

Resources are finite, population increases, consumption grows and all these result in the risk of volatile commodity prices, instability in global markets, and significant water, waste and carbon footprints. In February 2011, the short-term doubling in cotton prices justified one facet of many sustainability problems. Transforming the economy from one based on the principle that natural resources are infinite to the one acknowledging resource scarcity requires fashion companies to make sustainability a central concern to change products, processes, and business strategies in a radical way. The scale and urgency of sustainability integration must be promoted through a more authentic comprehension of quality, lifestyle and exclusiveness, involving respect for each other, the planet and societies (Bendell and Kleanthous 2007). Even though many fashion retailers claim that sustainability is embarked in corporate DNA and there are a number of examples of fashion retailers making progress, sustainability often remains vague and distinct from core business strategy rather than a fundamental component. A systematic approach is needed to integrate sustainability that must involve long-terms shifts in organizational structure, institutions and policy frameworks.

Box 1—True Fashion Impact

Significant environmental impacts are associated with the clothing life cycle. 132 million metric tonnes of coal is required for around 60 million tonnes of textile production (Ecotextiles 2011). The International Union for the Conservation of Nature reports that the dyeing and treatment of textiles generates between 17 and 20% of all industrial water pollution (Aiama et al. 2016). Additionally, the apparel industry is responsible for 22.5% of all insecticide use globally (Balch 2016). Energy inputs in polyester production are around 125 MJ/kg of the fibre and over two-thirds (70%) of the total emissions related to polyester products arise in the production phase (Balch 2016). However, about half of the emissions associated with the clothing life cycle come from consumer use phase, such as washing, tumble-drying, ironing and using. Statistics show that 400% more clothing is purchased today than it used to be done 20 years ago. Greenpeace reports that around 80 billion garments are produced worldwide, which equals to over 11 garments for each person every year. WRAP has found that (WRAP 2012) global footprint per household in the UK generates around 1.5 tonnes of CO_2-e emissions each year, which is equivalent to driving a car almost 6000 miles. Similarly, waste is a critical aspect affected by fashion consumption. More than 200,000 L each year is the global water footprint per household in the UK, which equals to filling over 1000 bathtubs to capacity.

3 Fashion Paradox

The main contradiction between production and consumption lies in the fact that today's consumers buy more than what they need. Economic growth has led to economic abundance. Consequently, consumers all over the world, regardless of their social segment, are wealthier than at any other time in history (Benammar 2008). However, this escalating economic abundance results in the continuous resource extraction and unavoidable inconsistencies. For example, the fashion industry generates a number of livelihoods worldwide, however it results in wasteful cycle of seasonal changes. Vintage, on the other hand, receives attention to appreciate the craftsmanship of couture while the idea of 'new' becomes an obsession to express individualities (Black 2012).

It must be acknowledged that sustainability in fashion is not a new trend. Eco trend reached its peak for the first time in the early 1990's. People showed a great motivation and enthusiasm to change fashion on a permanent basis by making everlasting and resilient items. However, one main characteristic of the entire system was underestimated back then, that is, the power of change (Teunissen 2008). The ecological collections developed in the mid-nineties mainly focused on materials, such as unbleached cotton. This way, eco fashion through garments such as trench coats and classic shirts was associated with minimalist, clean but not 'so aesthetic' looks (Teunissen 2008). The majority of consumers still associate environmental friendly clothes with unbleached, undyed and shapeless items (Jensen and Dahl 2008). Yet, aesthetic should be considered as equally important as the ecological aspect. This would bring a great challenge of making fabrics and producing garments looking glamorous, fashionable, sexy and conventional. As stressed by British designer Stella McCartney '*the important thing is to create beautiful luxurious things. Having people not notice that something is organic or faux leather could be considered one of the biggest challenges*'.

Previous endeavours of eco fashion were seeing pillars individually, however they are interrelated and a holistic perspective should be introduced to look at the system as a whole. Squaring a circle between product design, global SCs and sustainability would not seem optimistic (Brass and Bowden 2012). Then, people who design, produce, distribute, promote and consume fashion should be informed and be part of a larger conversation on sustainability. Fashion SCs involve a number of processes through the stages of raw material sourcing, fibre production, fabric production, apparel manufacture, transportation, and consumer use phases. Prior to the development of improvement scenarios, it is vital to examine details regarding environmental and social impacts each SC stage generates.

3.1 Raw Material Sourcing

Sourcing is already complicated without sustainability, as price, quality, time, vendor relations and geographical issues are complex enough. Nonetheless, sourcing must be pursued with consciousness. In terms of raw materials, cotton is the most widespread profitable non-food crop in the world. Cotton represents a dominant source of natural fibres used in clothing since 77% of natural fibre production comes from cotton (Carbon Trust 2011a). Cotton production provides income for more than 250 million people worldwide, employing about 7% of all labour in developing countries (WWF 2014). The price of cotton has dropped during the last 30 years, however the production costs are increasing. Consequently, wages are getting lower for farmers in places like India, Kyrgyzstan and West Africa. In

Uzbekistan, the entire cotton industry is built on forced child labour (Baptist World Aid Australia 2011). Cotton has also significant environmental impacts. 20,000 L of water is needed to produce 1 kg of cotton (WWF 2014), and further, global cotton production generates emissions of around 220 $MtCO_2$-e, accounting for approximately 0.8% of global CO_2 emissions. Around 30% of CO_2-e emissions associated with cotton production are embodied in the international trade of cotton as raw cotton (Carbon Trust 2011b) before the cotton is transformed into textiles, clothing and other final products. Additionally, cotton consumes 16% of all insecticides and 6.8% of all herbicides used worldwide (Organic Cotton 2015). A shift to organic cotton could be a way to reduce the environmental impact. Indeed, organic cotton does not require pesticides; however, yields of organic cotton are lower and extremely significant amounts of water are still needed, which presents the biggest drawback of organic cotton cultivation. Further, price and reliable supply emerge as other critical challenges for the execution. Consequently, on the whole, many challenges occur in raw material sourcing with respect to environmental and social facets.

3.2 Fibre Production

Fibres provide the initial source for textile products. Composition of an average garment includes natural fibres such as cotton, flax silk, and wool and manufactured fibres such polyurethane, polyamide, acrylic, polyester, and viscose. The amount of fibre used varies depending on the garment type. For example, cotton is the main fibre used for underwear production, while polyester accounts for one third of a dress and polyamide is the main component of swimwear. Cotton and polyester dominate global clothing fibre inputs, collectively supplying over 85% of all fibre used in clothing (Carbon Trust 2011a). Cotton meets 28% of fibre demand while man-made fibres meet a significant 58% of the demand. Polyester is a key synthetic fibre, and its high performance as well as its stability makes it the most popular polymer, e.g. approximately 11 million tons of polyester is produced each year, accounting for one half of the total annual production of all synthetic fibres. Polyester comes from finite resources and consume considerable amount of energy accompanied by toxic additives. Unluckily, polyester cannot be considered optimal since most polyester is manufactured while antimony is used as a catalyst. Antimony trioxide leaking from polyester fibres during the high-temperature dying processes emerges in the wastewater and cause chronic bronchitis.

3.3 Apparel Manufacture

Apparel manufacture techniques must be optimized through both technical and non-technical components. The former refers to products with high technical quality in terms of resistance, sensitivity, and wash ability while the latter aims at a better wardrobe management as a principle to avoid purchasing items. Process optimization is more linked with the upstream part, emphasising the interactions among manufacturers and buyers. Hence, this could be secured through a better fashion SC execution. Despite the significant nature of fabrics, manufacture phase generates a noteworthy amount of fabric waste, e.g. about 20% of the fabric swept off the floor of cutting room (Rissanen 2008). When this fabric waste ends up in the landfill, natural fibres decompose however chemicals and methane are released. Synthetics, on the other hand, take such long times to decompose. Fabric recycling is though is not suitable for all uses. Current recycling practices are actually downcycling leading to a reduction in the value and the life span of a recycled fibre becomes relatively shorter than the virgin ones. Innovative solutions are needed to mitigate the negative impact, e.g. switching 10% of cotton fabric to a 50:50 polycotton-blended fabric could reduce waste footprint around 1.7%, water footprint by 3% and carbon emissions by 0.4% (WRAP 2012).

3.4 Use Phase

Although widely accepted theories in product design suggest that 80% of a garments' environmental impact is determined by the material choice, this could differ based on a product type. Research shows that use phase of a pair of jeans contributes more than 60% of primary energy consumption, human toxicity and production of household waste (Bain et al. 2009). Additionally, the use phase accounts for 35 and 59% of carbon emissions and water eutrophication and between 10 and 34% of ozone layer depletion and water consumption (Bain et al. 2009). Behavioural patterns of washing, ironing and dry cleaning differ drastically depending on the person, product type and contexts. Firstly, products that easy to wash and do not require long washing cycles could be produced. This could be linked to producing garments that require less maintenance. Consequently, the way products are consumed could be unsustainable even if the products are produced in a sustainable way. Therefore, the actual behaviour of consumers counts significantly. It could be stated that an object cannot be sustainable if the behaviour it causes is not. Avoiding weak links, which require maintenance and reduce the lifespan, designing products that could be repaired and upgraded, increasing the functionality of the product, and increasing the lifespan through product quality could be listed as some ways to optimize the product life cycle. Design phase on this regard needs to explore opportunities to downgrade the environmental impact to be generated during the use phase.

Considering aforementioned facts, some might think that fashion is trivial, however fashion has become a source of the conversation. Early initiatives and actions by designers have begun to coincide with a wider political response to environmental and social actions employed in developed countries. A growing emphasis on sustainability can be a manifest throughout the fashion industry. Green is the new black has been a widely accepted and used as a metaphor for sustainable fashion that creates a meaningful correspondence between eco-friendliness and fashion, as well as a connection between short and long term goals (Kaiser 2008).

4 Advancing Sustainability Through Luxury

Luxury is not the opposite of poverty; it is the opposite of vulgarity – Coco Chanel

The concept of luxury gets its roots from the history. According to the Oxford Latin Dictionary, the term 'luxury' arises from the Latin word 'luxus', meaning exaggerated life, glamour, comfort and wealth (Brun and Castelli 2013). In the ancient world, luxury goods were associated with wealth, exclusivity and power. At the end of the 19th century, the second industrial revolution gave luxury's modern meaning of being enjoyable beyond the necessities of life. The leather goods of Hermes, the jewellery of Cartier, the luggage of Louis Vuitton and the raincoats of Burberry are some of the signature luxury products of that century. Succeeding the growth of the fashion business in the 20th century, superior quality, durability, performance and design have become the most relevant aspects in the luxury market. Till the 1970s, the era of exclusivity was experienced where mostly European family businesses centred on craftsmanship. The changing scene of luxury and further segmentation within the industry might make some argue that the exclusivity of luxury have been devalued due to socio-economic expansion of luxury brands across a number of product sectors, social classes and time zones (Bendell and Kleanthous 2007). Similarly, some fashion experts state that as family-owned fashion companies become multinational corporations, the industry has diminished its integrity and the history of craftsmanship.

In this setting, the highest quality product is expected to generate the most benefits for everyone involved in its production and distribution. Correspondingly, SC actors' knowledge and awareness of such benefits are fundamental for the experience. Nevertheless, social dynamics are changing and visible shifts in the luxury paradigm occur. Luxury brands are fragile to reputational damage since a big proportion of their brand value is derived from consumer loyalty and trust. Hence, luxury brands must seek ways to represent the greatest positive contribution. Caniato et al. (2009) suggest a set of critical success factors that are crucial to compete in the luxury market, including premium quality, craftsmanship, exclusivity, brand building, style and design, emotional appeal, country of origin, uniqueness, performance and innovation and lifestyle creation. Nonetheless, as Brun and Castelli (2013) recommend, the success factors must be revised and be extended by contemplating on recent market trends such as ethical aspects and sustainability. To this end, luxury could present a new yet a responsible business case through which profounder values are embodied in sourcing, manufacturing, distribution and retail of products.

4.1 The Quality to Foster Sustainability

The quality of any product has become more important than its look, and therefore the quality could be articulated to be an emergent as a distinctive feature not only for luxury but also for natural resources. The average party top, for example, is worn only 1.7 times before getting thrown away and this type of mass produced garments made of synthetic fabrics cannot be recycled, and they end up at landfill (Gibson 2012). Consequently, the new trend of sensible products with higher values of exclusivity and quality could purportedly nurture sustainability. This statement is in line with Prof. Dr. Michael Scott from Cambridge University who stressed that (2011) luxury is indeed a good tool to understand the value and priorities of diverse societies, present and past. Heritage and quality of luxury appeal because they do not cause global warming, dwindling natural resources, and pollution (Joy et al. 2012). Henceforth, luxury could largely contribute to the phenomena of democracy, social harmony and sustainability. Yet, sustainability should be translated as a guideline in luxury fashion to generate a corporate language. In today's challenging environment, luxury fashion companies are required to change their institutional logics to re-build aspiration, credibility and trust.

Incorporating elements of value such as higher ethical standards in sourcing, efficient energy and material use, low-impact manufacturing, distribution and retail, and the establishment of recycling and repair services could reduce negative impact while reshaping the brand value by refining environmental and social performance in today's resource-constrained state. Luxury brands adapt traditions to create ever-lasting products with style and to maintain the brand's heritage into the future. Accordingly, luxury brands could be at the forefront for sustainability and luxury could be redefined as a positive and a revolutionary road by endorsing distinctive attributes of longevity, superior performance and premium quality to handle the dangerously growing culture of disposability. Sustainable fashion through luxury could be defined as what could minimize the impact on the environment and on the societies while pursuing corporate goals. Sustainable 'luxury fashion' thus opts for business activities through exclusivity, quality and craftsmanship that might not impoverish the planet. In order to foster sustainability and achieve further improvements, integration along the luxury fashion SCs is needed and to this end, business model thinking must be employed and sustainability must be tapped into business culture of luxury.

With a final goal of advancing environmental and social sustainability, global luxury fashion companies develop and deploy some management capabilities. It is time to showcase some pivotal practices implemented by leading luxury fashion companies. Hermés has been advancing textile recycling by generating new fashion in Petit H line from waste caused by its operations. In terms of biodiversity, Stella McCartney takes many inspiring yet leading actions by which collections never feature leather, fur or any kind of animal product. Bottega Veneta, as another pioneer in luxury fashion, generates organic fertilizers out of leather cutting process. Gucci, similarly, pays a great attention to material sourcing, as PVC is fully eliminated from their collections. Prada is recycling and/or reusing more than 70% of its waste. Swarovski produces renewable energy in the form of hydropower. Bulgari, on the other hand, obtained 'Energrid' certification for renewable energy utilization at its Valenza and Solonghello manufacturing facilities. In addition to these cutting edge environmental friendly practices, luxury fashion brands advance social progress. The Kering Corporate Foundation, since its launch in 2009, battles with violence against women by focusing on three geographical areas in terms of combatting sexual violence in the Americas, domestic violence in Asia and harmful traditional practices in Western Europe. Additionally, LVMH is committed to advance gender equality and to promote the United Nations Women's Empowerment Principles. Besides, LVMH established the Institut des Metiers d'Excellence with a goal of preserving artisanship in traditional craftsmanship.

5 The Prerequisites for Sustainable Luxury

Luxury could create a profoundly positive value for a holistic ecology. While quality and technical performance are fundamental for luxury, a long history alongside heritage could further strengthen sustainability values. Premium quality, craftsmanship, superior technical performance and innovation could offset many of the issues concerning sustainability. Figure 1 illustrates the main steps on how to create *sustainable luxury* as a process to approach fashion with care and attention to live in a meaningful, pleasurable, and thoughtful way.

5.1 Design Better

Product design is where a considerable environmental impact of the clothing is determined. Changes in design practices could lead to significant improvements on how long the clothes could remain usable. In this sense, luxury products could present many opportunities within the clothing life cycle through increased longevity. Fabric quality is pivotal to extend the usable life of the items, as higher quality fabrics could endure wear and use better (Cooper et al. 2013). Therefore, embodying classic styles along with superior technical performance in luxury products could result in the use for longer periods. Modular design concepts could also empower the customization of clothing. Multifunctional concepts create

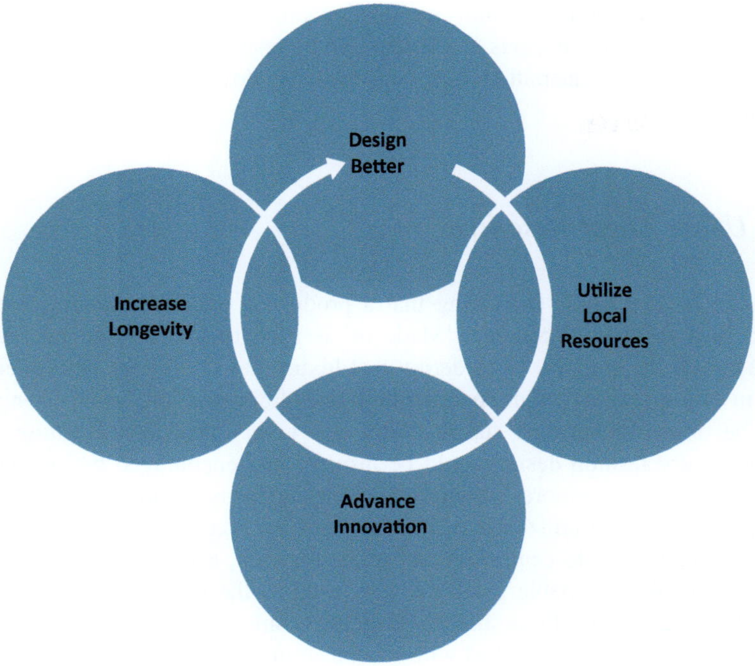

Fig. 1 Sustainable luxury principles

ultimate product versatility and adaptability. For example, designing products with detachable parts could enable consumers to use the product more than one way. Further, some components could be replaced and repaired easily. Every tone of re-used clothing and household textiles could provide net carbon savings. Though it is worth stressing that fashion designers and design teams need to work under great commercial pressure, and therefore producing clothing through accurately sourced materials, ensuring that every element of the collections is truly sustainable, manufacturing under fair trade principles are not always feasible. At the beginning of the design phase, strategic design decisions must be made to nourish ingredients and materials such as eco-intelligent constitutes. Processes, on the other hand, must be designed to reduce negative impact and reduce energy as well water use. Closing the loop on material flows could be secured in an ecological sound only if the materials are designed to be ecologically safe. Otherwise, the closed loop becomes polluted with toxic chemicals triggering health problems and a descending twist in value. 'Shelf quality' that aims at making the product look good in store must also be improved toward long-lasting quality. Sustainable solution could be further developed through technology-enabled fashion design and new product development applications such as utilizing smart fabrics, make-to-order clothing and digital textile printing (Loker 2008). When designers make purchase decisions in a responsible manner, they could not only provoke suppliers to pursue more environmental and ethical friendly production processes, but they would also be attributed to today's conscious consumers. Thus, responsibility and consciousness must be gradually transmitted from upstream to downstream levels of luxury fashion SCs.

5.2 Utilize Local Resources

'Made in' is a signature indicating that a product is planned, manufactured and packed in a mentioned country. 'Made in' is used to indicate the international uniqueness of the country. Considering that luxury has often been associated with made in origin, quality, high specialization and differentiation, some strong links could be observed between sustainability and luxury products. Creating a sustainable luxury fashion design could be inspired and encouraged by emphasizing local sourcing. Furthermore, carbon and energy intensity could be lowered across local production. The utilization of local sources for luxury products could indeed deliver an alternative to centralization, standardization, and identical products. To this end, anything available locally can be used to the best advantage and what cannot be produced locally is shared within the cosmopolitan society (Clark 2008). Though designers must be linked with local resources. These interactions could be formed within more transparent production systems in which less intermediation is ensured to preserve cultural and material values. Local production is concerned with creating a sector with sensitivity to place and scale in order to sustain communities and support jobs while preserving the quality of the environment (Fletcher 2008). Producing luxury locally could take a stance for distinctiveness, superior quality and social improvement. Through local production, luxury fashion could bring a wealth of new perspectives and practical solutions to many environmental and social problems. Utilizing local sources and emphasizing quality, craftsmanship and experienced labour during production are bound to result in less intermediation and more transparent production systems. Luxury fashion could further cultivate sustainability when the focus is shifted from fashion as an image to fashion as a material topic.

5.3 Advance Innovation

Virtual prototyping and 3D simulation could be further embarked as a good way to mark a progress in sustainable process development for luxury products. 3D prototyping provides design verification in the early stages of the design process. Development times could be shortened, and the use of physical could be dramatically minimized because this technology only produces the material when it is physically needed for the product. Thus, waste and emission reductions could be achieved. Physical prototypes and sales samples could be replaced with virtual 3D files that could provide substantial benefits. Adidas has been utilizing different types of 3D printers to produce prototypes. Nike's Flyknit technology for footwear is described as the lightest and one of the most supportive materials, which reduces waste through the one-piece knitted upper construction. Issey Miyake, on the other hand, refined his pleating techniques on 1988. A revolutionary technique was developed through which pleats are applied after the fabric is cut and sewn. Since 1999, Pleats Please Issey Miyake and other collections have been developed based on the A-POC method. A-POC, A piece of cloth, is a manufacturing method, which uses computer technology to create clothing from a single piece of thread in a single process. A-POC collection uses compute-enabled manufacturing to produce knitted or woven continuous tubes of fabric that could be cut into the desired silhouette by the consumer by following an outline, which indicates embedded layers already woven or knitted into the fabric. In this way, material waste is reduced in production while custom-made clothing is enabled. Thus, technological developments are indeed needed to embed in luxury production to help substantially reduce the environmental footprint.

5.4 *Increase Longevity*

The number of 'wash and wear' cycles the clothing goes through before its disposal is an important determinant. To this end, reductions in the number of wash and wear cycles for clothing could significantly lower the dominance of use-phase emissions. Furthermore, longevity of fashion items could lead to substantial emission reductions. If cloths have a longer usable life, the need to replace them could be lower, which could lead to reduction in the volumes discarded, subsequently resources consumed in the manufacturing could be fewer. Thus, it could be hypothesized that the provision of high quality and longevity could pose a great opportunity toward emission reduction associated with clothing over its life cycle. Research shows that extending the average life of clothes by just nine months could save €6.5 billion in resources utilized to supply, wash and dispose of clothing. Similarly, 33% longer lifetime of clothing, equivalent to 9 months, could save up to 10 $MtCO_2$-e carbon, and 400,000 tonnes of waste (WRAP 2012). It was agreed that the cut is the single most design important factor in extending a garment's life expectancy, and there are several cuts identified that could potentially increase the lifespan. In this sense, tailored pieces could last longer. Thus, high-end fashion products embodying classic styles with superior technical performance will tend to be longer lasting. Luxury apparels, on the other hand, make extensive use of luxury fibres—such as silk, fine wools and cashmere—, luxury fabrics—such as velvet, sating, taffeta, chiffon, crepe, organza, lace, and tulles—, elegant fabric designs and advanced textile technologies. Laser cutting, digital embroidery, and various technologies are applied to create highly technical yet desirable fabrics for luxury products. Yet, it must be stressed that these delicate fabrics are more vulnerable to damage during production and wear. Therefore, designers and material suppliers should always be in coordination for longer lifespan of the garments.

6 Conclusion

Luxury fashion inspires and pursues quality, longevity and exclusivity. Luxury has been established based on moral values and it is about having the very best. To this end, products causing environmental or social damage would not be considered the best by affluent consumers. What a luxury brand does today will be part of its heritage in fifty years. It must be noted that a luxury company's heritage from the 19th century will not remain the same, but it will be interpreted on the basis of contemporary values and the company's activities during the upcoming centuries. It is known that the world is experiencing a dark moment in terms of natural scarcity, social inequality and cultural shifts. However, luxury could create a better value for a holistic ecology by fulfilling societal and environmental needs. While quality is an essential success factor, a long history and heritage could further strengthen sustainability values. Sustainability cannot be secured without a joint effort throughout the chain. From cleaner production to responsible sourcing and conscious consumption, a long-term perspective should be embarked. Verifiable information on social and environmental performance should be further demanded throughout the entire value chain. Sustainable luxury could accordingly lead to a reduction in resource consumption while authentically improving quality, and durability.

Motivating current and potential suppliers on the significance of integrating sustainability into their management strategies is as equally important as deciding where to source or to produce garments. Influencing suppliers to create a sustainable SC requires commitment from both buyers and sellers. Therefore, a high level of commitment, communication, trust and sincere interest in encouraging and monitoring suppliers for sustainability management practices are required. Unfortunately, the fashion and apparel industry does not have a good track record of establishing long term relationships within supply chains. Suppliers indicate that buyers consider low cost a priority over building long-term and deeper relationships (Quinn 2008). But the way purchasing practices could have major impacts. When a buying firm employs aggressive price negotiations, short notice periods and late changes in the order, suppliers get negatively affected. When a buying firm cuts the production costs as a way to run a more lucrative business, most often suppliers are forced to ask for a lower price. However, this problem is often passed upstream. Consequently, unfair wages are paid to workers toward the end of the chain. Poor production planning and control requires more flexibility, which results in, excessive and even forced overtime performed by the workers. Nevertheless, such issues bring further and hidden problems, such as hidden subcontractors. For example, to keep costs down, a supplier might opt to sub-contract part of the order to another factory where standards could be much lower. In this case, buying firm, which placed the initial order, could not be even uninformed. Thus, last minute changes or unexpected modifications in the order could lead to dramatic consequences. Alignment of ethical strategies must be established and incentives must be provided to improve working conditions. Thus, an integrative business model is needed to link design, procurement, production and distribution with a better configuration of ethical practices in luxury fashion SCs.

Companies must move beyond the low hanging fruit philosophy, and therefore an inspection is needed to mitigate negative impacts and further integrate sustainability into critical issues such as SC configuration and design principles. To that end, significant resources are needed to implement radical changes in operational practices and to reengineer existing luxury fashion SCs. Embedding sustainability into every business decision could secure an economic stability while bringing positive organizational and operational outcomes. Sustainability integration could help companies get prepared for potential risks, act on opportunities and generate more value for its stakeholders. Sustainability could thus improve financial return in many ways, such as preventing brand damage, preserving 'license to operate', increasing operational efficiency and fostering ethical and environmental niche in the market. Further, effective occupational health and safety policies could help identify workplace risks so that potential negative impacts could be reduced and eliminated. Not only can injuries be reduced, but also absenteeism could be lowered. Undoubtedly, upstream suppliers take a dynamic stance to support buying fashion firms to change their inbound logistics processes. Internal sustainability practices are antecedents for external collaboration and SCs are therefore instrumental in complementing sustainability management capabilities of downstream partners by providing sustainable inputs and facilitating prevention processes.

Both industry and academia recently point out that multinational luxury fashion companies seek ways to reinforce their relationships with their SCs to achieve results, which will go beyond cost reduction. When companies establish stronger ties within their stakeholders, actors of SCs could better comprehend how the company functions and could contemplate on the direction the company might take in the future. This could enable working on innovative products with a better understanding of the societal expectations. On the contrary, when a relationship is weak, only temporal improvements could be achieved. Even though a supplier complies with its buying firm's code of conduct, if there is no real commitment for a managerial change the effort will likely dissipate. It is fundamental to embed sustainability into all stages of SCs, and in particular in the design process at macro and micro levels. Therefore, a new design thinking in luxury fashion to formulate innovative yet sustainable strategies must be considered pivotal.

Acknowledgements This research was conducted within the framework of the European Doctorate in Industrial Management (EDIM). This doctorate program is funded by European Commission's the Education, Audio-visual and Culture Executive Agency (EACEA), under Erasmus Mundus Action 1 programs.

References

Aiama D, Carbone G, Cator D, Challender D (2016) Biodiversity risks and opportunities in the apparel sector. Gland

Bain J, Beton A, Schultze A, Mudgal S (2009) Reducing the environmental impact of clothes cleaning

Balch O (2016) Apparel: research analysis. In: How brands can improve apparel value chains. http://innovation-forum.co.uk/perch/resources/if-management-briefing-apparel.pdf

Baptist World Aid Australia (2011) Fashion victims. In: Uzbekistan. http://www.baptistworldaid.org.au/assets/Be-Vocal-section/FV-Uzbekistan.pdf. Accessed 27 Nov 2015

Benammar K (2008) Abundance and scarcity. Concepts and rhetoric in ecology, economics, and eco-ethics. In: Brand J, Luyt T, Vos M (eds) Beyond green. d'Jonge Hond, Rotterdam, pp 40–52

Bendell J, Kleanthous A (2007) Deeper luxury. Surrey

Black S (2012) The sustainable fashion handbook. Thames & Hudson, London

Brass C, Bowden F (2012) Joining the dots: connecting design to a sustainable future. In: Black S (ed) The sustainable fashion handbook. Thames & Hudson, London, pp 108–112

Brun A, Castelli C (2013) The nature of luxury: a consumer perspective. Int J Retail Distrib Manage 41:823–847. doi:10.1108/IJRDM-01-2013-0006

Caniato F, Caridi M, Castelli CM, Golini R (2009) A contingency approach for SC strategy in the Italian luxury industry: do consolidated models fit? Int J Prod Econ 120:176–189. doi:10.1016/j.ijpe.2008.07.027

Carbon Trust (2011a) International carbon flows: clothing. London

Carbon Trust (2011b) International carbon flows: cotton. London

Clark H (2008) Slow+Fashion: an oxymoron or a promise for the future? Fash Theory J Dress Body Cult 12:427–446. doi:10.2752/175174108x346922

Cooper T, Hill H, Kininmonth J, Townsend K (2013) Design for longevity. In: Guide increasing act life cloth. http://www.wrap.org.uk/sites/files/wrap/Design for Longevity Report_0.pdf

Ecotextiles (2011) Estimating the carbon footprint of a fabric. In: Estimating the carbon footprint of a fabric. https://oecotextiles.wordpress.com/2011/01/19/estimating-the-carbon-footprint-of-a-fabric/. Accessed 12 Oct 2014

EURATEX (2015) The EU-28 textile and clothing industry. In: EU-28 Textile and Clothing Industry in the year 2014. http://euratex.eu/library/statistics/key-data/key-data-details/?tx_ttnews%5Btt_news%5D=4846&cHash=a43f98d261ba88233db94846a8f7e08e

Fletcher K (2008) Sustainable fashion and textiles: design journeys. In: Earthscan. http://katefletcher.com/publications/books/sustainable-fashion-and-textiles-design-journeys/. Accessed 15 Mar 2014

Gibson PC (2012) Sustainable fashion in the age of celebrity culture. In: Black S (ed) The sustainable fashion handbook. Thames & Hudson, London, pp 20–24

ILO (2016) Textiles, clothing, leather and footwear sector. In: Textiles, clothing, leather and footwear. http://www.ilo.org/global/industries-and-sectors/textiles-clothing-leather-footwear/lang–en/index.htm. Accessed 4 Jan 2016

Jensen TH, Dahl EH (2008) Guidelines. In: Brand J, Luyt T, Vos M (eds) Beyond green. d'Jonge Hond, Rotterdam, pp 148–158

Joy A, Sherry JF, Venkatesh A et al (2012) Fast fashion, sustainability, and the ethical appeal of luxury brands. Fash Theory J Dress Body Cult 16:273–296. doi:10.2752/175174112X 13340749707123

Kaiser S (2008) Mixing metaphors in the fiber, textile and apparel complex: moving toward a more sustainable fashion. In: Ulasewicz C, Hethorn J (eds) Sustainable fashion: why now? A conversation about issues, practices, and possibilities. Fairchild Books, pp 139–165

Loker S (2008) A technology-enabled sustainable fashion system: fashion's future. In: Ulasewicz C, Hethorn J (eds) Sustainable fashion: why now? A conversation about issues, practices, and possibilities. Fairchild Books, pp 95–133

Mosher M, Smith L (2015) Integrating sustainability into business: a guide for sustainability practitioners

Norman W, Macdonald C (2004) Getting to the Bottom of "Triple Bottom Line". Bus Ethics Q 14:243–262. doi:10.5840/beq200414211

Organic Cotton (2015) Environmental risks. In: Risks of cotton farming. http://www.organiccotton.org/oc/Cotton-general/Impact-of-cotton/Risk-of-cotton-farming.php

Quinn S (2008) Environmental stewardship and sustainable sourcing. In: Ulasewicz C, Hethorn J (eds) Sustainable fashion: why now? A conversation about issues, practices, and possibilities. Fairchild Books, pp 358–385

Research and Markets (2015) Apparel retail: global industry guide. Dublin

Rissanen T (2008) Creating fashion without the creation of fabric waste. In: Ulasewicz C, Hethorn J (eds) Sustainable fashion: why now? A conversation about issues, practices, and possibilities. Fairchild Books, pp 184–207

Teunissen J (2008) Autism in fashion. In: Brand J, Luyt T, Vos M (eds) Beyond green. d'Jonge Hond, Rotterdam, pp 18–40

WCED (World Commission on Environment and Development) (1987) Our common future. Oxford University Press, Oxford

WRAP (2012) Valuing our clothes. In: The true cost of how we design, use and dispose of clothing in the UK. Available via http://www.wrap.org.uk/sites/files/wrap/VoC FINAL online 2012 07 11.pdf. Accessed 14 Nov 2015

Wu Z, Pagell M (2011) Balancing priorities: decision-making in sustainable supply chain management. J Oper Manage 29:577–590. doi:10.1016/j.jom.2010.10.001

WWF (2014) Cotton. In: Sustainable agriculture. http://www.worldwildlife.org/industries/cotton. Accessed 10 Jan 2016

Sustainable Luxury in Hotels and Resorts: Is It Possible?

Gianna Moscardo

Abstract The increasing affluence of countries such as India, China and Brazil supports a growing global middle class interested in travel, consumption and luxury purchases. This trend presents some serious challenges for addressing a range of sustainability issues in tourism. Rising awareness of these sustainability issues has resulted in a number of luxury brands and businesses adopting sustainability practices. This paper begins by describing this response in tourism using an approach adapted from general sustainability analysis, the Drivers, Pressures, State, Impacts, and Responses (DPSIR) framework to understand the context for the development of the concept of sustainable luxury tourism. It then examines three cases of sustainable luxury tourism experiences exploring the extent and nature of the sustainability strategies adopted. This analysis particularly examines the trade-offs required between luxury and sustainability. Overall the results suggest that while it is possible to improve the overall environmental and social performance of luxury tourism there are a number of ongoing tensions between sustainability and luxury that create difficulties for the management of this form of tourism. This conclusion has important consequences for the sustainability of international tourism overall given its growing dependence on outbound travel from these emerging economies.

Keywords Luxury tourism · Sustainability · DPSIR

1 Introduction

This paper critically examines the concept of sustainable luxury tourism and seeks to address the question of whether such a phenomenon is possible. Although many seem to have assumed that there is such a thing as sustainable tourism, there

G. Moscardo (✉)
College of Business, Law and Governance, James Cook University,
Townsville, QLD 4811, Australia
e-mail: Gianna.moscardo@jcu.edu.au

M.A. Gardetti (ed.), *Sustainable Management of Luxury*,
Environmental Footprints and Eco-design of Products and Processes,
DOI 10.1007/978-981-10-2917-2_8

continues to be a debate about the validity of describing tourism as sustainable (Lansing and de Vries 2007). One perspective on this debate is provided by proponents of sustainable tourism who take for granted that it is possible to manage the negative impacts of tourism in such a way that this activity could be reasonably described as sustainable (Stabler 1997). Another perspective comes from those who think that while it is possible under certain, prescribed circumstances for tourism to make a contribution to sustainability in destination region, it is not possible to judge tourism as either sustainable or unsustainable (see Butler 1999 and Wall 1997). A third perspective argues that tourism can never be truly sustainable because it relies on non-essential consumption and inequality between hosts and guests in many locations (Lansing and de Vries 2007).

There are four main criticisms of the assumption that sustainable tourism is possible. Firstly, it is argued that discussions of sustainable tourism typically focus only on alleviating the negative impacts on the destination and do not consider impacts throughout the whole tourism system. Moscardo (2009) provides an example of this problem in an analysis of discussions of sustainable tourism and the Great Barrier Reef. In this case climate change was identified as a major threat to this environment but little consideration was given in discussions of sustainable tourism in this destination to the role that the long-haul travel undertaken by tourists to get to the destination might play in this threat. Secondly, there is often an assumption that some form of tourism is sustainable and the possibility that no tourism might be sustainable is rarely, if ever considered (Moscardo 2008; Sharpley 2000). Thirdly, discussions of sustainable tourism are typically driven by considerations of what markets want rather than what the destination needs (Moscardo 2008; Saarinen 2006). Finally, there is the issue that tourism in any form is still a form of unnecessary consumption. For many authors unnecessary consumption is at core of the current unsustainable economic system (Grigsby 2004).

Similar arguments have been made about the concept of sustainable luxury. There are those that assume that changes to the production and use of luxury goods can minimise negative impacts and thus support their sustainability (Bendell and Kelanthous 2007). Others believe that sustainable luxury is not possible, either because it depends upon unnecessary consumption and social inequality (Husic and Cicic 2009), or because the markets for luxury goods are unlikely to accept the changes necessary to minimise negative impacts (Mastrelli 2007).

Given that there are a number of forces encouraging growth in luxury tourism and that there is an increasing number of companies claiming to provide sustainable luxury tourism opportunities, it seems timely to critically examine this concept. This paper will use a framework adapted from the Drivers, Pressures, State, Impacts and Responses (DPSIR) approach to both describe the context in which sustainable luxury tourism exists and to outline the pressures contributing to the increased adoption of sustainability strategies by luxury tourism providers. According to Svarstad et al. (2008, p. 119) the DPSIR "is a widely accepted and commonly used framework for interdisciplinary indicator development, system and model conceptualization, and the structuring of integrated research programmes and assessments". The paper will then critically examine the integration of sustainability and luxury in three examples of self-labelled sustainable luxury tourism—Soneva Fushi Resort in the Maldives, Hotel Su Gologone in Sardinia, Italy, and Camp Okavango in Botswana.

2 Defining the Key Terms

While definitions of both the key terms of sustainability and luxury are subject to considerable debate and multiple definitions of each are available, there is greater consensus around the key features of sustainability than of luxury. The emrgnc website (EMRGNC 2003), for example, lists more than 100 definitions of sustainability and the related term sustainable development. A content analysis of these 100 definitions (Moscardo 2012) reveals a number of common and recurring themes including:

- The need to include environmental, economic and human or social issues in sustainability;
- Taking a long term or future orientation to decision-making;
- The importance of recognising generational equity and social justice issues as central to sustainability concerns;
- Recognising that individual and business actions take place within complex systems and require systems analysis and system thinking; and
- Awareness that consumption based on global disparities in wealth and power are at the core of sustainability issues.

In addition to these themes, discussions of sustainability definitions also highlight the distinction between strong and weak sustainability. One approach to this distinction is based on the idea of balancing multiple forms of capital such as natural, financial, human, and social to achieve sustainability (Lehtonen 2004). Within this approach a weak sustainability perspective is one that considers all forms of capital as equal, while a strong sustainability perspective recognises the critical importance of natural capital for human survival and that it cannot be substituted or replaced easily (Dietz and Neumayer 2007). An alternative view of weak and strong sustainability suggests that a weak sustainability approach is one that assumes that economic growth with only limited change to the current system is compatible with moving towards sustainability, while a strong approach requires fundamental change in the current economic and consumption system (Davison 2001).

For the purpose of this paper tourism can be seen as contributing to sustainability only if it is developed and managed in such a way as to address the issues raised in the previous sections. This paper also takes a strong sustainability perspective in that it recognises that natural capital cannot be replaced by other forms of capital and must be maintained or enhanced (Lawn 2012). In terms of the second perspective on weak and strong sustainability this paper is on the stronger side of the continuum suggesting that economic growth should not be the primary or sole goal of development and that considerable change may be needed in the nature of consumption to achieve sustainability. This paper does not, however, take the most extreme stance that all unnecessary consumption must be eliminated, because this is judged by the author to be very unlikely to happen in the short to medium term future.

The concept of luxury is more difficult to define than sustainability with several very different definitional approaches and a number of issues that need to be addressed. Firstly it is important to recognise a difference between luxury as a general concept—where one might say that luxury is having a little time to relax and read in a busy day—and luxury in the context of specific products and services (Beverland 2004). In the broadest sense luxury is associated with anything that is not essential, with something indulgent and with extravagance (Tynan et al. 2010). In the more focussed context of goods and services luxury includes a number of additional characteristics. These include high prices, exclusivity, iconic brands, perceptions of comfort, quality, uniqueness and scarcity, associations with prestige, affluence and high social status, conspicuous consumption, aspiration, personalised and intensive service, controlled distribution, and authenticity (Beverland 2004; Husic and Cicic 2009; Keller 2009; Tynan et al. 2010). As can be seen, the list includes some objective characteristics such as price, some consumer perceptions such as quality and rarity, some experiential dimensions such as service and comfort, and some consumer motivations such as conspicuous consumption. This range creates some conflict in discussions of luxury definitions. For example, some authors distinguish between old or true luxury, the very top tier in terms of expense and scarcity, and new luxury, less expensive, more experiential and more accessible (Tynan et al. 2010). This new luxury exists as part of the trend towards the democratization of luxury, with a range of luxury brands providing goods and services for a wider range of consumers (Husic and Cicic 2009; Tynan et al. 2010).

This paper will adopt the definition of luxury consumption provided by Tynan et al. (2010) that combines the characteristics of the main approaches to luxury definitions. In this context luxury consumption will be defined as goods and services that "are high quality, expensive and non-essential products and services that appear to be rare, exclusive, prestigious, and authentic and offer high levels of symbolic and emotional/hedonic values through customer experiences" (p. 1158). The symbolic and hedonic values include comfort and social status connected to conspicuous consumption. Luxury tourism can therefore be defined as tourism products and services that are promoted as, and perceived by tourists to be, associated with the characteristics included in the definition of luxury with the exception of high price. Price is a little more complicated in the case of luxury tourism as there exists the possibility that tourists can experience travel products and services that would be luxury in their countries for considerably less cost when they travel to destinations where lower wages and standards of living allow for the provision of luxury at a cheaper price.

3 Understanding the Context for the Rise of Sustainable Luxury Tourism—A DPSIR Approach

The DPSIR is an analytic approach for describing and analysing systems that are connected to various sustainability issues developed from earlier simpler pressure-state-response (PSR) approaches (Niemeijer and de Groot 2008). It is most commonly used to map out major aspects of human-environment systems to identify sustainability indicators that can then be used to assess and monitor either places, such as islands or coastal wetlands, or specific sustainability issues such as water usage or biodiversity (Svarstad et al. 2008). It can also be used to evaluate sustainability of specific issues or development options (Omann et al. 2009). The key elements and relationships of the approach are described in Fig. 1. It has been suggested that the DPSIR approach has been underutilized in tourism and could be a tool to support more critical thinking about sustainability in this sector (Rempel 2012). This paper adapts and uses the DPSIR approach following guidelines provided by Niemeijer and de Groot (2008) and the approach taken by Omann et al. (2009), to provide and explain the context within which sustainable luxury tourism has emerged. The following sections therefore examine the available relevant literature in terms of drivers, pressures, states, impacts, and sustainable luxury tourism as a response.

3.1 Drivers

A review of the literature on luxury consumption in general and luxury tourism in particular suggests three related key drivers:

- Economic growth in the emerging economies of countries such as China, India, Russia, Brazil, and Saudi Arabia resulting in a larger and more affluent middle class;

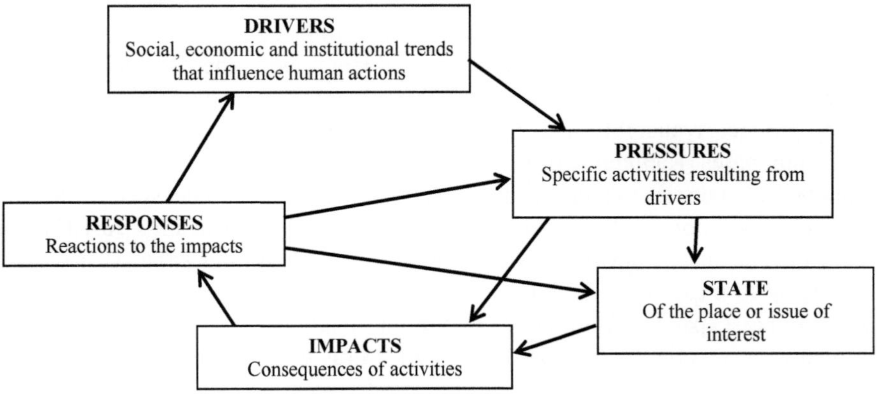

Fig. 1 Overview of the DPSIR approach

- The democratization of luxury; and
- The use of tourism, especially luxury tourism, as a regional economic development strategy by many governments.

There is considerable discussion in a number of areas about the rising affluence and economic power of these emerging economies (Court and Narasimhan 2010). These discussions often focus on the implications of a rapidly expanding middle class with a strong interest in both international travel and leisure retail spending (Kharas 2010). It is estimated, for example, that there are 50 million middle class households in India and 157 million in China (Kharas 2010). International travel is very attractive to these people, with international travel statistics revealing a continuing trend of high growth in outbound travel from these countries. According to the UNWTO (2015) China is now the largest source of international travellers in the world with double digit growth every year since 2004. China is now the largest source of international tourism expenditure at 292 US$billion, more than double that of the next country on the list, the USA. Russia is fifth in terms of international tourism expenditure ahead of countries like France and Canada, and Brazil is tenth. After China, the fastest growing sources of international tourists include India, countries from the Middle East such as Saudi Arabia and Qatar, and other Asian emerging economies such as Thailand (UNWTO 2015).

These global economic trends and the rapid rise of the middle class in BRIC countries drives not only growth in tourism but in luxury tourism as these consumers have also been identified as having high levels of interest in luxury consumption. Zhan and He (2012) claim that Asian consumers, especially middle class consumers in China and Japan, are particularly interested in luxury goods, accounting for 25% of the global market for luxury goods. Wiedmann et al. (2007) extend this argument providing evidence of increasing demand for luxury goods in a number of countries including parts of the former USSR, as well as Asia and Central and South America. This general interest in luxury is also evident in tourism (Yeoman and McMahon-Beattie 2006). Mohsin and Lockyer (2010), for example, argue that there has been significant growth in domestic demand for luxury hotels in India, while Park and Reisinger (2009) report high demand for luxury shopping as a critical part of tourism experiences for travellers from Asia.

This is a factor also connected to the second key driver—the democratization of luxury. The luxury democratization process has two main components—changes in luxury product availability and distribution and changes in consumer interest in luxury purchases (Kapferer and Bastien 2009). Many luxury brands have developed a number of strategies to increase their sales and market sizes. One such strategy is to develop additional related, but cheaper brands. Roberto Cavalli's Just Cavalli brand and Alexander McQueen's McQ lines are examples of this strategy. A second strategy is to expand the brand into related products and services. Armani, for example, has expanded from haute couture into ready to wear clothing as well as cosmetics, accessories, and homewares. A third strategy is to make luxury products more available through partnerships with department stores and other chains. This means that consumers can now more easily combine luxury and non-luxury purchases as part of their regular shopping. This encourages acceptance of more frequent indulgence in extravagant purchases and makes consumers comfortable with luxury products and services, and in turn this encourages more luxury consumption, including travel.

The democratization of luxury involves the creation of two types of luxury consumption—the more expensive and exclusive options sometimes referred to as traditional luxury, and the more widely available and less expensive versions, which are sometimes referred to as new luxury. There is, however, some confusion in the use of the label new luxury with some authors using it to refer to a shift in all luxury consumption towards greater interest in experiences rather than products (Yeoman and McMahon-Beattie 2006).

The third key driver identified for luxury tourism is that of government tourism development policies. A number of countries including China, Mexico and island nations have explicitly chosen to pursue luxury tourism as a development strategy (Scheyvens 2007; Brenner and Aguilar 2002; Wang and Wall 2007). Arguments for such policies include:

- A belief that luxury tourism is the best match for the interest of the fastest growing tourism markets;
- An expectation that this will attract higher spending tourists, which will support greater economic benefits for the local destination (Gulette 2007); and
- A desire to promote a more positive destination (Wang and Wall 2007).

3.2 Pressures

These three driving forces contribute to both overall tourism growth and increasing luxury tourism opportunities. This section will examine what these drivers encourage in terms of the nature of luxury tourism products and services. The section begins by examining models of, and research into, luxury consumption on general. A consistent conclusion in the existing literature on luxury consumption is that there exist several different clusters of motivations for this behaviour and these combine with consumer characteristics to produce different patterns or styles of luxury consumption. One common approach to organising motives is provided by Wiedmann et al. (2009) who classified luxury consumption motivations into five groups: price/cost; functional values including usability, uniqueness and quality; social values including conspicuous consumption, status seeking and prestige; individual or personal values such a hedonism or self-indulgence and materialism; and self-identity as a person of taste or as a connoisseur. A similar approach can be found in the work of Husic and Cicic (2009) who describe five groups of motives including those linked to Veblen's view that luxury is about conspicuous consumption, social status and prestige; luxury as a reflection of snobbery where the critical element is exclusivity and uniqueness; the bandwagon effect where luxury consumption is driven by an interest in being fashionable and linked to iconic brands; perfectionism where the emphasis is on perceived quality; and hedonism, which is focussed on indulgence, self-reward and comfort.

A number of studies have identified luxury market segments by the different levels of importance placed on these different types of motives. Mo and Roux (2009), for example, identified four luxury market segments. The first were those interested in indulgence and hedonism, the second were interested in social conformity combining aspects of conspicuous consumption and fashion, the third segment combined conspicuous consumption with snobbery, and the fourth group were focussed on quality and value as luxury drivers. A similar model is provided by Berthon et al. (2009) who combined different motives by with personal characteristics and experience to suggest four types of luxury consumption including:

- Evanescent conspicuous consumption—pursued by novice consumers aspiring to higher social status and influenced by fashion with an emphasis on social motives and iconic brands;
- Commercial conspicuous consumption—pursued by novice consumers with individual motivations and an emphasis on exclusivity;
- Ephemeral aesthetic consumption—pursued by expert consumers with a desire to be seen as experts or connoisseurs; and
- Classic aesthetic possession—pursued by expert consumers interested in beauty, perfection and endurance.

Other research has examined differences in luxury motivation linked to consumer characteristics such culture, age, wealth and luxury experience. Studies of Chinese luxury consumers consistently report a greater emphasis on the use of luxury purchases to signal social status and generate prestige (Zhan and He 2012; Bian and Forsythe 2012; Mo and Roux 2009). In addition there is evidence that luxury as conspicuous consumption to generate social status and prestige is more likely in newly affluent consumers (Atwal and Williams 2009). These results suggest that the emerging middle classes in BRIC countries are more likely to be interested in the conspicuous consumption and social status aspects of luxury.

The discussions of growth in luxury consumption and the emerging middle class in BRIC nations also highlight the importance of travel to these consumers with growth in demand for both luxury travel options and shopping for luxury goods as a key travel activity (Park and Reisinger 2009). As the use of luxury as a tourism development tool and demand for luxury tourism grows, research investigating luxury tourism consumption has begun to emerge. Moscardo and Benckendorff (2010), for example, analysed tourist reviews of luxury accommodations and found that overall tourists linked luxury to the provision of facilities such as large baths and rooms, hotel appearance and furnishings, comfort, privileged access to special locations, privacy, and high levels of personal service. Similar features have been listed in other papers suggesting that luxury tourism is associated with exclusivity, privileged access, intensive and extensive service, privacy, spacious personal facilities, quality in facility appearance, furnishings and service attributes such as food and dining options, comfort, and uniqueness (Ahn and Pearce 2013; Mohsin and Lockyer 2010).

3.3 State

These characteristics raise some serious questions for tourism businesses facing the challenge of moving towards greater sustainability. While it is not possible to provide a detailed assessment of the state of the global environment, it is possible to highlight some major areas of concern and examine the connection between these and luxury tourism. According to Ronchi et al. (2002), the three main challenges for achieving global sustainability are, managing resource use, limiting negative environmental impacts and achieving justice and equity in socio-economic development. Table 1 lists some of the key problems under each of these categories

Table 1 Global sustainability problems and characteristics of luxury tourism

Problem	Related human actions	Common luxury tourism characteristics
Managing resource use • Rapid decline in non-renewable energy sources • Excessive and inefficient use of freshwater • Use of arable land for other forms of development	Continued reliance on non-renewable energy in general and in particular for comfort and convenience Excessive use of freshwater for comfort and convenience Development of arable land for residential and other commercial activities—inefficient use of space	High usage of energy for features such as air-conditioning and refrigeration of luxury food and beverage High levels of freshwater consumption in swimming pools, baths and showers Large spaces for both individual accommodations and for whole properties to provide privacy and exclusive access
Environmental problems • Carbon gas emissions and climate change • Water, air and ground pollution • Waste disposal • Habitat destruction and bio-diversity loss	Use of inefficient and carbon based fuels Pollution and waste generated from production and usage associated with extensive consumption of non-essential goods and services Destruction of habitat to support production	Long distance travel to reach many locations involving large carbon emissions Habitat damage for building Provision of seafood in dining Larger waste issues associated with more intensive consumption Chemicals associated with cleaning and cosmetics provided

(continued)

Table 1 (continued)

Problem	Related human actions	Common luxury tourism characteristics
Equity in socio-economic development • Exploitation of lower standards of living to support cheaper production of goods and services for consumption elsewhere • Poor working conditions include lower pay and limited training • Economic leakage from foreign owned businesses • Limited distribution of economic activities to local community • Loss of local culture • Asymmetry in ownership of key resources increasing local dependency and vulnerability	Globalisation of the economy where cheaper production costs in the global South allow for increased consumption in the global North Reliance on post-colonial models of development	Often provide luxury tourism opportunities in countries with lower standards of living because of cheaper wages and land prices Multinational or foreign ownership, the need to import goods to support luxury tourist expectations and to use non-local staff and managers to meet international luxury tourism standards support high economic leakage and limited benefits for local communities Destruction or commodification of local culture

Sources Esty and Winston (2009), Moscardo (2012), Lansing and De Vries (2007), Ronchi et al. (2002)

linking them to human actions. A number of authors provide reviews of the relationships between tourism and these global sustainability problems highlighting tourism as a significant contributor to negative impacts in these areas (Buckley 2012; Dolnicar and Leisch 2008; Hall 2010; Weaver 2014).

Table 1 also links these sustainability problems to the common characteristics of luxury tourism sought by consumers. There is evidence that luxury tourism is associated with significantly higher levels of energy and water consumption, carbon emissions and waste production than other forms of tourism and/or residential use (Hadjikakou et al. 2015; Robbins and Gaczorek 2015). Lansing and de Vries (2007) provide examples where luxury tourist demands conflict with the actions required for sustainability including the use of large amounts of freshwater in luxury resort swimming pools in locations where local residents have very limited access to freshwater for basic daily needs, low wages paid to resort workers, and high levels of economic leakage from common luxury resort models. Ryan and Stewart (2009) in an analysis of a specific luxury eco-tourism option raise similar issues about the use of freshwater in swimming pools in desert climates and the difficulties in reconciling the expectations of luxury tourism consumers with sustainability objectives.

3.4 Impacts

This very brief review of the state of the global environment suggests that luxury tourism is likely to have significant negative impacts both on the visited destination and more broadly through the global nature of the tourism distribution system. Not surprisingly there is considerable discussion of tourism and sustainability and pressure from consumers, NGOs and governments to adopt more sustainable practices. This results in two important impacts of relevance to the present discussion. The first is pressure on luxury tourism businesses to adopt sustainability initiatives and actions (Butler 2008).

The second impact has been an increased focus in tourism development regulation on creating and managing all tourism businesses to be more environmentally responsible. In many remote and/or peripheral regions there is a trend towards limiting the size of tours and resorts, requiring self-sufficiency in energy and water usage, more effective and green waste management systems, and the use of sustainable building products and methods. These requirements can substantially increase the cost of starting and running tourism businesses which encourages many operators to adopt luxury tourism as a business model (Ryan and Stewart 2009). As a result there is now considerable overlap in many locations between eco-tourism and luxury tourism.

3.5 Sustainable Luxury Tourism as a Response

Figure 2 maps out the elements and relationships discussed in the previous sections as a total DPSIR cycle. This figure demonstrates how this system supports the rise of the concept of sustainable luxury tourism as a response to growing demand for luxury tourism coupled with increasing concerns about global sustainability. Proponents of the concept of sustainable luxury in general (Hennigs et al. 2013) and sustainable luxury tourism in particular (Ahn and Pearce 2013; Weinstein 2010) argue that it is possible to redesign and alter key aspects of luxury to limit negative impacts. Further, it is argued that luxury consumers are increasingly aware of, and concerned about, sustainability issues and so are actively seeking more responsible luxury options (Bendell and Kelanthous 2007). Related to this claim is the argument that luxury consumption is shifting from social to individual motives with a greater emphasis on unique experiences (Yeoman and McMahon-Beattie 2006). These arguments together are taken to suggest that luxury tourists may be prepared to accept changes in the key characteristics of the luxury tourism experience if they are linked to less negative impacts and they may accept less in the way of comfort, indulgence and tangible facilities in exchange for more engaging experiences (Moscardo and Benckendorff 2010).

An alternative argument is that luxury and sustainability are incompatible (Beckham and Voyer 2014). The available evidence supports this case with

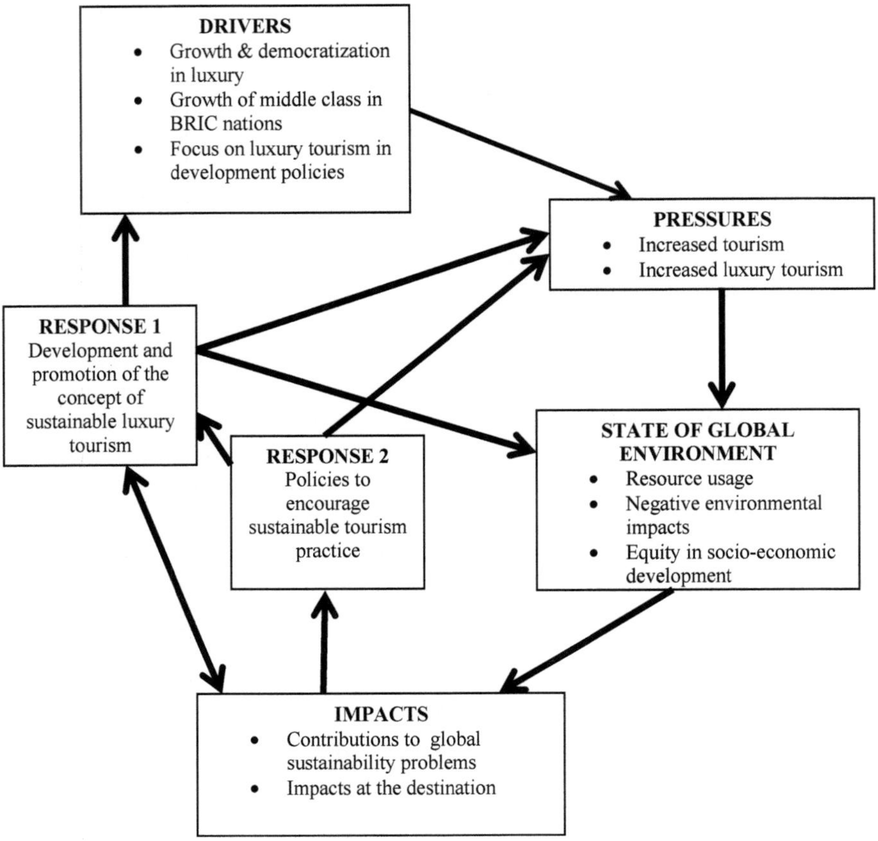

Fig. 2 The DPSIR analysis of sustainable luxury tourism

research reporting that luxury consumers do not link luxury with sustainability (Kapferer and Michaut-Denizeau 2014; Beckham and Voyer 2014), have negative attitudes towards the inclusion of sustainability elements in luxury goods (Achabou and Dekhili 2013) and do not use sustainability as a decision criteria in luxury purchases (Joy et al. 2012; Davies et al. 2012).

Preliminary research into tourism enterprises promoted as offering sustainable luxury tourism provides similar results. Moscardo and Benckendorff's (2010) study of tourist perspectives on luxury accommodation concluded that overall few tourists made any positive links between luxury and sustainability, with the exception of some who chose speciality forms of luxury accommodation such as historic inns or high-end bed and breakfast options. In a similar netnographic study of tourist reviews of two luxury Soneva resorts, Robbins and Gaczorek (2015) reported that few guests had clear ideas of sustainability in this context and most were focussed on the luxury elements of their stay rather than the sustainability of the resort. While very few reviews were negative, Robbins and Gaczorek (2015) reported that a perception that the sustainability strategies diminished the luxury aspects of the experience was a source of concern for some tourists.

Research evaluating the actual sustainability performance of luxury tourism businesses claiming to be sustainable also raises some challenges. Robbins and Gaczorek (2015), for example, noted that while the environmental performance of the two Soneva resorts studied was better than that of other luxury tourism offerings, resources use and waste production for guests was still much higher than that of local residents. Research into the environmental performance of luxury resorts in the Seychelles, French Polynesia and the Maldives reported that few had adopted sustainability strategies and those that had were generally reluctant to choose strategies that limited the comfort or activities of their guests (de-Miguel-Molina et al. 2014). Finally a sustainability assessment of a luxury resort with corporate social and environmental responsibility programs in the Dominican Republic found mixed results with positive performance in some environmental and economic areas, but less success in social issues (Sasidharan and Hall 2012).

4 Exploring Tensions Within the Concept of Sustainable Luxury Tourism

The existing literature reveals contradictory positions about the validity and potential of the concept of sustainable luxury tourism and several authors have suggested that attempts to introduce sustainability into luxury will create tensions that need to be both acknowledged and understood (Melissen et al. 2016). This is especially important given that the DPSIR analysis clearly shows increasing pressure on tourism businesses to both offer luxury and to manage sustainability. One way to explore these tensions and potential challenges in managing sustainable luxury tourism is through examination and comparison of case studies that provide different approaches to the concept of sustainable luxury tourism (Eisenhardt and Graebner 2007). The following sections report on such a study analysing three contrasting cases of sustainable luxury tourist accommodation—the Soneva Fushi resort in the Maldives, the Su Gologone hotel located in central Sardinia, and Desert Delta Safari's Camp Okavango in Botswana.

Each case chosen met the criteria of being actively promoted as sustainable luxury either by business or others in the tourism distribution system and/or having sustainability accreditation or recognition through awards. Soneva Fushi resort is part of group of hotels distinguished by their adherence to and promotion of the concept of "intelligent luxury". This phrase refers to the combination of sustainable, local, organic, wellness, learning, fun experiences with luxury (http://www.soneva.com/slow-life). This attempt to integrate sustainability and luxury is often cited as a leading example of successful sustainable luxury tourism (Ip-Soo-Ching and Veerapa 2013; Scheyvens 2011; Robbins and Gaczorek 2015). The Su Gologone Experience Hotel is a member of the Splendia Luxury and Character Hotels, Chic Retreats Small Boutique

Hotels and Luxury Villas (http://www.sugologone.it/en/) and has won a Conde Nast excellence award in 2014 for the best sustainable hospitality and corporate responsibility in Europe and the Mediterranean (http://www.johansens.com/inspiration/conde-nast-johansens-awards-for-excellence-winners-europe-and-mediterranean). It is included in a number of lists of sustainable hotels (http://italysustainabletravel.blogspot.com.au/2013/01/hotel-su-gologone.html, http://firstclass-holidays.org/en/green-luxury.html?attr[]=63) and has been part of several European Union sustainability programs. http://ec.europa.eu/environment/life/project/Projects/index.cfm?fuseaction=home.showFile&rep=file&fil=LIFE04_ENV_FR_000340_LAYMAN.pdf). The third case, Camp Okavango, has also been recognised by Conde Naste as a luxury tourism option in Botswana (http://www.luxurytravelmagazine.com/news-articles/botswana-safari-camps-make-cond-nast-traveler-gold-list-2014.php), has Green+ Tourism certification, and advertises an extensive range of sustainability initiatives and programs. Table 2 provides a summary description of each case against the key characteristics of luxury tourism described in the earlier sections. In addition to meeting the criteria of being an example of sustainable luxury tourism, the three cases were chosen because they were very different in terms of location, style, and the type of tourists likely to visit them.

The overall aim of the case study analysis was to explore the sustainability strategies adopted in each location and how tourists respond to both these sustainability programs and the overall luxury experience provided, in order to identify tensions and challenges to the integration of sustainability and luxury. It was not the aim of the case study analysis to provide a systematic sustainability assessment of each place, but rather to explore patterns in the areas that they chose to address in their sustainability programs and how their guests respond to these, if at all. The case analysis was based on examinations of websites, both the official websites of the three locations and other travel websites, news articles, available documents and annual reports, relevant academic papers where available, tourist reviews on trip advisor and in two cases, Su Gologone and Camp Okavango, personal site visits by the researcher.

Table 3 provides an overview of the sustainability initiatives associated with the case organised into the areas of impacts identified in the DPSIR analysis—energy and water usage, carbon emissions, habitat damage and biodiversity, pollution and waste disposal, employment conditions, economic leakage, local employment and financial support for local business, support for local culture and communities, issues of inequality and encouraging sustainability beyond the tourist experience. It is important to note that the strategies presented in Table 3 reflect the extent to which each business provides publicly available information in English and this varied considerably between the cases. It is likely that the table underestimates the range and extent of the strategies for Su Gologone. Even with this note of caution it is clear that the most comprehensive set of sustainability strategies and actions are those of Soneva Fushi, which is consistent with its profile as a leader in this area. Esty and Winston (2009) warn against setting unrealistic expectations for sustainability performance for businesses. It is easy to overlook significant advances when

Table 2 Key luxury characteristics of the three cases

Characteristic	Soneva Fushi Resort	Su Gologone	Camp Okavango
Advertised room rate for 2 people per night in April 2016 Average local monthly wage (average US monthly wage is USD$3263)	USD$2985 USD$543	USD$203 USD$1322	USD$1146 USD$616
Star rating and description in TripAdvisor	5 star luxury	4 star luxury	4 star luxury
Exclusivity	57 villas	68 rooms	12 safari style tents/suites for maximum of 24 guests
Privileged access to special locations	Situated on a private island *A jewel of an island—large by Maldives standards—with perfect beaches, a vibrant reef encircling the island, a colony of nesting turtles and its very own world class observatory. Located in the Baa Atoll, Maldives' only UNESCO World Biosphere Reserve*	Located in Supramonte in the mountains in Northern Sardinia close to the Gulf of Orosie and Gennergentu National Park and within walking distance of the archaeological site of the Su Gologone Spring	Camp Okavango sits at the heart of the Okavango Delta on Nxaragha Island providing access to the wildlife and environment of the Okavango Delta
Extensive and intensive levels of services	*In each of these memorable domains, a butler called Mr. or Ms. Friday from the classic island explorer's tale Robinson Crusoe will attend to your needs and assist with your holiday desires* Multiple communal spaces, restaurants, bars and leisure facilities	*The charm of this country hotel resides all in its unique blend of suggestive atmosphere, quality service, comfort and tradition* Multiple communal spaces, restaurants, bars and leisure facilities	*Their exquisite attention to detail, outstanding personalized service, and respect for the surrounding environment blend to create an outstanding guest experience on safari*
Privacy	Each room is its own villa, many with own swimming pool, located separately from others "*Accommodations look upon the Indian Ocean*	Each room has a private outdoor space whether it is an outdoor courtyard or balcony with some including	*Each of the guest suites have been built on individual raised wooden platforms and set beneath the thick Okavango Delta*

(continued)

Table 2 (continued)

Characteristic	Soneva Fushi Resort	Su Gologone	Camp Okavango
	from their own natural sandbox,... Thirteen encompass private seawater swimming pools too"	and additional outdoor bath tub	*vegetation. The en-suite bathrooms have double vanities, double shower and a separate water closet (with a view). The stack away sliding doors open onto a private open-air viewing deck offering beautiful views over the Okavango Delta*
Spacious accommodation	Villa sizes range from 182 to 873 m^2	Varying room sizes as each room is different. Approximately 30–45 m^2 as an average	Safari tents are approximately 20–25 m^2 plus outdoor deck
Quality appearance and furnishings	Smallest villa available is described as follows *A few barefoot steps from the Indian Ocean sit our original Maldivian retreats amidst thriving jungle. Recline outside on the plush daybed and padded sun loungers or dine by candlelight on your private stretch of sand. Inside the cozy hideaway, climb between unbleached organic cotton sheets on the plumped up bed (king or twins as you prefer). The generous bathroom extends from covered bath with bespoke amenities to an outdoor rain shower in the verdant garden*	*A romantic refuge of great comfort: elegant in their whitewashed walls and clear-blue doors, fitted with antique furniture, traditional Sardinian fabrics and chests, precious paintings and profiled by old beams of juniper wood*	*East African-style safari tents, individually situated on raised teak platforms with en-suite facilities and private viewing decks*
Comfort	*High ceilinged bedrooms, plush pillow top beds made with organic cotton sheets and open-air garden bathrooms*	*A romantic refuge of great comfort*	

(continued)

Table 2 (continued)

Characteristic	Soneva Fushi Resort	Su Gologone	Camp Okavango
Uniqueness	Range of special activities available to guests	*The hotel guests are surrounded by artistic works of big contemporary artists and many genial artisan, skilful in working wood, clay and iron*	*Desert and Delta Safaris offers unique wildlife, nature and cultural experiences Each room has its own unique view into the lush Okavango forest from its own private deck*
Quality Food and dining options	Multiple dining options and European Chefs	*Here there is a gastronomic itinerary composed of the fragrances and flavours of tradition, all of which are a delight to discover, starting with the no-carbon-footprint breakfast, with Oliena honey, jams, cheeses and Sardinian charcuterie, as well as the typical breads and sweets baked daily in the kitchen. The lunches and dinners have simple, rich, unforgettable perfumes*	*Exquisite cuisine*

Note Items in italics are taken directly from Soneva Fushi, Su Gologone, and Camp Okavango Desert and Delta safari websites. Other information comes from marketing material. Average income figures from http://www.numbeo.com/cost-of-living/ checked on the 31st March 2016

identifying gaps and areas still to be addressed and so it is important to acknowledge that all three cases have a wide range of initiatives across a range of different sustainability dimensions.

The second key stage of the case study analysis included guest perspectives taken from online reviews available on the TripAdvisor websites for each location. Firstly 20 reviews were selected at random for each location, then the total set of reviews were searched for those that included the words luxury or luxurious and where more than ten reviews met this criteria, a random selection of ten included in the sample. The reviews were then searched for words related to sustainability and again where more than 10 reviews met this criteria a random selection of ten were included in the sample. Finally the reviews were searched for examples that included words related to both luxury and sustainability. In all cases there were reviews that included discussions of luxury but only guests reviewing Soneva Fushi included words related to sustainability. So this meant a sample of 30 reviews each for Camp Okavango and Su Gologone and a sample of 50 reviews for Soneva Fushi.

Table 3 Summary of sustainability strategies adopted in the three cases

Sustainability aspect	Soneva Fushi	Su Gologone	Camp Okavango
Energy usage	Limit use of air-conditioning using design and insulation to maintain comfortable temperature Currently reliant on diesel generators but building solar panels, wind generators and moving to bio-fuel with targets to eliminate non-renewable energy usage	Part of an EU program to reduce energy usage Suggested that they use alternative energies but details not available Limited use of air-conditioning	Solar generated electricity No air-conditioning or heating - use design to maintain comfortable temperatures
Water use	Have own desalination plants to produce water for the resorts Produce own filtered water in reusable glass bottles on site for guest use to cut down on importation of plastic bottled water Recycle water within the resort Donate part of water sales to Water Charity and Thirst Aid projects	Part of an EU program to reduce water usage	Range of water conservation strategies including encouraging to refill rather than replace water bottles
Carbon emissions	Mitigate long haul travel by guests with carbon offsets Converting to renewable energy sources Provide bicycles for guests on the island	Advertise non-carbon breakfasts	
Habitat damage and biodiversity	Support research into local ecosystems Supported legislation for shark protection in their area Contribute to forest and mangrove restoration projects		Work with Department of Wildlife and National parks to support habitat preservation
Pollution	Treat waste water before releasing it Use biodegradable cleaners, pesticides and packaging	Source many products from own or local organic gardens and farms	Primary waste-water flows into a series of above-ground, sealed water-treatment tanks. The end product is potable water that is used to irrigate gardens

(continued)

Table 3 (continued)

Sustainability aspect	Soneva Fushi	Su Gologone	Camp Okavango
Waste Disposal	Reduce waste through refillable containers, recycle paper and other materials, use food scraps as fertiliser in organic herb and vegetable gardens Limit the use of plastic Use local and/or recycled materials for construction		Any combustible litter is burnt in a controlled area while all glass, plastic and cans are stored in animal-proof cages and removed by road to Maun for recycling
Employment Conditions	Have gender equity programs Encourage extended family involvement Training programs for staff across different employment areas Provide healthcare and accommodation	Employ local people as staff but also local artisans for hotel guest experiences	All guides are local Batswana, recruited from nearby villages Desert and Delta have developed training programmes that has produced numerous leaders from local community who are employed in management positions throughout company
Economic leakage, local employment and support for local businesses	Employed local craftsmen for some construction Target local people as employees Support local health care initiatives Provide support of a local charity working with disabled residents	Use food products of the Oliena territory Supports local artists and artisans with their work being displayed and sold on the property	Locally based procurement programme purchases from local companies or alternatively from local distributers wherever possible Desert and Delta Safaris' parent company, Chobe Holdings Limited, is the only tourism company listed on the Botswana Stock Exchange and provides a mechanism for the citizens of Botswana to take a stake in and play a real and meaningful role in the development of the country's tourism resource
Support for local culture and communities	Environmental education programs for local communities Support local schools and play grounds	Strong focus on local traditional arts, crafts and food through all aspects of the hotel. Use of local products and	HIV/AIDS wellness programme, with specific outreach to staff, their families and the communities within the areas that it operates in,

(continued)

Table 3 (continued)

Sustainability aspect	Soneva Fushi	Su Gologone	Camp Okavango
	Contribute to a number of community projects	traditional techniques in building and furnishing	which in turn has been acknowledged and commended by the International Labour Organisation as leaders of industry in this field
Tackling inequality	Contribute to a number of international community development and aid projects	Nothing reported	Support for local community initiatives tackling poverty, support for disadvantaged children to go to school, support for local craft cooperatives and funding to support infrastructure development in local villages
Encouraging sustainability beyond the tourist experience	Offer incentives to guests to volunteer in local community projects Provide information about sustainability projects on their website and offer guest sustainability education on site	Offer classes in local arts, crafts and cooking Provide information on local environments and biodiversity	Habitat and wildlife conservation messages are often included in wildlife viewing activities

These reviews were content analysed focussing on the features guests associated with luxury and a positive experience, guest awareness of, and interest in, sustainability issues, any linkages between sustainability and luxury, and concerns or problems with these linkages.

Across all three cases guest views on luxury were consistent with what might be expected from the characteristics reported in the literature. Luxury in this tourism context was associated with conspicuous and extravagant consumption, personalised service, comfort, space, quality of facilities and privacy. For Camp Okavango guest comments included statements such as "*outstanding customer service on every level from the room attendants, guides, food attendant, to management everyone is at your beck and call*" and "*the tent … featuring teak furniture giving the ambience of a traditional safari tent but maintained that element of luxury that set it apart*". For Su Gologone comments included "*it feels like an exclusive, elegant, but remarkably unpretentious club*" and "*the room was spacious and decorated with local elements, the bed and pillows were very comfortable, and the huge bathroom added a luxurious feel*". Guest comments about luxury for Soneva Fushi highlighted facilities and extravagance "*the villas are very spacious and well laid out, we had about 4 showers in our garden*", and "*it was over the top amazing*" (referring to the number of Jacuzzis, private pools and televisions screens in a villa), quality of food "*the food is spectacular*", and privacy "*looks like your own private island*". Even when the reviews did not explicitly include words related to luxury it was clear across all three locations that positive experiences were consistently linked to these luxury features and/or the privileged access to special environments these places provided for their guests. "*In the bush walks I felt like a privileged guest allowed to be in the same terrain where elephant, lechwe, giraffe, cape buffalo had grazed*" (Camp Okavango). "*The setting is simply gorgeous, adjacent to an alp-like sheer granite mountain, overlooking green valley below*" (Su Gologone). "*We were blown away with set up on arrival to the secluded and breath taking beach*" (Soneva Fushi).

As previously noted only reviews for Soneva Fushi specifically included words related to sustainability although many of the reviews for Su Gologone referred to their connection to local arts, crafts and culture. Guest discussions of sustainability for Soneva Fushi were characterised by two main themes. The first was the dominance of environmental issues in guest reviews. Where sustainability was specifically mentioned the majority of comments were about the environmental programs especially those related to energy, water and waste. Very few guests mentioned the economic and social programs. The second major theme was one of happiness that the various sustainability initiatives did not interfere with guest comfort or service. For example, "*it is amazing that "going green" can be done in such luxurious way*", "*the sustainability concept of Soneva Fushi is credible and not presented in an obtrusive or pretentious way*", and "*our accommodation was luxurious and spacious ... the outdoor bathroom was amazing ... Soneva walks the talk of being sustainable and the complete absence of anything plastic was impressive*".

Across all three locations the available negative comments were consistently linked to perceptions of poor service, lower quality facilities, lack of privacy and/or lack of comfort, service and facilities. For example At Su Gologone problems with limited air-conditioning, the number of steps to different rooms, and limited wi-fi were common sources of complaint. Similar issues were identified for Camp Okavango with complaints about a lack of wi-fi, insufficient electricity, and no air-conditioning. For Soneva Fushi guest complaints focussed on repetition of menus, lack of privacy, and concerns over the construction of new buildings. While none of the reviews explicitly linked a sustainability strategy to a complaint, it is clear that guests were sensitive to any action that detracted from the luxury elements that they expected.

5 Tensions and Challenges in Sustainable Luxury Tourism

The case study analysis suggested that it is possible to provide tourists with a satisfactory luxury experience while implementing some sustainability strategies that can improve a businesses' overall sustainability performance. This does not, however, mean that the there is any commonality between the concepts of sustainability or luxury or that these improvements to performance are sufficient to argue that luxury tourism can be sustainable or make a positive contribution to sustainability more generally. The analysis identified a number of tensions and challenges that suggest it is very difficult to argue for the validity of the concept of sustainable luxury tourism.

It is clear that tourists see luxury as conspicuous, privileged and extravagant consumption and few connect luxury with sustainability. Guests appear to be willing to accept sustainability initiatives as long as these initiatives do not detract from their comfort, alter their activities or access, or limit their actual consumption. This tension between consumption and sustainability exists at both a practical and more philosophical level. At the philosophical level it is clear that luxury tourism depends upon consumption beyond the normal and thus relies on encouraging actions that move away from sustainability and that exploit global North-South differences in standards of living and wages. This avoidance of changes to the features of consumption runs the risk that the luxury tourism experience may contribute to a false sense of complacency about the issues and encourage greater consumption at home and in further travel experiences. Yeoman and McMahon-Beattie (2006), for example, note that there is spill-over effect from experiencing luxury while traveling to greater consumption of luxury goods at home.

At the practical level it sets real limits to the extent to which these places can improve their sustainability performance. For example, while water conservation efforts at Soneva Fushi are admirable, it is clear that guests expect and enjoy having access to multiple baths, private swimming pools and limitless showers and so at some point the resort will be unable to generate significant improvement in their water consumption performance. As noted elsewhere (Lansing and de Vries 2007; Robbins and Gaczorek 2015) having a better performance than other comparable businesses is laudable, but it does not mean that the performance is good enough to be classified as sustainable. Similarly, all three cases, especially Soneva Fushi and Camp Okavango, rely on long haul travellers and this makes a significant contribution to their carbon emissions. Soneva Fushi attempts to balance this by investing 2% of room tariffs into carbon offsetting actions, but it is not clear if this is actually enough and there are no reported strategies to change the markets to those who would travel less distance.

The issue of long haul travel and carbon offsets is linked to the second challenge identified in the analysis which is the lack of attention paid to managing tourists' direct impacts on the local destination. This is especially relevant to Soneva Fushi where many guest reviews highlighted their ability to engage in special activities in fragile environments such as dining or sleeping on a sand bar. Guests at all locations seem unaware of the potential negative environmental impacts of these special activities and it is not clear how many of these sorts of special experiences can be provided without longer term cumulative impacts on the physical settings. This issue of avoiding changes to guest activities has been noted as a problem for luxury resorts in a range of locations (de-Miguel-Molina et al. 2014).

The third tension identified is that between the environmental and social dimensions of sustainability. Although all three cases appeared to have programs linked to the social and economic dimensions of sustainability, guests were either unaware of, or not interested in, these. While all three cases highlight their employment of locals, a number of features of these claims are problematic. It might be argued, for example, that this is more a matter of convenience and cost for the business rather than a genuine example of corporate responsibility (Cherapanukorn and Focken 2014). Possibly guests may feel much more uncomfortable if they consider too deeply the circumstances and daily lives of local residents as it is likely to be a stark contrast with their luxury consumption. For example, the smallest guest room available at the Soneva Fushi resort at 182 m^2 is more than five times larger than an average Maldives home, which has an average size of 36 m^2 typically for 10 people with 16% having no electricity and the majority using electricity only for minimal lights at night and wood or oil for cooking (based on figures provided by Nations Encyclopedia 2012). Resort guests have access to multiple swimming pools and most have access to two bathrooms, while 68% of the residents rely totally on rainwater and most use much less water per day (Nations Encyclopedia 2012).

The overall impression from the guest reviews that did mention sustainability is that the sustainability actions provide more of a reason for feeling good about ones' self than an opportunity to make a genuine contribution to sustainability. In the absence of more intensive guest engagement in sustainability practices at these places it is unlikely that these places are making much of a change in sustainability beyond the experience.

Finally, Moscardo (2008) also points out that tourism development is rarely directly compared to alternative forms of development and is often not assessed in terms of opportunity cost. In the present case it is possible that the development of a luxury resort, regardless of its environmental impact, may not have been as sustainable an option for the destination as some other form of tourism or some other activity. Work by Scheyvens (2007) and Sandbrook (2010) suggest that local benefits from tourism may be much higher for more budget forms than for luxury tourism. These alternative tourism approaches are linked to more stable and equitable development with a wider range of opportunities for local involvement and control.

6 Conclusions

The case study analysis suggests that it is possible for a luxury tourism business to reduce negative and increase positive impacts on the destination. The overall assessment does not, however, suggest that luxury tourism can actively contribute to sustainability. At core luxury tourism is driven by motives that run counter to an effective sustainability business philosophy. Luxury tourism is based on conspicuous consumption, extravagant amenities and facilities and personal indulgence, and a belief that these should not be sacrificed. The full costs of such luxury consumption in many countries are hidden and the long term costs of not changing consumption behaviours are ignored or hidden in the luxury tourism experience. At a practical level this tension between maintaining the luxury elements and trying to limit the wider costs of this type of consumption is likely to be resolved in favour of the luxury elements in order to maintain the business thus invalidating claims of sustainability.

This overall conclusion has important implications for tourism. The increasing demand from various emerging economics for both travel in general and for luxury tourism more specifically presents a significant threat to any attempts to improve the sustainability of tourism overall. At a practical level, given the real limits that exist in terms of the ability of luxury tourism providers to improve their individual business sustainability performance, the significant increases in tourism from these economies is likely to far outweigh any positive outcomes that can be achieved. If tourism is to improve its sustainability then it must seek fundamentally different models of tourism development and provision. This requires academic research in a number of areas including a better understanding of how to encourage tourists to make different choices, more awareness of cross-cultural aspects of travel decision-making, better sustainability assessment tools, and more innovation in the development of different tourism destination development options.

References

Achabou MA, Dekhili S (2013) Luxury and sustainable development: is there a match? J Bus Res 66(10):1896–1903

Ahn YH, Pearce AR (2013) Green luxury: a case study of two green hotels. Coll Publishing 8(1):90–119

Atwal G, Williams A (2009) Luxury brand marketing—the experience is everything! Brand Manage 16(5/6):338–346

Beckham D, Voyer BG (2014) Can sustainability be luxurious? a mixed-method investigation of implicit and explicit attitudes towards sustainable luxury consumption. Adv Consum Res 42:245–250

Bendell J, Kelanthous A (2007) Deeper luxury. World Wildlife Fund, UK, Godalming

Berthon P, Pitt L, Parent M, Berthon J-P (2009) Aesthetics and ephemerality: observing and preserving the luxury brand. Calif Manage Rev 52(1):45–66

Beverland M (2004) Uncovering "Theories-in-Use": building luxury wine brands. Eur J Mark 38 (3/4):446–466

Bian Q, Forsythe S (2012) Purchase intention for luxury brands: a cross-cultural comparison. J Bus Res 65:1443–1451

Brenner L, Aguilar AG (2002) Luxury tourism and regional economic development in Mexico. Prof Geogr 54(4):500–520

Buckley R (2012) Sustainable tourism: research and reality. Ann Tourism Res 39(2):528–546

Butler RW (1999) Problems and issues of integrating tourism development. In: Pearce DG, Butler RW (eds) Contemporary issues in tourism development. Routledge, London, pp 65–80

Butler J (2008) The compelling "Hard Case" for "Green" hotel development. Cornell Hospitality Q 49:234–244

Cherapanukorn V, Focken K (2014) Corporate social responsibility (CSR) and sustainability in Asian luxury hotels: policies, practices and standards. Asian Soc Sci 10(8):198–209

Court D, Narasimhan L (2010) Capturing the world's emerging middle class. http://www.mckinseyquarterly.com/Capturing_the_worlds_emerging_middle_class_2639. Accessed 25 Aug 2012

Davies IA, Lee Z, Ahonkhai I (2012) Do consumers care about ethical-luxury? J Bus Ethics 106(1):37–51

Davison A (2001) Technology and the contested meaning of sustainability. Suny Press, New York

de-Miguel-Molina B, de-Miguel-Molina M, Rumiche-Sosa ME (2014) Luxury sustainable tourism in small island developing states surrounded by coral reefs. Ocean Coast Manage 98:86–94

Dietz S, Neumayer E (2007) Weak and strong sustainability in the SEEA: concepts and measurement. Ecol Econ 61:617–626

Dolnicar S, Leisch F (2008) Selective marketing for environmentally sustainable tourism. Tour Manage 29:672–680

Eisenhardt KM, Graebner ME (2007) Theory building from cases: opportunities and challenges. Acad Manage J 50(1):25–32

EMRGNC (2003) Defining sustainability: a hundred perspectives. http://www.emrgnc.com.au/SustainabilityDefinitions.pdf. Accessed 15 Nov 2010

Esty DC, Winston AS (2009) Green to gold, 2nd edn. Wiley, Hoboken

Grigsby W (2004) Buying time and getting by: the voluntary simplicity movement. State University of New York, Albany

Gulette GS (2007) Migration and tourism development in Huatulco, Oaxaca. Curr Anthropol 48(4):603–611

Hadjikakou M, Miller G, Chenoweth J, Druckman A, Zoumides C (2015) A comprehensive framework for comparing water use intensity across different tourist types. J Sustain Tourism 23(10):1445–1467

Hall CM (2010) Changing paradigms and global change: from sustainable to steady-state tourism. Tourism Recreation Res 35(2):131–143

Hennigs N, Wiedmann KP, Klarmann C, Behrens S (2013) Sustainability as part of the luxury essence: delivering value through social and environmental excellence. J Corp Citizensh. (52):25–35

Husic M, Cicic M (2009) Luxury consumption factors. J Fashion Mark Manage 13(2):231–245

Ip-Soo-Ching JM, Veerapa NK (2013) Permaculture: insights into kitchen gardens of environmental and eco-tourism operator in Thailand. Acta Hortic 1088:633–636

Joy A, Sherry JF, Venkatesh A, Wang J, Chan R (2012) Fast fashion, sustainability, and the ethical appeal of luxury brands. Fashion Theory 16(3):273–295

Kapferer JN, Bastien V (2009) The specificity of luxury management: turning marketing upside down. J Brand Manage 16:311–322

Kapferer JN, Michaut-Denizeau A (2014) Is luxury compatible with sustainability? luxury consumers' viewpoint. J Brand Manage 21(1):1–22

Keller KL (2009) Managing the growth tradeoff: challenges and opportunities in luxury branding. J Brand Manage 16(5/6):290–301

Kharas H (2010) The emerging middle class in developing countries. Working paper no. 285. OECD Development Centre, Paris

Lansing P, de Vries P (2007) Sustainable tourism: ethical alternative or marketing ploy? J Bus Ethics 72:77–85

Lawn P (2012) The economics of sustainable development. In: Moscardo G, Lamberton G, Wells G, Fallon W, Lawn P, Rowe A, Humphrey J, Wiesner R, Pettitt B, Clifton D, Renouf M, Kershaw W (eds) Sustainability in Australian business. Wiley, Milton, pp 99–132

Lehtonen M (2004) The environmental-social interface of sustainable development: capabilities, social capital, institutions. Ecol Econ 49:200–211

Mastrelli T (2007) Luxury with a conscience. Hospitality Des 29(6):87

Melissen F, van Ginneken R, Wood RC (2016) Sustainability challenges and opportunities arising from the owner-operator split in hotels. Int J Hospitality Manage 54:35–42

Mo T, Roux E (2009) Exploring Chinese consumer values and attitudes towards luxury brand consumption. Centre de Recherche en Marketing de Bourgogne (CERMAB), University of Bourgogne. http://leg2.u-bourgogne.fr/CERMAB/z-outils/documents/actesJRMB/JRMB14-2009/Mo%20Roux.pdf. Accessed 31 Mar 2016

Mohsin A, Lockyer T (2010) Customer perceptions of service quality in luxury hotels in New Delhi, India: an exploratory study. Int J Contemp Hospitality Manage 22(2):160–173

Moscardo G (2008) Sustainable tourism innovation: challenging basic assumptions. Tourism Rev Int 8(1):4–13

Moscardo G (2009) Challenges for tourism development and management on Australia's great barrier reef coast. In: Dowling R, Pforr C (eds) Coastal tourism development—planning and management issues. Cognizant Communications, New York, pp 252–264

Moscardo G (2012) Sustainability, economy and society. In: Moscardo G, Lamberton G, Wells G, Fallon W, Lawn P, Rowe A, Humphrey J, Wiesner R, Pettitt B, Clifton D, Renouf M, Kershaw W (eds) Sustainability in Australian business. Wiley, Milton, pp 1–34

Moscardo G, Benckendorff P (2010) Sustainable luxury: oxymoron or comfortable bedfellows. In: Proceedings of the 2010 international tourism conference on global sustainable tourism, Mbombela, Nelspruit, South Africa, 15–19 Nov 2010

Nations Encyclopedia (2012) Maldives. http://www.nationsencyclopedia.com/economies/Asia-and-the-Pacific/Maldives.html. Accessed 26 Aug 2012

Niemeijer D, de Groot RS (2008) A conceptual framework for selecting environmental indicator sets. Ecol Ind 8:14–25

Omann I, Stocker A, Jager J (2009) Climate change as a threat to biodiversity: an application of the DPSIR approach. Ecol Econ 69(1):24–31

Park K-S, Reisinger Y (2009) Cultural differences in shopping for luxury goods: Western, Asian, and Hispanic tourists. J Travel Tourism Mark 26:762–777

Rempel JM (2012) Exploring the causal nexus of tourism impacts on quality-of-life. In: Uysal M, Perdue R, Sirgy MJ (eds) Handbook of tourism and quality-of-life research. Springer, Dordrecht, pp 583–606

Robbins D, Gaczorek J (2015) Luxury and sustainability in tourism accommodation—an exploration of how to reconcile apparently incompatible objectives using a case study approach. In: Hay R (ed) Conference proceedings of BEST EN Think Tank XV: the environment-people nexus in sustainable tourism: finding the balance, James Cook University, Townsville, pp 166–186. http://www.besteducationnetwork.org/13790. Accessed 3 Mar 2016

Ronchi E, Federico A, Musmeci F (2002) A system-oriented integrated indicator for sustainable development in Italy. Ecol Ind 2(1–2):197–210

Ryan C, Stewart M (2009) Eco-tourism and luxury—the case of Al Maha, Dubai. J Sustain Tourism 17(3):287–301

Saarinen J (2006) Traditions of sustainability in tourism studies. Ann Tourism Res 33:1121–1140

Sandbrook CG (2010) Local economic impact of different forms of nature-based tourism. Conserv Lett 3:21–28

Sasidharan V, Hall ME (2012) Dominican resort tourism, sustainability, and millennium development goals. J Tourism Insights 3(1):5–26

Scheyvens R (2007) Poor cousins no more: valuing the development potential of domestic and diaspora tourism. Prog Dev Stud 7(4):307–325

Scheyvens R (2011) The challenge of sustainable tourism development in the Maldives: understanding the social and political dimensions of sustainability. Asia Pac Viewpoint 52 (2):148–164

Sharpley R (2000) In defense of (mass) tourism. In: Robinson MR, Long P, Evans N, Sharpley S, Swarbrooke J (eds) Reflections on international tourism; environmental management and pathways to sustainability. Business Education Publishers, Sunderland, pp 269–284

Stabler MJ (1997) An overview of the sustainable tourism debate and the scope and content of the book. In: Stabler MJ (ed) Tourism and sustainability: principles to practice. CAB International, Wallingford, pp 1–21

Svarstad H, Petersen LK, Rothman D, Seipel H, Watzold F (2008) Discursive biases of the environmental research framework DPSIR. Land Use Policy 25(1):116–125

Tynan C, McKechnie S, Chhuon C (2010) Co-creating value for luxury brands. J Bus Res 63(11):1156–1163

UNWTO (2015) UNWTO tourism highlights 2015 edition. http://www.e-unwto.org/doi/pdf/10.18111/9789284416899. Accessed 11 June 2016

Wall G (1997) Sustainable tourism—unsustainable development. In: Wahab S, Pigram JJ (eds) Tourism development and growth. Routledge, London, pp 33–49

Wang Y, Wall G (2007) Administrative arrangements and displacement compensation in top-down tourism planning—a case from Hainan Province, China. Tour Manage 28:70–82

Weaver DB (2014) Asymmetrical dialectics of sustainable tourism: toward enlightened mass tourism. J Travel Res 53(2):131–140

Weinstein J (2010) Green luxury no oxymoron. Hotels 44(2):30

Wiedmann KP, Hennigs N, Siebels A (2007) Measuring consumers' luxury value perception: a cross-cultural framework. Acad Mark Sci Rev 7(7):333–361

Wiedmann KP, Hennigs N, Siebels A (2009) Value-based segmentation of luxury consumption behavior. Psychol Market 26(7):625–651

Yeoman I, McMahon-Beattie U (2006) Luxury markets and premium pricing. J Revenue Pricing Manage 4(4):319–328

Zhan L, He Y (2012) Understanding luxury consumption in China: consumer perceptions of best-known brands. J Bus Res. 65:1452–1460

Diamonds, Development, and Dollars: A Case Study on a Newly Founded Gemological Lab in Botswana Aiming at Sustainable Community Empowerment

Thomaï Serdari

Abstract Drawing on a theoretical framework published elsewhere (Serdari 2016), this paper presents the development of a model gemological lab in Botswana (designed and implemented by Ms. Giovanna Federico, Ph.D. whom the author interviewed). The work poses questions on the nature of diamonds mining and its challenging relationship with natural resources and showcases best practices in reinforcing a sustainable value chain management by focusing on the following: production processes; employee relationships; employee training; and other related organizational issues. In other words, this case study touches on the idea of sustainable diamond mining through social and environmental performance. Luxury items such as quality diamonds are valued not only according to the Gemological Institute of America (GIA) 4Cs (Color, Clarity, Cut, Carats) but also based on the entire production process as previously published by the author in her piece on luxury production. The proposed case study showcases how luxury production processes reinforce initiatives on sustainability particularly in relation to economic empowerment. It also suggests that a fifth C, standing for Creator, would advance sustainability in the diamond industry by slowing down the process.

Keywords Luxury · Sustainable luxury · Green luxury · Luxury business model · Diamonds · Diamonds mining · Diamonds processing · Slow luxury

1 Introduction

On March 16, 2016, journalist Kati Chitrakorn, writing for the *Business of Fashion*, the leading digital authority on the global fashion industry in news, analysis, and business intelligence, reported from the annual Baselworld in Basel, Switzerland. The preeminent event for luxury watchmakers and jewelers, Baselworld is known as the quintessential industry fair where the most venerated brand names and luxury

---

T. Serdari (✉)
Leonard N. Stern School of Business, New York University, New York, USA
e-mail: tserdari@stern.nyu.edu

M.A. Gardetti (ed.), *Sustainable Management of Luxury*,
Environmental Footprints and Eco-design of Products and Processes,
DOI 10.1007/978-981-10-2917-2_9

houses unveil their innovations in the field of horology and their feats in gem-studded jewelry designs. Baselworld is the apex of luxury.

According to Chitrakorn, this year's Baselworld was slightly different. Executives of the luxury brands exhibiting at the fair have been under tremendous pressure in the $80 billion diamond market where "slowing demand in key markets such as the United States, China and the Middle East has sent a ripple through the global diamond pipeline, driving prices down by nearly 25 percent in 2015, according to consulting firm Bain & Company." Chitrakorn correctly identifies the culprit as the weaker demand for diamonds "coupled with a growing supply glut." In other words, luxury retailers greedy in capturing customers' attention and dollars for short-term gains have overproduced and overstocked diamond merchandize, falling into their own trap: diamonds, the commodity they have been marketing as 'luxury,' are not as a powerful aspirational item for younger consumers who have reversed expected growth patterns for diamond retailers internationally. In addition to a bruised bottom line for retailers, this has resulted in an unprecedented cut back in labor forces anywhere from Russia, Canada, Australia and Southern African countries such as Angola, Botswana, DR Congo, South Africa, and Zimbabwe. It has also hurt the diamond pipeline midstream with repercussions felt on the trading floor in Antwerp.

Consumers' attention and desires are split amongst several fast growing luxury categories, such as travel, electronics, fine wines and art. To compete in the luxury market, diamond retailers may want to consider their role in promoting transparency on how value is created through the diamond pipeline; in adding value through knowledge sharing and locals' empowerment in the countries in which they mine; and in treating quality diamonds as luxury items through operations that follow a luxury business model as opposed to commoditizing their most valuable asset.

What follows is a case study on the role gemological labs play in enabling local labor forces to regain proper ownership of the value chain in diamond mining. The case is based on a lab in Botswana, which Giovanna Federico, Ph.D. GIA (GIA stands for Gemological Institute of America), interviewed for this paper, assisted in organizing while she also trained the staff. Our conversation began informally because of our common interest in the luxury market. It soon became clear that the diamond industry lacks transparency in a market where the luxury consumer demands it across products categories. This prompted the connection between the diamond industry value chain and the theoretical framework of the absolute luxury business model that has already been published elsewhere. It initiated questions on the definition of sustainability within the context of the diamond trade. In other words, sustainability within that context relates as much to management of natural resources and minimal impact on the environment as to empowering local labor forces and communities to fully participate and contribute in the exploitation of their countries' natural wealth. This has already happened in other luxury industries, such as silk, crystal, porcelain, and leather industries in France or textiles and fashion industries in Italy. A Batswana artisan's personal story and approach to work should matter as much as that of a French artisan. It should be celebrated as eloquently as that of a European luxury worker and it should somehow appear in the classification of diamonds beyond GIA's 4Cs (Color, Clarity, Cut, Carats).

The diamonds market will become less volatile when its major players understand the term 'sustainability' as one that expands beyond the discourse of natural resources preservation or climate change and that encompasses fair management and development of human resources as well as economic empowerment of upstream labor in a post-colonial world.

2 Sustainable Diamonds in the Context of Luxury Studies

Sustainable diamond mining is reshaping the diamond industry today. This is mainly achieved through social and environmental performance of companies engaged in the mining, cutting, polishing, and trading of diamonds. In the context of luxury studies, sustainable diamond mining raises interesting questions that have to do with the marketing of quality, as opposed to industrial, diamonds, the branding of diamonds marketed directly to consumers, and the transparency in the trading of diamonds both on the wholesale and retail level.

While discussions on provenance of quality diamonds have intensified since the late 1990s when the United Nations had to intervene to resolve the repercussions of Angola's civil war, the trade of diamonds is *selectively* transparent. Consumers purchase diamonds that have been valued according to the GIA's 4Cs (Color, Clarity, Cut, Carats) and that may come with a certificate of non-conflict provenance. However, both these classification systems do not take into account broader issues of sustainability both on the social or environmental level and how these may impact the pricing of the specific pieces, especially as consumers grow more and more aware of how the concept of luxury directly relates to the question of a sustainable future for our planet. Today's consumers are demanding greater levels of transparency prior to the purchase of luxury goods.

The importance of value added based on sustainable value chain management has not been explored in the literature, especially as value can impact pricing of luxury goods at the retail level. To illustrate how value added can be created, this case study will examine the formation of a model gemological lab in Botswana. The production processes organized in it perfectly follow the template of the absolute luxury business model published by one of the authors elsewhere. A combination of knowledge published in the academic press as well as observations recorded directly at the site, this case study attempts to bridge the gap between academia and industry and aims at establishing a new framework that may aid the consumer when purchasing diamonds that have been branded as luxury.

3 Luxury as Aspiration

A number of authors have defined the concept of luxury, a concept so elusive that allows for various interpretations via distinct methodological approaches. Historians, philosophers, linguists, scholars of literature or practitioners among others have all contributed specific points of view in articulating definitions that are based on empirical evidence. Taking into account published works that are truly milestones in the field of luxury studies (Veblen 1899; Sekora 1977; Berry 1994; Berg 2005; Kapferer and Bastien 2009; Gardetti and Girón 2014) an overarching definition of luxury can be articulated as follows: "Luxury is a universal certainty that implies scarcity, beauty, and culture. It is a multi-sensory experience that impacts the individual on a sensorial/physical, emotional, and intellectual level." (Serdari 2016) In addition to summing up all previous definitions, the aforementioned explanation of luxury fits well with Miguel Angel Gardetti's work at the Center for Studies on Sustainable Luxury in Argentina where the notion of *sustainable* luxury has been introduced and studied as the new paradigm that prefaces the future of luxury in a newly re-organized global economy supported by local luxury enterprises as opposed to global multinational brands (Gardetti and Torres 2013). A similar disruption has begun in the diamond mining industry as well where the monopoly of DeBeers yielded to the formation of several multi-national firms competing for market share and where finally, local governments (especially in Africa where states had been excluded from profits rendered by their own natural resources) are major stakeholders in mining, cutting, and polishing companies. In other words, the shift from the global to the local has already taken place and what remains to be discussed is how to best prepare people at the local level to participate in the creation of their nation's wealth, how to contribute in efforts to preserve their natural environment, and how to best demonstrate that the passion and character of particular cultures can be preserved in the aura of the final product.

4 The Luxury Business Model

Business models are designed to generate revenues for a company and achieve profits. While there are myriad of ways to organize a firm's business model, most are based on the premise that significant reduction of costs increases profits. For firms that compete in the luxury market business models are more complex because they require a different strategic approach to organizing the firm's infrastructure, personnel, raw materials, production methods, distribution channels etc. The best way to illustrate a luxury business model is as shown in Fig. 1 (Serdari 2016). This illustration explains the elements that characterize production of luxury goods and alludes to the manipulation of the perception of luxury in the market, where specific strategies can be applied to further enhance consumers' perception of aspirational luxury goods (Kapferer and Bastien 2009). Reading the business model from left to

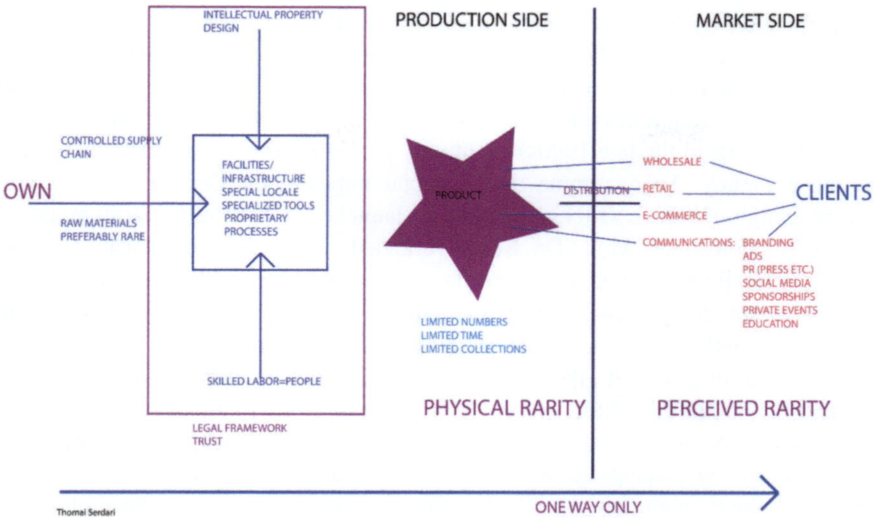

Fig. 1 The luxury business model (© Thomaï Serdari)

right, one realizes that control over each stage of production and distribution is key for luxury firms, starting with the sourcing of raw materials (which often requires vertical integration, a strategy successfully implemented by luxury companies *Hermès* and *Chanel*). The luxury firm should own its infrastructure, preferably in its original location where it employs workers whose tenure is usually long-term so that the firm takes advantage of their tacit knowledge. A legal trust protects the luxury firm from "corporate raiders" but also reinforces its creative competitive advantage by safeguarding its intellectual property. The number of items produced is limited and decided by the firm; these products may appear in limited editions or they may appear for a limited amount of time. The luxury company has full control over all distribution and communications channels, whether physical or virtual. In other words, the luxury business model is one that operates based on the idea of scarcity (sourcing, operations, production) and the idea of rarity, either real or perceived (distribution, communications, sales). The moment a firm deviates from this model, it actually exposes itself to risk of intellectual infringement; inability to access raw materials; inability to access appropriate labor force; loss of control over availability of product, pricing, and communications. The challenge luxury firms face is one of great capital costs early on and before the products enter the market combined with the delicate act of balancing the production of ultimate luxury products (star products) alongside accessible luxury items.

In diamonds sales, the idea of luxury is deliberately cultivated in consumers' perception. *De Beers* (a group of companies that has maintained a leading role in the diamond exploration, diamond mining, diamond retail, diamond trading and industrial diamond manufacturing sectors) deliberately controlled that perception for the majority of the twentieth century. De Beers monopolized all wholesale channels and even penalized those miners who left the cartels to deal directly with wholesalers and cutters. The company methodically stockpiled and then flooded the market with the particular quality of diamonds produced in that mine (Cadieux and Conklin 2005). Regardless, the strategy that De Beers practiced for over 60 years (from the 1930s to the late 1990s) eventually placed a tremendous financial burden on the company. To overcome that financial impasse, De Beers partnered with luxury conglomerate LVMH and together launched a new retail venture going directly to consumer with "De Beers" branded diamonds. In securing a retail presence, De Beers further reinforced its position of power in shaping consumers' opinion about the perceived rarity of diamonds and continued doing so as the market of branded luxury goods kept growing from the mid-1990s to the present (Berghaus et al. 2014). As the illustration clarifies, on the market side, businesses can rely on a multitude of tools (such as new technological platforms for commerce, communications in the form of ads, sponsorships, private events, education, social media, and even pricing) to strengthen the "luxury halo" effect created around certain products, regardless of whether these are truly luxury or simply premium goods that have been branded as luxury. From diamonds that could be worn only by kings and royalty in pre-modern times, the twentieth century experienced a massification of the luxury trade (Berg Fashion Library of Dress 2016). Consumers' perception of diamonds was marked by De Beers's strong messaging that "diamonds are forever," a promise that both resonated with the romantic views of marriage and eternal love in relationships and with the implied allusion to timelessness, a quality attributed to luxury goods and responsible for their future value. In other words, De Beers (along with several other players in the luxury market) strategically manipulates the perceived rarity of its products when in fact several of those are made with industrial diamonds (a commodity and by-product of cutting quality diamonds). While this has been extensively discussed (Cadieux and Conklin 2005; De Koninck and Debaere 2007) it also shows that if we are to speak of "luxury" diamonds we need to move away from the market side of the luxury business model of Fig. 1 and concentrate on the production side.

It is this part of the model that will inform our discussion of mining processes in the context of sustainable, luxury diamonds. Here we see that a business that produces luxury goods maintains facilities in a special geographic location and has remained in the same location for years, if not centuries. This continuity of presence relates to special natural resources found in that area and implies a unique relationship with the people of that area, who, most probably, contribute their labor. A special intra-generational bond is created between the company and the locals, who, through continued exposure to the proprietary processes of production, acquire two types of knowledge: explicit knowledge of the industry and its traits and tacit knowledge of production processes (Adamson 2013). For several luxury producers this means that if they were to move their facilities elsewhere they would miss the tacit knowledge their workers acquired over the years and while this cannot be quantified, the final product would suffer in quality and perhaps be inferior to the original luxury good. Additionally, as in all businesses, a legal trust is created to safeguard the business against competitors (and corporate raiders) as well as to protect the proprietary processes in place, including any type of intellectual property developed herein. Ideally, a true luxury producer owns the supply chain and therefore has control over the quality of the raw materials processed in the business.

In all luxury enterprises, the product is driving both the business and the brand. The "star" product, whether this is a crocodile handbag or an exceptional piece of jewelry, is produced in limited numbers (luxury producers follow a strategy of controlled/limited supply) or is available for a limited time or in limited collections. Therefore, it is the type of product, its numbers, and the time during which it becomes available that establish its veracity as "luxury."

Now that we have examined the specifics of the luxury business model, we will proceed with a general background of diamond mining processes and how they impact the environment before presenting a detailed report from Botswana's model gemological lab and its potential in ascribing to the luxury business model.

5 Diamond Mining and Sustainability

While initially discovered in meteorites by Erofeev and Lachinov in 1888, diamonds occur in Kimberlites in the upper sections of the Earth's crust, forming pipes, dikes, veins, and sills. Kimberley and De Beers, international mining and exploration companies operating in Africa, Australia, and Europe, discovered the first Kimberlite diamonds in South Africa (Roberts 1972; Robertson 1974). According to Orlov (1977), whose seminal work on diamond mineralogy remains the most referenced source in the field, Kimberlites are located in the tectonically stable center of continents known as shields or platforms. Because of the high alkali composition of the rock of which Kimberlites are composed, scientists believe that Kimberlite rocks are intrusions formed by the alkalitic magma from the upper mantle pushed into the upper areas of the planet's crust. Kimberlite pipes in particular occur where the earth's crust in undergoing bending as the magma is pushing upwards. This phenomenon, known as Kimberlite volcanism, occurred during the Precambrian and early Paleozoic times (Orlov 1977).

Kimberlite pipes and their distinct formations define the methods of diamond extraction, namely open pit and underground mining. In open pit mining the diamond rich material is loosened by removing layers of earth gradually and while the newly formed benches move downward in a circular motion. The deeper the open pit, the narrower it gets. While the upper strata of the pit are softer and easier to extract, the deeper the pit the harder the surface becomes. Mining in open pit often requires use of explosives (Legrand 1980). This may impact the stability of the pit and certainly permanently alters the environment. The stability of the mine depends on the width of the circular benches around it as well as on their height. In other words, the sloping of the open pit mine must be smooth to allow for maximization of extraction and prevent the pit from becoming too narrow too soon. Profitability of pit mining is measured in carats per ton of material. The value of the pit is therefore decreasing the deeper and narrower the pit becomes (Legrand 1980).

To circumvent steep or unstable pits that are both dangerous and costly, underground mining is utilized. This process relies on building complex underground tunnels that perforate the host rock underneath the open pit and give access to the remaining Kimberlite deposit of the pipe that had originally formed the open pit mine.

While the diamond industry has adopted strict regulations to mitigate the impact of mining processes on the environment, the local social fabric, and its economy, not all diamond mining is regulated. Occurrence of informal alluvial mining is prevalent in diamond-rich and poverty stricken regions, particularly in Africa. This method of mining is known as 'artisanal' mining, which, in our artisanal luxury obsessed Western market, may be misleadingly interpreted as positive (Van Bockstael and Vlassenroot 2011). This could not be farther from the truth. "Alluvial diamonds is the term used to describe diamonds that have been removed from the primary source (Kimberlite) by natural erosive action over millions of years, and eventually deposited in a new environment such as a river bed, an ocean floor or a shoreline" (Diamondfacts.org). Alluvial diamond mining is a small-scale operation undertaken by local miners, often with no other employment prospects, who extract diamonds from alluvial deposits by digging and sifting through mud, gravel, and sand using shovels, sieves, and their bare hands. The implications of alluvial diamond mining for the diamond trade are tremendous. The activities associated with this method are outside any formal legal and regulatory framework (Van Bockstael and Vlassenroot 2012). According to the International Institute for Environment and Development, while the productivity of this type of miners is limited and low compared to formal diamond mining, it still presents great dangers in terms of their own safety and the resulting impact on the environment (IIED 2006). The deposits exploited are marginal, the labor involved strenuous and dangerous, and the return on the miners' investment is minimal. Most importantly, once an area has been disturbed for mining it is not restored in its original state simply because the rural labor force does not have the time (or funds) to invest in such operation. Environmentally, alluvial mining has the greatest impact in specific regions where the balance of its microclimate is permanently altered because of shifting grounds and water drainage. Additionally, the difficulty of alluvial mining presents labor challenges that are often solved by employing children, a practice that is well known to local governments who have little power in intervening, especially because alluvial mining is mostly migratory (Huang 2002). This may also explain why alluvial mining camps foster greater potential for incidence of prostitution and spreading of HIV/AIDS, further heightened because of the transitory nature of artisanal operations. In spite of their dangerous labor conditions and their dedication to a lifestyle with a range of inherent dangers, migrant alluvial diamond miners are not educated about the true value of the rough diamonds they mine and are, therefore, vulnerable to exploitation. Finally, alluvial mining diamonds are not documented. Both their origin and trade routes are untraceable and as such they may be easily used to finance local conflicts or terrorist actions. This is perhaps the most misleading interpretation of the term 'artisanal' in the context of diamond mining. 14% of the world's rough diamonds stems from artisanal, informal alluvial diamond digging (Diamondfacts.org). Artisanal luxury, seen as wholesome and authentic in Western markets, stands for 'conflict' or 'blood' in the context of the diamond economy and perfectly contradicts the humanistic value inherent in luxury (Serdari 2016).

Setting alluvial mining aside, the main question about sustainability and diamond mining sets forth a two-prong discussion: how to minimize its impact on the environment and how to maximize the benefits to the communities and countries involved in the mining processes.

Environmental impact is managed by the guidelines set forth in ISO 14001 (the most important standard within the ISO series developed and published by the International Organization for Standardization that "define, establish, and maintain en effective quality assurance system for manufacturing and service industries.") ISO 14001 specifically defines the framework that "a company or organization can follow to set up an effective environmental management system." The standard is revisited each year in order to achieve an up-to-date roadmap for companies that must audit, manage, correct, and communicate their environmental systems both to minimize occurring damages and to safeguard against biggest challenges such as climate change.

The specificity of the diamond mining processes and its direct correlation to the geographic location where they incur call for additional attention beyond the prescribed measures outlined by ISO 14001. Diamond mining companies are cognizant of the gaps in regulations that may lead to operational oversights especially because most develop scientific processes internally but deploy them internationally in locations with special environmental conditions and microclimates.

The areas of impact that are monitored through the ISO 14001 standard are:

Land exploration: Impact is minimized by reusing vehicle tracks, clearing out minimal amounts of soil during drilling and sampling, and replacing topsoil from exploration sites.

- Open pit and underground mining: The large quantities of surplus rock, sand, and soil extracted from the pit usually accumulate in the immediate vicinity of the mining area. They need to be managed while mining is in progress and rehabilitated once the mine closes. It is expected that the topsoil will return to its previous state when mining operations have ceased. Prominent companies with open pit mines are BHP Billiton and Rio Tinto; De Beers owns open pit and underground mines.
- Coastal and alluvial mining: In addition to processes applicable to pit mining, removal of plants and soil and construction of sea walls may be required when mining on beaches. Environmental impact relates both to restoration of land and plant life. Standards of sustainable diamond mining in coastal areas require that special techniques are used to monitor how much deposit is removed because of the wind and the waves and how to restore plant life in areas that suffer from low rainfall (American Museum of Natural History 2006). De Beers is in coastal and alluvial mining.

- Marine mining: Diamonds are accessed beneath the seabed. All seabed matter that is removed before mining needs to be replaced in its original position and state. While restoration is not immediate, research has shown that over a period of a few years, vegetation and fish life return to the mined seabed area. De Beers started offshore (marine) diamond mining in 1961.
- Energy use, air quality and pollution: Diamond mining uses heavily two forms of energy, electricity and hydrocarbons (diesel, marine gas, oil and petrol). Therefore, it contributes significantly to the greenhouse effect with emissions of gases and various chemical, synthetic and natural, substances. In other words, there is no mining process that does not negatively impact the planet's climate. This is perhaps the only area that cannot change in the near future. As a result, diamond mining can never be fully sustainable and cannot be marketed a 'green luxury.'
- Waste and recycling: The types of waste from diamond mining are similar to other large scale industrial operations and include oil, paper, scrap metal, batteries, tires, plastic, and glass. Recycling materials, including separating and recycling oil, is the predominant effort in minimizing environmental damage due to waste.
- Water consumption: Diamond mining uses large quantities of water for extraction. This poses tremendous challenges, especially in African regions where water is scarce. A related issue to excessive water consumption is prevention of water pollution. Therefore, initiatives on conservation of water (reduction of use, recovery, reuse and recycling) are as important and carefully monitored as are initiatives on topsoil and waste monitoring to prevent contamination of bodies of water. Diamond mining companies are heavily invested in researching and developing technology that relies on alternative water sources. However, this would be another area that would challenge initiatives to market diamonds as 'green luxury.'
- Disruption of biodiversity: Diamond mining activities disturb plant, animal and fish life. Mining companies today strive to minimize their impact on these ecosystems through all stages of operations, from land exploration to closing of the mine. Seeds of native species around the mining site are harvested and replanted, often in collaboration with the Millennium Seed Bank (run by The Royal Botanic Gardens at Kew Gardens in London, UK). Wildlife protection is more challenging than seed conservation. There is no organization similar to the Millennium Seed Bank, which implies that efforts for wildlife protection and restoration are left to the mining companies with various degrees of commitment and success. Mining company Rio Tinto has been monitoring environmental impact of operations in the Northwest territories of Canada for the last fifteen years mainly because fish is an important source of food for the local indigenous population. There are a number of Nature Preserves to develop breeding programs for rare and endangered species (such as antelope, disease-free buffalo, white rhino calves, African elephants, wild dog, and black rhino).

In summary, the diamond mining industry presents a complex case of an agent that has the potential to both irreparably harm the environment as well as to fund research and initiatives that may prove useful beyond the immediate applications on diamond mining sites. Quality diamonds are the result of heavy capital investments by mining companies and cannot be labeled as 'green luxury' even though environmental protection and sustainability efforts are the highest priority of multinationals that invest substantial time, efforts, and funds in positively impacting the environment.

6 Report from Botswana

According to Wall Street Journal former reporter and current US News editor Vanessa O'Connell, "Botswana has been attractive to mining companies because it remains the world's largest producer of gem-quality diamonds since the 1980s. The Jwaneng Diamond Mine is the world's richest by value of recovered diamonds." The journalist outlines other reasons why Botswana has retained its appeal: it is among Africa's least-corrupt countries; its government regulates labor conditions in the mines, which the government owns in a 50–50 partnership with international mining company De Beers; Botswana guarantees provenance of its gems and appeases consumers' concerns over "conflict diamonds" sold to fund wars (Wall Street Journal Oct. 26, 2009).

Botswana's population is relatively small compared to other African countries and its numbers have further diminished after the country was hit by the AIDS epidemic. It is an underdeveloped country with minimal manufacturing infrastructure which puts Botswana at a disadvantage compared to other countries competing for diamond cutting and polishing operations. "A parcel of diamonds polished [in Botswana] costs about $100 per carat compared with $30 in India," O'Connell asserts. In the last ten years however, the government took a stronger stance in upgrading the country's status amongst competitors. After renegotiating its mining contract with De Beers, Botswana "licensed 16 international cutting firms willing to build factories [there]." (Wall Street Journal Oct. 26, 2009) There was an important requirement for this deal to go through and that was that the companies would have to train locals to polish diamonds. This agreement would then entitle the companies to the right to buy rough diamonds in Botswana, a market that had been completely closed to them.

At first, the workers were selected on the basis of their education (they had to have a high school diploma); their math aptitude; and their handiwork, tested in how accurately and how fast they could "place 40 tiny metal sticks into the holes set in a board." (Wall Street Journal Oct. 26, 2009) It takes about four to six months for workers with the basic qualifications outlined above to become proficient in bruting (bruting is the art of cutting a diamond round) and faceting rough stones. Of those hired, only about 70% make it to the stage of "qualified worker," which also means that they can finish one polished diamond per day, a considerably low level of productivity compared to what one encounters in India or Vietnam.

O'Connell reveals that Tiffany's, which took a majority stake in Rand Diamonds, an international cutting firm approved by Botswana's government to do business there for the right to buy rough diamonds "doesn't tell its customers where individual diamonds are mined or polished, and declined to say how much of its overall inventory is now sourced and polished in Botswana." According to Tiffany's CEO, Mr. Kowalski, "We [Tiffany] want the focus to be on the quality of the diamond ring, not how it came to be." (Wall Street Journal Oct. 26, 2009)

How the diamond ring came to be, however, is very much what should define its value—without discounting the quality of the final product, it is the entire production process that adds value to quality diamonds.

Finally, and while Tiffany's has been able to capitalize on its substantial capital investments in Botswana through Rand Diamonds with diamond merchandize representing a 47% of Tiffany's sales, their expat supervisors in Botswana's gemological labs prefer to perfect the work themselves rather than impart their knowledge to the locals. In other words, big diamond retailers remain secretive about how and where they are sourcing their diamonds. Even when they invest in local infrastructure in diamond rich countries, they limit knowledge transfer to the local laborers to the absolute minimum. On the surface, it seems that Botswana's government involvement in the mining industry is a propitious omen for the locals' future. Yet, the relationship between several international diamond mining players and the locals remains one-sided and practically opaque.

This was also confirmed during the author's interview with Giovanna Federico (see Appendix) who was hired by New York-based company Gemological Science International to set up a gemological lab in Botswana, specifically to carry on with the operations of cutting and polishing diamonds. Well acquainted with the precise sequence of operations prescribed for diamond mining (sorting for quality and size, cutting, polishing, grading based on the Gemological Institute of America four Cs) Federico confirmed that while the technology of diamond mining in its various forms (discussed in the early part of this paper) has been developed in Western countries, the reality of implementing it on site and specifically in the underdeveloped countries of Africa is far from what one finds in scientific textbooks. Intermittent supply of electricity, which in turn impacts connectivity over the Internet, is the gravest challenge because it directly disrupts the lab operations. On the other hand, these difficulties do not stop the locals from wanting to participate in the global economy, mainly because they have been exposed to it thanks to the mining industry that attracts large numbers of foreigners who visit Botswana to work. Powerful unions formed to negotiate with the multinational mining companies often mitigate the tension between foreigners and local labor. The Batswanas desire more involvement in the diamond industry and more exposure to various aspects of diamond processing. There is definitely a clash of cultures.

On the other hand, and while environmental and humane conditions are starkly improved compared to the early years of mining, local workers still feel that they are not given enough knowledge that will make them employable to above ground operations, in labs or trade. Federico also refers to the highly specialized instruments that are used in gemological labs and which require advanced skills in handling. These skills can only come with training in which most labs are lacking. Therefore, there is a great need of investment in training and development, which when present, ensure operations of the highest standards but also prepare the lab personnel for a variety of roles within the diamond industry. The four fully trained local employees that Federico mentored during her stay in Botswana, were, by the time she left, highly desirable staff either at the lab or the office and were responsible for a flow of about 150 diamonds per day. This points to a novel idea on further grading diamonds by adding a fifth C to GIA's 4Cs: C as in Creator.

7 The (Missing) Fifth C: A New Framework for Determining Diamond Value

Africa's wealth of diamond supplies has failed to repay its countries because their local populations have not been included at the later stages of diamond processing (downstream). In other words, African natives (from Botswana, DR Congo, South Africa, Zambia, and other diamond rich territories) have been treated as commodities while international mining companies try to maximize the amount of diamonds that are extracted, cut, and polished before moving them to the big trading centers (mainly Antwerp, Belgium but also Mumbai, India and Shanghai, China) (De Koninck and Debaere 2007).

In spite of local governments' influence in recapturing their countries' natural wealth, as for example was the case with the Batswana government and its agreement with international mining companies to carry the exploitation of the local mines as equal partners, a large part of the diamond trade remains either unregulated (as discussed in the context of alluvial mining) or based on personal attention and trust (De Koninck and Debaere 2007). The implications of the 'hand-shake' economy is that the industry remains deliberately opaque in an effort to maximize the profits occurred through wholesale of cut and polished diamonds and later retail distribution of diamond merchandise (Cadieux and Conklin 2005).

The question therefore is whether the diamond industry can claim its place within sustainable luxury and if so, whether this can be accomplished through training, development, and economic empowerment of local labor forces, whose role within the diamond industry will not be exclusively linked to underground mining or to industrial tasks on the excavation site, especially because as has already been noted these are very precarious roles that may soon disappear because of the rapid technological advancements in the field. If underground mining becomes safer when automated systems are introduced, the displaced labor force can still contribute in the value chain of diamond production by applying time and expertise at downstream operations.

While diamonds that are free of "blood" or "conflict" associations reach the consumer with a certificate stating that "the diamonds herein invoiced have been purchased from legitimate sources not involved in funding conflict and in compliance with United Nations resolution. The seller hereby guarantees that these diamonds are from areas that are free of conflict, based on personal knowledge and/or written guarantees provided by the supplier of diamonds" (De Koninck and Debaere 2007: 15), no specifics are provided about the country of origin or place of processing. In other words, where did the stone originate and where did it travel before reaching the supplier and finally the retailer? How could a customer be certain that Batswanas were actually part of the value chain that brought the brilliant diamond in her hands?

Adding a new field for 'Creator,' introduces the fifth C in GIA's framework of 4C's. If that field is to be filled out correctly through every single step of processing, the retailer should be able to provide the customer with a new certificate that clearly states the Clarity, Cut, Color, Carats, and Creator for each diamond. The Creator field can be further broken down to subfields for: country of origin; name of mine; processing of rough; cutting; and polishing, all of which are activities that truly differentiate 'blood-free' diamonds from conflict diamonds. Other than the country of origin and the name of mine that are place specific (and obviously in the same country) all other activities may take place in other parts of the world (Vietnam and India are market leaders in the cutting and polishing of stones). Gemological labs have names and people who work in them have identities, personal stories, and skills.

What the fifth C would actually do is slow down the process of diamond production, an idea that contradicts how commodities are processed but actually reinforces the absolute luxury business model discussed earlier. Marketing alone has proven unable to sustain the appeal of diamonds for customers whose interest is equally intrigued by adventurous expeditions, luxurious travel, or innovative appliances. The luxury business model provides a framework that would allow businesses to reorganize their activities around main points of interest, for example matching the location of the mine with the location of the gemological lab. In that case, the Creator field can impart significant information about the cultural background of the people who worked on cutting the diamonds or those who polished them and so on, a practice similar to the one communicated by luxury houses in the leather accessories business where artisans inscribe their initials on each product. On one hand, the consumer would gain insights into the main agents who made the purchased stone as beautiful as it appears to be. On the other hand, and because the fifth C slows down the process of production, diamond retailers may avoid overproduction and focus instead on supplying the market with controlled numbers of diamond products, a model much closer to how true luxury firms operate.

8 Conclusions

This case study attempted to explain the predicament the diamond industry faces on a regular basis with varied intervals of peaks and valleys. This stems from oversupply of diamonds (both quality and industrial) of often dubious origins and unclear path to the consumer. The experience gained by working closely with the local labor force in Botswana revealed the need for stronger inclusion of local workers in upstream processes, better opportunities in professional development, and higher standards in execution. With these criteria in place, production of diamonds may begin to actually repay the countries of origin for their natural resources and help them face challenges in other vital domains, such as water and electricity supply, restoration of the natural environment, and preservation of continuous source of income that may finally help these countries catch up with the developed world in health and education. Sustainable luxury is slow. It is not about fast processes or great numbers of supply. On the contrary, sustainable diamonds that ascribe to the 5Cs generate wealth for all the stakeholders involved in the creation process and retain, in addition to monetary value, humanistic value, the most intangible and yet most fundamental component of luxury.

Appendix: Interview with Giovanna Federico, Ph.D. Gemological Institute of America Certified Gemologist

TS: Giovanna, you were recently invited to travel to Botswana and assist in setting up a gemological lab. What type of lab is it?

GF: This new lab, by Gemological Science International originally from New York, is focusing on cut and polished diamonds exclusively. It is set up so that employees analyze each single gem and generate a report stating its unique characteristics.

TS: How does this relate to the entire value chain of diamond mining?

GF: While the process may differ depending on where the mine is located we can safely assume that operations follow a prescribed sequence from mining the rough diamonds that are extracted from their natural Kimberlite. Subsequently, rough diamonds are sorted by quality and size. Then they are cut. This is a very important step because each rough diamond requires a different cut. That allows the rough to be cut according to its natural characteristics. By doing so, this process emphasizes the natural beauty of the stone. Cutting is followed by polishing, which enhances the diamond's sparkle. It is only after all these steps have been completed that the gems reach the gemological laboratory. There, they are analyzed and assigned a certificate that lists the diamond's grade. This grading system is based on the "Four Cs" standard (clarity, cut, color and carat weight).

TS: You have long experience in these processes. Why do you think that Botswana offers a challenge to mining companies? Why would they need an expert like you to assist in organizing the lab? What is different about Botswana?

GF: Botswana is a very interesting place. It is a country with great ambition. Locals are business-oriented and eager to learn about new processes and anything that will connect them quickly to the global economy. Locals are accustomed to seeing a lot of foreigners in Botswana thanks to the diamond industry that is thriving there. We all know that Botswana has traditionally rendered great numbers of excellent quality diamonds and a few of substantial number of carats. The diamond industry, therefore, is fundamental in moving the local economy forward. Simply put, it is the main source of income for a large majority of the population. This is not to say that the relationship between Batswanas and foreigners, especially foreign business entities, is not occasionally tense, especially when things are not handled properly, or when Batswanas feel shortchanged. Employees at these foreign mining companies are well aware of the power they have through organized, strong unions. In my experience, the local population is of a moderate temperament and peaceful attitude. But we cannot forget how close their reality is to their recent experience of the apartheid that dominated their lives for a very long time. Also, in spite of the general progress in business, the constant flow of foreigners and Batswanas' exposure to diversity, Botswana is still a developing country with tremendous challenges.

TS: This means that arriving in Botswana to set up procedures in a gemological lab may have been a cultural shock for you.

GF: There are certainly enormous differences in culture, in daily modes of life, and education between myself and the people I came to train. I try to always take these into consideration to create a positive and productive atmosphere conducive to learning.

TS: Other than cultural differences, what are the challenges present in Botswana in particular?

GF: The biggest challenge is shortage of water. Most of the country's rivers have dried up—the entire country is in a state of drought. The government is working with neighboring countries facilitating the flow of water back into Botswana from a variety of sources. This is no small feat.

TS: What are the consequences of the water shortage for daily life and businesses?

GF: When I arrived [in March of 2015] and throughout my stay [that ended in June of 2015] the water supply was interrupted from 9 am to 6 pm four times a week both in residential and business addresses. There are private people and businesses with funds to invest in large water tanks to be installed in proximity to their residencies or offices.

TS: I understand there can be no diamond mining without water.

GF: Yes, that is correct. Diamond mining relies heavily on the use of water, especially upstream when the Kimberlite is crushed and washed in order to extract the rough diamonds. Water is necessary at various stages of the mining process.

TS: Could you tell us more about the process itself?

GF: I had the chance to visit one of the biggest mines in South Africa on a different trip. The engineer who took me to 740 m underground explained to me how the business is organized. Mining companies invest a vast amount of money in research primarily and, subsequently, digging and exploring. Once the Kimberlite pipes have been located and confirmed, tunnels are dug out. These tunnels need to be supplied with water and electricity. Ironically, diamonds, which represent one of Africa's important natural resources, require such large quantities of other natural resources (water and electricity) that are pretty scarce in that entire region. To come back to the actual mine, I noticed that the tunnels were excavated from above, laterally, and underneath as well. It is a very complex engineering process. Once the tunneling system has been completed, the developer builds an entire village around the mine. This is a full-service urban capsule with shops, restaurants, and medical facilities all of which are designed to accommodate the miners and their families.

TS: How are the conditions underground?

GF: Conditions underground have improved from the past. Today, developers build lounges and resting areas below ground for the hundreds of people who work there daily.

TS: The industry has turned a little bit more humane…

GF: Yes, definitely. It is an industry for the modern world. However, in Botswana the challenges are numerous. In addition to water shortage, there is a shortage of electricity. This greatly affects the operations below and above ground. Electricity is cut at least once a week. Networks require electricity as well. At the lab, our Internet connection would be interrupted and force the staff to abandon their tasks or repeat work that they had started. This happened multiple times per week, every week. Therefore, we need to think of these businesses within the context of a third world economy and appreciate the extra effort that goes into cutting, polishing, and grading these diamonds that, we know, will end up at a retailer's window on Fifth Avenue.

TS: You spoke of the Internet and computers. How automated are these gemological labs?

GF: Automated systems are already being considered as an integral part of the process. However, while labs rely on them above ground, underground automation has not happened yet. Mining companies are investing heavily in research of automated systems that will eventually replace the miners who work underground. This will have life changing consequences for the locals and this is why I believe in helping them acquire knowledge that will make them employable in operations above ground. The human and social repercussions of this development are not been discussed openly. I believe that they present both a threat against Batswanas' livelihood and a remarkable opportunity for them to be prepared for more sophisticated, and certainly more comfortable, types of employment above ground.

TS: What were the conditions you faced at the lab when you first arrived?

GF: At that point, the lab was very basically developed. There was one staff person, a local young lady who had not been fully trained. Her work was inconsistent and relied on a very rudimentary technological system. Gemological labs required sophisticated machinery both to identify the traits of each gem and to allow for a seamless workflow that allows each expert to be fully employed all the time. None of this existed before my arrival in March 2015.

TS: Could you give us more details on this type of technology?

GF: The machinery required is very specific. A lab must be equipped with the following:

1. Sarine scan: it's a scanning instrument that reports measurements of each single diamonds This information will be part of the final report when the diamond leaves the lab.
2. A diamond scale: This is a very sensitive and precise instrument that focuses only on the carat weight of the stone. Periodically it needs to be re-calibrated and cleaned.
3. A high-resolution scope to analyze the internal characteristics of the diamond. It allows gemologists to assess the stone's clarity grade.
4. To assess the color grade a lab needs the grading light cabinet, a special box which mimics natural daylight.
5. A fluorescence box determines the level of fluorescence or the lack of it in a diamond.

These are the basics of a diamond-grading lab. In other departments, such as the one determining the stone's chemical composition and the origin of color, there are other types of essential equipment, without which a lab cannot be considered reputable. Yet, not all labs are following these standards and, unfortunately, the staff at these other labs are not able to determine the grades of the stones and their more complex traits. They limit their authority in determining authenticity. This is not nearly enough for diamonds that enter the global trade.

TS: Could you explain how you modernized this lab?

GF: I studied the systems that had been installed and made sure I knew how to troubleshoot anything related to technology. I also supervised staff while they were grading diamonds. I organized the lab's operations and procedures while also assigned and monitored electrical functions. I designed and implemented new office systems for maintaining the lab's efficiency. This included making sure that procurement and set up of equipment (Sarin machine, scopes etc.) was timely and up to standards. I supervised the staff by actively measuring their results against industry standards and by making the necessary adjustments. I also acted as a liaison between the lab and headquarters providing them with summary reports and up to the minute information on how the systems supported or prohibited the staff from doing their work. In the process, I recruited a number of new staff, which I found to be the most satisfying part of my assignment. I trained them on each of the machines and explained to them the nuances of color, clarity, polish, and symmetry in grading the stones. At the end of the process, in June 2015, I left the lab with fully trained staff who could be employable in any lab, even in the most high-tech facilities in developed economies. Beyond their basic training, the employees were invited to participate in a series of educational workshops I designed and organized as well as encouraged to develop personal networks and to participate in professional societies. I was personally interested in both leaving an exemplary lab behind at the end of my visit and in having trained people who would be able to seek new opportunities within the international community of the diamond mining industry and trade. When I left Botswana, I had four fully trained locals who were working at the lab and the office and were generating a workflow of about 150 diamonds on a daily basis.

References

Adamson G (2013) The invention of craft. Bloomsbury Press, London

American Museum of Natural History (2006) The nature of diamonds: alluvial mining. Accessed online on 4 Apr 2016

Berg Fashion Library Online. Accessed online on 4 Apr 2016

Berg M (2005) Luxury and pleasure in eighteenth-century Britain. Oxford University Press, Oxford

Berghaus B, Müller-Stewens G, Reinecke S (2014) The management of luxury: a practitioner's handbook. Kogan Press, New York

Berry J (1994) The idea of luxury. Cambridge University Press, Cambridge

Cadieux D, Conklin D (2005) De Beers and the global diamond industry. Harvard Business School Cases, Cambridge, MA

Chitrakorn K (2016) Distress in the $80 Billion Diamond Market. In: Business of fashion: the daily digest. Accessed online on 8 Apr 2016

De Koninck V, Debaere P (2007) Is the diamond capital of the world losing its sparkle? Diamond trading and cutting in a changing global economy. Harvard Business School Cases, Cambridge, MA

Diamondfacts.org. Accessed on 2 Apr 2016

Gardetti MA, Girón ME (2014) Sustainable luxury and social entrepreneurship: stories from the pioneers. Greenleaf Publishing, Sheffield

Gardetti MA, Torres AL (eds) (2013) Sustainable luxury. The Journal of Corporate Citizenship 52, Dec 2013

Huang Y (2002) Note on conflict diamonds: why are civil wars, like diamonds, forever?. Harvard Business School Cases, Cambridge, MA

International Institute of Environemnt and Development (2006) Artisanal and small scale mining. Accessed online on 20 Mar 2016

Kapferer J-N, Bastien V (2009) The luxury strategy: break the rules of marketing to build luxury brands. Kogan Press, New York

Legrand J (1980) Diamonds, myth, magic, and reality. Crown, New York

O'Connell V (2009) Diamond Industry Makeover Sends Fifth Avenue to Africa. Wall Street Journal, 26 Oct. http://www.wsj.com/articles/SB125650986946206903. Accessed online on 8 Apr 2016

Orlov Y (1977) The mineralogy of the diamond. Wiley, New York

Roberts B (1972) The diamond magnates. Hamish Hamilton, London

Robertson M (1974) Diamond Fever: South African Diamond History 1966-9 from Primary Sources. Oxford University Press, London

Sekora J (1977) Luxury: the concept in western thought. Johns Hopkins University Press, Baltimore

Serdari T (2016) Steidl: printer, publisher, alchemist: the field of luxury production in Germany. Luxury: History Culture Consumption 2(2):33–51

Van Bockstael S, Vlassenroot K (eds) (2011) Diamonds, rice, and a 'Maggi Cube': artisanal diamond mining and livelihoods in Liberia. Academia Press, Gent, NL

Van Bockstael S, Vlassenroot K (eds) (2012) A farmer's best friend? Artisanal diamond mining and rural change in West and Central Africa. Academia Press, Gent, NL

Veblen T (1899) The theory of the leisure class. Prometheus Books. New York

More Than Skin Deep: A Service Design Approach to Making the Luxury Personal Care Industry More Sustainable

R. Wakefield-Rann

Abstract The core values of both luxury and sustainability are at odds with a consumer culture characterised by cheap, disposable products and undervalued natural resources. Although some product categories within the luxury goods sector have upheld the values of quality and durability, others, such as personal care, have come to rely on materials and processes that are harmful to ecosystems and human health. The luxury personal care industry trades on qualities of purity, freshness, beauty and the 'natural'. However, the industry remains unsustainable through its continued use of single-use plastic packaging and particular synthetic chemical additives. For this to change, the way in which personal care products are delivered and administered must be fundamentally redesigned. This chapter presents a case study of luxury personal care company LUSH, and examines how its innovative approach to service design could provide a genuinely sustainable model for luxury personal care companies, and potentially the broader industry. The central elements of this model include local production, 'naked' products, short expiry dates, and innovative retail design.

Keywords Service design · Sustainable luxury · Personal care products · Social practices · Endocrine disruption · Plastic pollution

1 Introduction

The luxury goods sector is under increasing pressure to reform its core practices and values to align with sustainability principles. The concept of luxury has been accused of being incompatible with sustainability, as it has come to be associated with waste, greed and social inequality. Reports on business sustainability have also found that luxury companies have lagged behind other sectors in improving their

R. Wakefield-Rann (✉)
Faculty of Design, Architecture and Building,
University of Technology Sydney, Sydney, Australia
e-mail: rachaelwakefieldrann@gmail.com

M.A. Gardetti (ed.), *Sustainable Management of Luxury*,
Environmental Footprints and Eco-design of Products and Processes,
DOI 10.1007/978-981-10-2917-2_10

business practices to make them more sustainable (Bendell and Kleanthous 2007). However, there is a compelling case that core luxury values are aligned with the values of sustainability, including quality, durability, expert craftsmanship, and the importance of precious natural resources. For a business to be genuinely sustainable, it must emphasise quality over quantity and a reduced rate of consumption: two of the core values that characterise luxury. Some luxury brands have begun to successfully make this connection and transition to a genuine model of sustainable luxury (Tutty 2016). Yet, as this chapter will demonstrate, a number of brands have attempted to become sustainable while maintaining a business model that still depends on cheap resource procurement, disposable products and/or packaging and rapid turn-around. A business cannot be considered sustainable unless it accounts for its social and environmental impacts across the supply chain, in a way that is able to meet the needs of current consumers without inhibiting the capacity of future generations to meet their needs.

Despite the commensurate vales of luxury and sustainability, it has been particularly difficult for luxury personal care brands to make the necessary changes to become sustainable. Unlike other luxury product categories, personal care products (PCPs) are non-durable and are intended to be consumed on a regular basis. These product qualities contribute to two key barriers currently preventing personal care businesses from becoming more sustainable: their dependence on disposable plastic packaging, and their use of certain chemicals for product preservation, fragrance and texture. This is not to say personal care businesses do not face other sustainability challenges, but that these obstacles have been the most difficult to overcome. A number of brands have made positive efforts to source more ingredients from organic and fair-trade producers, use more recycled packaging, reduce their carbon footprint and exclude some harmful chemicals from their products (Kessler 2015; Matusow 2010). However, their continued use of plastic and harmful chemicals means the core requirements of genuine sustainability have not yet been met by any medium or large-scale personal care brands.

This chapter contends that the key sustainability challenges faced by personal care companies cannot be resolved unless the daily personal care routines of consumers can be reformed. Everyday personal care routines for most adult individuals in wealthy industrialised countries will generally involve combinations of products, including: anti-bacterial liquid soap and body wash, shaving gel, exfoliators, face cleanser, shampoo, conditioner, hair spray, deodorant, toothpaste, dental floss, mouth wash, moisturizer, face wipes, and for many, numerous kinds of cosmetics, make-up removal wipes and liquid, nail-polish and anti-aging creams. This sheer number of products, and the amount of materials, energy and chemicals required to produce them, are unprecedented in history and result in significant human and ecological harm. For a brand to be sustainable it cannot contribute to and proliferate these practices. It must offer a new way of practicing personal care that does not require an excess of chemicals, energy, materials and packaging.

Sustainability has been difficult to achieve because current market expectations regarding the shelf life of products, their price, sensual qualities, and ease of use necessitate the use of durable plastic containers and numerous chemicals. The majority of PCPs sold are packaged in disposable plastic containers (Matusow 2010). These not only require non-renewable resources such as petroleum to manufacture, the products and their packages often contain chemicals that have been identified as carcinogens, mutagens, allergens or endocrine disrupting chemicals (EDC) (Liboiron 2015; Kessler 2015; Zoeller et al. 2012). EDCs, for example, interfere with the development and functioning of the hormonal system, which has been correlated with numerous health conditions in animals and humans, including obesity, diabetes and cancer, among others (Diamanti-Kandarakis et al. 2009; Schug et al. 2013). Plastic packaging and the plastic microbeads often contained in PCPs also end up in the environment, where they have been shown to cause various forms of harm, including choking and starving animals via ingestion, and altering the reproductive development of marine species.

This chapter proposes that to become sustainable, luxury personal care companies must employ a new service design model that enables the removal of disposable plastic packaging and toxic chemicals from their products. Service design has the unique capacity to effectively reform both the organisational structures of luxury personal care businesses and the expectations consumers have of their products. 'Service design' refers to the way that people, infrastructures, communications and material components of a service are arranged to produce certain outcomes within an organisation, or to orient customer behaviour a desired way (Blomkvist et al. 2010). Unlike product design, service design delivers a constellation of ingredients that work adaptively in unison to support an outcome. In the case of luxury PCPs, a service design approach offers the ability to re-evaluate and re-design better ways of coordinating supply chain logistics, organisational structures, digital customer interfaces, customer support, retailing environments and product design in support of more sustainable outcomes. Service design could therefore enable a new way of selling and practicing personal care that attends to the key pleasures and practical functions people require from their products, without relying on plastic packaging or toxic preservatives, fragrances and colors.

To determine how luxury personal care businesses can successfully develop a more sustainable service design model, the complex role of PCPs in people's daily lives must first be understood. The first section of this chapter will therefore examine the numerous functions that PCPs commonly fulfill. If a business fails to recognise the essential needs or aspirations consumers seek to satisfy through PCPs, such as cleanliness, beauty and relaxation, it is not likely to succeed on the market. To determine the most significant targets for sustainable change within businesses, the subsequent section will analyse the types of harm associated with current PCP use. This is important because the criteria currently used by personal care companies to define and measure harm have excluded significant variables, most notably plastic packaging and toxic chemicals.

The following section will examine how the adoption of new models for designing, manufacturing and retailing personal care by luxury businesses could have the potential to improve the sustainability of the personal care industry more broadly. There are two reasons for this: first, as public awareness of the negative impacts of PCPs has increased, demand has increased for products that exclude potentially harmful ingredients (Transparency Market ResearchTMR 2015). This has prompted more consumers to purchase luxury products, which are perceived to be of greater quality and more trustworthy (Bendell and Kleanthous 2007). Second, luxury brands play a significant role in establishing the aspirations and standards of other market sectors, from premium to budget (Kapferer 2010). This means that the sustainable practices adopted by luxury companies could catalyze change across the industry.

To demonstrate how this may be achieved, the final section will examine the service design model employed by luxury personal care brand LUSH. The LUSH model offers a new way of delivering personal care that has broad appeal, does not compromise on luxury and does not require synthetic plastics and chemical additives. Notable elements of their approach include the use of 'naked', or package-free, products, 'delicatessen' style retailing, and customer connectivity with producers and product manufacturers along the supply chain.

2 Why We Use Personal Care Products

Consumer motivations for buying personal care products are intimately linked to contemporary cultural definitions of beauty and cleanliness, and the personal hygiene practices we perform to achieve them. The way hygiene and beauty are measured and understood changes over time, necessitating different products and routines. Hygiene is defined as the "conditions or practices conducive to maintaining health and preventing disease, especially through cleanliness" (OED 2016). However, the modern practices that have evolved to achieve cleanliness and prevent disease are actually contributing to the rise of certain diseases and the contamination and depletion of the ecosystems upon which we depend.

Historically, ideas about how the body becomes infected by disease, and the associated programs of hygienic reform, have transformed the meaning of cleanliness and how it should be achieved. For example, in sixteenth century Europe it was widely believed that the body was porous and able to be penetrated by heat and water. As such, bathing was seen as important to restore perceived imbalances in bodily 'humours' (Shove 2003: 86). This transmuted into a belief that opening the pores to water and the air could let in various maladies and allow vital substances to seep out. As a result, bathing came to be socially frowned upon. Failure to conform to contemporary hygiene norms, even if they contradict those that have recently preceded them, consistently results in individuals being labelled as disgusting and morally reprehensible (Smith 2007). Conversely, if the prescribed hygiene practices are performed well, they can actively contribute to what Bourdieu refers to as 'beauty capital', benefiting the individual socially and professionally (Ross-Smith and Huppatz 2010; Shove 2003).

In addition to disease prevention, insurance against social exclusion, and a desire to enhance one's 'beauty capital', PCPs are now expected to nourish and restore the mind and soul. Although rationales for hygiene and personal care rituals are still well grounded in discourses of germ eradication and aesthetic standards, they have been overlaid by narratives espousing the benefits of bathing for relaxation and stress relief (Shove 2003: 105). PCPs, particularly in the luxury category, offer to reconnect urban consumers to nature through the use of scent, often invoking forests, fruit, flowers and beaches. The role of personal care rituals in replenishing the mind and spirit is supported by the function of the modern bathroom as one of the few places many people can be truly alone. Bathrooms are often the only room in modern houses that have a lock (Penner 2013). In this sense, the bathroom is both a sanctuary in which products are administered for relaxation and a place where the body is cleansed and prepared according to social expectations.

The multifunctional role that PCPs play in peoples' lives means that sustainable reform programs based simply on reducing the number of products currently consumed is unlikely to produce the required reduction in harmful activity. The social and personal consequences would be too great. The role of PCPs in enabling people to simultaneously meet social standards of cleanliness and appearance, while allowing them to feel relaxed and pampered cannot be ignored. Rather, the challenge will be to create, and make available, an alternative set of products that cater to the diverse functions served by current PCPs without the associated harm to humans and ecosystems.

3 Forms of Harm

To adequately assess how luxury personal care companies can become more sustainable, the nature and extent of the harm they cause must be adequately characterised. To do this, the following section examines the scientific literature that addresses the negative impacts of PCPs on the environment and people. This examination will reveal the types of change necessary for the luxury personal care industry to become more environmentally sustainable and safe and why the majority of current interventions are insufficient.

3.1 Ecological Impacts

The impact of PCP production, use and disposal on global ecosystems alters the course of numerous environmental processes, some of which cause explicit harm, while others are more challenging to quantify and trace. Two forms of harm will be discussed here: harm linked to product packaging made from plastic polymers, and harm associated with the chemicals that are added to plastics and the products they contain. These are not the only aspects of the PCP industry that cause damage to ecosystems and communities, there are also concerns relating to resource use and greenhouse gas emissions produced by globalised PCP supply chains (Sayer et al. 2012; Pereira de Carvalho and Barbieri 2012). These issues are already well documented in other papers (Sayer et al. 2012; Tan et al. 2009). The impact of plastic and chemical pollution from PCPs on ecosystems and human health has been largely neglected in existing attempts to develop sustainable PCPs. The following paragraphs will demonstrate why these issues must be addressed for a genuinely sustainable PCP industry to exist.

The plastic packaging used in the majority of hygiene products is cited in the literature as the product component likely to pose the greatest long-term risk to global ecosystems (Liboiron 2015: 9; Gabrys 2013). There are now numerous scientific studies demonstrating that plastic polymers from product packaging are responsible for choking and entangling animals (Moore 2013) and limiting oxygen transfer in soils (Barnes et al. 2009). As plastic enters waterways and marine environments it breaks down into microplastics (5 mm or less), which are ingested by marine organisms such as fish and birds, and even animals as small as mussels and plankton (Avio et al. 2015; Browne et al. 2008; Cole et al. 2013).

In addition to plastic packaging, many PCPs contain synthetic plastic polymers in their formulas in the form of microbeads. The 2015 United Nations Environment Program (UNEP) report *Plastic in Cosmetics: Are we polluting the environment through our personal care?,* noted that plastic ingredients are included in a range of product types, including: "deodorant, shampoo, conditioner, shower gel, lipstick, hair colouring, shaving cream, sunscreen, insect repellent, anti-wrinkle creams, moisturizers, hair spray, facial masks, baby care products, eye shadow (and) mascara" (Leslie 2015). Plastic ingredients from these products regularly enter the environment from urban areas because they do not decompose in wastewater treatment systems and are released into waterways via raw sewage, treated effluents, and as sludge applied as fertiliser to agricultural land.

Research has also found that chemicals that are added to, or accumulate in, plastic as it moves through the environment, leach out into animal bodies when ingested (Rochman et al. 2013). This is problematic because most plastic chemical additives, or plasticizers, are classified as endocrine disrupting chemicals (EDC). The endocrine disrupting effects of these plasticizers have been observed in mammals, birds, reptiles, fish, and molluscs in both wild populations (Vos et al. 2000) and in laboratory settings (Oehlmann et al. 2000). A review conducted by Vos et al. (2000) notes that the causal links between impaired reproduction and EDCs are now well documented for a number of species, resulting in local or regional population changes. Examples include: severe population declines in certain raptor species in Europe and North America due to egg-shell thinning and impaired reproduction and immune function in Baltic grey and ringed seals.

Despite the existence of strong correlative and laboratory evidence of harm caused by EDCs, the literature highlights that regulation and remediation are challenging for a number of reasons. First, the precise causative pathways through which organisms are affected by plastics and their EDC additives are difficult to trace and quantify. A key reason for this is that EDCs can often have stronger effects at low doses (Vandenberg et al. 2012). Second, plastics are ubiquitous, which makes it difficult to regulate and monitor the distribution of harmful EDCs (Bushnik et al. 2010; Bergman et al. 2013; Liboiron 2015). As a result, it is no longer possible to establish uncontaminated control groups for experimental scientific research. Evidence of impact, as measured against such control groups, is required within the current regulatory paradigms of most countries to definitively establish that a substance is causing harm and act on it. Consequently, the World Health Organization (WHO) and the UNEP, claim that approximately 800 chemicals that are known or suspected to be endocrine disruptors are still produced and circulated in everyday consumer goods, including PCPs (UNEP/WHO 2013).

In addition to the chemicals used in plastic packaging, a number of studies have found that chemicals used in the PCPs themselves, include EDCs (Wolff et al. 2010; Ferrer et al. 2012; Routledge et al. 1998; Stoker et al. 2010; Clayton et al. 2011). Other chemicals in hygiene products have been linked to different forms of harm, such as carcinogenic formaldehyde, and quaternium-15 and DMDM hydantoin, the 'donor' preservatives that enable it to be released (Kessler 2015; Health and Services 2014).

Many of these problematic chemicals are washed down drains with products such as body wash and shampoo. There is research to suggest that they could potentially be adversely affecting the chemistry of marine ecosystems, particularly if they are picked up and carried by plastic polymers (Leslie 2015). A 2015 study by Carbajo et al. examined the aquatic toxicity of eight preservatives frequently used in PCPs, finding that they all showed considerable toxicity in the studied biological assessments, with differing levels of potency (Carbajo et al. 2015). In addition to ecological impacts, PCP chemicals can have an impact on humans through the ingestion of effected marine products or direct application to the body (Diamanti-Kandarakis et al. 2009). The following section will briefly examine what is known about the impact of chemicals in PCPs on human health.

3.2 Health Impacts on Humans

An increasing body of research suggests that chemicals in many hygiene products can have endocrine disrupting effects in humans via direct use. There is strong correlative and laboratory evidence to suggest that EDCs are involved in a number of health conditions, including feminization of male foetuses, early-onset puberty and menopause and senility, obesity, diabetes, stunted brain development, miscarriage, and cancer, among others (Grün and Blumberg 2009: 20; Reuben 2010: 40; Bergman et al. 2013; Di Renzo et al. 2015). In Nordic countries, research suggests that the adverse effects on male reproductive systems from regular exposure to EDCs, including the use of hygiene products, could be costing up to €36 million for each year of exposure (Olsson 2014: 55). However, assessing harm to humans is beset with the same issues faced when establishing harm in ecological settings. In both cases there is difficulty in determining linear, clear, repeatable causation pathways between a chemical and a specific form of harm (Liboiron 2015). For this reason, the harm caused by EDCs is commonly compared with that caused by smoking: both are strongly correlated with particular health effects, but neither can be shown to directly *cause* harm (Pasqualotto et al. 2004; Liboiron 2015: 3).

Despite the strong correlative evidence that EDCs are involved in promoting certain diseases, the experimental conditions required to prove this, and support decisive action, cannot be obtained (Liboiron 2015). The issue is not lack of correlative evidence between EDCs and health effects, but that the evidence does not provide a definitive picture of which chemicals produce which effects and at what levels. Endocrine disruption can manifest differently in different bodies depending on the hormones and receptors affected and the age of the person (Huang et al. 2015). Young children, pregnant women and foetuses are more susceptible to certain manifestations of endocrine disruption (Diamanti-Kandarakis et al. 2009; Reuben 2010). In addition, there are no test bodies against which exposure effects can be measured, as all bodies already have active hormones, and all bodies tested globally contain traces of synthetic EDCs (Meeker et al. 2009; Bushnik et al. 2010). This presents a problem for regulation and other forms of action to mitigate potential harm.

One of the most important considerations highlighted in the literature is that when assessing the potential harm of hygiene products no chemical product can be examined in isolation, as aggregate exposures, or what is commonly termed 'the cocktail effect', influences how chemicals interact and behave in bodies (Loretz et al. 2008; Meeker et al. 2009). Consider how many PCPs the prototypical modern woman in an industrialised country may use each day. She might begin her day with a shower using exfoliating body wash, shampoo, conditioner, and foaming face cleanser, to be followed immediately by the application of moisturiser, make-up, hairspray and deodorant. Preservatives and fragrances commonly used in all of these product types have been identified as EDCs (Biesterbos et al. 2013; Kessler 2015). The effects of EDCs in PCPs are also influenced by EDCs used in other everyday products, for example water bottles and food tins containing BPA. Gosens et al. (2014) argue that the aggregation of exposure to a substance from different sources via different pathways is not adequately addressed in common risk assessments of chemical substances, leading to a significant underestimation of risk.

Many of the chemicals in consumer products that have been tested and banned are replaced by other untested chemicals that behave in the same way, and are often subsequently also banned (Lakind and Birnbaum 2010; Blum 2016). Johnson & Johnson are currently in the process of reformulating hundreds of their products to find substitutions for a number of ingredients that have attracted safety concerns (Kessler 2015). Because the company wishes to retain the colour, scent and texture of many of these products they are seeking chemicals that behave in the same way as the old ones. However, one of their research development managers acknowledged that tinkering with the formula of a product often has unintended consequences, and that fixing one problem often creates another (Kessler 2015). This process of substituting one harmful chemical with another that behaves in a near identical way has been likened to a game of "whack a mole" by the former commissioner of the US Food and Drug Administration (FDA) (Kennedy 2007).

In response to the enduring problems associated with chemical substitutions in consumer products, scientists are increasingly advocating a ban on entire classes of chemicals. For example, the Green Science Policy Institute, is calling for a 50% reduction over the next five years in the use of six families of chemicals used in consumer products: highly fluorinated chemicals, flame retardants, organic solvents, certain metals, antimicrobials, bisphenols, and phthalates (GSPI 2016). The latter three of these classes are commonly used in PCPs. A number of scientists, including The Endocrine Society, are also calling for the Precautionary Principle to be instituted more rigorously in business practices, which would require industry, and society more broadly, to ask if we need particular chemicals, given their potential for harm (Zoeller et al. 2012; Diamanti-Kandarakis et al. 2009).

Even if problematic classes of chemicals were banned, the way that PCPs are manufactured and used must be fundamentally redesigned if equally damaging substitutions are to be avoided. The dominant structure of PCP supply chains at present necessitates the use of strong yet flexible packaging to ensure easy product transport and use, and preservatives to prolong shelf life (Kessler 2015). Liquid products generally require more preservative than solid or dry ones to prevent microbial growth (LUSH 2016b). If the dominant supply chain model is to be maintained, it is likely that the problem of substitution will continue, as its functionality depends on products continuing to behave in the same way. The following section will address public and PCP industry responses to these concerns and why the luxury sector is important for broader change across the PCP industry.

4 How Luxury Can Improve the Sustainability of the Personal Care Industry

The luxury sector of the personal care industry has the capacity to affect positive change beyond the regular luxury consumer market. Consumers who have not traditionally participated in the luxury market are increasingly turning to luxury PCPs because they are perceived to be more trustworthy, unadulterated, pure and concerned with quality than other product categories. The demand for luxury products, particularly those with 'organic' or 'natural' attributes, has increased as public awareness about the health and environmental impacts of chemicals in PCPs has improved. According to a market report published by Transparency Market Research, the demand for organic PCPs globally is expected to experience a growth rate of 9.6% from 2012 to 2018 (TMR 2015). In a comprehensive survey reported by Statista the most commonly reported reason for purchasing luxury PCPs was "I believe luxury personal care products are better for my body" (Statista 2012). This trend towards purchasing luxury PCPs for health reasons is represented in numerous places around the world (Bendell and Kleanthous 2007).

Consumers are also turning to the luxury sector because it is increasingly perceived as an avenue to express one's deepest values, including care for the environment. According to recent WWF report *Deeper Luxury* on the sustainability performance of luxury brands, one's capacity to purchase products that reflect personal values and aspirations for a better world are now key defining features of personal success (Bendell and Kleanthous 2007). They report that this is not only the case for Western luxury consumers but also the wealthy and middle classes of Latin America, Asia and Eastern Europe. It is therefore logical to infer that as luxury brands increasingly position themselves as avenues through which one can express ethical values, consumers would view luxury PCPs as a viable solution to their sustainability and health concerns. The increasing market share of luxury personal care brands, due to their perceived health and ethical benefits, has afforded the luxury sector a greater capacity to reduce the harmful impacts of personal care practices.

The reach of luxury brands has also been extended with the rise of 'new luxury'. Luxury brands are broadening their product ranges with 'masstige' (prestige for the masses) products that are more accessible to middle-class consumers (Truong et al. 2009; Kapferer and Bastien 2009). This masstige approach is now one of the major growth strategies for luxury brands (Truong et al. 2009). While this strategy has increased the influence of luxury brands, it has made it harder to appeal to consumers on the basis of exclusivity. Rather, value is more often being added through the improvement of environmental and social performance (Bendell and Kleanthous 2007). This confluence of broad appeal and a greater concern for the values that brands represent, mean that the luxury sector of the PCP market is well positioned to lead change.

The increasing popularity of luxury PCPs and their association with sustainable values potentially provide important opportunities to reduce the harmful impacts of PCP production and use. However, the types of reforms that are currently being made to create more sustainable and healthy brands have largely been tokenistic or misguided. The following section examines the ways in which PCP brands have been reforming their products and business models to become more safe and sustainable, and why these attempts have not been sufficient.

5 Addressing Ineffective Sustainability Strategies

In recognition of the increased demand for natural products, many luxury and other PCP brands have changed aspects of their production, branding and retailing strategies. The methods employed by companies to increase the 'green' credentials of their brands can be loosely grouped into five broad strategies: first, the substitution of chemical preservatives or fragrances that are thought to be harmful, with others that serve the same function; second, the addition of organic ingredients; third, the use of post-consumer recycled packaging or 'eco-packaging'; fourth, marketing the 'natural' attributes of products; and fifth, supporting charities or initiatives that align with the values they aspire to represent.

The removal of particular chemicals that have attracted public disapproval is one of the key ways companies attempt to improve their health and sustainability credentials. For example, Avon have reportedly vowed to stop using the anti-bacterial chemical triclosan, while Johnson & Johnson have vowed to eliminate both triclosan and Diethyl phthalate (DEP) from all products (Kessler 2015). It is now common to see brands labeled as paraben-free, such as Jurlique and Clinique, or exclude the chemicals from some of their products, such as Estee Lauder, Lancôme and Elizabeth Arden, as part of their natural marketing strategies.

The addition of organic ingredients is another common strategy used. By sourcing some organic ingredients companies are able to use the term in their marketing. A number of brands, including L'Occitane, Burt's Bees, Aveda and Dr. Hauschka have glossaries for their natural or organic ingredients on their websites, but omit details about their synthetic ingredients. The terms 'natural' and 'organic' have no legal definitions, so can be used in marketing without adhering to specific criteria. Some companies seek to add credibility to their brand by joining labelling schemes that require them to meet certification criteria (Hartlieb and Jones 2009). This provides more certainty for the consumer that companies can be held accountable to their claims. However, many of these schemes only relate to one aspect of sustainability, such as the organic production of raw materials or the use of sustainable palm oil (Alves 2009; Teoh 2010). This, plus the lax certification criteria of some schemes, means that the inclusion of sustainability certifications on labels can be deceptive to consumers (Alves 2009: 6).

A number of brands also make appeals to the concept of recycling to improve their sustainability credentials. Burt's Bees, L'Occitane, and Estee Lauder, among many others, advertise that at least a portion of their products are sold in post-consumer recycled plastic. The first problem with recycling is that people often do not recycle recyclable products. The second is that the process of recycling is still very resource intensive: it consumes energy, requires virgin materials, and creates pollution (Liboiron 2009). At best it can marginally minimise the impact of a product, and at worst, it can encourage the use of disposables even further. The notion that a product is recyclable, or made from recycled materials, gives the impression that the product causes less environmental damage. This allows our unsustainable reliance on disposable products to continue without acknowledging the high energy and material costs still associated with the routine use of recycled, or recyclable, single-use products. Plastics are particularly inefficient to recycle as there are many types of plastic, each of which has a different melting point, set of plasticizing chemicals, and density (MacBride 2011). This makes recycled plastics harder to use as raw stock material. As a result, the recycling of recovered plastic cannot be considered a sustainable solution to plastic waste.

The terms ‘Sustainable’ and ‘Eco-packaging’ also represent a form of ‘green washing’. Forecasters have predicted that the sustainable packaging sector will grow faster than the packaging industry overall (Matusow 2010). They also predict that the fastest-growing segment of the sustainable packaging sector will be plastic-based packaging. Davide Nicosia of design firm NiCE Ltd. promoted the fact that they had developed a way to reduce the amount of plastic in a shampoo bottle by 13%, as a way to achieve a more sustainable product (Matusow 2010). The remaining 87% plastic is still an unsustainable amount of plastic for a disposable product to contain, and therefore should not be considered ‘Sustainable’ packaging.

Companies use green marketing strategies not only to promote or exaggerate natural elements of their products, but to grow the market demand for products with these attributes. Since a number of large companies have invested in natural product ranges, or acquired brands situated in the natural and organic sector, it is in their interest to further promote ‘natural’, ‘organic’, ‘sustainable’ and ‘toxic-free’ qualities as key attributes that consumers should consider when purchasing PCPs (TMR 2015; GVW 2015). Examples of companies that have sought to increase their presence in the natural product market, and thereby expand it, include L’Oreal, which acquired Kiehl’s in 2000 and The Body Shop in 2013, and Clorox which acquired Burt’s Bees in 2007 (GVW 2015).

Another common strategy, employed by luxury personal care companies in particular, is to associate the brand with a social or environmental cause that is unrelated to the operation of the business. An exemplary case is *Crème de la Mer* by Le Mer, an Estee Lauder brand, which donates a portion of sales revenue to National Geographic to promote ocean conservation. However, *Crème de la Mer* is packaged in small jars with plastic lids. The product is therefore contributing to the production and distribution of disposable plastic packaging, one the greatest threats to the world’s oceans. This type of strategy is not necessarily harmful, but it has the capacity to be if the brand using it is undermining the values it is espousing.

The identification of these strategies does not suggest that personal care brands are not genuinely committed to making their products more safe and sustainable. The ethos of many luxury brands such as L’Occitane and Aveda have resulted in them making changes that are less tokenistic, and have positive impacts on the environments and communities from which ingredients are sourced. However, their continued use of plastic packaging precludes their products from being classified as genuinely sustainable and safe.

The substitution of problematic ingredients, the sustainable sourcing of some materials, and the use of recycled or recyclable materials are not sufficient interventions to make a product sustainable. To understand how more meaningful change may be created, the following section will examine how an intervention at the service design level might provide opportunities for meaningful change to occur.

6 Intervening in Personal Care Practices

Current expectations about what products should be able to do, and the ways collective personal care routines are performed, necessitate the use of plastic containers and chemical additives. As certain products enter our daily lives, become normal and alter the way we conduct certain activities, they shape our expectations around what it means to do those activities well (Shove 2003: 79–91). For example, the introduction of liquid soap and body wash in plastic pump-packs and squeeze bottles is replacing the use of bar soaps in daily washing practices around the world (Mintel 2015). The introduction of this new technology has influenced what we expect from washing, via the marketing of these products and how we have come to use them. Liquid soaps are touted by marketers to be more hygienic than soap bars, as they are contained in sealed, impenetrable packages and cannot accumulate dirt and other particles. They also do not leave a sticky residue where they are stored. Additives to liquid soap and body wash also influence our expectations about what a body-cleansing product should accomplish. They will often include exfoliating microbeads, vitamins, anti-bacterials, anti-perspirants, moisturisers, fragrances that evoke nature, and foaming surfactants to make users feel like they are cleaning more effectively. The normalisation of multi-functional body care products has created new expectations around the functions that products should fulfil and ultimately what it means to be clean (Shove 2003: 79–91). To create more sustainable PCPs, our routines and the functions that PCPs perform within them, must be reassessed.

The majority of PCP brands proclaiming to be sustainable are intended to integrate into current routines are consequently unable to challenge and resist some of the most unsustainable elements of personal care practices. Influential systems theorist Donella Meadows proposed a hierarchy of effective points to intervene in a system. According to this hierarchy, interventions involving chemical substitutions and the addition of organic ingredients could be placed at the bottom, as they involve tinkering with numbers, or 'diddling with details', but do not affect the positive feedback loops that reinforce unsustainable behaviours (Meadows and Wright 2008: 148). What is required, then, are products that encourage people to practice personal care differently, but in a way that is attractive and viable for individuals currently engaged in unsustainable consumption practices.

7 LUSH: Fresh, Handmade Cosmetics

The British luxury cosmetics company LUSH have pioneered a new service model for the manufacturing and retailing of PCPs. LUSH are a private company founded in Dorset, England in 1994 that exclusively produces handmade PCPs (Buchwalter 2007). They currently have over 900 stores globally in 49 countries (LUSH 2016a). LUSH occupy a unique place in the luxury personal care market, as they appeal to traditional values of superior quality, beauty, bespoke products, and indulgence, they are also known for their social and environmental activism.

The reason LUSH has been selected as a case study, rather than a smaller luxury company connected to a single location, is specifically because they have demonstrated their global appeal and the capacity to influence trends in multiple cultural contexts. This means that they, or other luxury brands following this model, could have a greater potential to effect change in collective practices and conventions.

It must be noted that LUSH still engages in some of the practices that have been identified as concerning throughout this chapter. Despite selling 46% of their products without packaging, they still use containers for the remainder (LUSH 2016e). Like other brands, they have attempted to reduce their impact by only providing packages made from post-consumer recycled plastic, recyclable or biodegradable materials, and incentives are provided to return recyclable packages to stores for recycling. For example, LUSH Australia states that customers can bring back 5 empty black pots to receive a free 'Fresh Face Mask' (LUSH 2016e). They also continue to use synthetic preservatives, fragrances and colours in a number of products. This is because they have made the decision to continue selling product types, such as liquid gels, moisturisers, perfume and make-up that require these ingredients to function as intended. However, LUSH's innovative service model is not undermined by the continued sale of these products, because it is not dependent on them, and could remove them if desired.

Concerning products notwithstanding, this chapter proposes that there are four notable elements of LUSH's service design which could provide a basis for luxury personal care companies to become genuinely environmentally sustainable and safe without compromising on luxury and ease of use.

7.1 *Service Elements*

The first important element of LUSH's Service Design is the use of packaging-free, or 'naked', products. Nearly half of LUSH's products have no packaging. This amounts to over 100 products including soaps, bath bombs, shampoo bars, bubble bars, massage bars, body butters and solid facial cleansers. According to LUSH's Fresh Thinking catalogue, their first shampoo bar weighed 55 g, was one fifth the size of a 250 g bottle of shampoo but lasted three times as long. These bars not only omit plastic, but often do not require anti-microbial preservatives because they do not contain water. LUSH have since increased their packaging-free range and in 2007 launched a global campaign called Get Naked. This involved employees removing as much clothing as they dared in protest against extra packaging. This promotional activity was reinforced through a mini-website called The Naked Truth, which provided information on the harms associated with packaging and offered a "buy two, get one free" deal for all products without packaging. The use of this strategy demonstrates that 'naked' products are perceived as something that could actually improve the popularity of their brand, rather than detract from it.

Far from being a disincentive, some research suggests that products with less packaging actually reinforce consumer trust. According to research conducted by Di and Arbajian (2014) the exclusion of packaging enables people to conduct a sensory evaluation of a product rather than having to rely on the claims made on its packaging. The sensory recognition of products is being increasingly promoted as an important component of brand recognition and success (Lindstrom 2005). The ability to experience a product before buying offered by packaging-free retailing means the removal of packaging is not necessarily a concession to sustainability, but an asset.

The second important element of LUSH's service design model is their use of an attractive 'delicatessen style' retailing environment. The interactive sensory experience offered by naked products is amplified at the store level. Stores are designed to have different sections that invite customers to experience products in various ways. Unwrapped slabs of colourful soaps are stacked in one area, ready for sales staff to cut off custom-sized chunks on a wooden board, which are then wrapped in greaseproof paper. In another corner there may be a bar of fresh skin care products where facemasks can be ladled out of pots. When LUSH began, its target market was young, however the 'delicatessen' approach appeals to a wider market and now their fastest growing sector is the 35–50 age group (Buchwalter 2007). This is potentially because the 'deli' approach allows customers to have products custom-sized and conduct a thorough sensory evaluation of the products before selecting which one they want, much like choosing the nicest looking apple at the grocer. This element of choice holds significant appeal and potential to expand retail models that do not rely on packaging.

Third, LUSH have a relatively transparent approach to ingredient sourcing and customer relations. Unlike a number of the other companies discussed in this chapter, LUSH publishes their full ingredient list online, and attempts to justify their use (LUSH 2016d). This is not to suggest that all elements of their business practice are transparent, only to illustrate their transparency is superior to other comparable brands. With respect to their strong ethical stance against animal testing, they not only refuse to buy raw products from companies that test on animals, but have actively helped raw material suppliers convert to no-animal testing policies throughout their business (Buchwalter 2007). This approach of actively lobbying producers to change practices that contradict company values should be an integral aspect of an ethical business. LUSH also allow themselves to be held accountable by the public if they fail to meet their ethical, or any other, standards. This is done through the use of online forums such as the LUSH North American online forum, which has a section titled Rants and Raves, under which customers can express their thoughts on any aspect of the brand without it being censored (Buchwalter 2007).

Fourth, LUSH involves customers in the production of their products. There are two notable ways this is done. The first involves the identification of the person that handmade the product. The majority of LUSH products are handmade within relatively close proximity to the point of sale, partially because of the preservation requirements of some products. Every item has a sticker on it with an image of the person who made it, the date it was made and when it will expire. According to Buchwalter (2007), customers get to know the products made by certain makers and even become fans of particular producers. The second is a newer initiative called LUSH Kitchen which provides customers with an insight into the production process via short videos regularly uploaded online, and even allows them to order the exact products they see being created (LUSH 2016c). This is part of a digital strategy, created in collaboration with Method, that won a User Experience Award in 2014 (UXAwards 2014). Both of these methods serve to reconnect customers with producers in innovative ways that do not depend on enterprises being small scale and locked into a single location. By engaging individuals with the contents of their products and how they are made, they become more aware and better equipped to scrutinise such processes.

7.2 Further Steps

There are crucial areas for improvement that must be addressed for the LUSH service design model to function in a comprehensively sustainable and safe way.

The first is the discontinuation of liquid products that require plastic packaging and synthetic preservatives. The continued use of these products not only has direct environmental impacts, but perpetuates personal care practices that necessitate their use.

The second consideration is the local sourcing of ingredients. Although much of the product production takes place in close proximity to stores, raw ingredients are still sourced globally. This not only has an impact in terms of resources and greenhouse gas emissions associated with transport, but in terms of the packaging required for transporting goods long distances. Companies could therefore emulate the model proposed but focus on developing formulas that are based on ingredients that can be sourced or grown local to the stores.

The final point that must be considered is the accessibility and convenience of the products. In order to have the greatest impact, products should not only be available to the wealthy. This means that product prices should be less than standard luxury products, without compromising on luxury attributes, as with other brands that fit the new 'democratised' luxury model.

7.3 Management Implications

The way that businesses redesign their supply chains, retailing strategies and products to meet the basic criteria for sustainability specified in this chapter can be successfully conducted in a number of ways. The LUSH model does, however, suggest some valuable ways that particular management structures and strategies could be used. LUSH have noted that their flat organisational structure has contributed to their capacity to be responsive to local needs while ensuring that decisions reflect the core ideals and values of the head management team (LUSH 2016a). Although a close connection is maintained between retail employees and the guiding values of company directors, LUSH has allowed the makers of their handmade products a considerable degree autonomy over how they are made. This contributes to a sense of bespoke product creation and fosters customer connection and loyalty to individual makers.

8 Conclusions

Plastics and manufactured chemicals afford luxury PCPs the temporal qualities and multi-functionality needed for individuals to conveniently clean, adorn, pamper and relax themselves according to current cultural norms and expectations. These qualities also determine the post-consumer life of these products as they breakdown in environments and bodies, and integrate into these natural systems in complex ways. Their biological effects can be seen over generations, and their ecological effects over centuries.

The way that sustainability and harm are understood and measured influences where the boundaries around the problem are drawn and what is defined as a matter of concern. If companies were to acknowledge that the continued use of plastics and particular classes of chemicals are crucial elements of environmental and human health problems, they must also acknowledge that change cannot simply be achieved by substituting certain ingredients. Change must involve a restructuring of the dominant ways of manufacturing, transporting, retailing and using PCPs in everyday life.

The luxury ethos provides a promising basis for an intervention to reverse the positive feedback loops that reinforce unsustainable behaviours. The luxury sector has the power to and to transform the values that underpin how and why personal care practices are performed. If this new model is to be sustainable, it must also be widely accessible. Changes made in businesses that only a few have access to will minimise damage, but cannot hope to alter collective personal care practices at the scale required. The increasing appeal of luxury across a broader market and the tendency for practices in the luxury sector to flow into other sectors makes luxury businesses a promising intervention point for change across the industry.

The version of the LUSH service design model presented in this chapter offers a new way of provisioning personal care that has global reach, yet does not necessitate long-lived synthetic plastics and their chemical additives. The use of 'naked' products, sensory engagement, 'delicatessen' style retailing, and connectivity with raw ingredient and product producers, enhance product appeal, consumer trust and connectivity along the supply chain. The transformative capacity of this model depends on it being implemented in a way that maintains a commitment to sustainability, while recognising the pressure on individuals to conform to particular hygiene and beauty norms, and the time pressures that govern personal care routines within modern life.

Luxury personal care brands have an important opportunity to set new sustainability standards and practices for the personal care industry and consumers. This chapter has provided insight into the shortcomings of current attempts to implement sustainable business models and establish the basic criteria required for any personal care business to be considered sustainable. Looking to the future of luxury personal care, the best way for each business to meet sustainability criteria and implement a new service design model will differ depending on their current supply chain structure, organisational structure, scale and customer base. Further research must therefore be undertaken by each business to ensure the service design model they adopt is appropriate for the operational context, without compromising on essential sustainability requirements.

References

Alves I (2009) Green spin everywhere: how greenwashing reveals the limits of the CSR paradigm. J Glob Change Governance 2(1):1–26

Avio CG, Gorbi S, Milan M, Benedetti M, Fattorini D, d'Errico G, Pauletto M, Bargelloni L, Regoli F (2015) Pollutants bioavailability and toxicological risk from microplastics to marine mussels. Environ Pollut 198:211–222

Barnes DK, Galgani F, Thompson RC, Barlaz M (2009) Accumulation and fragmentation of plastic debris in global environments. Philos Trans R Soc Lond B: Biol Sci 364(1526):1985–1998

Bendell J, Kleanthous A (2007) Deeper luxury: quality and style when the world matters. WWF, UK

Bergman Å, Heindel JJ, Jobling S, Kidd K, Zoeller TR, Organization WH (2013) State of the science of endocrine disrupting chemicals 2012: summary for decision-makers. World Health Organization, Geneva

Biesterbos JW, Dudzina T, Delmaar CJ, Bakker MI, Russel FG, von Goetz N, Scheepers PT, Roeleveld N (2013) Usage patterns of personal care products: important factors for exposure assessment. Food Chem Toxicol 55:8–17

Blomkvist J, Holmlid S, Segelström F (2010) This is service design research: yesterday, today and tomorrow

Blum A (2016) Tackling toxics. Science 351(6278):1117

Browne MA, Dissanayake A, Galloway TS, Lowe DM, Thompson RC (2008) Ingested microscopic plastic translocates to the circulatory system of the mussel, mytilus edulis (L.). Environ Sci Technol 42(13):5026–5031

Buchwalter C (2007) Living a LUSH life. Available: http://www.carriebuchwalter.com/uploads/2/7/6/6/2766408/cb_lush_paper.pdf

Bushnik T, Haines D, Levallois P, Levesque J, Van Oostdam J, Viau C (2010) Lead and bisphenol A concentrations in the Canadian population. Health Rep 21(3):7

Carbajo JB, Perdigón-Melón JA, Petre AL, Rosal R, Letón P, García-Calvo E (2015) Personal care product preservatives: risk assessment and mixture toxicities with an industrial wastewater. Water Res 72:174–185

Clayton EMR, Todd M, Dowd JB, Aiello AE (2011) The impact of bisphenol A and triclosan on immune parameters in the US population, NHANES 2003–2006. Environ Health Perspect 119 (3):390

Cole M, Lindeque P, Fileman E, Halsband C, Goodhead R, Moger J, Galloway TS (2013) Microplastic ingestion by zooplankton. Environ Sci Technol 47(12):6646–6655

Di Y, Arbajian P (2014) Understanding alternative choices of handmade cosmetics in the postmodern consumer society

Di Renzo GC, Conry JA, Blake J, DeFrancesco MS, DeNicola N, Martin JN, McCue KA, Richmond D, Shah A, Sutton P (2015) International federation of gynecology and obstetrics opinion on reproductive health impacts of exposure to toxic environmental chemicals. Int J Gynecol Obstet 131(3):219–225

Diamanti-Kandarakis E, Bourguignon J-P, Giudice LC, Hauser R, Prins GS, Soto AM, Zoeller RT, Gore AC (2009) Endocrine-disrupting chemicals: an endocrine society scientific statement. Endocr Rev 30(4):293–342

Ferrer A, Hidalgo C, Kaps R, Kougoulis J (2012) Revision of European ecolabel criteria for soaps, shampoos and hair conditioners. Market analysis for the joint research center of the European Union & the institute for prospective technological studies

Gabrys J (2013) 12 Plastic and the work of the biodegradable. Accum: Mater Polit plast:208

Gosens I, Delmaar CJ, ter Burg W, de Heer C, Schuur AG (2014) Aggregate exposure approaches for parabens in personal care products: a case assessment for children between 0 and 3 years old. J Eposure Sci Environ Epidemiol 24(2):208–214

Grün F, Blumberg B (2009) Endocrine disrupters as obesogens. Mol Cell Endocrinol 304(1):19–29

GSPI (2016) Six classes. http://greensciencepolicy.org/topics/six-classes/. Accessed 20 Feb 2016

GVW (2015) Organic personal care market worth $15.98 billion by 2020. http://www.grandviewresearch.com/press-release/global-organic-personal-care-market. Accessed 24 Jan 2016

Hartlieb S, Jones B (2009) Humanising business through ethical labelling: progress and paradoxes in the UK. J Bus Ethics 88(3):583–600

Health UDo, Services H (2014) 13th report on carcinogens. Polycyclic aromatic hydrocarbons—national toxicology program http://ntp.niehs.nih.gov/ntp/roc/content/profiles/polycyclicaromatichydrocarbons.pdf

Huang H-B, Chen H-Y, Su P-H, Huang P-C, Sun C-W, Wang C-J, Chen H-Y, Hsiung CA, Wang S-L (2015) Fetal and childhood exposure to phthalate diesters and cognitive function in children up to 12 years of age: Taiwanese maternal and infant cohort study. PLoS ONE 10(6):e0131910

Kapferer J-N (2010) All that glitters is not green: the challenge of sustainable luxury. Eur Bus Rev:40-45

Kapferer J-N, Bastien V (2009) The specificity of luxury management: turning marketing upside down. J Brand Manage 16(5–6):311–322

Kennedy D (2007) Toxic dilemmas. Science 318(5854):1217

Kessler R (2015) More than cosmetic changes: taking stock of personal care product safety. Environ Health Perspect 123(5):A120

Lakind JS, Birnbaum LS (2010) Out of the frying pan and out of the fire: the indispensable role of exposure science in avoiding risks from replacement chemicals. J Eposure Sci Environ Epidemiol 20(2):115–116

Leslie H (2015) Plastic in cosmetics, are we polluting the environment through our personal care?

Liboiron M (2009) Recycling as a Crisis of Meaning. eTopia

Liboiron M (2015) Redefining pollution and action: the matter of plastics. J Mater Cult:1359183515622966

Lindstrom M (2005) Broad sensory branding. J Prod Brand Manage 14(2):84–87

Loretz L, Api A, Babcock L, Barraj L, Burdick J, Cater K, Jarrett G, Mann S, Pan Y, Re T (2008) Exposure data for cosmetic products: facial cleanser, hair conditioner, and eye shadow. Food Chem Toxicol 46(5):1516–1524

LUSH (2016a) Fresh thinking: summer catalogue Australia

LUSH (2016b) Liquid soap vs. solid soap. https://www.lushusa.com/Liquid-Soap-vs.-Solid-Soap/Article_Liquid-Soap-Vs-Solid-Soap,en_US,pg.html?fid=tips-tricks. Accessed 23 Mar 2016

LUSH (2016c) LUSH kitchen. https://uk.lush.com/products/kitchen. Accessed 22 Feb 2016

LUSH (2016d) LUSH: fresh, handmade cosmetics. https://www.lush.com.au. Accessed 22 Feb 2016

LUSH (2016e) Reducing landfill. https://www.lush.com.au/shop/info/23/naked. Accessed 5 Jan 2016

MacBride S (2011) Recycling reconsidered: the present failure and future promise of environmental action in the United States. MIT Press, Cambridge

Matusow J (2010) Sustainable packaging: the beauty Industry's Perfect Storm? http://shows.beautypackaging.com/articles/2010/04/sustainable-packaging-the-beauty-industrys-perfect. Accessed 27 Jan 2016

Meadows DH, Wright D (2008) Thinking in systems: A primer. chelsea green publishing, Vermont

Meeker JD, Sathyanarayana S, Swan SH (2009) Phthalates and other additives in plastics: human exposure and associated health outcomes. Philos Trans R Soc Lond B: Biol Sci 364 (1526):2097–2113

Mintel (2015) Annual soap & shower market overview—International—2015

Moore C (2013) Rapidly increasing plastic pollution aquaculture threatens marine life. Tul Envtl LJ 27:205

OED (2016) Hygiene. http://www.oxforddictionaries.com/definition/english/hygiene. Accessed 31 Mar 2016

Oehlmann J, Schulte-Oehlmann U, Tillmann M, Markert B (2000) Effects of endocrine disruptors on prosobranch snails (Mollusca: Gastropoda) in the laboratory. Part I: bisphenol A and octylphenol as xeno-estrogens. Ecotoxicology 9(6):383–397

Olsson M (2014) The cost of inaction: a socioeconomic analysis of costs linked to effects of endocrine disrupting substances on male reproductive health. Nordic council of ministers, Copenhagen

Pasqualotto FF, Lucon AM, Sobreiro BP, Pasqualotto EB, Arap S (2004) Effects of medical therapy, alcohol, smoking, and endocrine disruptors on male infertility. Rev Hosp das Clínicas 59(6):375–382

Penner B (2013) Bathroom. Reaktion books, UK

Pereira de Carvalho A, Barbieri JC (2012) Innovation and sustainability in the supply chain of a cosmetics company: a case study. J Technol Manage Innov 7(2):144–156

Reuben S (2010) Reducing environmental cancer risk: what we can do now. President's cancer panel, Annual report 2008–2009

Rochman CM, Hoh E, Hentschel BT, Kaye S (2013) Long-term field measurement of sorption of organic contaminants to five types of plastic pellets: implications for plastic marine debris. Environ Sci Technol 47(3):1646–1654

Ross-Smith A, Huppatz K (2010) Management, women and gender capital. Gender Work Organ 17(5):547–566

Routledge EJ, Parker J, Odum J, Ashby J, Sumpter JP (1998) Some alkyl hydroxy benzoate preservatives (parabens) are estrogenic. Toxicol Appl Pharmacol 153(1):12–19

Sayer J, Ghazoul J, Nelson P, Boedhihartono AK (2012) Oil palm expansion transforms tropical landscapes and livelihoods. Glob Food Secur 1(2):114–119

Schug T, Abagyan R, Blumberg B, Collins T, Crews D, DeFur P, Dickerson S, Edwards T, Gore A, Guillette L (2013) Designing endocrine disruption out of the next generation of chemicals. Green Chem 15(1):181–198

Shove E (2003) Comfort, cleanliness and convenience: the social organization of normality, vol 810. Berg, Oxford

Smith V (2007) Clean: a history of personal hygiene and purity. OUP, Oxford

Statista (2012) Consumers' motivational factors for purchasing luxury personal care products in the United States in 2012. http://www.statista.com/statistics/251649/us-consumers-reasons-to-buy-luxury-personal-care-products/). Accessed 17 Jan 2016

Stoker TE, Gibson EK, Zorrilla LM (2010) Triclosan exposure modulates estrogen-dependent responses in the female wistar rat. Toxicol Sci:kfq180

Tan K, Lee K, Mohamed A, Bhatia S (2009) Palm oil: addressing issues and towards sustainable development. Renew Sustain Energy Rev 13(2):420–427

Teoh CH (2010) Key sustainability issues in the palm oil sector. a discussion paper for multi-stakeholders consultations (Commissioned by the world bank group)

TMR (2015) Organic personal care products market—global industry analysis, size, share, growth, trends and forecast, 2014–2020. Transparency market research

Truong Y, McColl R, Kitchen PJ (2009) New luxury brand positioning and the emergence of masstige brands. J Brand Manage 16(5):375–382

Tutty J (2016) 2016 Predictions for the luxury industry: sustainability and innovation. Positive luxury

UNEP/WHO (2013) State of the science of endocrine disrupting chemicals—2012. WHO-UNEP, Geneva

UXAwards (2014) 2014 UXies. http://userexperienceawards.com/ux-awards-2014-winners/. Accessed 7 Mar 2016

Vandenberg LN, Colborn T, Hayes TB, Heindel JJ, Jacobs DR Jr, Lee D-H, Shioda T, Soto AM, vom Saal FS, Welshons WV (2012) Hormones and endocrine-disrupting chemicals: low-dose effects and nonmonotonic dose responses. Endocr Rev 33(3):378–455

Vos JG, Dybing E, Greim HA, Ladefoged O, Lambré C, Tarazona JV, Brandt I, Vethaak AD (2000) Health effects of endocrine-disrupting chemicals on wildlife, with special reference to the European situation. Crit Rev Toxicol 30(1):71–133

Wolff MS, Teitelbaum SL, Pinney SM, Windham G, Liao L, Biro F, Kushi LH, Erdmann C, Hiatt RA, Rybak ME (2010) Investigation of relationships between urinary biomarkers of phytoestrogens, phthalates, and phenols and pubertal stages in girls. Environ Health Perspect:1039–1046

Zoeller RT, Brown T, Doan L, Gore A, Skakkebaek N, Soto A, Woodruff T, Vom Saal F (2012) Endocrine-disrupting chemicals and public health protection: a statement of principles from the endocrine society. Endocrinology 153(9):4097–4110

Part III
Sustainable Luxury Business Practices: Different Aspects

Promoting Sustainable Management: World Wildlife Fund's Hybrid Strategy to Change the Luxury Industry

Graham Bullock

Abstract Scholars generally characterize activist organizations as pursuing either "reformist" or "radical" agendas and using either collaborative or confrontational strategies to engage with firms. This paradigm oversimplifies the behavior of these organizations and ignores the possibility that they may pursue hybrid approaches that combine a range of strategies. This chapter explores how the non profit WWF used such an approach to engage the luxury industry in its 2007 *Deeper Luxury* report. The chapter analyzes the report's descriptive and injunctive normative statements that positively engage the industry and its controversial rating system that directly confronts the top ten luxury firms. The chapter also examines the sustainability-related documents of the ten rated companies, and finds that their responses to WWF's normative demands varied significantly, with L'Oreal, LVMH, and PPR demonstrating the greatest responsiveness. It concludes that this hybrid approach had mixed results in promoting sustainable management within the luxury industry, and may require collaboration among organizations with different skills and reputations.

Keywords Luxury · Sustainability · Industry norms · Nongovernmental organizations (NGOs) · Environmental assessment · Corporate social responsibility (CSR)

1 Introduction

In 2007, the United Kingdom's World Wildlife Fund office (WWF-UK) released a report entitled, "*Deeper Luxury*: Quality and Style When the World Matters." Written by two advisors to the organization and clearly displaying WWF's panda logo on its cover, this report called on luxury companies to "put sustainability

G. Bullock (✉)
Political Science and Environmental Studies, Davidson College, Box 7134, Davidson, Charlotte, NC 28035-7134, USA
e-mail: grbullock@davidson.edu

M.A. Gardetti (ed.), *Sustainable Management of Luxury*,
Environmental Footprints and Eco-design of Products and Processes,
DOI 10.1007/978-981-10-2917-2_11

issues at the heart of their business strategy and to benchmark their progress" (Bendell and Kleanthous 2007). It maps out the threats and opportunities facing the industry, challenges the myths of "shallow luxury," and grades the performance of the ten largest global luxury companies. L'Oreal, Hermes and LVMH received the highest grades in the group (C+) while Bulgari and Tods received the lowest (F). The 45-page report also discusses cutting-edge examples of sustainably-designed luxury goods, and presents a ten point plan to "unlock the latent commercial potential of sustainable luxury brands" (Bendell and Kleanthous 2007).

Why did WWF-UK, the oldest national organization of the largest multi-national conservation non-governmental organization (NGO) in the world (Banks 2010), decide to write this report on the luxury industry? What does this report reveal about WWF-UK's strategy for engaging the industry, and how have the companies responded to the claims and recommendations made in the report? How effective was it as an engagement tactic for WWF-UK? This chapter addresses these questions within the context of scholarly research on non profit strategy and the relationship between business and society, and shows how the *Deeper Luxury* report is an intriguing and provocative example of a non profit organizational tactic that reflects a hybrid organizational strategy. This hybrid strategy combines confrontational and advocacy elements with collaborative and operational features (Parker 2003), and creates both risks and opportunities for non profit organizations attempting to engage the private sector.

2 From Organizational Macro-Strategies to Campaign Micro-Tactics

Past research has focused on describing and evaluating such non profit strategies at the level of the organization. As Moore (2000) outlines, non profit organizational strategies should take into account the public value they expect to create, their sources of credibility and legitimacy, and their operational capacities. Scholars have developed increasingly sophisticated metrics that take into account these different factors and attempt to measure the operational effectiveness of non profit organizations (Sowa 2004; MacIndoe and Barman 2013). Researchers have also conducted case studies of particular non profit strategies, such as partnerships with government, cause-related marketing, and collaborations with industry (Liston-Heyes and Liu 2010; Ber and Branzei 2010; Pinkse and Kolk 2012; Austin and Seitanidi 2012; Cornforth et al. 2014). While some research has investigated the factors that lead stakeholder groups to target particular firms (Hendry 2006), less work has been done exploring why these groups focus on particular industries and economic sectors.

Other scholars have developed useful typologies and classifications of these organizational strategies (Bosso 1995; Frooman 1999; Hond and Bakker 2007; Kourula and Laasonen 2010; Schiller and Almog-Bar 2013). Elkington and Beloe

(2010) categorize environmental non-governmental organizations, for example, as either "polarizers," which "aim to achieve change by disrupting the status quo through confrontation," or "integrators," which "aim to achieve change through constructive partnerships with businesses, governments, and other stakeholders." This classification is similar to Hoffman and Bertels (2010: 62) distinction between "bright greens" that work within the market system and "dark greens" that seek to radically change that system. Den Hond and Bakker (2007) propose that "reformative" activist groups are more likely to focus on the reinstitutionalization of new norms and utilize tactics of material and symbolic gain for targeted firms, while "radical" activist groups are more likely to emphasize the deinstitutionalization of existing norms and utilize tactics of material and symbolic loss for targeted firms.

Less attention has been paid in the literature to studying and evaluating the specific tactics non profit organizations use in their campaigns to engage companies and how they fit into these broader strategic frameworks and typologies. Such analyses are particularly important in understanding how the non profit sector engages with the private sector on issues relating to their mission. Specifically, why do non profit organizations choose to focus on particular industries, how do they choose to engage them, and how do those industries respond to their method of engagement? Such a focus on micro-level tactics can improve our understanding of more macro-level organizational strategy, and show how the "on-the-ground" operations and tactics of non profits do not always fit neatly into the idealized typologies described in the literature. It can also reveal the difficulties and complexities organizations face in implementing such strategies, and highlight opportunities to improve upon them in the future.

Van Huijstee and Glasbergen's (2010) work on Rabobank is one of the few studies that effectively utilizes the theoretical frameworks discussed above to analyze the specific tactics used by non-governmental organizations. They explore how two environmental organizations used contrasting strategies to engage with Rabobank, a Dutch for-profit financial services provider. World Wide Fund for Nature Netherlands, a typical "engager" that uses symbolic gain strategies, developed a partnership with Rabobank to launch a climate-neutral credit card, while Friends of the Earth Netherlands, a typical "confronter" known for its symbolic damage approach to companies, published a report on the role of Dutch banks in climate change that inflicted "substantial symbolic damage" via attention from both the media and Parliament. The authors found that both strategies were effective, and complemented each other in ways that successfully deinstitutionalized and reinstitutionalized norms around bank policies that relate to climate change.

In light of this result, is it possible for one organization to develop a tactic that effectively creates both symbolic damages and gains and enables it to act as an engager and confronter simultaneously? Following the work of van Huijstee and Glasbergen (2010) but instead focusing on a single organization attempting to implement such a hybrid corporate engagement strategy, this chapter analyzes WWF-UK's *Deeper Luxury report* as a micro-level tactic that reflects the broader strategic orientation of the organization towards the luxury industry. In order to address the research questions outlined above, the chapter explores the nature of the luxury industry and its complicated relationship to sustainability efforts, conducts a detailed content analysis of the *Deeper Luxury* report, and examines the publicly-available sustainability-related documents produced by the ten companies evaluated in the report. It utilizes a variety of both qualitative and quantitative content analysis methods to both document the different forms of normative statements found in the report and explore its use of both normative change and performance measurement to engage the luxury industry. The analysis contributes to existing literature on the luxury industry, non profit organizational strategy, and corporate social responsibility, and has important implications for the relationship between business and society and efforts to extend corporate citizenship initiatives into new sectors and issue areas.

The chapter shows how WWF-UK, in order to attract the attention of one such sector, the luxury industry, used its *Deeper Luxury* report to engage in a two-pronged hybrid strategy that focused on both changing industry norms and measuring industry performance. Such a strategy has the potential to overcome the limitations of each approach in isolation and serve as a compromise response to the perception of "sustainable luxury" as an oxymoron. It also enables WWF-UK to play the role of both an integrating "bright green" and a polarizing "dark green" by highlighting both the best and worst practices in the industry. WWF-UK plays the integrator role through the report's emphasis on normative change and its greater prevalence of positive, prescriptive statements that encourage the industry in a more sustainable direction. By also giving the luxury companies relatively poor grades that frame the overall industry's performance as lackluster and deficient, WWF-UK uses the report's performance measurement component to simultaneously play the polarizer role.

While the use of both normative "carrots" and measurement "sticks" may have encouraged some luxury companies, such as L'Oreal, PPR, and LVMH, to expand their sustainability-related reports and programs, it appears to have had a more limited effect on other companies, such as Swatch, Bulgari, and Tods. These mixed results reveal important challenges facing such hybrid strategies—they are difficult to implement and organizations such as WWF who traditionally play the role of "good cop" may not be as effective at simultaneously playing the role of "bad cop." Advocacy organizations should therefore consider collaborating with other organizations that can complement their skills and reputations in pursuing such strategies, and be prepared to pursue more purely confrontational or collaborative approaches if they fail.

This chapter is divided into five parts. The first part discusses the opportunities and challenges associated with encouraging a more sustainable luxury industry in light of both WWF-UK's stated reasons for publishing the report and insights from the literature on luxury. The second part analyzes the two components of WWF-UK's strategy—encouraging normative change and measuring performance. The third part describes an industry association's initial response to the report, and the fourth part provides insights into the industry's response over time by presenting data from a content analysis of the sustainability-related documents published by the ten companies that WWF-UK evaluated. The fifth section of the chapter discusses the implications of this case for advocacy organizations, the luxury industry, and future research on "sustainable luxury."

3 Why Luxury?

3.1 WWF's Stated Reasons

Content analysis of the *Deeper Luxury* report reveals several reasons why WWF decided to focus on the luxury industry. These include the size of the industry, its global expansion, and its democratization—"luxury brands have become more accessible, making it harder to appeal to consumers on the grounds of exclusivity" (Bendell and Kleanthous 2007). It also cites the wide range of environmental and social impacts created by the industry—from the mining of gold to the unethical treatment of supply chain workers. The authors mention the increased attention to these impacts by the media, celebrities and Hollywood as an additional motivation for their work, and specifically reference the 2007 film *Blood Diamond.* The report also claims to be filling an important knowledge gap as the first "systematic analysis of the luxury brands' social and environmental responsibilities, performance, and opportunities" (Bendell and Kleanthous 2007). Perhaps the most important reason, however, was the perceived influence of the sector—"luxury brands have the power to influence consumer aspiration and behavior by editing consumer choices…and by influencing how, when, and for how long consumers use their choices" (Bendell and Kleanthous 2007). WWF-UK saw the value of directing this power towards encouraging more sustainable consumption. The importance of this factor resonates with Hendry's (2006) finding that environmental non-governmental organizations often target individual companies they perceive as having a significant impact on the environment.

3.2 Insights from the Literature: Why not Luxury?

These reasons resonate with the broader literature on sustainable luxury. Kapferer (2010) asserts that the luxury industry has attracted the attention of sustainability advocates for four reasons—luxury is highly visible, fundamentally irrational, focused on excess, and directly related to social inequality. He concludes that these factors are related more to the symbolic position of the sector than its objective environmental impact or its "economic weight." The disproportionate importance of brand value and reputation to luxury companies has also been cited as a reason for the attention they have received because it makes them inherently vulnerable to "name and shame" strategies that put their brand equity at risk (Klein 1999; Vogel 2008; Kapferer 2010). Celebrities may also be sensitive to criticisms of the brands they have endorsed (Kapferer 2010; Qureshi 2012), which is an idea that may attract NGO attention and one that the WWF report emphasizes. Others have asserted that it is the high profitability of the industry that demands their attention to sustainability (Kahn 2009). In two experimental studies, Boenigk and Schuchardt (2014) also show that luxury brand partnerships improve attitudes towards the participating non profit organizations and enables them to raise more donations, particularly from wealthy individuals.

Another question to ask, however, is why *not* to focus on luxury? One reason is its complexity—"luxury" industry is not clearly defined, includes a wide range of heterogeneous products and companies, and has a diverse set of impacts. A luxury brand traditionally was thought of as being selective, exclusive, scarce, sophisticated, and tasteful, but the democratization of luxury has broadened that definition—now there is both "mass" luxury and "inaccessible" luxury (Tungate 2009). The concept of luxury has several layers of meaning, from anything that is perceived as desirable and more than necessary or ordinary to specific goods and commodities that have high levels of price, quality, aesthetics, or rarity (Heine 2010). Thus luxury covers an enormous range of products, such as fine dining, cosmetics, fragrances, leather goods, textiles, watches, jewelry, shoes, furniture, cars, hotels, real estate, and boats (Cheuvreux 2012).

Furthermore, these products are produced by companies that have different corporate structures and use different supply chain strategies (Caniato et al. 2011). The social and environmental impacts of these companies and products are wide-ranging and vary by product—the primary concern about luxury automobiles, for example, is likely to be greenhouse gas emissions, while the primary concern about luxury leather goods is likely to be toxic waste produced during the tanning process (Cheuvreux 2012). Such diversity may make it difficult to offer assessments and recommendations to the "luxury industry" that are actionable and relevant for specific products. It also makes it difficult to focus on a specific social or environmental problem associated with the industry, and monitor progress towards solving that problem.

The definitions cited above point towards an additional reason for environmental organizations not to focus on the sustainable consumption of luxury goods, which are by definition unnecessary and non-functional. Why should responsible consumers buy them at all? Many commentators have noted the apparent paradox of "sustainable luxury." Belli (2008) cites "solar-powered tanning beds, solar-powered yachts and sustainable caviar" as evidence that green consumerism has become the "ultimate guilt-reducer, a way to enjoy all the ease and comfort life has to offer without feeling that you're actually doing very little to contribute." Uren (2012) asserts that "luxury can often be shorthand for excessive consumption," while Kapferer (2010) describes the intrinsic tradeoffs between the quality and creativity that luxury requires and the self-restraint and frugality that sustainability demands. The *Deeper Luxury* report notes that "luxury brands face criticism that they fuel unnecessary consumption and provoke discontent among poorer members of society," and cites the fact that the Mayor of Beijing banned luxury billboards and the Indian government taxes luxury goods at 114% as evidence of this criticism (Bendell and Kleanthous 2007).

This darker side of luxury and the complexity of the industry make the choice to focus on luxury brands more interesting and controversial. Rather than focus on a narrower category (e.g. jewelry, watches) or encourage a broader disengagement from luxury goods because of their extravagant and resource-intensive nature, WWF-UK decided to pursue the opportunities that the broad but unique nature of the luxury industry presented. In particular, the industry's power to influence consumption norms, the firms' vulnerability to reputational risk, and the momentum created by the *Blood Diamond* movie were likely the strongest factors driving their decision to publish the report. This decision did not come without risks, as their framing of "sustainable luxury" could have been perceived as a cognitively dissonant oxymoron that was too broad and incoherent to translate into specific actions and metrics. It also threatened to demoralize WWF-UK's employees and volunteers, as Boenigk and Schuchardt's (2014) research shows non profit partnerships with luxury brands can create "identification conflicts" and negative attitudes among the non profit organization's employees and volunteers. These tradeoffs associated with engaging the luxury industry may explain the hybrid strategy that WWF-UK utilized in its *Deeper Luxury* report.

4 The Demands of *Deeper Luxury*

So what did WWF-UK demand of the luxury industry, in terms of such actions and metrics? How did they seek to convince readers of the report that sustainable luxury is not an oxymoron, but a realistic and important goal for both luxury producers and consumers? A content analysis of the report reveals that the authors pursued two specific strategies—changing the norms associated with luxury and measuring the social and environmental performance of luxury companies. This section summarizes these two approaches, while the next two sections analyze the companies' response to them.

4.1 Changing of Norms

The concept of norms is used extensively in the social sciences to analyze human behavior. While conceptualizations of the term vary widely, Interis (2011) provides a useful definition of a norm as "a voluntary behavior that is prevalent within a given reference group." Building on the literature on norms, he argues there are two main types of norms—*descriptive norms*, which are simply "what people do," and *injunctive norms*, which are "what people should do." The latter involves some threat of sanctioning, either by oneself ("*internal*") or by others ("*external*"), if the norm is not followed. Norms can also have different polarities, from the *prescriptive* (what one should do) pole to the *proscriptive* (what one should not do) pole (Jasso and Opp 1997; Wilson et al. 2001).

As Sunstein (1996) notes, "Changes in norms may be the cheapest and most effective way to make things better…[and] promote human well-being." Changing and activating norms is a common strategy of social and environmental activists, who often act as "norm entrepreneurs" (Sunstein 1996). Normative change is a key component of both the deinstitutionalization and reinstitutionalization processes discussed by den Hond and Bakker (2007). But how do activists specifically implement this strategy? How do they promote certain norms and discourage others? The types of norms discussed above provide a useful framework for exploring these questions and for understanding WWF-UK's approach to changing norms in its *Deeper Luxury* report. These normative dimensions highlight strategic choices made by the organization, and have implications for the ultimate effectiveness of the initiative.

In order to analyze the authors' normative change approach, I used the textual coding software Atlas.ti to code the report for the different types of norms outlined above. Only normative statements that are directly relevant to the luxury industry and sustainability were coded. Passages that make statements about current luxury

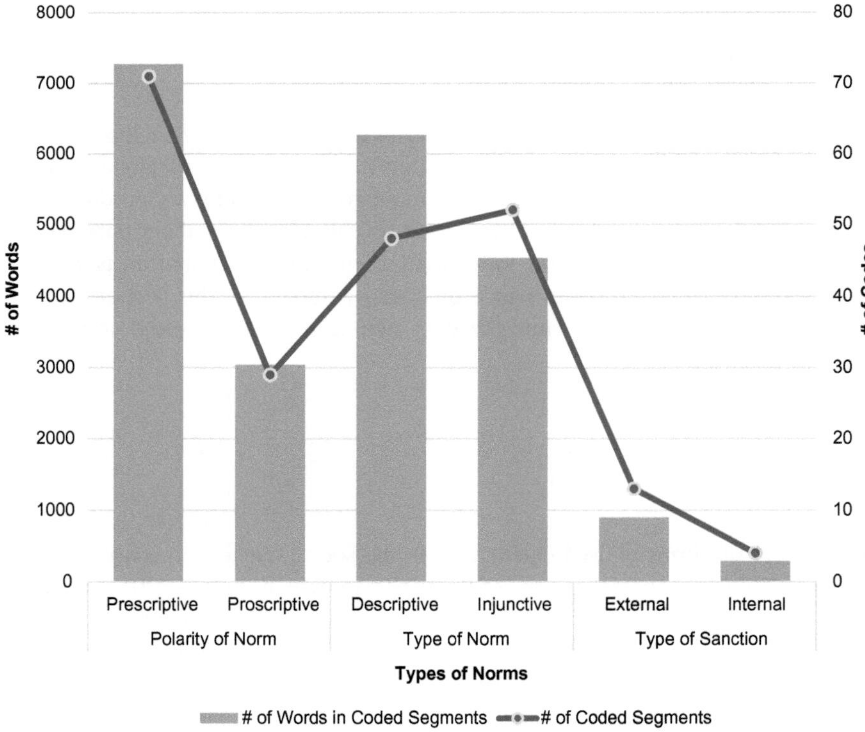

Fig. 1 Count of norm codes and number of words in coded quotations

industry, consumer, or celebrity practices relating to sustainability were coded as descriptive norms, while passages that make statements about the expected behaviors of these luxury actors were coded as injunctive norms. In this coding scheme, I depart from Interis's (2011) conception of injunctive norms as requiring an explicit sanctioning mechanism (as many strong "injunctive" statements do not necessarily imply a specific sanction), and define injunction as any statement that has a clear "should" or "ought" implication.

Figure 1 shows both the number of coded segments and the total number of words coded by type of norm. The number of text segments coded as injunctive norms (52) is greater than those coded as descriptive norms (48), but the number of words in the descriptive norm texts is greater (6265) than those in the injunctive norm texts (4529). The greater number of words in the descriptive norm text segments may in part be due to extensive descriptions in Chap. 7 of the report of seven "future makers"—such as Osklen in Brazil, John Hardy in Bali, and Tesla in the US—that "embody aspects of the future of luxury by excelling in key aspects of their social and environmental performance" (Bendell and Kleanthous 2007).

The nearly equivalent number of descriptive and injunctive codes, however, reveals the balanced strategy that WWF-UK has taken in the report. While certain sections focus on one type of norm, in general the authors weave injunctive and descriptive norms together throughout the report. As an example, in their section on gold, the authors describe the support of the No Dirty Gold Campaign from Cartier, Piaget, Tiffany & Co and Van Cleef & Arpels—a descriptive norm—and then immediately state that the "luxury watch and jewellery industry can and should… position ethically sourced gold as a more attractive alternative for the discerning customer"—an injunctive norm (Bendell and Kleanthous 2007). Such a balanced approach between describing and demanding preferred behaviors was not matched, however, by a similar balance in the polarity of the report's normative statements. In coding for polarity, I defined a prescriptive norm as any behavior or policy presented as acceptable or positive, and a proscriptive norm as any behavior or policy presented as unacceptable or negative. Every normative statement made was coded as either prescriptive or proscriptive, and many more prescriptive text segments (71) and words (7274) were found than proscriptive ones (29 text segments, 3043 words). Thus the authors focused on positive norms more extensively than negative ones. This apparent hesitancy to make negative demands on the luxury industry corresponds with the fact that normative sanctions are rarely mentioned in the document. While sanctions imposed by external actors are more frequently referenced (13) than sanctions that the actor might impose on itself (4), neither are common.

4.2 *Measuring of Performance*

The second major mechanism employed in *Deeper Luxury* to engage the luxury industry is related to but conceptually distinct from normative change. It is the development and distribution of measurements of corporate environmental and social performance. This approach is a form of information-based governance, which uses the provision of information to exert power and stimulate collective action (Bullock 2015a). The effectiveness of information-based governance depends on how individuals and organizations perceive four critical dimensions of the information provided—its source, content, the methods by which the information was developed, and the interface by which the information is delivered to its audience (Bullock 2011). This section examines each of these components of the *Deeper Luxury* rating system.

The report's ratings have two primary sources—Ethical Investment Research Service (EIRIS) and Covalence. EIRIS is a 28-year-old non profit organization that provides "responsible investment research" on nearly 3000 companies (EIRIS 2013a, b). While EIRIS has a detailed Conflict of Interest Policy and Code of Conduct designed to insure its independence (EIRIS 2008a, b), it does not disclose its client list and the data it uses is self-reported by the companies they are assessing. Covalence is a 12-year-old company that publishes the EthicalQuote Index, a rating of 2800 companies that "integrates thousands of news pieces gathered online and classified according to 50 criteria inspired by the Global Reporting Initiative" (Covalence SA 2013a). No Code of Conduct and Conflict of Interest Policy could be found on its website, and its clients include many of the companies it evaluates, including Coca-Cola, Dell, and IBM (Covalence SA 2013b). Its data come from online search engines, individual websites, and submissions from interested parties, which are weighted equally and not assessed for their validity (Covalence SA 2012).

WWF-UK used data from these two sources to develop ratings of the ten companies with the largest market capitalizations in 2006 that derive a significant portion of their revenue from personal luxury items—Bulgari, Coach, Inc., Hermes, L'Oreal, LVMH (owns over 50 brands, including Louis Vuitton and Hennessy), PPR (owner of the Gucci Group), Richemont (owner of Cartier and Mont Blanc), Swatch Group Ltd., Tiffany & Co, and Tods. The rating assesses both what the companies claim about their own environmental, social, and governance performance and what others are saying about their performance (in 2006). While the 95 specific criteria used in the assessment are not provided, four general categories of environment, human rights, corporate governance, and stakeholder relations are listed, along with example criteria, such as "equal opportunity" and "environmental policies." In its limited explanations for the grades each company received, the report identifies additional criteria that individual companies did particularly well or poorly on, such as "health and safety in the workplace" and "environmental management" (Bendell and Kleanthous 2007).

Half of the grade is based on 50 criteria derived from self-reported company data compiled by EIRIS, while the other half is based on 45 criteria derived from 512 news stories about the ten companies found by Covalence. The EIRIS criteria were divided into the four equally-weighted categories listed above, which were aggregated into a numerical score from 0 to 50. The Covalence criteria were also aggregated to create a numerical score from 0 to 50, although nothing is said about how they were weighted. The combined score was then converted into letter grades, with any score below a 50 receiving an F, which is the grade Bulgari and Tods received. Other grades given were three Ds (Swatch, PPR, Richemont), one D+ (Tiffany), one C (Coach), and two C+s (LVMH, Hermes).

These low grades present all of the companies in a poor and unflattering light. They were published as bar graphs on a WWF-UK website and in a 45-page PDF report. The report also includes the final 0–100 scores, the two 0–50 sub-scores, and 1–10 rankings of the companies. The "positive news" and "negative news" scores from Covalence were plotted in a two-dimensional graph that shows L'Oreal having nearly three times as many good news stories as the other companies. The EIRIS sub-scores were presented in two figures—one vertical bar graph showing the overall sub-scores for self-reported performance and one stacked vertical bar graph showing scores for the four sub-categories in different shades of blue (Bendell and Kleanthous 2007).

5 The Industry's Initial Response

How did the industry respond to the report's ratings and normative demands? The immediate response from the most relevant industry association, The Council for Responsible Jewellery Practices (CRJP), was unequivocally negative. The CRJP is an "international, not-for-profit standards setting and certification organization established to reinforce consumer confidence in the jewellery industry by advancing responsible business practices," and at the time listed 81 members, including 32 jewellery retailers, 17 jewellery manufacturers and wholesalers, 17 gold and diamond-related firms, eight trade associations, and five mining companies. Three of the companies rated by *Deeper Luxury*—Tiffany & Co., Richemont, and Bulgari—were members of the Council in 2007.

The CEO of the CRJP wrote a letter soon after the release of the report to the CEO of WWF-UK that was highly critical of its contents. The letter states that "the report and WWF-UK, as its publisher, deserve the strongest condemnation for the completely undeserved and unwarranted criticism the report contains regarding a number of CRJP members," and that the CRJP was "profoundly concerned that WWF, a NGO previously respected by the CRJP and its members, would publish something with the exceptionally unprofessional content and tone of this report" (Miller 2007). The letter specifically criticizes WWF-UK for its "fatally flawed scoring methodology," its lack of consultation with the companies named in the report, its "glaring omissions" of the work being done by the CRJP's membership (including with WWF itself), its "poorly conducted research," and its failure to compare the progress of the jewellery sector with other consumer good sectors (Miller 2007). These claims are evidence of the strongly negative reaction that the report—and in particular its rating methods—elicited from at least some representatives of the luxury industry.

6 Deepening Over Time?

Despite this initially combative response, has the luxury industry deepened its engagement with sustainability issues since the report's publication? In this section, I analyze the corporate responses to the normative strategies employed by WWF in the report by conducting a quantitative content analysis of the companies' 2001–2012 sustainability-related documents.

6.1 Changes in Sustainability Reporting Practices

In the report, WWF provides a 10-Point Plan of action for the luxury industry. Two of these prescriptive, injunctive norms—"understand your brand" and "understand your consumers"—are only tangentially related to sustainability. Four of these norms relate primarily to sustainability performance measurement, reporting, and improvement—"get your house in order," "C(S)R: Handle with care!," "Communicate," and "Measure, monitor, and report regularly." While it is beyond the scope of this chapter to directly analyze the sustainability performance of these ten companies, several aspects of their sustainability reporting can be measured. Such an approach builds on past analyses of corporate sustainability reports and websites in the literature (Chapple and Moon 2005; Rodriguez Bolivar 2009; Perego and Kolk 2012; Fifka and Drabble 2012; Weber and Marley 2012) and focuses on three particular aspects of these reports. The first is to what extent the companies have published information on their corporate social responsibility policies or performance. The second is to what extent those documents provide data for the indicators recommended by the Global Reporting Initiative (GRI), which was one of the few specific policy recommendations made by WWF-UK. The third

Table 1 Luxury companies' 2001-2012 sustainability-related documentation

Company	Publication	# of pages	GRI	# of GRI indicators	Third-party audit
L'Oreal	Since 2004	52	Yes	98	Yes
Hermes	since 2005	17	–	–	–
LVMH	Since 2001	50	Yes	42	Yes
Coach	Undated	12	–	–	–
Tiffany & Co.	Since 2010	73	Yes	126	Yes
Swatch	Since 2004	1	–	–	–
PPR	Since 2006	61	Yes	107	Yes
Richemont	Since 2005	55	Yes	93	–
Bulgari	None	–	–	–	–
Tods	None	–	–	–	–

is whether those reports have been audited by an independent third party, and the fourth is the number of pages dedicated to sustainability issues in the companies' most recent set of documents. The data for these metrics are presented in Table 1.

The level of performance measurement and reporting varies significantly across the ten companies rated by WWF-UK. L'Oreal, LVMH, Swatch, and Richemont have been reporting their performance since before *Deeper Luxury* was published, while Coach and Tiffany began reporting afterwards and Bulgari and Tods provide no such reporting. The length of these documents also varies substantially—from the relatively few pages provided by Swatch (1 page), Coach (12), and Hermes (17), to the 50 or more pages provided by PPR, Richemont, Tiffany, L'Oreal, and LVMH. Reference to the GRI also varies—half of the companies (Hermes, Coach, Swatch, Bulgari, and Tods) make no such reference, while the other half does (L'Oreal, LVMH, Tiffany, PPR, and Richemont). Only four companies (LVMH, PPR, Tiffany, and L'Oreal) had their reports audited by independent third parties.

6.2 Changes in Motivation, Innovation, and Communication Practices

The remaining four recommendations provided in WWF-UK's 10-Point Plan are to "motivate" its employees, "innovate" new sustainable products, "collaborate" with a range of different actors, and "sign up your consumers to the sustainability journey." Three of these four injunctive norms contain a root keyword—motivate, innovate, collaborate—that can be used to measure the extent to which these three injunctive norms have been referenced by the top ten luxury companies (the fourth relating to consumers is less amenable to this form of analysis and should be a subject of future research). I used the coding software MaxQDA 11 to auto-code every instance of these keywords (i.e. any word with the roots motivat*, collaborat*, and innovat*) in the companies' 2001–2012 sustainability-related documents. These included corporate social responsibility, sustainability, and environmental reports; integrated annual reports (only the sections focused on sustainability were included in the analysis); and supplemental sustainability documents. Such auto-coding can produce measurement error, but that error should be consistent across companies, types of documents, and codes. It also avoids human coding error and subjective decisions about what counts as motivation, innovation, and collaboration.

As Fig. 2 shows, the most common keyword found in the companies' documents was **innovation**. The WWF report encourages companies to innovate across a wide range of contexts, and in particular to "think of new ways for your existing products or services to do useful jobs while enhancing people's perceptions of themselves as environmentally or socially responsible." L'Oreal has clearly focused on this theme; its 2011 report begins by stating, "If 2011 had to be summarised in two words, it would be the year of 'sustainable innovation.' We innovated in every

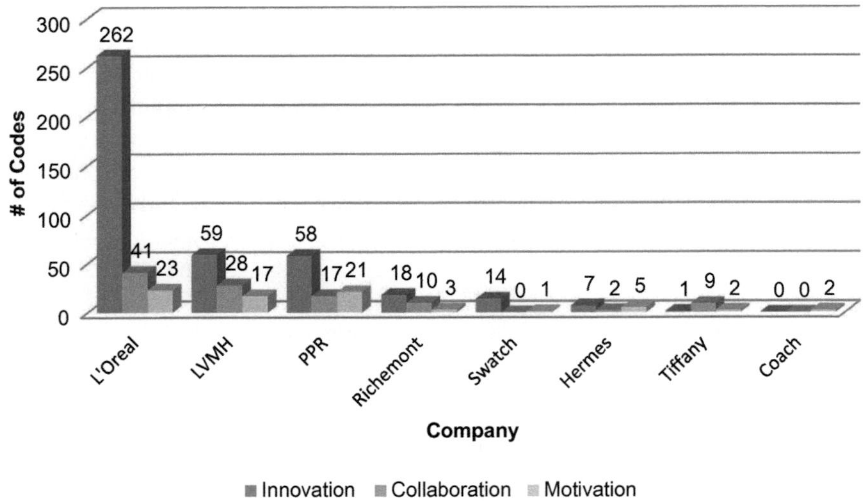

Fig. 2 Frequency of innovation, collaboration and motivation keywords in luxury company sustainability-related documents (2001–2012, by company). *Note* Tods and Bulgari are not included because no sustainability-related documents could be found on their websites

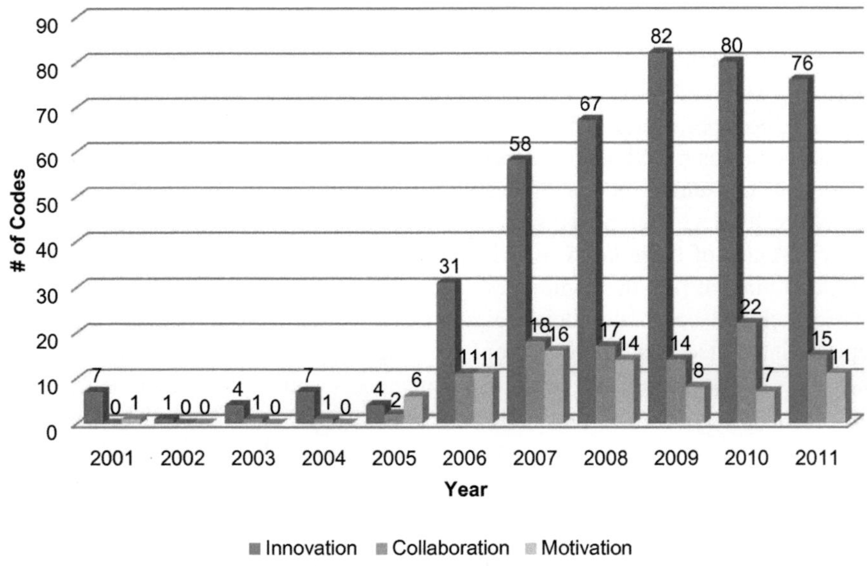

Fig. 3 Frequency of innovation, collaboration and motivation keywords in luxury company sustainability-related documents (2001–2011, by year). *Note* 2012 is not included because only one report (Richemont) had been released for this year when the data was collected in February 2013. Data for Coach is also not included because its documents were undated

field, setting ourselves ambitious goals in terms of respect for the environment and for all stakeholders…" It uses the word over 250 times in its reports since 2005—more than all of the other companies combined. LVMH (59) and PPR (58) are the next most likely to reference innovation, followed by Richemont (18) and Swatch (14). As Fig. 3 shows, the use of the term increased dramatically between 2005 and 2009, and then slightly declined in 2010 and 2011.

The next most common norm to be cited is **collaboration**. WWF encouraged companies to "explore possibilities to team up, both internally and externally," among staff in "multi-functional, multi-skilled teams" and with third parties such as strategic partners, celebrity representatives, non-governmental organizations, competitors, and industry associations. L'Oreal is also the most frequent user of the keyword "collaborate" (41 times), followed by LVMH (28) and then PPR (10). An example from LVMH's 2011 report: "Aware of the advances that can be made through holding discussions and collaborating with others, particularly in environmental matters, LVMH makes a point of forming key partnerships with national and international non-profit organizations, local communities and educational institutions." The use of the term also increased dramatically between 2005 and 2007, but has since leveled off.

The least common of the three norms to be mentioned is **motivation**. The term is most commonly found in L'Oreal's documents (23 times), although PPR (21) and LVMH (17) are a close second and third. In its "Motivate" recommendation, WWF-UK suggested that luxury companies "reward and congratulate staff for meeting environmental and social objectives and targets," and these companies apparently have picked up on this norm. In its 2011 report, PPR claims that "we are motivating ourselves and our customers to be aware of the planet's fragility," while LVMH states that "motivating employees begins with training" and that a total of 16,726 h of training were offered in 2007. Figure 3 shows an increase in the use of the term between 2004 and 2007, a decrease between 2007 and 2010, and an increase again in 2011.

7 Conclusion

7.1 A Two-Pronged Hybrid Strategy

This chapter has described the two-pronged strategy that WWF-UK used to engage the luxury industry in its *Deeper Luxury* report. The report is a micro-level tactic designed to both change industry norms and measure industry performance, and enabled the environmental organization to both confront the industry and offer it an olive branch of collaborative possibilities. This tactic represents a broader hybrid strategy that combines approaches used by "polarizers" and "dark greens" with those of "integrators" and "bright greens" (Elkington and Beloe 2010; Hoffman and Bertels 2010). WWF-UK plays the integrator role through the report's emphasis on normative change, while playing the polarizer role through its hard-hitting ratings of the industry's performance on sustainability issues. Both roles have elements that soften their polarizing or integrating functions; for example, the report makes several negative, proscriptive normative statements that puts an edge to its otherwise relatively supportive language, while the report's grades differentiate between the companies and does not fail all of them, demonstrating that WWF-UK is capable of recognizing some examples of positive performance.

This hybrid strategy is perhaps not surprising for WWF, as van Huijstee and Glasbergen (2010) characterize WWF as a hybrid NGO that combines advocacy and operational activities, and Banks (2010) states that "the organization does not hesitate to criticize bad actors," such as the Asia Pulp and Paper Company. However, rating the top ten companies in an entire industry is a relatively unusual exercise for WWF, and other scholars have generally considered it to be an "integrator" and "light green" environmental NGO (Elkington and Beloe 2010; Hoffman and Bertels 2010). The extensive section on WWF's history of working with business in *Deeper Luxury* reveals its historical orientation towards collaboration and its underlying goal of partnering with the firms that it is critiquing. As Banks (2010) notes, WWF has strong and numerous ties to the business community through its board of directors, and "partners with corporations on a range of activities, such as philanthropic conservation efforts, development of issue-specific best practices, and transformative organizational change."

WWF-UK may have needed to take such a hybrid approach because of the oxymoronic, darker side of luxury and the perception that many forms of luxury are inherently unsustainable. While not endorsing this view of luxury, the report's title encapsulates WWF-UK's compromise response to it. "Deeper" luxury may refer to an intermediate level of depth that goes beyond a "deep" luxury that is merely philanthropic and environmentally and socially responsible to one that actively seeks to change society's norms around consumption. It does not embrace, however, the "deepest" form of luxury, which seeks to actively avoid luxury goods and services and focuses instead on leisure, natural beauty, social equity, and simple, ethical living. It is arguably WWF's historical orientation towards collaboration that precludes such a stance, but even more confrontational organizations have often avoided this interpretation of responsible consumption. Greenpeace's Greener Electronics initiative, for example, does not discourage consumers to reduce their use of electronics, but instead encourages companies to make greener products. The attempt to balance engagement and confrontation—to play both the good cop and the bad cop—is not unique to WWF, but apparent in many organizations, from the Rainforest Action Network to Environmental Defense (Krill 2010; Ruta 2010; Lyon 2010).

WWF-UK may have also pursued normative change and performance measurement simultaneously in order to compensate for the weaknesses of each approach. Calls for a systemic normative shift in the luxury industry may have been necessary, but can sound like empty rhetoric without specific ways to measure such a shift. In this sense, they may overly emphasize the symbolic gains and the reinstutionalization process without first threatening symbolic losses and beginning the process of deinstitutionalization (Hond and Bakker 2007). Sustainability metrics can communicate clear priorities that serve to ground such normative statements, but their robustness and validity can be limited, especially when available expertise and data are limited and the rated industry is particularly broad and heterogeneous (as the luxury industry is). Bold and positive normative statements presented in a colorful report might serve to distract readers from these limitations.

7.2 Implications and Lessons Learned

This hybrid strategy appears to have been only partially successful. While Jem Bendell claims that the report "appeared in over 50 newspapers and magazines worldwide in the month of its release," its use as a Google search term disappeared only four months later (Bendell 2011: 145; Google 2013). In March 2013, there were less than 10,000 Google hits for "deeper luxury," compared to 300,000 for "green luxury" and 213,000 for "sustainable luxury." In terms of corporate responsiveness to the demands made in the report, the results are decidedly mixed. As of early 2013, two of the ten companies had not produced any publicly-available sustainability-related documents, and three had only published a few pages about their sustainability performance. Five had made no reference to the GRI, and six had not had their reports independently audited by a third party. Nevertheless, several companies have increased the amount and complexity of their reporting dramatically since 2007, and have extensively referenced several of the sustainability norms emphasized in *Deeper Luxury,* and especially the idea of sustainable innovation. L'Oreal is the leader in referring to all three of these norms, followed by PPR and LVMH, while Hermes, Coach, Tiffany, and Swatch make few or no mentions of them.

While these are imperfect proxy metrics of effectiveness, the initiative clearly does not have a mainstream following and has not stimulated comprehensive or widespread changes in the luxury industry. Such a result does not prove that hybrid confrontational/collaborative strategies are inherently ineffective, but it does demonstrate that they can be difficult to implement successfully. This paper has focused on how such a strategy is reflected in one particular tactic used by WWF-UK; future research should explore how these hybrid strategies are implemented in other contexts and through other mechanisms. For example, interviews with non profit employees employing such a hybrid strategy could further reveal the reasoning and dynamics driving its implementation, and contribute to updating the literature's conceptualization of non profit organizational strategy at the macro-level and its understanding of how it connects to the on-the-ground tactics used by these organizations.

Towards that end, this case provides several initial insights and lessons learned regarding these hybrid strategies and tactics. First, it is important to create transparent, credible, and iterative ratings. WWF-UK does not provide the specific performance criteria and the weights of those criteria used in the assessment, making it difficult for companies to know how to improve their grade. It uses Covalence data in problematic ways that reduce its credibility. For example, it does not take into account the absolute number of "good news" or "bad news" stories, only their relative proportions. So even though L'Oreal had almost three times as much good news as any other company, it received a lower Covalence score than Hermes and Coach, which received perfect 50 scores but had two of the four lowest number of good news stories. The weighting of every Covalence source as equal is also highly questionable. Furthermore, the ratings were only published once. Performance scorecards should provide regular, iterative assessments of organizations so that the public can monitor their performance over time (Gormley and Weimer 1999).

Second, a single actor playing both the good cop and bad cop roles may not be believable. The intended audience may sense the actor's underlying orientation and not find either its threats or its rewards credible—a key factor in driving organizational effectiveness (Moore 2000; Brown 2008; Bullock 2015b). Audiences may have accepted WWF's attempts to positively encourage the luxury industry to become more sustainable, but found its confrontational and controversial ratings as cognitively dissonant and out-of-character for the organization. Third, building on Moore's (2000) emphasis on operational capacity, it is difficult to do any one of these strategies well, and the risk of doing them all poorly is high. WWF, for example, seems to have been partially successful with its normative approach, but its performance measurement component had many weaknesses. It may therefore be more effective for organizations with different expertise and orientations to collaborate to implement such hybrid strategies, as WWF and Friends of the Earth ultimately did in their successful interactions with Rabobank in the Netherlands (van Huijstee and Glasbergen 2010). And fourth, it is important to issue specific policy recommendations and performance standards so that companies can be held accountable. Most of WWF-UK's admonitions, such as "get your house in order," are too vague to effectively implement and assess.

Finally, organizations should be prepared to take a more purely confrontational or collaborative approach if the hybrid approach fails. Direct confrontation in particular may be an important niche for organizations to fill in specific issue areas, such as sustainable luxury, where stronger responses may be appropriate. Whether such an approach will resonate with the public depends on the extent to which the industry responds to the demands from the more moderate strategy. Such a response must go beyond the use of terms such as collaboration, motivation, and innovation that is documented in this chapter, and show significant, relevant, and continuous progress in not only environmental and social performance but also in encouraging more sustainable consumer norms and behaviors. Future research should focus on transparently, rigorously and iteratively identifying and measuring these outcomes, and the extent to which different luxury companies are contributing to them. A long-term commitment to such research by industry, academia, and advocacy organizations such as WWF will enable luxury consumers to know the real depth of their favorite brands' commitment to sustainability.

Acknowledgements I would like to acknowledge the assistance of Emma Park in conducting the research for this chapter. As an undergraduate at Davidson College, Emma helped collect and analyze the corporate social responsibility reports and related documents discussed in the chapter. She was also very helpful in thinking about WWF's strategy and the industry's response to its Deeper Luxury report.

References

Austin JE, Seitanidi MM (2012) Collaborative value creation: a review of partnering between nonprofits and businesses. Part 2: Partnership processes and outcomes. Nonprofit Volunt Sector Q 41:929–968. doi:10.1177/0899764012454685

Banks M (2010) World wildlife fund. Good cop/bad cop: environmental NGOs and their strategies toward business. Resources for the Future Press, Washington, D.C., pp 171–183

Belli B (2008) Green living deluxe: how the environmental movement went way, way upscale. E 19

Bendell J (2011) Evolving Partnerships: a guide to working with business for greater social change. Greenleaf Publishing

Bendell J, Kleanthous A (2007) Deeper luxury: quality and style when the world matters. World Wildlife Fund (UK)

Ber MJL, Branzei O (2010) (Re)forming strategic cross-sector partnerships relational processes of social innovation. Bus Soc 49:140–172. doi:10.1177/0007650309345457

Boenigk S, Schuchardt V (2014) Nonprofit collaboration with luxury brands: positive and negative effects for cause-related marketing. Nonprofit Volunt Sector Q. doi:10.1177/0899764014551280

Bosso CJ (1995) The color of money: environmental groups and the pathologies of fund raising. In: Interest group politics, 4th edn. CQ Press, Washington, D.C., pp 101–130

Brown LD (2008) Creating credibility: legitimacy and accountability for transnational civil society. Kumarian Press

Bullock G (2015a) Independent labels? The power behind environmental information about products and companies. Polit Res Q 68:46–62. doi:10.1177/1065912914564685

Bullock G (2011) Green grades: the popularity and perceived effectiveness of information-based environmental governance strategies. Ph.D. Dissertation, University of California, Berkeley

Bullock G (2015b) Signaling the credibility of private actors as public agents: transparency, independence, and expertise in environmental evaluations of products and companies. Bus Polit 17:177–219

Caniato F, Caridi M, Castelli C, Golini R (2011) Supply chain management in the luxury industry: a first classification of companies and their strategies. Int J Prod Econ 133:622–633

Chapple W, Moon J (2005) Corporate social responsibility (CSR) in Asia a seven-country study of CSR web site reporting. Bus Soc 44:415–441. doi:10.1177/0007650305281658

Cheuvreux (2012) Luxury goods—sustainability sector profile. Cheuvreux Credit Agricole Group

Cornforth C, Hayes JP, Vangen S (2014) Nonprofit-public collaborations: understanding governance dynamics. Nonprofit Volunt Sector Q. doi:10.1177/0899764014532836

Covalence SA (2013a) About. In: ethicalquote.com. http://www.ethicalquote.com/index.php/about-us/. Accessed 26 Feb 2013

Covalence SA (2013b) Clients. In: ethicalquote.com. http://www.ethicalquote.com/index.php/clients/. Accessed 26 Feb 2013

Covalence SA (2012) Methodology. In: ethicalquote.com. http://www.ethicalquote.com/docs/CovalenceEthicalQuoteMethodology.pdf. Accessed 26 Feb 2013

EIRIS (2013a) About us. In: eiris.org. http://www.eiris.org/about_us.html. Accessed 26 Feb 2013

EIRIS Our Clients (2013b) In: eiris.org. http://www.eiris.org/managers/our_clients.html. Accessed 26 Feb 2013

EIRIS (2008a) EIRIS independence policy. In: eiris.org. http://www.eiris.org/about-us/documents/independencepolicy-summary.pdf. Accessed 26 Feb 2013

EIRIS (2008b) EIRIS code of conduct. In: eiris.org. http://www.eiris.org/about-us/documents/codeofconductpolicy.pdf. Accessed 26 Feb 2013

Elkington J, Beloe S (2010) The twenty-first century NGO. Good cop bad cop: environmental NGOs and their strategies toward business. Resources for the Future Press, Washington, D.C., pp 17–47

Fifka MS, Drabble M (2012) Focus and standardization of sustainability reporting—a comparative study of the United Kingdom and Finland. Bus Strategy Environ 21:455–474. doi:10.1002/bse.1730

Frooman J (1999) Stakeholder influence strategies. Acad Manag Rev 24:191–205. doi:10.2307/259074

Google (2013) Google trends. In: google.com/trends. http://www.google.com/trends/explore. Accessed 4 Mar 2013

Gormley WT, Weimer DL (1999) Organizational report cards. Harvard University Press

Heine K (2010) Luxury & sustainability: implications of a consumer-oriented concept of luxury brands (abstract). Istanbul

Hendry JR (2006) Taking aim at business what factors lead environmental non-governmental organizations to target particular firms? Bus Soc 45:47–86. doi:10.1177/0007650305281849

Hoffman A, Bertels S (2010) Who is part of the environmental movement? Good cop/bad cop: Environmental NGOs and their strategies toward business. Resources for the Future Press, Washington, D.C., pp 48–72

Hond FD, Bakker FGAD (2007) Ideologically motivated activism: how activist groups influence corporate social change activities. Acad Manag Rev 32:901–924. doi:10.2307/20159341

Interis M (2011) On norms: a typology with discussion. Am J Econ Sociol 70:424–438

Jasso G, Opp K-D (1997) Probing the character of norms: a factorial survey analysis of the norms of political action. Am Sociol Rev 62:947–964. doi:10.2307/2657349

Kahn J (2009) Luxury-goods makers embrace sustainability. The New York Times

Kapferer J-N (2010) All that glitters is not green: the challenge of sustainable luxury. Eur Bus Rev

Klein N (1999) No logo: taking aim at the brand bullies. Picador, New York

Kourula A, Laasonen S (2010) Nongovernmental organizations in business and society, management, and international business research review and implications from 1998 to 2007. Bus Soc 49:35–67. doi:10.1177/0007650309345282

Krill J (2010) Rainforest Action Network. Good cop/bad cop: environmental NGOs and their strategies toward business. Resources for the Future Press, Washington, D.C., pp 208–220

Liston-Heyes C, Liu G (2010) Cause-related marketing in the retail and finance sectors an exploratory study of the determinants of cause selection and nonprofit alliances. Nonprofit Volunt Sector Q 39:77–101. doi:10.1177/0899764008326680

Lyon TP (2010) Good cop/bad cop: environmental NGOs and their strategies toward business. Resources for the Future, Washington, D.C.

MacIndoe H, Barman E (2013) How organizational stakeholders shape performance measurement in nonprofits exploring a multidimensional measure. Nonprofit Volunt Sector Q 42:716–738. doi:10.1177/0899764012444351

Miller J (2007) CRJP asks WWF-UK to withdraw "deeper luxury" report. In: diamonds.net. http://www.diamonds.net/News/NewsItem.aspx?ArticleID=19927&ArticleTitle=CRJP+Asks+WWF-UK+to+Withdraw+%27Deeper+Luxury%27+Report. Accessed 10 Feb 2013

Moore MH (2000) Managing for value: organizational strategy in for-profit, nonprofit, and governmental organizations. Nonprofit Volunt Sector Q 29:183–208. doi:10.1177/089976400773746391

Parker AR (2003) Prospects for NGO collaboration with multinational enterprises. Globalization and NGOs: transforming business, government, and society. Greenwood Publishing Group, Santa Barbara, pp 81–105

Perego P, Kolk A (2012) Multinationals' accountability on sustainability: the evolution of third-party assurance of sustainability reports. J Bus Ethics 110:173–190. doi:http://dx.doi.org.ezproxy.lib.davidson.edu/10.1007/s10551-012-1420-5

Pinkse J, Kolk A (2012) Addressing the climate change—sustainable development Nexus: the role of multistakeholder partnerships. Bus Soc 51:176–210. doi:10.1177/0007650311427426

Qureshi A (2012) On the green carpet. Financ Times

Rodriguez Bolivar MP (2009) Evaluating corporate environmental reporting on the internet: the utility and resource industries in Spain. Bus Soc 48:179–205. doi:10.1177/0007650307305370

Ruta G (2010) Environmental defense fund. Good cop/bad cop: environmental NGOs and their strategies toward business. Resources for the Future Press, Washington, D.C., pp 184–194

Schiller RS, Almog-Bar M (2013) Revisiting collaborations between nonprofits and businesses: an NPO-centric view and typology. Nonprofit Volunt Sector Q 42:942–962. doi:10.1177/0899764012471753

Sowa JE (2004) No longer unmeasurable? a multidimensional integrated model of nonprofit organizational effectiveness. Nonprofit Volunt Sector Q 33:711–728. doi:10.1177/0899764004269146

Sunstein CR (1996) Social norms and social roles. Columbia Law Rev 96:903–968. doi:10.2307/1123430

Tungate M (2009) Luxury world: the past, present and future of luxury brands. Kogan Page Limited, London and Philadelphia

Uren S (2012) Can luxury be sustainable? Green Futures, 37–37

van Huijstee M, Glasbergen P (2010) NGOs moving business: an analysis of contrasting strategies. Bus Soc 49:591–618. doi:10.1177/0007650310365516

Vogel D (2008) Private global business regulation. Annu Rev Polit Sci 11:261–282. doi:10.1146/annurev.polisci.11.053106.141706

Weber J, Marley KA (2012) In search of stakeholder salience: exploring corporate social and sustainability reports. Bus Soc 51:626–649. doi:10.1177/0007650309353061

Wilson KL, Lizzio AJ, Zauner S, Gallois C (2001) Social rules for managing attempted interpersonal domination in the workplace: influence of status and gender. Sex Roles 44:129–154. doi:10.1023/A:1010998802612

Part IV
Sustainable Management in the Luxury Value Chain

Luxury Products and Services and the Sustainable Value Chain: Six Management Lessons from Gucci

John Armitage, Joanne Roberts and Yasmin K. Sekhon

Abstract This chapter investigates the case for the re-moralisation of the idea of luxury and thus for the re-moralisation of luxury products and services through a consideration of luxury products and services and the 'sustainable value chain', a concept that incorporates a company's production system and its impact on its stakeholders from shareholders, customers and suppliers to local communities and national governments. Following a discussion of the idea of luxury and the nature of luxury products and services, the idea of sustainability is considered before the sustainable value chain is explored in the context of the production of luxury products and services. The sustainable practices of the Italian luxury brand Gucci are then explored and six management lessons derived from Gucci's experience are outlined for all luxury businesses.

keywords Luxury · Luxury products and services · Sustainability · Sustainable value chain · Gucci

1 Introduction: Luxury Products and Services

Luxury is often associated with expensive, elegant, and refined products and services of the highest quality as well as with a rich, comfortable, and sumptuous lifestyle. Such products and services are also related to excessive quantity and viewed by many as superfluous, unnecessary, an indulgence or a waste of resources. These latter characteristics account for the negative moral view of luxury products and services that was generally accepted in the West until the eighteenth century. Indeed, luxury products and services were commonly regarded as morally abhorrent, even sinful; they were, therefore, subject to control through sumptuary laws that regulated their permitted consumption (Adams 2012; Hunt 1996).

J. Armitage · J. Roberts (✉) · Y.K. Sekhon
Winchester Luxury Research Group, Winchester School of Art,
University of Southampton, Park Avenue, Winchester, Hampshire SO23 8DL, UK
e-mail: j.roberts@soton.ac.uk

M.A. Gardetti (ed.), *Sustainable Management of Luxury*,
Environmental Footprints and Eco-design of Products and Processes,
DOI 10.1007/978-981-10-2917-2_12

The 'de-moralisation' of luxury products and services occurred following the arguments propounded by Hume (2012 [1752]) and Smith (1981 [1776]) concerning the beneficial outcomes related to the production and consumption of luxury products and services. In his highly influential book, *The Idea of Luxury*, Berry (1994) traces the historical development of the concept of luxury and, by implication, luxury products and services, in the West. He defines luxury as the antonym of necessity, given that it is distinct from basic needs, which are non-intentional and universal. Hence, for Berry, the idea of luxury occupies the realm of wants and desires.

However, luxury products and services, which may be regarded as socially non-necessary by some people, may be 'needed' by other people, either in a specific instrumental sense or because such products and services are the object of intense desire or identification. Consequently, not all unnecessary products or services are luxury products and services for everyone. For instance, Veblen's (1899) concept of conspicuous consumption, which is so often associated with luxury products and services, can, according to Berry's perspective, be interpreted as the instrumental consumption of luxury products and services since its purpose is to signal one's social status to others. Therefore, the consumption of some luxury products and services may be necessary for certain individuals to maintain their social position. But, for Berry (1994: 41, original italics), '*luxuries are those goods that admit of easy and painless substitution because the desire for them lacks fervency*'. What is clear from Berry's (1994) analysis, then, is that, ultimately, the idea of luxury and thus luxury products and services cannot be objectively defined because they depend on cultural, social, and individual contexts and meanings.

From a business and marketing perspective, however, Chevalier and Mazzalovo (2012) argue that luxury products and services can be defined if they meet three criteria. Firstly, they must have a strong artistic content; secondly, they must be the result of craftsmanship; and, thirdly, they must be international. The link between art, craftsmanship and luxury products and services is not new. Works of art and luxury products and services entail craftsmanship, require high levels of skill, time, and expensive materials. Consequently, consumption of luxury products and services has been the preserve of wealthy individuals and institutions. Nevertheless, changing income levels and techniques of production have made luxury products and services increasingly available to a wider range of individuals since the 20th century (Thomas 2007). Chevalier and Mazzalovo's (2012) suggestion that, for something to be a luxury product or service, it must also be international, is a consequence of the cultural and economic processes of globalization that have taken place since the mid-1980s. Moreover, in a diverse world, the view that luxury products and services must be international is to some extent in conflict with the idea of luxury products and services as socially constructed. Yet, this insistence on the international dimension is an insistence that is embedded in a business perspective on luxury, which is primarily concerned with market size and with the expanding geographical reach of luxury brands as a means to produce sustainable profits, especially among the large luxury sector conglomerates, including, for instance, LVMH (Moët Hennessy Louis Vuitton SE), Kering, and Richemont.

Luxury products and services have also been classified in terms of their accessibility by Allérès (1990), who identifies three levels of luxury products and services, namely, inaccessible (exclusive and/or unique luxury products and services), intermediate (expensive replicas of unique luxury products and services), and, accessible (factory produced and in large production runs). Today, we are witnessing the increasing democratization of or access to luxury products and services, and a proliferation of terms associated with them, such as 'new luxury' or 'mass luxury'. According to Kapferer and Bastien (2012), this explosion of labels is the result of, on the one hand, the efforts of traditional brands to trade up to luxury brands, and, on the other hand, the drive for profits among luxury brands and businesses by offering luxury products and services to a wider global market. Such changes also reflect the fragmentation of the luxury production process, such that the design of luxury products and services involve significant artistic inputs and craftsmanship, but the final products and services can be mass produced in low cost locations without any loss of quality (Thomas 2007; Tokatli 2014). This shift to 'mass luxury' has been accompanied by the emergence of the idea of 'meta-luxury' (Ricca and Robins 2012) or 'über luxury' (Quintavalle 2013) to distinguish between mass produced luxury products and services and those luxury products and services that are exclusive, because they are the result of high levels of skill and craftsmanship, and because their cost of production renders them accessible only to 'Ultra High Net Worth Individuals' [individuals with a net worth of at least $30 million USD (Wealth-X and UBS 2014)].

Additionally, the meaning of luxury products and services varies through time and space, and across economic, social, and cultural contexts. For instance, in the 1980s, a mobile telephone was a luxury product and service, but, today, it is regarded as a necessity in the developed and developing countries, with 96 mobile telephones per 100 world inhabitants (ITU 2015). Furthermore, the possession of an Internet-connected computer may be regarded as a luxury product and service in present day developing countries, yet a computer is seen by many as a necessity in the developed nations. So, luxury products and services are not fixed temporally, spatially, or economically, but, rather, are relative and socio-culturally constructed concepts, products, services, and practices.

Clearly, luxury is a complex idea, but it is also manifested in a very real form in the global marketplace where luxury products and services may be defined by high price. The characteristics of demand for luxury products and services differ from those of normal products and services. Luxury products are often referred to as 'Veblen goods'. According to Veblen's (1899) theory of conspicuous consumption, as the price of a luxury product increases, the utility that is gained from its consumption also increases, because such an increase in price allows consumers to signal their own increasing status. For this reason, a Veblen good is not necessarily of a higher quality than a normal product, as its status as a luxury product or service depends on the perceptions of consumers, of their ability to denote social standing through, for instance, recognizable luxury brand logos. While there is much anecdotal evidence to suggest that luxury goods and services do display the characteristics of Veblen goods (Bagwell and Bernheim 1996), it would be inappropriate to intimate that the price of all luxury products and services is merely determined by a form of consumer demand that is underpinned by the utility gained from implying high social standing. Though the accessible luxury products and services delineated by Allérès (1990) may be differentiated from normal products and services, largely by higher prices, with consumers willing to pay more to display their social position, for those luxury products and services that Allérès defines as inaccessible, in the sense of being exclusive unique products and services, their high price relative to normal products and services is likely to be determined by the greater costs of production. For even if the high price of such inaccessible luxury products and services will attract consumers who gain utility from communicating their social station, the price of such luxury products and services will primarily be based on the high costs of production rather than solely on demand derived from the desire of consumers to demonstrate their elevated social position.

From a business viewpoint, products and services acquire luxury status from the perceptions of consumers, or the high production costs, or a combination of the two (Roberts and Armitage 2015). Nevertheless, whatever the source of the status of luxury products and services, it is important that managers of luxury businesses remember the socio-cultural foundations of the meaning of luxury products and services. This is because, as socially constructed concepts, products, services, and practices, what is defined as luxury and therefore what are recognized as luxury products and services can shift rapidly due to factors beyond the commercial domain.

Moreover, the complex meaning of luxury, luxury products, and luxury services is ever more important because the luxury global market is expanding rapidly. The value of the luxury products and services market exceeded €850 billion in 2014, having grown at a rate of 7% on the previous year (D'Arpizio et al. 2014: 3). The de-moralisation of luxury products and services that began in the eighteenth century has unleashed a seemingly ever growing demand for expensive, elegant and refined products and services, which, especially from the standpoint of sustainability, can be viewed as contributing to the wasteful use of resources and detrimental to the alleviation of poverty and inequality. Also, the rising market for luxury products and services seems at odds with the current period of global economic instability stemming from the global financial crisis of 2008 (Armitage and Roberts 2016). To repay debts incurred during the financial crisis, for example, governments across the world have implemented austerity policies and cut public spending on social services. In contrast to the austerity experienced by the majority, the richest 1% of the world's population continues to increase its share of wealth (Dorling 2014; Piketty 2014). Besides, growing concerns over climate change and the depletion of resources has implications for the growth of consumption, and particularly consumption of luxury products and services. Yet, perhaps there is a case for what Christopher J. Berry calls the 're-moralisation' of the concept of luxury? Berry's (2016: 47) idea of 're-moralisation', of the notion that there is a feasible and sustainable approach to luxury in the modern world, identifies three grounds for such a re-moralisation of the concept of luxury in the current era, namely ethical, social and ecological. Among these, he sees the ecological argument as having the most potential for changing moral attitudes towards consumption because an ecological crisis will force a change of consumption patterns. But is the ever rising demand for luxury products and services incompatible with sustainability? Is the production of luxury products and services an uneconomical use of resources and harmful to the mitigation of privation and inequity? And just what are the connections between climate change, the diminution of resources, and the consumption of luxury products and services? In what follows, we investigate the case for the re-moralisation of the idea of luxury and thus for the re-moralisation of luxury products and services through a consideration of luxury products and services and the 'sustainable value chain', a concept that incorporates a company's production system and its impact on its stakeholders from shareholders, customers and suppliers to local communities and national governments. We then examine the sustainable practices of the Italian luxury brand Gucci, before drawing out six management lessons that Gucci's experience offers for all luxury businesses. But let us begin with the concept of sustainability.

2 Sustainability

The concept of sustainability has grown in popularity since the 1980s. The Brundtland Report of 1987 (World Commission on Environment and Development 1987: 8), which defined global sustainability as 'development that meets the needs of the present without compromising the ability of future generations to meet their needs,' was a key initiative stimulating interest in this idea (Kuhlman and Farrington 2010). Today, sustainability also involves rising expectations for corporate social and environmental performance. For, alongside Corporate and Social Responsibility (CSR), Corporate Social Performance (CPS), and Corporate Environmental Reporting (CER), all of which have become increasingly accepted in the corporate world since the 1990s (Carroll 1999), sustainability is now a key issue for businesses, including businesses in the luxury sector. Initially focused on the environment in response to global proposals and agreements, including the 1992 United Nations Framework Convention on Climate Change (UNFCCC) and the Kyoto Protocol of 1997, the notion of sustainability has expanded to include a range of international economic and community schemes, including the United Nations' (UN) Global Compact, which seeks to create a sustainable and inclusive global economy that delivers lasting benefits to people (UN 2016). Most recently, in September 2015, and building on previous programmes on sustainability and development, including the Millennium Development Goals (UN 2015a), the UN established 17 global Sustainable Development Goals (SDG), with the aim of achieving them by 2030 (Box 1).

Box 1 Sustainable Development Goals

- Goal 1. End poverty in all its forms everywhere
- Goal 2. End hunger, achieve food security and improved nutrition and promote sustainable agriculture
- Goal 3. Ensure healthy lives and promote well-being for all at all ages
- Goal 4. Ensure inclusive and equitable quality education and promote lifelong learning opportunities for all
- Goal 5. Achieve gender equality and empower all women and girls
- Goal 6. Ensure availability and sustainable management of water and sanitation for all
- Goal 7. Ensure access to affordable, reliable, sustainable and modern energy for all
- Goal 8. Promote sustained, inclusive and sustainable economic growth, full and productive employment and decent work for all
- Goal 9. Build resilient infrastructure, promote inclusive and sustainable industrialization and foster innovation
- Goal 10. Reduce inequality within and among countries
- Goal 11. Make cities and human settlements inclusive, safe, resilient and sustainable
- ***Goal 12. Ensure sustainable consumption and production patterns***
- Goal 13. Take urgent action to combat climate change and its impacts*
- Goal 14. Conserve and sustainably use the oceans, seas and marine resources for sustainable development
- Goal 15. Protect, restore and promote sustainable use of terrestrial ecosystems, sustainably manage forests, combat desertification, and halt and reverse land degradation and halt biodiversity loss
- Goal 16. Promote peaceful and inclusive societies for sustainable development, provide access to justice for all and build effective, accountable and inclusive institutions at all levels
- Goal 17. Strengthen the means of implementation and revitalize the global partnership for sustainable development

*Acknowledging that the United Nations Framework Convention on Climate Change is the primary international, intergovernmental forum for negotiating the global response to climate change.

Source: United Nations (2015b: 16)

Evidently, the concept of sustainability, particularly insofar as it is driving increasing expectations about corporate and environmental performance in the production and consumption of luxury products and services, is of critical environmental, economic, and communal importance. Certainly, as can be observed in Box 1, above, the UN's global Sustainable Development Goals have numerous implications, not only in terms of assuaging poverty and promoting good health, gender equality, inexpensive energy, hardy infrastructure building, safe cities, and so on, but, crucially for our purposes, as Goal 12 puts it, in guaranteeing 'sustainable consumption and production patterns', inclusive of such patterns in the luxury sector. In what ways, then, might we conceive of luxury products and services not only within the context of ever expanding demand for such products and services but also within the context of wider Sustainable Development Goals that demand urgent action, for instance, to resist climate change (Goal 13), oceanic degradation, terrestrial desertification (Goal 14 and 15), and growing social inequalities (Goal 10) while simultaneously promoting well-being, empowerment and opportunity for all (Goals 3–5)?

Following Hart and Milstein (2003: 56), we argue that a sustainable luxury enterprise 'is one that contributes to sustainable development by delivering simultaneously economic, social, and environmental benefits—the so-called triple bottom line.' The term 'Triple Bottom Line' (TBL) was coined in 1994 by John Elkington, the founder of the consultancy company SustainAbility (SustainAbility and UNEP 1994; Elkington 1997). Elkington (1997) argues that companies engaged in the production of luxury and other products and services should prepare three bottom lines: firstly, the traditional profit and loss account (profit); secondly, a measure of how socially responsible the company has been (people); and, thirdly, a measure of how environmentally responsible the company has been (planet) (*The Economist* 2009).

Of course, the motivation for company level sustainable practices varies from sector to sector, from that of a self-understood moral obligation to a grudging response to external government regulation (Hart and Milstein 2003). Generally, however, sustainability is increasingly viewed as a business opportunity since it offers scope for increased efficiency, lower costs, and lower risk; for growing revenues and market share through innovation; and for image and reputation enhancement, all of which provide a positive means of setting a company apart from its competitors. Sustainability can also be a way for luxury and other companies to attract and secure resources for the future (Fawcett et al. 2015). But, as Fawcett et al. (2015) have demonstrated, the adoption of sustainability does tend to be popular with business in times of economic growth but, in times of economic decline, it is inclined to be viewed by many companies *as* a luxury, such that sustainability has so far failed to become a mainstream practice. Similarly, in prosperous periods, luxury consumers especially are willing to pay more for their products and services than when the economy is in recession. Yet, if stimulated by conspicuous consumption, high end luxury consumers are content to pay a sustainability premium to display their wealth through the acquisition of luxury products and services. The truly wealthy, not surprisingly, are unaffected by economic downturns and thus continue to purchase sustainable and other luxury products and services. But, for luxury companies, we contend, the key to long term success is to truly embrace sustainable business ideas and practices.

Indeed, and although global sustainability is a complex concept and practice, following Hart and Milstein (2003: 59), we suggest that sustainable business concepts and practices can create value by reducing the level of material consumption and pollution; by operating at greater levels of transparency and responsiveness; by the development of new, disruptive, technologies that shrink the size of the human footprint on the planet; and, by meeting the needs of those at the bottom of the world income pyramid in a way that facilitates inclusive wealth creation and distribution.

The ever present challenge of producing luxury products and services in an era of climate change, resource depletion and widespread inequality, therefore, requires ensuring that sustainable development remains a key part of luxury business thinking, discourses, and practices. The secret is to produce luxury products and services with full awareness of the environmental and social challenges facing the world's population and to embed sustainable development, not only in luxury business philosophy, but, also in the practices associated with the sustainable value chain.

3 Luxury Products and Services and the Sustainable Value Chain

Before we can discuss luxury products and services and the sustainable value chain, it will be helpful if we first of all clarify the term 'value chain'. The concept of the value chain describes the complete range of activities that luxury and other firms and workers perform to bring a product from conception to end-use and beyond. Such activities include Research and Development (R&D), design, production, sales and marketing, distribution, support to the final consumer, and recycling (Gereffi and Fernandez-Stark 2011: 4). Popularized by Porter (1985) as a means of conceptualising the production system of an organization as a set of subsystems, each with inputs, transformation processes, and outputs, the idea of the value chain offers opportunities to further the firm's competitive advantage. Incorporating sustainability into the conventional value chain involves thinking beyond the value created for the end user or customer. Indeed, it is necessary to consider the 'chain' or the wider context within which an activity occurs, focusing on the full range of stakeholders, including all those individuals, groups and organizations that are affected by the value chain activity, from, for instance, workers, customers and suppliers to shareholders, local communities and governments. Fundamentally, a sustainability approach requires an appreciation of the environmental and social impacts of the value chain, enabling businesses and society to address the challenges associated with the lifecycle of products and services (World Business Council for Sustainable Development 2011: 3). Similarly, a sustainable supply chain needs to reflect the longer term ramifications of the creation of value for all supply chain stakeholders at a local, national and global level. Consequently, supply chain sustainability involves the management of environmental, social and economic impacts, and the encouragement by management of good governance practices, throughout the lifecycles of products and services. The objective of supply chain sustainability is, firstly, to create, protect and grow long-term environmental, social and economic value for all stakeholders involved in, and affected by, bringing products and services to market and, secondly, to promote broader sustainable development objectives (United Nations Global Compact and Business for Social Responsibility 2010).

It is important to note that the value chain and the supply chain are often used as complementary ways of analysing an organization's activities. Actually, sometimes, the terms are used interchangeably. Supply Chain Management (SCM), which also emerged in the 1980s, can be defined as a way of managing the flow of products and services from suppliers to the end user (Cooper et al. 1997). SCM evolved throughout the 1990s as companies increased the efficiency of their operations by means of the implementation of Information and Communication Technologies (ICTs). For instance, the large American retailer, Wal-Mart, using SCM through point-of-sale data to enable continuous replenishment of inventory, was able to significantly drive down costs (Sherer 2005). Due to their focus on quality and design excellence, luxury companies were initially slow to take up SCM practices. However, by the mid-2000s, luxury companies, including LVMH, Versace, Girogio Armani, Burberry and Prada were adopting ICT systems to improve the efficiency of their supply chains. Indeed, Louis Vuitton's former CEO, Yves Carcelle, recruited managers formerly employed by Toyota, and therefore familiar with SCM practices, to improve the efficiency of Louis Vuitton's supply chain (Passariello 2006).

Nevertheless, there are important differences between a value chain approach and a SCM approach when analysing business processes. Whereas a SCM approach emphasises the costs and efficiencies of supply, a value chain approach focuses on the *customer* as the source of value, with value flowing from the customer, in the form of demand, to the supplier (Ramsay 2005). Thus, SCM concentrates on integrating supplier and producer processes, improving efficiency and reducing waste, while a value chain focuses on creating value in the eyes of the customer (Feller et al. 2006).

Notably, a value chain may extend not only beyond the boundaries of the business organization to include a wide range of suppliers but also beyond the boundaries of the nation state. In fact, from a Global Value Chain (GVC) standpoint, the world economy is a complex network of suppliers and buyers that are integrated and led by Multinational Enterprises (MNEs) (Lee and Gereffi 2015: 321). United Nations Conference on Trade and Development (UNCTAD 2013: 133–135), for example, estimates that 80% of world trade now passes through GVCs, with the developing country share of global value-added trade doubling from 20 to 40% between 1990 and 2010. Moreover, in a study of 39 countries, Jiang and Milberg (2013) estimate that GVCs have generated 88 million jobs. Similarly, MNEs, through their leadership of GVCs, have great scope to spread sustainable development practices across multiple supplier organizations in widely varied locations. Fawcett et al. (2015), for instance, argue that the implementation of sustainable business practices across a firm's value chain can lead to significant efficiency gains as well as reputation enhancement.

Having elucidated the term value chain concerning the organization, the global economy, MNEs, trade, jobs, and sustainable business practices, the key question now is: just what are the links between the consumption of luxury products and services, climate change, resource depletion, the well-being of workers and communities and the sustainable value chain?

We argue that the consumption of luxury products and services and a commitment to sustainability are not incompatible or contradictory. For example, one of the important links between sustainability and luxury products and services is, as Gardetti and Torre (2015), put it, a return to 'the essence of luxury with its traditional focus on thoughtful purchasing and a return to artisanal manufacturing, to the beauty of quality materials and to respect for social and environmental issues'. Correspondingly, Kapferer (2010: 42) argues that there is a convergence between luxury products and services and sustainability because '[d]urability is at the heart of sustainable development as well as luxury. Durability is the enemy of the fashion industry and of the mass market industry, based on planned obsolescence. On the contrary, luxury is the business of lasting worth'. Such remarks may be true of meta luxury products and services but they are not true for mass luxury products and services. In mass markets, luxury clothing, for example, is made in the same outsourced, off-shored, factories that are used by lower cost retailers (Thomas 2007). The business model of luxury companies distributing such products and services is thus based on a rapidly revolving fashion cycle, with obsolescence acting as an essential element that ensures continually growing sales, profitability, and return on shareholders' investments.

However, numerous studies reveal that sustainability is a low priority for many luxury consumers (Davies et al. 2012; Achabou and Dekhili 2013; Beckham and Voyer 2014). For instance, Achabou and Dekhili (2013) examined the propensity of consumers to consider recycled materials in luxury purchases through an empirical study of a French luxury clothing company. They found that incorporating recycled materials in such products and services negatively affects consumer preferences, indicating an incompatibility between recycling and luxury clothing. Such research suggests that, despite the growing concerns of consumers regarding sustainability, the environmentally responsible behaviour of the brand is of secondary concern to consumers of luxury products and services whose primary focus is on the intrinsic quality of the products and services.

How, then, can luxury companies make the case for the re-moralisation not only of the idea of luxury but also for the re-moralisation of luxury products and services through the sustainable value chain?

First, luxury companies can make the argument for the re-moralisation of the idea of luxury and luxury products and services through publicly embracing the sustainable value chain. Second, luxury companies can communicate their social and environmental activities more effectively to all their stakeholders (Joy 2015). So, for instance, LVMH adopted an environmental charter as far back as 2001, and has been auditing its carbon footprint since 2004. But are LVMH's customers aware of its charter? Luxury companies would thus be well advised to follow the general trend of producing detailed corporate responsibility reports concerning all of their business activities, as exemplified by Tiffany & Co. (2014), which already produces separate corporate responsibility and sustainability reports, in addition to its regular annual reports.

Yet, it is the luxury group Kering that has made most progress in measuring its impact on the environment. Certainly, Kering was one of the first companies to publish an economic valuation of the environmental impacts caused by greenhouse gas emissions and water consumption along the entire supply chain of Puma, one of its sportswear brands (Confino 2011). Moreover, Kering has made a commitment to the difficult task of integrating both its social and economic impacts into its future accounts.

Consequently, Kering is leading the way among luxury companies in terms of sustainability, having developed an innovative methodology, the Environmental Profit & Loss (EP&L) tool, which makes the seemingly invisible impacts of its business visible, quantifiable, and comparable (Kering Group 2016). Traditionally, a number of metrics have been used to assess the environmental impact of a business, from tonnes and hectares to cubic meters, depending on what is being measured, whether it is land employed or water consumed in the production process or the quantities of pollutants emitted across the supply chain. The use of a variety of units of measurement has made it challenging for businesses to coherently evaluate the financial impact of their activities on the environment. However, Kering's EP&L tool permits an overarching measurement to be applied, facilitating the assessment and comparison of activities, including risks and opportunities, across the supply and value chains. Likewise, Kering (2013) openly shares the methodology that underpins its EP&L tool, thereby encouraging its adoption by other businesses.

In these ways, Kering seeks to understand what impact its decision-making has on its own sustainable practices. Furthermore, the knowledge gained through the use of the EP&L tool is employed to inform corporate decisions with the aim of minimising Kering's risk and strengthening the long-term presence of the Group and its individual brands. The EP&L tool measures the impact of activities throughout the value chain, from the acquisition of raw materials, processing, manufacturing, and assembly to operations and retail. Indeed, in September 2015, Kering was recognized as an industry trailblazer in the 2015 Dow Jones Sustainability Indices (DJSI) at the World and Europe levels, leading the Textiles, Apparel and Luxury Goods sector for the second consecutive year (Kering Kering 2016a). Additionally, in 2016, Kering was acknowledged as one of the world's most sustainable corporations, according to the Corporate Knights' Global 100 index, where it is ranked 43rd overall, and second within the index's Textile, Apparel and Luxury Goods sector (Corporate Knights 2016).

But what is the secret of Kering's success regarding its management decision-making concerning sustainability, future planning, and branding? How does Kering's EP&L tool impact on its value chain? Let us consider the case of the Italian luxury brand Gucci, which, as a wholly owned subsidiary of Kering, accounted for over 52% of Kering's luxury related revenue in 2014 (Kering 2014), before we go on to draw out the management lessons that can be learned from this luxury industry leader.

4 Gucci and Sustainability

In this section, our focus is not only on Gucci's luxury products and services but also on its approach to sustainability, to its production processes, to its resources, to its customers, and, above all, to its re-moralisation of its own luxury products and services by way of its sustainable value chain.

Founded in 1921, Gucci is of course an Italian luxury fashion and leather goods brand, which has developed its business practices to ensure that the sustainability agenda is at the heart of its corporate performance. According to Kering (2014: 42), Gucci's parent company, 'Gucci's goal is to continue enhancing brand value while delivering long-term growth across product categories and maintaining high standards of environmental and social sustainability.' Thus, Gucci, as a renowned fashion producer and leader, is an organization that is also highly committed to offering its consumers sustainable luxury products and services. Gucci's holistic approach to sustainable development goals is, moreover, reflected in its internal and external structures, from the use of sustainable tools through to its triple bottom line.

However, as we will show, Gucci does not view sustainability as a business threat but as a business opportunity through which it can both challenge its competitors in the marketplace and secure sustainable resources to build luxury brand value. Gucci's embrace of sustainable business practices is, then, creating value for the company even in an era of climate change, diminishing resources and widespread inequality. Hence, Gucci has incorporated the idea of sustainable development into its marketing discourse and activities. In 2011, for example, and as part of a public relations and sustainability exercise, Gucci launched a design competition, where eighteen participants were asked to design a handbag using bamboo with minimal environmental impact (Spagnoli Gabardi, n.d). The winner's design was successful in fulfilling Gucci's sustainability agenda as well as in generating excellent public relations for the company since the competition highlighted its innovative, sustainable, and customer friendly practices. The use of bamboo was also extended to Gucci's watch and bracelet lines. Gucci thus produces luxury products and services that are aware of climate change but, within its own business context, also embedded within its marketing discourses and sustainable value chain.

Similarly, Gucci's luxury products and services, with their commitment to sustainability, are constituted from value created at every stage of the supply chain. One Gucci initiative, for instance, which parallels its customer's growing interest in sustainable consumption (Hoeffler et al. 2010; Brosius et al. 2013), is its acquisition of sustainable leather through carefully chosen suppliers and the development of eco-friendly handbags and eyewear. Indeed, in 2013, Gucci launched a range of leather handbags sourced from Rainforest Alliance certified ranches in the Brazilian Amazon (Rainforest Alliance 2013). The Rainforest Alliance supports farming communities in Brazil by encouraging low environmental impact and sustainable practices, including, for example, guaranteeing the good treatment of livestock and the promotion of the rights and wellbeing of ranch workers. Furthermore, Gucci signed a moratorium with Greenpeace whereby the leather used by Gucci was not sourced from cattle raised on former Rainforests in the Amazon (Campione 2013). The 'Gucci for Green carpet challenge' 'zero deforestation' handbags, designed in collaboration with Livia Firth and launched in 2013, also demonstrates how a luxury brand that emphasizes heritage, exclusivity, and contemporary Italian craftsmanship is able to incorporate traditional sustainable practices into its products and services (Karmali 2013). Moreover, each handbag comes with its own passport, detailing its journey through the supply chain, from the sourcing of leather, to its design, manufacture, and final creation (Rainforest Alliance 2013).

Gucci's sustainable product range also extends to its eye and footwear. For instance, in 2012, it launched sunglasses made from liquid wood, a biodegradable, eco-friendly material that is an alternative to plastic, which is often used in the manufacture of eyewear (Kering 2016b; Spagnoli Gabardi, n.d). Hence, the production of these sunglasses had a lower environmental impact than many others because the material used was natural rather than refined. More recently, Gucci launched 'Sustainable Soles', including a ballerina flat female shoe that is made from bio plastic together with a men's range featuring bio rubber soles (Spagnoli Gabardi, n.d).

Additional to its sustainable product lines, in 2007, Gucci launched its certification process, which is applied throughout its supply chain. This certification means, for instance, that Gucci will ensure that its jewellery materials are ethically mined. To achieve this level of sustainability, Gucci has diversified its supply chain process such that, rather than relying on third party mining companies, it has taken ownership of mines and is intensely involved in the extraction of precious metals. Hence, Gucci's 'operational capabilities are assisting the organization's efforts to reduce its environmental impact throughout its supply chain' (Darnall et al. 2008: 30).

However Gucci's sustainable practices also extend throughout its value chain, such that its customers are given the opportunity to *participate in* Gucci's eco-friendly programme through, for example, the recycling of packaging. Customers can thus return their packaging, using a pre-paid envelope, to a Gucci centre where it will be used to make new products. Indeed, since 2010, Gucci has ensured that all of its packaging is produced either from recycled materials or paper certified as 100% sustainable and recyclable by the Forest Stewardship council.

Gucci's commitment to the sustainable value chain, then, places the customer centre stage whilst also maintaining trade, jobs, and the development of sustainable business practices. Making the critical connection between the consumption of Gucci luxury products and services, climate change, and declining resources through the sustainable value chain, the company's approach to sustainability is one that incorporates a value led customer focus across all parts of its business strategy. This method not only lowers Gucci's environmental impact but also adds value to Gucci's customer's experiences of the company's luxury products and services. The all-encompassing approach adopted by Gucci thus signals a long-lasting commitment to the well-being of all the company's stakeholders. It is a method that is welcomed by customers because, as their own expectations of conspicuous (Douglas and Isherwood 1979; Dittmar 1992, 2008) or inconspicuous consumption (Erkhardt et al. 2015) of luxury brands rises, Gucci, through its sustainable value chain and product ranges, offers its customers the opportunity to sometimes display but always enjoy ethical and sustainable luxury products and services.

Understanding the importance of the social issues associated with sustainable business practices, Gucci considers the wider social impact of its own business processes. In 2004 Gucci committed to achieving SA8000 certification, an international standard from Social Accountability International that aims to support companies in the development and management of social accountability systems, which address CSR issues including child labour, health and safety conditions and working hours and compensation. Gucci went on to gain SA8000 certification in 2007 for the supply chain of its leather and jewellery goods (Massimiliano et al. 2016). For Gucci, the importance of looking after its workforce includes ensuring diversity, providing rights and equal opportunities for its staff as well as respect for health and safety. In addition Gucci takes a longer-term view of its decisions by implementing policies and procedures that encourage continuous improvement.

Furthermore, Gucci has been actively involved in philanthropy concerning issues relating to women's and girl's education. Having forged a partnership with the United Nations Children's Emergency Fund (UNICEF), Gucci has made charitable donations and created 'UNICEF' products, where money from sales has been passed on to the charity. An example of the latter was Gucci's Nice shopper bag, where 25% of the retail sales price of each bag purchased was donated to UNICEF's "Schools of Africa" and "Schools of Asia" programmes, providing education for vulnerable children (UNICEF 2013). Additionally, Gucci has invested in developing partnerships with female celebrities including Beyonce Knowles-Carter, Salma Hayek Pinault, Halle Berry and Jennifer Lopez, such as the launch in 2013 of the Chime for Change global campaign to raise funds for projects promoting education, health and justice for girls and women around the world. Through crowdfunding, this campaign has funded over 400 projects associated with its 144 non-profit partners in 86 countries, reaching hundreds of thousands of girls and women globally (http://www.chimeforchange.org/). Moreover, through its holding company, Gucci also supports the Kering Corporate Foundation, which was launched in 2009, and which embodies the social commitment of the group (http://www.keringfoundation.org/kering-foundation). The Foundation combats violence against women through community based projects and initiatives with the aim of giving women increased independence. Many of these activities are global in nature and aim to empower women everywhere. Hence, Gucci places a strong emphasis on socially led initiatives, which extend beyond practices focused on the environment, giving a greater level of societal and shared value amongst its stakeholders.

In this way, Gucci has enhanced its social and economic luxury brand value through the sustainable value chain and extended its *symbolic* social value beyond heritage, craft and other traditional luxury brand values, such that the purchase of a sustainable Gucci product or service now indicates a more thoughtful and solicitous act that concerns itself with issues of durability and meta luxury, rather than ones based purely on conspicuous consumption and social status (Vickers and Renand 2003: 472; Han et al. 2010; Choo et al. 2012).

This last is critical since Gucci's most remarkable achievement here is that it has re-moralised its luxury products and services through the implementation of its own sustainable value chain. To attain this, Gucci has not only embraced its own sustainable value chain *internally* but has also actively publicized its commitment to a sustainable value chain *externally* through a range of marketing and public relations activities, including the handbag design competition noted above. Hence, Gucci communicates to all of its stakeholders its commitment to the sustainable value chain. So for instance, armed with the knowledge of Gucci's sustainable business practices, customers are empowered to incorporate issues concerning sustainability into their luxury purchasing decisions and suppliers are integrated into networks of sustainable production.

Of course, to provide convincing information about Gucci's sustainable business practices, it is necessary to have a robust assessment of the impact of its activities on the environment and society. Hence, the use of the EP&L tool helps Gucci to display its own reduced influence upon the earth through quantifying such effects in monetary terms, by measuring, for example, the potentially environmentally harmful chemicals used in its own production processes and those of its suppliers. The knowledge gained through the adoption of the EP&L tool allows Gucci to measure and implement a 'cradle to grave' approach to the management of its 'ecobalance', including its choice of product lines as well as the materials used in their manufacture. Thus, Gucci's measurement of its environmental footprint highlights the company's ecological liabilities and allows Gucci to implement sustainable supply and value chains, inclusive of its sourcing of sustainable materials, the recycling of its unwanted packaging, and for the whole to be easily evaluated. However, Gucci has yet to establish methods to measure the full impact of its social initiatives, focusing merely on documenting the number of projects, partnerships and individuals supported. Such social initiatives have the potential to have lasting value with significant long-term benefits but cannot be wholly quantified in monetary terms.

As this section has illustrated, through its adoption of sustainable business practices across its supply and value chain, Gucci is at the forefront of the luxury sector in terms of developing sustainable consumption and production patterns. We turn now to a discussion of what other luxury businesses can learn from Gucci's adoption of a sustainable approach to its value chain.

5 Six Management Lessons from Gucci: Discussion and Conclusions

Our exploration of Gucci's sustainable supply and value chain yields six key lessons for managers of luxury businesses. The *first* of these management lessons is that luxury customers are sympathetic to a corporate sustainable luxury strategy.

Even though luxury products and services produced through a sustainable value chain may be more costly than those produced in non-sustainable ways, customers appreciate the added value of contributing to initiatives seeking to protect of the environment, improve the welfare of workers, and raise the well-being of all communities touched by the production, distribution and consumption of luxury products and services. For those customers principally motivated by conspicuous consumption, the added cost of sustainably produced luxury products and services adds utility through its ability to indicate increased status.

Yet, the relationship between the consumption of luxury and sustainability goes further than a desire to protect the environment, to support safe working conditions, enhance the welfare of communities, or even engage in new types of conspicuous consumption. Thus, the *second* management lesson that other luxury companies can learn from Gucci concerns the cultivation of symbolic value associated with sustainable business practices. To accomplish this, Gucci embeds environmental and social awareness in every aspect of its luxury products and services through an emphasis on the symbolic values of the quality and the excellence of the brand, of its social eminence and worth. Consequently, luxury businesses must encourage in their customers the symbolic values associated with social and environmental responsibility, durability and meta luxury with a view to making sustainable luxury consumption a high social signifier for their customers. For, through its sustainability initiatives, Gucci has added symbolic value to its luxury products and services, making its customer's consumption of them more than simply a hedonic experience but also, and somewhat paradoxically, an act of luxury consumption that is embedded in a socially responsible agenda, which promotes and contributes to the global sustainable development agenda. In this sense, Gucci is engaged in the re-moralisation of luxury.

Therefore, the *third* management lesson that other luxury companies can learn from Gucci is the necessity to embrace their own sustainable value chain *internally* within the company. It is through an internally consistent commitment to sustainability that it is possible to convey a strong and convincing message concerning the businesses' devotion to sustainability to external stakeholders. Relatedly, the *fourth* management lesson that other luxury companies can learn from Gucci is, then, the necessity of publicizing *externally* the fact that they have adopted their own sustainable value chain internally. So, for instance, luxury customers want to see more than LVMH's near invisible charter or sections added to the end of a luxury company's annual report. By communicating its commitment to sustainability to all of its stakeholders, Gucci not only sets an example for other luxury companies to follow but also broadcasts its ethical and environmental awareness, thereby adding to the symbolic value of its products and services.

As a result, the *fifth* management lesson that other luxury companies can learn from Gucci is the need to demonstrate in a systematic and convincing fashion their own impact upon the environment and their contribution to the achievement of the UN's sustainable development goals. As noted, Gucci's EP&L tool helps the company measure and display its own reduced influence upon the environment through quantifying such effects in monetary terms. Consequently, other luxury companies need to use instruments similar to the EP&L tools employed by Gucci to measure their environmental footprint and to assess their efforts to improve sustainability. Gucci is also active in documenting its positive contribution to sustainability and social responsibility through its own website and reports as well as those of Kering, its parent company. Moreover, through initiatives like Chime for Change, Gucci activity promotes its contribution to sustainable development through social media, connections with top celebrities, and live high profile events, such as the Chime for Change concert held in London in 2013, featuring performances from a host of female artists, including Beyonce Knowles-Carter and Jennifer Lopez (http://www.chimeforchange.org/).

Hence, the *sixth* and final management lesson that other luxury companies can learn from Gucci is that customer consumption of sustainable Gucci luxury products and services has not proven problematic for the company but, on the contrary, beneficial to it. Consequently, the experience of Gucci underlines that the pursuit of sustainable business practices can add to, rather than diminish, a luxury company's competitive advantage. For not only do sustainable business practices attract favourable publicity and thereby act as a means of promoting Gucci's socially responsible approach but they also contribute to the symbolic value of its products and services, thereby enhancing their luxury status.

In conclusion, if luxury companies wish to become not merely leaders in sustainability but also a luxury industry leader like Gucci, they would be well counselled to accept the ideas associated with luxury products and services and the sustainable value chain discussed in this chapter. The six management lessons drawn from Gucci's implementation of sustainability across its supply and value chains can be adopted, adapted, and extended by other luxury businesses. For the need for sustainable business practices is becoming ever more pressing, such that businesses like Gucci, and its parent group, Kering, are now shaping sustainable consumption and production patterns in the luxury sector. In so doing, they are encouraging a re-moralisation of luxury that gives recognition to the environmental and social costs involved in the production of exclusive and exquisite products and services. Through openly sharing the methodology that underlies the EP&L tool employed by Gucci, Kering is similarly promoting the implementation of sustainability across the luxury sector and beyond. Or, as François-Henri Pinault, the CEO and Chairman of Kering, succinctly put it: 'sustainable business is smart business' (Godelnik 2013).

References

Achabou MA, Dekhili S (2013) Luxury and sustainable development: is there a match? J Bus Res 66:1896–1903

Adams WH (2012) On luxury: a cautionary tale. Potomac Books, Washington DC

Allérès D (1990) Luxe, stratégies-marketing. Economica, Paris

Armitage J, Roberts J (2016) Critical luxury studies: defining a field. In: Armitage J, Roberts J (eds) Critical luxury studies: art, design, media. Edinburgh University Press, Edinburgh

Bagwell LS, Bernheim BD (1996) Veblen effects in a theory of conspicuous consumption. Am Econ Rev 86(3):349–373. http://www.jstor.org/stable/2118201

Beckham D, Voyer BG (2014) Can sustainability be luxurious? A mixed-method investigation of implicit and explicit attitudes towards sustainable luxury consumption. Adv Consum Res 42:245–250

Berry CJ (1994) The idea of luxury: a conceptual and historical investigation. Cambridge University Press, Cambridge

Berry CJ (2016) Luxury: a dialectic of desire? In: Armitage J, Roberts J (eds) Critical luxury studies: art, design, media. Edinburgh University Press, Edinburgh

Brosius N, Fernandez KV, Cherrier H (2013) Reacquiring consumer waste: treasure in our trash? J Publ Policy Mark 32(2):286–301

Campione C (2013) Gucci "bag" a new kind of eco-luxury, Blogpost, Greenpeace international, 13 Mar. Available at: http://www.greenpeace.org/international/en/news/Blogs/makingwaves/gucci-bag-a-new-kind-of-eco-luxury/blog/44319/. Accessed 21 Mar 2016

Carroll AB (1999) Corporate social responsibility: evolution of a definitional construct. Bus Soc 38(3):268–295

Chevalier M, Mazzalovo G (2012) Luxury brand management: a world of privilege, 2nd edn. Wiley, Singapore

Choo HJ, Moon H, Kim H, Yoon N (2012) Luxury customer value. J Fashion Mark Manag Int J 16(1):81–101

Confino J (2011) Puma world's first major company to put a value on its environmental impact. The Guardian Professional Network, The Guardian, 16 May. Available at: http://www.theguardian.com/sustainable-business/puma-value-environmental-impact-biodiversity. Accessed 2 Mar 2016

Cooper MC, Lambert DM, Pagh JD (1997) Supply chain management: more than a new name for logistics. Int J Logistics Manag 8(1):1–14

Corporate Knights (2016) 2016 Global 100 results: the results for the 2016 global 100 most sustainable corporations in the world index. Corporate Knights: the magazine for clean capitalism, 20 Jan, Winter Issue. Available at: http://www.corporateknights.com/reports/2016-global-100/2016-global-100-results-14533333/. Accessed 2 Mar 2016

D'Arpizio C, Levato F, Zito D, de Montgolfier J (2014) Luxury goods worldwide. Market study, fall-winter 2014: the rise of the borderless consumer. Bain & Company: http://www.bain.com/publications/articles/luxury-goods-worldwide-market-study-december-2014.aspx. Accessed 22 Jan 2016

Darnall N, Jolley GJ, Handfield R (2008) Environmental management systems and green supply chain management: complements for sustainability? Bus Strat Env 17:30–45. doi:10.1002/bse.557

Davies I, Lee Z, Ahonkhai I (2012) Do consumers care about ethical-luxury? J Bus Ethics 106 (1):37–51

Dittmar H (1992) The social psychology of material possessions: to have is to be. Harvester Wheatsheaf, Hemel Hempstead

Dittmar H (2008) Consumer culture, identity and well-being: the search for the 'good life' and the 'body perfect'. Psychology Press, Hove

Dorling D (2014) Inequality and the 1%. Verso, London and New York

Douglas M, Isherwood B (1979) The world of goods: towards an anthropology of consumption. W. W. Norton, New York

Elkington J (1997) Cannibals with forks: the triple bottom line of 21st century business, Capstone

Erkhardt GM, Belk RW, Wilson JAJ (2015) The rise of inconspicuous consumption. J Mark Manag 31(7–8):807–826

Fawcett SE, Brockhaus S, Knemeyer MA, Fawcett AM (2015) Sustainability as strategy: caught in the luxury trap. Supply Chain Manag Rev, 16–25. Available at: www.scmr.com. Accessed 3 Mar 2016

Feller A, Shunk D, Callarman T (2006) Value chains versus supply chains. BPTrends. Available at www.BPTrends.com. Accessed 4 Mar 2016

Gardetti MA, Torre AL (2015) Introduction. In: Gardetti MA, Torre AL (eds) Sustainable luxury: managing social and environmental performance in iconic brands. Greenleaf Publishing, Sheffield

Gereffi G, Fernandez-Stark K (2011) Global value chain analysis: a primer. Center on Globalization, Governance & Competitiveness, Durham, NC. Available at: www.cggc.duke.edu/pdfs/2011-05-31_GVC_analysis_a_primer.pdf. Accessed 1 Mar 2016

Godelnik R (2013) 5 lessons from Kering, the sustainability dark horse. Triple pundit: people, planet, profit. Monday, 12 Aug. Available at: http://www.triplepundit.com/2013/08/5-lessons-kering-unknown-sustainability-leader/. Accessed 26 Mar 2016

Han YJ, Nunes JC, Dreze X (2010) Signaling status with luxury goods: the role of brand prominence. J Mark 74:15–30

Hart SL, Milstein MB (2003) Creating sustainable value. Acad Manag Executive 17(2):56–67

Hoeffler S, Bloom PN, Keller KL (2010) Understanding stakeholder responses to corporate citizenship initiatives: managerial guidelines and research directions. J Publ Policy Mark 29(1):78–88

Hume D (2012 [1752]) Political discourses. Forgotten Books, London

Hunt A (1996) Governance of the consuming passions: a history of sumptuary law. St Martin's Press, New York

ITU (2015) World telecommunication/ICT indicators database 2015 (19th edn/December 2015). Available from: https://www.itu.int/en/ITU-D/Statistics/Pages/publications/wtid.aspx. Accessed 4 Mar 2016

Jiang X, Milberg W (2013) Capturing the jobs from globalisation: trade and employment in global value chains. Capturing the gains, working paper 2013/30, University of Manchester. Available at: www.capturingthegains.org/publications/workingpapers/wp_201330.htm. Accessed 1 Mar 2016

Joy A (2015) Fast fashion, luxury brands, and sustainability. European Financial Review, 22 June. Available at: http://www.europeanfinancialreview.com/?p=4589. Accessed 4 Mar 2016

Kapferer J-N (2010) All that glitters is not green: the challenge of sustainable luxury. Eur Bus Rev, 40–45

Kapferer J-N, Bastien V (2012) The luxury strategy: break the rules of marketing to build luxury brands. Kogan Page, London

Karmali S (2013) Gucci launches first zero-deforestation handbag line. Vogue, 5 Mar. Available at: http://www.vogue.co.uk/accessories/news/2013/03/05/gucci-creates-first-zero-deforestation-ethical-handbag-collection—green-carpet-challenge. Accessed 23 Mar 2016

Kering (2013) Environmental profit & loss: methodology and 2013 group results. Available at: http://www.kering.com/sites/default/files/document/kering_epl_methodology_and_2013_group_results_0.pdf#page=24. Accessed 23 Mar 2016

Kering (2014) Activity report 2014. Available at: http://www.kering.com/sites/default/files/ra2014_kering_en.pdf. Accessed 5 Mar 2016

Kering (2016a) Kering recognised as one of the global 100 most sustainable corporations in the world by Corporate Knights, Press Release, 21.01.2016. Available at: http://www.kering.com/sites/default/files/press_release/pr_kering_global100_21.01.16.pdf. Accessed 2 Mar 2016

Kering (2016b) Gucci's sustainable products. Available at: http://www.kering.com/en/sustainability/achievements/guccis_sustainable_products. Accessed 23 Mar 2016

Kering Group (2016) Environmental P&L. Available at: http://www.kering.com/en/sustainability/epl. Accessed 2 Mar 2016

Kuhlman T, Farrington J (2010) What is sustainability? Sustainability 2(11):3436–3448. doi:10.3390/su2113436

Lee J, Gereffi G (2015) Global value chains, rising power firms and economic and social upgrading. Crit Perspect Int Bus 11(3/4):319–339

Massimiliano B, Perego P, Ravagli R (2016) The evolution of corporate social responsibility in Gucci: from risk management to stakeholder engagement. In: Lindgreen A, Kotler P, Vanhamme J, Maon F (eds) A Stakeholder approach to corporate social responsibility: pressures, conflicts, and reconciliation. Routledge, London

Passariello C (2006) Louis Vuitton Tries modern methods on factory lines. Wall Street J, 9 Oct. Available at: http://www.wsj.com/articles/SB116035785257486463. Accessed 23 Mar 2016
Piketty T (2014) Capital in the twenty-first century. Harvard University Press, Cambridge
Porter ME (1985) Competitive advantage: creating and sustaining superior performance. The Free Press, New York
Quintavalle A (2013) Über luxury: for billionaires only. In: Hoffmann J, Coste-Maniere I (eds) Global luxury trends. Palgrave Macmillan, Hounsmill, Basingstoke, pp 51–76
Rainforest Alliance (2013) Gucci goes sustainable: new handbag line made with leather from rainforest alliance certified™ Ranches, Press Release. Rainforest alliance, 4 Mar. Available at: http://www.rainforest-alliance.org/newsroom/press-releases/gucci-goes-sustainable. Accessed 21 Mar 2016
Ramsay J (2005) The real meaning of value in trading relationships. Int J Oper Prod Manag 25:549
Ricca M, Robin R (2012) Meta-luxury: brand and the culture of excellence. Palgrave Macmillan, London
Roberts J, Armitage J (2015) Luxury and creativity: exploration, exploitation, or preservation? Technol Innov Manag Rev 5(7):41–49
Sherer SA (2005) From supply-chain management to value network advocacy: implications for e-supply chains. Supply Chain Manag 10:77
Smith A (1981 [1776]) An enquiry into the nature and causes of the wealth of nations. Liberty Classics, Indianapolis
Spagnoli Gabardi C (n.d) Green Gucci: cleaner than the rest. ELUXE magazine. Available at: http://eluxemagazine.com/magazine/green-gucci/. Accessed 23 Mar 2016
SustainAbility with the United Nations Environment Programme (UNEP) (1994) Company environmental reporting: a measure of the progress of business and industry towards sustainable development. Technical report 24, United Nations Environment Programme, Paris
The Economist (2009) Triple bottom line: it consists of three Ps: profit, people and planet. The Economist, 17 Nov, Online extra. Available at: http://www.economist.com/node/14301663. Accessed 2 Mar 2016
Thomas D (2007) Deluxe: how luxury lost its lustre. Penguin, London
Tiffany & Co. (2014) Sustainability report. Tiffany & Co., New York. Available at: https://www.tiffany.co.uk/CSR. Accessed 4 Mar 2016
Tokatli N (2014) 'Made in Italy? Who cares!' Prada's new economic geography. Geoforum 54:1–9
UNCTAD (2013) World investment report 2013: global value chains: investment and trade for development. Unities Nations, New York and Geneva
UNICEF (2013) Gucci shopper bag sold in support of UNICEF, 27 Nov, UNICEF United States Fund. Available at: https://www.unicefusa.org/stories/gucci-shopper-bag-sold-support-unicef/7467. Accessed 23 June 2016
United Nations (2015a) The millennium development goals report 2015. Unities Nations, New York and Geneva. Available at: http://www.un.org/millenniumgoals/2015_MDG_Report/pdf/MDG%202015%20rev%20%28July%201%29.pdf. Accessed 2 Mar 2016
United Nations (2015b) Transforming our world: the 2030 Agenda for sustainable development, sustainabledevelopment.un.org, A/RES/70/1. Available at: https://sustainabledevelopment.un.org/content/documents/21252030%20Agenda%20for%20Sustainable%20Development%20web.pdf. Accessed 1 Feb 2016
United Nations (2016) United Nations global compact. Available at: https://www.unglobalcompact.org/what-is-gc/mission. Accessed 2 Mar 2016
United Nations Global Compact and Business for Social Responsibility (2010) Supply chain sustainability a practical guide for continuous improvement. UN Global Compact Office and Business for Social Responsibility. Available at: http://www.bsr.org/reports/BSR_UNGC_SupplyChainReport.pdf. Accessed 23 June 2016
Veblen T (1899) The theory of the leisure class: an economic study in the evolution of institutions. Macmillan, New York and London
Vickers JS, Renand F (2003) The marketing of luxury goods: an exploratory study—three conceptual dimensions. Mark Rev 3(4):459–478

Wealth-X and UBS (2014) World ultra wealth report 2014, Wealth-X and UBS: http://www.worldultrawealthreport.com/home.php. Accessed 22 Jan 2016

World Business Council for Sustainable Development (2011) Collaboration, innovation, transformation: ideas and inspiration to accelerate sustainable growth—a value chain approach. World Business Council for Sustainable Development. Available at: http://wbcsdservers.org/wbcsdpublications/cd_files/datas/business-solutions/consumption/pdf/CollaborationInnovationTransformation.pdf. Accessed 23 June 2016

World Commission on Environment and Development (1987) Our common future (Brundtland Report). Oxford University Press, Oxford

Author Biographies

Dr. John Armitage is Professor of Media Arts and Co-Director of the Winchester Luxury Research Group at Winchester School of Art, University of Southampton, United Kingdom. He is the co-editor with Joanne Roberts of *Critical Luxury Studies: Art, Design, Media* (Edinburgh University Press, 2016), and with Joanne Roberts and Jonathan Faiers of *The Luxury Reader* (Bloomsbury, forthcoming), and he is currently completing a single authored book titled *Luxury and Visual Culture* (Bloomsbury, forthcoming).

Dr. Joanne Roberts is Professor in Arts and Cultural Management and Director of the Winchester Luxury Research Group at Winchester School of Art, University of Southampton, United Kingdom. Her research interests include knowledge management, innovation, creativity and luxury. Joanne has published articles in a wide range of international journals, including *Journal of Management Studies*, *Journal of Business Ethics* and *Research Policy*. Additionally, she has authored and edited a number of books. Her latest sole-authored book is *A Very Short, Fairly Interesting and Reasonably Cheap Book about Knowledge Management* (Sage, 2015).

Dr. Yasmin K. Sekhon is an Associate Professor in Luxury Brand Management and Director of Postgraduate Programmes at Winchester School of Art, University of Southampton, United Kingdom. Her areas of research include the consumption patterns of first and second-generation immigrants, the consumption of luxury brands, the meaning and significance of luxury across cultures as well as the study of materialism in cross-cultural settings. She has published in a number of academic journals from *Consumption, Markets and Culture Journal*, *Academy of Marketing Review*, *Non-profit and Voluntary Sector Quarterly*, *International Journal of Market Research* amongst others. She is also on the editorial board of the *International Journal of Market Research.*

Luxury Fashion Brand Sustainability and Flagship Store Design. The Case of 'Smart Sustainable Stores'

Elisa Arrigo

Abstract In recent years, the relevance of sustainable practices has increased in the luxury environment as socio-environmental problems have become more severe and consumers started to react sensitively towards the conduct of companies and the eco-friendliness of their offer. Several luxury companies have thus realized that only by reinforcing their sustainability's values and improving their awareness among external stakeholders, could they safeguard their brand equity. Although many companies have already embraced sustainability in their development strategies, very little is known on luxury sustainability in academic research. In particular, a research gap has emerged on the role played by luxury flagship stores in disseminating externally the corporate sustainability of luxury fashion brands. Hence, this study aims to close the above-mentioned research gap by focusing on sustainability issues in store retailing research. To address this gap, apart from the literature review on luxury sustainability and flagship store format as a brand communication channel, a case study has been carried out on the 'Smart Sustainable Store' policy of the Kering Group. For the first time, findings show how sustainable design and management of the flagship store help to minimize the environmental impact of the store's operations by optimizing energy efficiency, and can make stakeholders more aware of the luxury fashion brand's commitment towards the environmental sustainability by safeguarding and improving its reputation.

Keywords Luxury fashion brand · Fashion retailing · Store design · Sustainability

E. Arrigo (✉)
Department of Economics, Management and Statistics (DEMS),
University of Milano-Bicocca, Milan, Italy
e-mail: elisa.arrigo@unimib.it

M.A. Gardetti (ed.), *Sustainable Management of Luxury*,
Environmental Footprints and Eco-design of Products and Processes,
DOI 10.1007/978-981-10-2917-2_13

1 Introduction

Sustainable development is on the agenda of several market actors such as governments, the business world, consumers, and non-governmental organizations (NGOs), and refers to promoting a society that can persist over generations through the conservation of natural resources. In 1987, the Brundtland Commission Report defined 'sustainability' as 'meeting the needs of the present generation without compromising the ability of future generations to meet their own needs' and, to date, many studies have examined whether by incorporating sustainability issues into the business strategies, this could represent a source of competitive advantage (Reuter et al. 2010; Markley and Davis 2007). Thus, corporate sustainability has emerged as a new managerial model based on stakeholder relationships, focused on the capacity of companies to strategically integrate economic, social and environmental issues into business operations. Sustainability contributes to the triple bottom line (Elkington 1997) via its favourable impact on the company's relationships with important stakeholders and can represent a key driver of innovation for many companies (Nidumolu et al. 2009).

In the luxury industry, companies have realized that luxury, whilst typically highly profitable, faces ethical and socio-environmental issues (Gardetti and Muthu 2015; Gardetti and Torres 2014). According to Kapferer (2010), luxury is so concerned about its sustainability, firstly due to it being a very "visible" sector with many people attracted by luxury brands and following their activities constantly. Furthermore, "no sector reveals the world social inequality as much as luxury" (Kapferer 2010, p. 42) by being accessible only to rich people and, consequently, luxury is considered as a factor of social tension.

Many luxury corporations have already developed new management approaches to incorporate economic, social and environmental sustainability considerations into their business practices. Sometimes, this has involved creating specific environmental task forces to monitor their sustainability policies. Luxury companies have understood, in fact, that only by reinforcing the values and the practices of sustainability, could they preserve their reputation and clearly distinguish themselves from everyday low-cost fashion brands.

Despite this quite rapid development of sustainable practices in businesses, academic research on sustainable luxury is still at an embryonic stage. This paper therefore aims to contribute to filling this gap by investigating the role of the flagship store format in translating the corporate sustainability of a luxury brand. The overall research question was 'Can luxury brands communicate their sustainability through the design and management of their flagship store?'. In order to address this research question, a descriptive case study of the 'Smart Sustainable Store' at the Kering Group, one of the most important luxury groups, has been carried out.

The paper is organised as follows: following the introduction, Sect. 2 analyses the theoretical background of luxury brands and sustainability, and the flagship store format. Section 3 proposes a new role of flagship store design and management in interpreting the sustainability commitment of a luxury brand. Section 4 describes the methodology and examines the case of Smart Sustainable Stores; Sect. 5 contains the results and discussion; and, finally, Sect. 6 draws together conclusions with future research directions.

2 Theoretical Background

2.1 *Sustainability and Luxury Brands*

Fashion and Luxury are among the most polluting industries in the business world because of their heavy consumption of natural resources such as water and energy used in manufacturing fabrics or their use of dangerous chemicals for dying and tanning. Luxury brands have recently become a target for public criticism; in fact, the global development of corporate activities worldwide that luxury brands have realized in recent years has exposed them to accusations of damaging the ecological environment.

Although the luxury industry is relatively small in terms of number of companies, it has an impact due to its level of sales and power to influence. In 2014, the overall luxury market exceeded €850 billion, showing a healthy growth of 7% and has a predicted 9% annual growth rate up to 2020 (Bain & Co., 2015). Further, luxury means excess: the best design, the best craftsmanship, the best retailing (the nicest locations and store concepts), and the best customer service occur in the luxury industry and, therefore, luxury brands frequently drive the world of marketing strategies (Ko and Megehee 2012). Thus, the fact that luxury companies adopt sustainable practices could have a positive influence on companies pertaining to other sectors.

In order to review the relationship existing between sustainability and luxury, it is useful to begin by considering the key-features of a luxury brand. In the managerial academic literature, several definitions of luxury brands have been proposed, most of which refer to the following elements: excellence, high transaction value, exclusivity, artisanship, and premium price (Jackson 2004; Keller 2009; Hennigs et al. 2012, Heine 2010; Chevalier and Mazzalovo 2012).

Fionda and Moore (2009) explored the critical dimensions of creating and maintaining a successful luxury brand over time, and identified nine key attributes, requiring a consistent and coherent approach to be efficiently managed. The first attribute refers to the development of a clear brand identity as a necessary pre-requisite for building good relationships with customers. Again, marketing communication is another key attribute to staying in touch with the market by creating brand awareness. Further, a third element is represented by the integrity of a luxury product depending on aspects of quality, artisanship, and functionality. Next, the design signature (iconic products, creative directors, and recognisable style), the history and heritage of the brand are recognized as being suitable to establish the perception of a luxury product as authentic in the customer's mind. Moreover, the typical exclusivity of luxury offerings, carefully controlled by companies through the system of limited edition sales or company-managed stores, constitutes another key feature. The exclusivity of luxury goods is additionally communicated through their high price that represents another key element of luxury brands and is frequently the first one to catch the customer's attention. Obviously, the corporate culture pervading the luxury company and the level of expertise accumulated over time are other elements necessary in building a successful luxury brand. Lastly, the environment in which a brand meets the buyers and the customer service provided in this space are essential attributes of luxury value propositions. In fact, a luxury store is the place in which buyers can fully live the brand experience of luxury since it provides an insight into the brand's lifestyle and values.

Jackson (2004) similarly considered the point of sale as one of the key dimensions of luxury brands together with global recognition, critical mass, core competence, high product quality and innovation, powerful advertising, and superb customer service.

Besides all these described features needed to create a successful luxury brand, Keller (2009) argued that, in general, luxury brands face three important trade-offs reflecting the challenges faced by managers in handling their development over time: exclusivity—accessibility; classic—contemporary images; and acquisition—retention.

They are considered as trade-offs since the strategies moving in the direction of improving exclusivity, classical image and customer retention can be opposed by those of accessibility, modernization and acquisition of new customers. In fact, luxury brands need to be perceived as up-market and high-class whilst, at the same time, they need to expand the customer base, and subsequently to increase their accessibility in order to generate sufficient profits. At this point, the exclusivity-accessibility paradox arises as the prestige of a luxury brand gets eroded if too many people are able to acquire it (for the rarity principle). Thus, luxury brands should look towards maximising their profits without selling their products too widely (Dall'Olmo Riley and Lacroix 2003). Furthermore, luxury brands have a heritage and tradition that sometimes are not able to be appreciated by a generation of young customers, so brands should evolve by incorporating more contemporary elements (classic-contemporary paradox). Finally, luxury brands are engaged in searching for new customers but also in maintaining existing ones (acquisition-retention paradox). These paradoxes make the management of luxury brands a complex activity putting pressure on marketing managers to find new evolving strategies that are consistent in the long-term, and sustainability could represent one of these particular strategies.

It is argued by some people that luxury and sustainability are irreconcilable by definition, since luxury implies excess, and production and consumption driven by anything apart from satisfying basic needs, and this represents the main cause of environmental and social problems. On the contrary, Kapferer (2010) suggests that luxury and sustainability tend to converge; in fact, every luxury brand has two attributes that move towards the direction of a sustainable development.

Firstly, both luxury brands and sustainable development focus on rarity. Natural resources are scarce and similarly the highly perceived value of luxury depends on a product's obvious rarity (rare materials, rare pearls, and rare artisanship) that determines its distinctiveness. Likewise, this rarity safeguards environmental sustainability since mass production is more damaging to the planet rather than artisanal goods, which produce less waste compared to the level of waste derived from the energy consumption of industrial manufacturing systems.

Secondly, luxury brands are characterised by their feature of durability, aligning with the concept of environmental sustainability by limiting the consumption of natural resources over time. Luxury goods are usually long lasting, and in the case, for example, of Louis Vuitton bags or Bulgari jewels, are passed down from generation to generation. In this way, waste is reduced and the environment is preserved.

It is possible, therefore, to affirm that the preservation of environmental resources determines the scarcity and, consequently, the exclusivity of luxury products. Luxury depends on sustainability and, equally, sustainability finds in the luxury concept a potential supporter since the durability and rarity of luxury brands make sustainability a way of evoking the brand's eternalness and emphasizing the brand awareness. Sustainable business practices can thus improve an existing luxury brand's image by increasing its perceived value and exclusivity, whilst achieving higher levels of sales and building customer loyalty (Guercini and Ranfagni 2013).

Hennigs et al. (2013) consider sustainability as part of the essence of luxury and propose an interesting framework of luxury sustainability value based on individual value perception. A superior customer value's creation can be obtained through the enhancement of significant and emotional brand associations; thus, the capacity of a luxury brand to reach an excellent level of sustainability depends on the customer's expectations and perceptions about luxury value.

In their framework, Hennigs et al. (2013) suggest that customers measure the performance of a luxury brand based on four dimensions: financial, functional, individual and social value. The first dimension, financial, refers to the value of a product expressed in monetary terms beyond other costs that a customer expends to acquire the product, such as the time taken to select it etc. In this case, since customers perceive luxury offerings as scarce and very expensive, sustainable and responsible consumption is encouraged. Moreover, the functional dimension of sustainable value is linked to aspects of quality, reliability and durability and different key areas are relevant in developing a perspective of sustainable luxury brand management such as the packaging or business operations. In order to address a functional sustainability, the entire supply chain processes should encourage sustainable strategies. Therefore, individual sustainable value proceeds from the basic consideration that customers buy luxury products in order to satisfy and reflect their personal values and aspirations, and a clear commitment to sustainability is not only desired by them, but also a requirement. Finally, the perceived utility obtained by customers through the use or consumption of luxury products is tied to an acknowledgment of their special social status. If a luxury brand is marked by social and environmental responsibility, this contributes to enhancing its customer relationships since customers are proud of buying products from a sustainable company. In conclusion, Hennigs et al. (2013) suggest that luxury brands need to modify their understanding of excellence, which is no longer associated with fashion or glamour but with engagement and sustainable business management.

Climate change, the increasing waste and the abuse of many natural resources are some of the major challenges that luxury companies face nowadays. In response to these challenges, luxury companies try to make an efficient use of resources, minimising their reliance on unsustainable resources and reducing the waste and pollution of their productive and distributive processes. They encourage the use of alternative energy sources and some companies have recently embraced new business models based on sustainable supply chains where their sourcing, manufacturing and distribution processes are redefined in order to safeguard the environment and improve efficiency (Kapferer and Bastien 2012).

2.2 The Flagship Store Format

Store management is increasingly relevant in current retailing research as the point of sale represents the place where luxury retailers can show their value propositions, besides spreading their brand values trying to positively influence the customer experience (Brakus et al. 2009).

As Nobbs et al. (2012) point out, luxury retailing displays three peculiarities with respect to the retailing of other product categories: frequently it constitutes a direct channel for companies that manufacture their products; these companies supply on the global market, and very often luxury retailers use direct investment as a strategy for foreign market entry. Consequently, store management is of key significance for luxury brands operating worldwide.

In previous research, many academics (Kent 2003; Clow and Baack 2010; Doyle and Broadbridge 1999) assimilated the store as a marketing communication tool, because, as suggested by Doyle and Broadbridge (1999), the store environment and the internal atmosphere characterize the touch point between customers and the retailer, the store being considered as a channel of brand communication. In fact, according to Gottdiener (1997), the retail atmosphere is a major part in creating a memorable and attractive customer experience.

The seminal work of Bitner (1992) on servicescape proposed that the physical retail environment assumes three strategic functions for marketing strategies: it offers a complete description of the retailer's total offering; through its internal layout, it helps customers to look around freely and comfortably; it acts as a signal of differentiation, by positioning the retailer at the desired level. Thus, the design of the physical surroundings in the retail environment is recognized to affect customer behaviour and purchasing aims. In fact, the physical environment, in this case the point of sale, includes many cues used by customers to value the retailer's capabilities and the quality of its offering (Orth et al. 2012). So the store can be considered as very powerful in communicating the luxury brand values to customers.

The complex mix of cues controlled by retailers inside the store and able to influence positively the customer responses include lighting, colour, layout, wall décor, quality of materials, sound, etc. The design of the retail store environment is therefore a crucial element in the luxury brands' marketing strategy, as it contributes to developing a sense of prestige and excellence (Hines and Bruce 2007). Luxury retailers can choose among many retail store formats to display and sell their products and the flagship store format is exactly designed to take the branding concept to an extreme level.

Kozinets et al. (2002) proposed one of the first descriptions of flagship stores founded on three key features:

- supply only a single brand;
- are owned by their manufacturers; and
- have, as their primary purpose, building the brand image.

This latter feature supports Fernie et al.'s (1998, p. 373) claim, that the flagship store's function is simply "*to act as publicity vehicles for the ranges and are not required to show a typical return on investment*".

A further enlargement of definitions followed the first produced by Kozinets et al. (2002). For example, Mikunda (2006, p. 228) considers the flagship store as "*the principal store of a retail chain*", Diamond (2005, p. 12) as "*the most important in a chain*", and Frings (2009, p. 458) as "*the largest and most representative store in a chain organisation*". Many features have been recognised as influential in making flagship stores the "*touch point for stakeholders*" (Nobbs et al. 2012). Flagship stores provide a sense of tangibility to the luxury brands and offer a space where luxury retailers can leverage the impact of other tools and events connected to the brand reinforcement, such as advertising, promotions or other related marketing activities. Consumers go to the flagship store not only to purchase items, but also to experience the luxury brand in an environment completely controlled by the retailer (Kozinets et al. 2002).

As suggested by Manlow and Nobbs (2013), the key dimensions of a flagship store can be derived from the definitions offered by Kozinets et al. (2002), Mikunda (2004), Jackson (2004), Diamond (2005), Varley (2007), Mores (2007), Frings (2009), and Kent and Brown (2009). In particular, four main features emerge from these definitions: the flagship store is located in a large outlet in a prominent area; it offers the widest and most in depth product assortment; it provides a high quality store environment; and finally, it serves to communicate the brand position and values, as flagship stores are becoming the company's prime mass medium.

Particularly interesting is the framework of the international luxury flagship store proposed by Moore and Doherty (2007) where four core dimensions are identified: strategic purpose; distribution hierarchy; language of flagship; and the location and place.

The strategic purpose of a flagship store is easily understandable from the huge investment required for its establishment that is often far superior to the return generated from sales. For this reason, luxury retailers usually open flagship stores only in the most lucrative foreign markets and consider them as a market-entry strategy (Moore et al. 2010). In this case, flagship stores can serve as an important mediator in the development and management of collaborative relationships between the luxury retailer and its partners within a foreign market (Moore and Doherty 2007).

Additionally, flagship stores display a considerable size and a strong visual impact; generally, their height makes them stand out from other stores in order to reveal in concrete terms the magnitude of the luxury brand. Flagship stores aim to attract wealthy customers, generating interest and curiosity towards a luxury brand, and are used as a vehicle of brand communication in order to spread brand identity values. Frequently, to add intangible luxury associations to the flagship store, the most prestigious architects and designers are involved in its planning and design (Kent 2007).

With reference to the distribution hierarchy, flagship stores provide customers with a special assortment of products together with an excellent standard of service. Often, in all countries, new collections are presented firstly in the flagship store and, only later, in the other outlets. In any case, certain items remain exclusively available in the flagship store in order to increase both the sense of prestige and customer traffic.

As described before, the store atmosphere is strictly tied to the luxury perception and brand positioning and, for this very reason, Moore and Doherty (2007) coined the expression "the language of the flagship". They referred to the many factors inside the flagship store that contribute to creating a luxurious experience for the customer, such as: the colour, the scale, the olfactory devices, the lighting, the furnishing materials, the sales assistants, etc.

Finally, the "place and location" dimension can be examined on two levels: at the macro level, flagship stores are placed in leading international cities and, at the micro level, they are concentrated in streets visited by wealthy customers or tourists and, frequently, their allure and prestige is further improved by the status and heritage of the building that they occupy. Moreover, by placing a flagship store inside a store agglomeration, luxury retailers benefit from each other's individual luxury brand positioning and, together, they enable the development of a luxury sense of place in the urban area delimited by their locations (Arrigo 2015).

Many luxury retailers confirm also the trend of creating a single flagship store in every overseas city or market with the aim of improving the sense of exclusivity and prestige that are enhanced by also adding unique features to distinguish it from others. This strategy is efficient because it enables luxury retailers to concentrate the entire investment on the new market entry in a unique store and, simultaneously, preserving full control over the branding strategies within that offshore market.

In their study on the meaning of the flagship stores for customers and managers, Manlow and Nobbs (2013) investigated the relationship existing between the flagship store format and the customer brand experience and they found that the store design constitutes a key component in the process realized by customers to associate an identity with each brand. In particular, Brengman and Willems (2003) examined the influential features of store design that consumers attribute to a brand's personality. For example, they show that customers perceive as sophisticated spacious, expensive and attractive stores and this contributes to creating specific customer brand associations.

Customers buy luxury goods to fulfill a dual need for status and a desire to live through pleasurable experiences and consequently the design of luxury flagship stores takes into account all the senses in order to create an emotional response in consumers (Riewoldt 2002). Luxury brands need to create a "sensorial universe" capable of reaching different emotional contact points so that customers can live a whole brand experience (Brakus et al. 2009).

3 The Flagship Store as Propagator of Corporate Sustainability

Higher consumer ecological sensibility, increased attention toward social responsibility and more competition have created growing research interest in the topic of sustainability along the supply chain and, in particular, on the role of retailers. In fact, since they occupy a crucial position between manufacturers and consumers by connecting upstream and downstream supply chain dynamics, retailers have the power both to establish a respect for sustainable standards inside their supply chain (Ogle et al. 2004) and to raise awareness on sustainability issues among the downstream buyers by influencing their consumption (Lehner 2015).

Wiese et al. (2012) provided a comprehensive overview of past and current research on sustainability in retailing and studied the role played by retailers in the supply chains through a broad analysis of sustainability considerations in different research areas. Their findings show that retail research is only partially focused on sustainability and that this latter topic has received attention only in specific sectors such as transport, agriculture or areas such as business administration, while it is relevant to all research areas and industry sectors.

The present study attempts to contribute to the area of luxury sustainability research in retailing by investigating the role played by the flagship store in translating and communicating the corporate sustainability of a luxury brand. As argued by Jackson (2004), the store is a key dimension of luxury brands constituting a real physical space to develop a relationship between a luxury retailer and its customers. Moreover, according to the notion of servicescape proposed by Bitner (1992), the store can be considered as a signal of differentiation and positioning of a luxury retailer because it includes many cues (Orth et al. 2012) used by customers to evaluate the luxury brand.

Particularly, the flagship store format is designed to provide a space in which customers can live a whole brand experience (Brakus et al. 2009), whilst according to Mores (2007), flagship stores are becoming the prime mass-medium used by retailers to communicate with their customers.

An international flagship store can furthermore be examined on four core dimensions identified by the location and place, strategic purpose, distribution hierarchy, and the language of the flagship (Moore and Doherty 2007). By focusing on the latter dimension, i.e. the language of the flagship, Nobbs et al. (2012) distinguish three key aspects that luxury retailers should consider in designing their flagships:

- the architecture;
- the provision of value-added services; and
- the organization of entertainment events (that the authors define as "third space").

In fact, the authors claim that the services provided to clients and events organized inside the flagship store should be as excellent and sophisticated as the luxury offerings. Only in this way, can luxury retailers trigger peoples' curiosity and persuade them to visit the store and, once inside, convince customers to stay there for a long period having more time to analyze and purchase luxury products.

Architecture plays an important role since it interprets materially the identity of a luxury brand by expressing the brand style and values that are considered and analyzed by the architects during the design of the flagship. Frequently, the boundaries of retailing, art and architecture disappear to convey visually the luxury brand, and luxury products become part of the entire composition not separated by the retail design.

In this study, for the first time, it is proposed that all these three key aspects of the language of the flagship store can be related to the corporate sustainability so that the flagship store can be considered as a translator of luxury brand sustainability. From the projection of the architectural elements up to the customer service provided by shopping assistants in the flagship store together with internally organized events, all of these elements can be managed in a sustainable way.

According to Hennigs et al. (2013), the capacity of a luxury brand to reach a level of sustainable excellence depends on the customer's expectations and perceptions about the luxury value that he/she measures using four dimensions: financial, functional, individual and social value. The authors highlight how the social value that customers obtain by acquiring a luxury product refers to the acknowledgment of a special social status and, consequently, if the luxury brand is a reflection of a form of economic, social and environmental sustainability, this contributes to enhancing customer relationships, as customers will be proud of associating themselves with such values.

Therefore, in the present paper, it is suggested that luxury fashion retailers can manage the three factors (architectural design, customer service and events' organization) in disseminating in an external way their commitment to social and environmental sustainability. Obviously, in order to achieve this objective, luxury retailers need to train their retail assistants who interact directly with customers, in order to make them fully aware of the sustainability practices realized in the manufacturing and distribution processes of the supply chain, so that they can clearly communicate this information to customers. For example, they should explain the provenance of luxury products in order to indicate that the cotton or wool used in the products comes from certified sources. In so doing, retail assistants could make customers more satisfied about their purchases since they could gratify their self-respect needs.

Furthermore, luxury fashion retailers should organize events or activities inside the flagship store to disclose their commitment to social and environmental sustainability in a more relaxed way, by entertaining buyers with conferences, videos and pictures that show actual sustainable actions and encouraging them to become more interested in social projects involving the luxury brand.

Finally, also the architectural design of the flagship store should comply with a set of standards aimed at reducing CO_2 emissions and waste through an efficient consumption of natural resources such as energy and water across all store operations. For example, luxury fashion retailers could light flagship stores with electricity produced from renewable sources thereby increasing energy efficiency, or they could limit the carbon emissions per store.

To explore more deeply the proposed role of the flagship store in delivering the sustainability commitment of a luxury brand, a descriptive case study is provided in the following section.

4 Methodology

Case study research is performed to support the theoretical assertions about the nature of reality and offers the opportunity for a holistic view of a process as opposed to a reductionist-fragmented view (Gummesson 2000). This methodology is recommended for compensating for a lack of established theory and/or where accepted principles have not been recognized and draws on secondary data, on direct observation of the phenomenon and on interviews with people involved. As Yin (2014) suggested, the case study method can be useful when the investigator has little control over events and when the focus is on contemporary phenomenon. Based on a literature review on the validity of case study research, this approach appears to be an efficient and valuable method to investigate and interpret luxury retailers' actions related to the communication of their commitment to sustainability.

The company chosen for the case study was the Kering Group, the owner of many well-known global luxury brands including Gucci, Bottega Veneta, Saint Laurent, and Pomellato. Kering was selected as it is recognized as one of the world's most sustainable corporations:

- it is the only luxury company to be included in the "2016 Corporate Knights' Global 100 Index" published at the World Economic Forum; and
- in September 2015, Kering was recognised as an industry leader in the Dow Jones Sustainability Indices (DJSI) World and Europe, heading up the Textiles, Apparel and Luxury Goods sector for the second consecutive year.

Thus, it can be considered a strong prototype of a global company committed towards sustainability in the luxury industry. In particular, this study will focus on the Saint Laurent luxury brand, pertaining to the Kering multi-brand portfolio, which has introduced a comprehensive and exemplary program to reduce the environmental impact of its flagship stores.

Data collection was based on secondary sources and interviews with luxury managers and experts. The Kering Group pursues a disclosed policy concerning its corporate sustainability by making available a lot of information and data both on the actions implemented from a sustainable perspective and on achieved results. More specifically, the following secondary sources were used for the case study: Annual Report (2014); Environmental Profit & Loss (E P&L) Report (2014); Environmental Reporting Methodology Memo (2014); the progress report on 2016 sustainability; and a section of the official corporate website dedicated to sustainability (www.kering.com/en/sustainability).

4.1 The Case of Smart Sustainable Stores Within the Kering Group

Within the Kering Group, sustainability means "*making the invisible visible*" and "*businesses need a new approach to see the reality behind their supply chains, to understand their environmental impacts, so that they can reduce them*". With these words, on the sustainability webpage of the company website (www.kering.com/en/sustainability), the Kering Group presents its commitment to sustainability. Sustainability creates value and represents long-term differentiation, new business development, an incentive for innovation, efficiency opportunities and a key factor in attracting and retaining the best employees. The corporation has designed a specific program in order to monitor and evaluate its impact on nature, entitled the Environmental Profit and Loss (E P&L) account that measures costs and benefits generated by each brand's environmental impact both within its own operations and across the supply chain.

In April 2012, the Kering Group announced a set of key sustainability targets to achieve by 2016 and the internal Sustainability Department defined detailed policy guidelines through the value chain of the Luxury and Sport & Lifestyle brands. During the last company's Annual General Meeting, Marie Claire Daveu, the Chief Sustainability Officer and Head of International Institutional Affairs said: "*We are focused and committed to enhancing our own business and be a business model that can illustrate and leverage change in our industry, while being transparent and collaborative in the process. Transparency and engagement with our stakeholders around the world is paramount and will help us continue to challenge ourselves to reach our sustainability ambitions.*"

In order to manage the sustainability issues, each brand has its 'Sustainability Lead' and the group-brand coordination is ensured through regular meetings to coordinate the sustainability strategy, to monitor its advances and to share best practices developed at brand level.

The fixed sustainability objectives relate to sourcing of raw materials, including alternatives, paper and packaging, water use, waste and carbon emissions and hazardous chemicals, while offsetting the remaining CO_2 emissions and supporting suppliers. Similarly, the indicators used to monitor the environmental impact are split into ten categories as specified in the Environmental Reporting Methodology Memo:

- Energy consumption;
- Water consumption;
- Waste production;
- Paper consumption;
- Packaging consumption;
- Consumption of raw materials;
- Transport (BtoB, BtoC and company cars);
- Air pollution;
- Environmental management; and
- General data on the site (surface, turn over, etc.) enabling ratios calculation.

Kering has built a reliable and secure web-based platform in order to gather, validate and combine its extra-financial data; the platform is able to allocate the indicators on the specific data of the sites and brands. Thus, the Sustainability Leads of each brand manage their network of users and, along with the Kering Sustainability Department, define their list of indicators and the main reporting guidelines (Environmental Reporting Methodology Memo 2014). For example, since every brand handles its own logistics, the transportation flows are extremely diversified, and as a consequence, so are the indicators used to measure the environmental impact of each brand.

Due to the implementation of the E P&L program, Kering is becoming more efficient in its energy consumption, procurement and management. In 2014, the main environmental impact derived from the production (49%) and processing (25%) of raw materials inside the supply chain, and the rest from the stores, warehouses and offices (7%). In the same period, the Group also increased the use of renewable energy by approximately 50% and as a result of using solar panels, it was able to save 384 tons of CO_2 (E P&L Report, 2014) by becoming more sustainable.

In order to explore the role of the flagship store in delivering the sustainability commitment of a luxury brand, the policy concerning the store management was analyzed. To reinforce its environmental management policy referring to store management, Kering created the 'Smart Sustainable Store' handbook that sets out the environmental best practice for retail locations by covering the management of energy, waste, water, paper and packaging, logistics and deliveries and store maintenance. To evaluate each store's efficiency, Kering measures the impact on air pollution, greenhouse gas emissions, land use, waste, water consumption, and water pollution.

The case study focuses on the Saint Laurent brand, one of the luxury brands within the Kering Group. Founded in 1961, as Yves Saint Laurent, it is one of the world's most prominent fashion houses. It was the first couture house to launch in 1966, the concept of women's luxury prêt-à-porter and over the years, its revolutionary styles have become a cultural and artistic reference for the fashion industry. Nowadays, Saint Laurent is a global brand in the vanguard of fashion, designing and marketing a broad range of women's and men's products, leather goods, shoes, and jewelry. Through a license agreement with L'Oréal, it also sells fragrances and cosmetics. From 2012, the creative and image director of the luxury brand has been Hedi Slimane.

Saint Laurent represents one of the most important luxury fashion brands of the Kering Group not only in terms of total sales, where it amounts to 12% of Group revenues (data referred to 2015) but also in its sustainability policy. In 2015, the brand Saint Laurent achieved 974 million euros in revenue and had a retail network of 142 directly operated boutiques including flagship stores in Paris, London, New York, Hong Kong, Shanghai, Beijing and Los Angeles. Saint Laurent is also present in selected multi-brand boutiques and department stores around the world.

The luxury fashion brand implements various actions to improve its environmental sustainability such as efforts to recycle waste and unused materials by establishing a partnership with two French NGOs giving a second life to fabrics used in old collections. In 2014, the percentage of certified (FSC or PEFC) or recycled paper exceeded 90% and many best practices were developed to implement environmental projects aimed at reducing the impact of transport and logistical operations. Moreover, Saint Laurent set up the *Saint Laurent's Règles d'Or*, ten golden rules for a good environmental practice throughout its stores, with a focus on waste management. This requires managers to ensure that all waste paper, cardboard packaging, glass, plastic bottles, tins and ink cartridges are sorted and, then, recycled.

5 Results and Discussion

According to the Smart Sustainable Store policy of Kering, Saint Laurent has introduced a broad program to reduce the environmental impact of its stores. Originally, Hedi Slimane, the creative and image director, placed emphasis on the simplicity concept through expensive materials, generous proportions and architectural clarity, as design features of the Saint Laurent architectural store concept. Later on, Saint Laurent prepared a guide outlining green building best practices, based on the LEED (Leadership in Energy and Environmental Design) requirements, particularly covering the lighting, heating, and air conditioning systems. This latter represents a certification system developed by the U.S. Green Building Council to encourage the construction of energy and resource-efficient buildings.

Over a period of time, environmental management best practices were implemented at an operational level across all regions, with an application rate of 83% at the end of 2014 (excluding department stores). With the support of Kering, Saint

Laurent is gradually switching all of its French energy contracts over to green electricity. Over the 2012–2014 period, the luxury fashion brand has been able to improve the energy efficiency of its stores by 27% and three flagship stores (38 Faubourg in Paris, Old Bond Street in London and 326 Rodeo Drive in Beverly Hills) have been awarded the highest possible level of LEED certification.

Beside the sustainable architectural design of its flagship stores, Saint Laurent uses many communication channels to raise employee awareness of sustainability issues such as dedicated newsletters, talks from outside experts as part of organized events. Recently, the luxury fashion brand has even scheduled one-hour training modules as part of new store opening programs in Europe.

From 2013, Saint Laurent has organized an annual sustainability conference, and for its second edition, Livia Firth, the founder of the Green Carpet Challenge and creative director of EcoAge, was invited to share her experiences with 150 employees on the integration of sustainability in fashion (2014 Kering Group Annual Report).

From an examination of the approach in which the Saint Laurent brand deals with its flagship stores, looking at the management process from the design phase up until the daily running based on the smart sustainable store program fixed by Kering, it is possible to summarize that:

- the architectural design of Saint Laurent flagship stores has been realized in order to achieve a greater environmental sustainability by reducing the consumption of energy and carbon emissions;
- Saint Laurent provides training on sustainability issues to its employees including those in direct contact with customers (i.e. retail assistants) so that they can make aware these latter about the luxury brand's sustainability policies; and
- Finally, it organizes events focused on the topic of sustainability such as its annual Sustainability conference, to spread awareness about Saint Laurent's commitment.

According to the literature review and, in particular, to the proposed role of the flagship store format as propagator of a luxury brand's sustainability, results have shown that Saint Laurent manages its flagship stores as branded spaces (Brakus et al. 2009; Jackson 2004) where to enter into contact with customers. The store architecture, the layout design, the reduced carbon emissions and energy consumption, combined with a training of shopping assistants about the social and environmental sustainability policies followed by the luxury fashion brand, and the organization of events to entertain buyers, are all elements that characterise the language (Moore and Doherty 2007) of the Saint Laurent flagship stores. Furthermore, as stated by Nobbs et al. (2012), the above-mentioned factors are able to communicate outside the excellent offer and service level provided inside the luxury flagship stores. In fact, through an optimal customer service, the organization of specific events, and a very carefully designed architecture able to interpret materially the identity and values of the luxury fashion brand, it is possible to assert that Saint Laurent flagship stores act as a communication mass medium, according to Mores (2007).

In particular, for the first time, this study shown how all these elements can play a key role also in signalling outside the luxury fashion brand's commitment towards social and environmental sustainability. Thus, the study extends previous research on the flagship store format by showing that flagship store design and management do not simply disseminate an excellence, superiority, and high quality of luxury products but, since these elements can be all related to the concept of sustainability, they make the flagship store an effective translator of the luxury fashion brand's corporate sustainability.

From a sustainable design of the architectural elements up to information delivered inside the flagship store about the realized sustainable policies through the shopping assistants during their normal working hours or in occasion of special events, all of these elements are easily communicable to the customers and understood by them. In this way, customers become aware of the luxury fashion brand's sustainability and can feel proud of themselves in acquiring one of its products. In fact, as stated by Hennigs et al. (2013), the ability of a luxury brand to reach a certain level of sustainable excellence depends on the customer's expectations and perceptions about the provided luxury value measured through the dimensions of financial, functional, individual, and social value. When a customer buys a luxury product, he/she can gain a social value related to the reflection of the luxury brand's social and environmental sustainability and this contributes to both enhancing the customer satisfaction and strengthening the customer-brand relationship.

The Saint Laurent commitment towards an environmental sustainability has been efficiently transferred outside since, based on publicly available data, the brand has been the first luxury brand to be awarded the LEED Platinum Certification for multiple freestanding stores. This achievement demonstrates how innovative and sustainable solutions help to minimise the use of natural resources such as water consumption, optimise energy efficiency and, at the same time, communicate the right message of being committed to environmental sustainability.

6 Conclusion and Future Research Directions

The pursuit of a sustainable development policy is an imperative for many companies nowadays. However, luxury companies have been frequently criticized according to public opinion, with luxury products representing the opposite of sustainability, since they imply excess, and signal inequality by emphasizing the differences that exist between the rich and the poor.

In reality, as detailed above and with reference to Kapferer (2010), luxury and sustainability tend to converge around features such as the rarity and durability of luxury offerings and many companies have developed various sustainable strategies to contribute towards limiting their damage to the planet. However, academic research in this field remains underdeveloped; this study aimed at making a contribution to the field of luxury retailing research on sustainability by examining the role played by the flagship store format in the environmental sustainability policy of a luxury fashion brand.

The study's findings confirm that a sustainable design and management of a flagship store helps luxury fashion brands to minimize the consumption of natural resources, by optimizing energy efficiency, but also can contribute to making stakeholders aware of the luxury brand's commitment toward sustainability. Thus, for the first time to the Author's knowledge, it has been proved how flagship stores can act as propagator of luxury brand sustainability. In fact, the study of the Saint Laurent brand has highlighted how by establishing environmental best practices in its flagship stores, based on the Smart Sustainable Store policy, the luxury fashion brand was able to achieve two important results. Firstly, it reduced its impact on the natural environment by limiting the consumption of paper, waste and electricity and, secondly, it improved its reputation as a sustainable brand being awarded the LEED Platinum Certification for multiple freestanding stores. Moreover, the Kering Group, the holding company, was the only luxury corporation to be included in the "2016 Corporate Knights' Global 100 Index" recognized as an industry leader in the Textile, Apparel & Luxury Goods sector in the "2015 Dow Jones Sustainability Indices (DJSI)".

These awards and acknowledgements have surely contributed to showing the worldwide commitment of the Saint Laurent brand and the Kering Group to environmental sustainability by enhancing their awareness of both the idea of a sustainable brand and a sustainable corporation. In conclusion, it is possible to state that flagship store design plays a key role in managing the development of a luxury fashion brand over time and, in particular, in improving stakeholder awareness about its corporate sustainability.

The purpose of this research was to better understand the match between current practice and existing theory in the field of sustainable luxury retailing. However, it is clear that further case study research may help to make the conclusions more valid and generalizable, with research into other global luxury brands being investigating their individual flagship store management policies. Moreover, further research could focus on examining other communication media and tools used by luxury brands to make customers aware of their commitment to sustainability with the aim of providing a framework for sustainable communication best practice.

References

Arrigo E (2015) The role of the flagship store location in luxury branding. An international exploratory study. Int J Retail Distrib Manage 43(6):518–537. doi:10.1108/IJRDM-08-2013-0158

Bitner MJ (1992) Servicescapes: the impact of physical surroundings on customers and employees. J Mark 56(2):57–71

Brakus JJ, Schmitt BH, Zarantonello L (2009) Brand experience: what is it? How is it measured? Does it affect loyalty? J Mark 73(3):52–68. doi:10.1509/jmkg.73.3.52

Brengman M, Willems K (2003) Determinants of fashion store personality: a consumer perspective. J Prod Brand Manage 18(5):346–355. doi:10.1108/10610420910981828

Chevalier M, Mazzalovo G (2012) Luxury brand management: a world of privilege. Wiley, New York

Clow KE, Baack DE (2010) Integrated advertising, promotion and marketing communications, Global edn. Pearson, New York

Dall'Olmo Riley F, Lacroix C (2003) Luxury branding on the internet: lost opportunity or impossibility? Mark Intell Planning 21(2):96–104. doi:10.1108/02634500310465407

Diamond E (2005) Fashion retailing: a multi-channel approach. Prentice Hall, Englewood Cliffs

Doyle SA, Broadbridge A (1999) Differentiation by design: the importance of design in retailer repositioning and differentiation. Int J Retail Distrib Manage 27(2/3):72–81. doi:10.1108/09590559910258571

Elkington J (1997) Cannibals with forks. The triple bottom line of 21st century. Capstone

Fernie J, Moore CM, Lawrie A (1998) A tale of two cities: an examination of fashion designer retailing within London and New York. J Prod Brand Manage 7(5):366–378. doi:10.1108/10610429810237637

Fionda AM, Moore CM (2009) The anatomy of the luxury fashion brand. J Brand Manage 16 (5/6):347–363. doi:10.1057/bm.2008.45

Frings GS (2009) Fashion: from concept to consumer. Prentice Hall, Englewood Cliffs

Gardetti MA, Muthu SS (eds) (2015) Handbook of sustainable luxury textiles and fashion. Springer, Singapore

Gardetti MA, Torres AL (eds) (2014) Sustainability luxury: managing social and environmental performance in iconic brands. Greenleaf Publishing, Sheffield

Gottdiener M (1997) The theming of America: dreams, visions, and commercial spaces. Westview Press, Boulder

Guercini S, Ranfagni S (2013) Sustainability and luxury: the Italian case of a supply chain based on native wools. J Corp Citizensh 52:76–89. doi:10.9774/GLEAF.4700.2013.de.00008

Gummesson E (2000) Qualitative methods in management research. Sage, Thousand Oaks

Heine K (2010) The personality of luxury fashion brands. J Glob Fashion Mark 1(3):154–163. doi:10.1080/20932685.2010.10593067

Hennigs N, Wiedmann KP, Klarmann C, Behrens S (2013) Sustainability as part of the luxury essence: delivering value through social and environmental excellence. J Corp Citizensh 52:25–35. doi:10.9774/GLEAF.4700.2013.de.00005

Hennigs N, Wiedmann KP, Klarmann C (2012) Luxury brands in the digital age—exclusivity versus ubiquity. Mark Rev St Gallen 1:30–35

Hines T, Bruce M (2007) Fashion marketing. Contemporary issues, 2nd edn. Elsevier, New York

Jackson T (2004) A contemporary analysis of global luxury brands. In: Bruce M, Moore CM, Birtwistle G (eds) International retail marketing. A case study approach. Elsevier, New York, pp 155–169

Kapferer JN (2010) All that glitters is not green: the challenge of sustainable luxury. Eur Bus Rev:40–45

Kapferer JN, Bastien V (2012) The luxury strategy: break the rules of marketing to build luxury brands, 2nd edn. Kogan Page, London

Keller KL (2009) Managing the growth tradeoff: challenges and opportunities in luxury branding. J Brand Manage 16(5/6):290–301. doi:10.1057/bm.2008.47

Kent T (2003) 2D23D: management and design perspectives on retail branding. Int J Retail Distrib Manage 31(2/3):131–142. doi:10.1108/09590550310465503

Kent T (2007) Creative space: design and the retail environment. Int J Retail Distrib Manage 35 (9):734–745. doi:10.1108/09590550773273

Kent T, Brown R (2009) Flagship marketing: concepts and places. Routledge Taylor & Francis Group

Ko E, Megehee CM (2012) Fashion marketing of luxury brands: recent research issues and contributions. J Bus Res 65(10):1395–1398. doi:10.1016/j.jbusres.2011.10.004

Kozinets RV, Sherry JF, DeBerry-Spence B, Duhacheck A, Nuttavuthisi K, Storm D (2002) Themed flagship brand stores in the new millennium: theory, practice, prospect. J Retail 78 (1):17–29. doi:10.1016/S0022-4359(01)00063-X

Lehner M (2015) Translating sustainability: the role of the retail store. Int J Retail Distrib Manage 43(4–5):386–402. doi:10.1108/IJRDM-02-2014-0013

Manlow V, Nobbs K (2013) Form and function of luxury flagships: an international exploratory study of the meaning of the flagship store for managers and customers. J Fashion Mark Manage 17(1):49–64. doi:10.1108/13612021311305137

Markley MJ, Davis L (2007) Exploring future competitive advantage through sustainable supply chains. Int J Phys Distrib Logistics Manage 37(9):763–774. doi:10.1108/09600030710840859

Mikunda C (2004) Brand lands, hot spots & cool spaces: Welcome to the third place and the total marketing experience. Kogan Page Publishers

Mikunda C (2006) Brand lands, hot spots & cool spaces. Kogan Page, London

Moore CM, Doherty AM (2007) The international flagship stores of luxury fashion retailers. In: Hines T, Bruce M (eds) Fashion marketing. Contemporary issues, 2nd edn. Elsevier, New York, pp 277–296

Moore CM, Doherty AM, Doyle SA (2010) Flagship stores as a market entry method: the perspective of luxury fashion retailing. Eur J Mark 44(1/2):139–161. doi:10.1108/03090561011008646

Mores CM (2007) From Fiorucci to the guerrilla stores: shop displays in architecture, marketing and communications. Windsor Books, Oxford

Nidumolu R, Prahalad CK, Rangaswami MR (2009) Why sustainability is now the key driver of innovation. Harvard Bus Rev:57–64

Nobbs K, Moore CM, Sheridan M (2012) The flagship format within the luxury fashion market. Int J Retail Distrib Manage 40(12):920–934. doi:10.1108/09590551211274928

Ogle JP, Hyllegard KH, Dunbar BH (2004) Predicting patronage behaviors in a sustainable retail environment adding retail characteristics and consumer lifestyle orientation to the belief-attitude-behavior intention model. Environ Behav 36(5):717–741. doi:10.1177/0013916504265442

Orth UR, Heinrich F, Malkewitz K (2012) Servicescape interior design and consumers' personality impressions. J Serv Mark 26(3):194–203. doi:10.1108/08876041211223997

Reuter C, Foerstl K, Hartmann E, Blome C (2010) Sustainable global supplier management: the role of dynamic capabilities in achieving competitive advantage. J Supply Chain Manage 46 (2):45–63. doi:10.1111/j.1745-493X.2010.03189.x

Riewoldt O (ed) (2002) Brandscaping: worlds of experience in retail design. Birkhäuser Verlag AG, Basel, Boston and Berlin

Varley R (2007) Retail product management. Routledge, London

Wiese A, Kellner J, Lietke B, Toporowski W, Zielke S (2012) Sustainability in retailing–a summative content analysis. Int J Retail Distrib Manage 40(4):318–335. doi:10.1108/09590551211211792

Yin R (2014) Case study research. Design and methods, 5th edn. Sage, Thousand Oaks

Vogue or Vague: Sustainability Performance Appraisal in Luxury Fashion Supply Chains

Hakan Karaosman, Alessandro Brun and Gustavo Morales-Alonso

Abstract Pressure over sustainability is constantly growing. Luxury goods companies are thus required to prioritize their corporate goals and to integrate sustainability into upstream supply chains (SCs). Nevertheless, it is getting difficult to find sustainable partners as a consequence of globally dispersed fashion SCs (FSCs). In order to commit to business sustainability, a luxury goods company must address and appraise not only its own but also its suppliers' social and environmental performance. While there have been efforts in assessing environmental sustainability, to date, there are still gaps in the current literature in terms of social sustainability assessment and to what extent social sustainability could be integrated into contemporary decision-making processes. Prior research stress that having an integrated approach to investigate interactions among social, economic, and environmental dimensions is more practical than applying deep yet disconnected investigation in only one dimension. Nonetheless, it is important to underline that sustainability indicators do not ensure the same impact on all industries; henceforth sector-specific assessment frameworks need a further investigation. This study therefore attempts to synthesise both current and novel components in a comprehensive framework to appraise sustainability performance. The main contribution of this study is the proposition of *a 360 Degrees Performance Appraisal model* to evaluate the impact of SC operations on sustainability. Hence, this chapter provides an understanding of (i) how SC of a luxury goods company must be configured toward sustainability, (ii) how sustainability performance must be assessed through incorporation of a wide range of stakeholders, and (iii) how sustainability could be

H. Karaosman (✉) · A. Brun
Department of Management, Economics and Industrial Engineering,
Politecnico di Milano, 20156 Milan, Italy
e-mail: hakan.karaosman@polimi.it

A. Brun
e-mail: alessandro.brun@polimi.it

H. Karaosman · G. Morales-Alonso
Department of Industrial Management, Business Administration and Statistics,
Universidad Politecnica de Madrid, 28006 Madrid, Spain
e-mail: gustavo.morales@upm.es

M.A. Gardetti (ed.), *Sustainable Management of Luxury*,
Environmental Footprints and Eco-design of Products and Processes,
DOI 10.1007/978-981-10-2917-2_14

further advanced in luxury FSCs. Not only could this tool provide an integrated approach to value sustainability by encompassing all related stakeholders associated, but it could also help luxury fashion companies monitor, interpret and further improve their suppliers' and sub-contractors' sustainability performance.

Keywords Sustainability performance assessment · Luxury goods industry · Self-assessment · Fashion supply chains · Sustainable fashion supply chains

1 Introduction

This century was nurtured under the mark of industrialization. Velocity has become frantic, lifestyles have changed and disconnection between production and consumption has become threatening. The world population has recently reached a major figure of 7.4 billion. Accordingly, fundamental needs of individuals such as food and clothing escalate drastically. The global textile and apparel industry had a total revenue of €1169 billion in 2014 with a significant annual growth rate of 3.8% between 2010 and 2014 (Research and Markets 2015). According to the European Apparel and Textile Confederation, the overall size of the industry in the EU-28 represented a turnover of €165 billion in 2014 and with investments of €4 billion (EURATEX 2015). The textile and apparel manufacturing industry, which directly employs over 60 million people, has become the world's third-biggest industrial industry, following automotive and electronics manufacturing (Ditty 2015). The luxury industry, on the other hand, experienced a lucrative period by exceeding €1 trillion in 2015 with a growth rate of 14%. The personal luxury goods, which is the core of luxury, ballooned to more than 250 € billion in 2015 (D'Arpizio et al. 2015). Nevertheless, despite the growth and high profit margins, the global fashion market is affected by macroeconomic and socio-political events. Supply network complexity, fragmentation of production, geographical dispersion, demand unpredictability, economic crisis, rising labour costs in emerging countries, social sustainability issues, time-based competition, and the mounting sensibility to environmental issues challenge the overall value proposition within the industry.

Let's think of boiling-hot water in a kettle. Water in the kettle gets heated till it starts boiling. The boiling water generates steam and pressure inside the kettle. As the water keeps boiling, more steam is generated. Nonetheless, after a certain point steam generated has no place to go within the kettle. When the water boils long enough to generate lots of steam and pressure, the steam starts escaping through the hole in the top of the kettle. This happens slowly first, however since more steam is generated faster than it could escape, pressure pushes the steam to flow through the hole even faster. Ultimately the steam rushes through the hole so fast, which produces vibrations and making the kettle whistle. This could be a depiction for today's sustainability need. Natural resources and social assets are overused and the growing pressure on earth is stuffed down. Since the water has been boiling for so long, there is an urgent need to integrate sustainability within energy and labour intensive industries before it causes the earth to whistle. In this competitive environment, fashion companies must change their institutional logics to re-build aspiration, credibility and trust. The increasing concern of consumers with respect to environmental and social issues must be considered one of the greatest cultural shifts of the 21st century; therefore, it is time to interrogate how and to what extent the luxury goods industry could label and fully integrate sustainability when the world matters. The justification of sustainability is difficult to quantify in global fashion supply chains (FSCs) where there appears to be multi tier suppliers in which information and material flow take place. However, it is the time to act in a responsible manner and look beyond traditional roots toward sustainability integration and performance appraisal.

Luxury fashion was designed to last, to be truly timeless and to be unique beyond forever. Yet, the fashion world has moved so far and segmentation within the luxury left us wondering whether it is still timeless. The fashion world has moved so fast, yet it now needs irreplaceable, enduring and sustainable designs. Today's luxury could be the consciousness of the future. A consciousness to capture the way tomorrow's needs meet today's contemplations.

Supply chain management (SCM) in the fashion industry refers to the management of physical, financial and logical flows in networks encompassing critical intra-and inter-organizational relationships. The main contribution of this study is the proposal of *a 360 Degrees Performance Appraisal model* to evaluate the impact of supply chain (SC) operations on sustainability through a comprehensive framework consisting of existing and novel elements. This chapter thus provides an understanding of (i) how SC of a luxury fashion company must be configured toward sustainability, (ii) how sustainability performance must be assessed, and (iii) how sustainability could be further advanced in luxury FSCs. Not only could this tool provide an integrated approach to value sustainability by incorporating all related stakeholders associated within a SC, but it could also help luxury fashion companies monitor, interpret and further improve their suppliers' and sub-contractors' sustainability performance. In this vein, Sect. 2 explains the real cost of fashion to illustrate why sustainability matters for fashion. Section 3, subsequently, describes how sustainability could truly meet luxury fashion. Section 4 displays the current position where sustainability stands within the luxury goods industry. Lastly, Sect. 5 delivers a resolution by providing a model to measure sustainability performance to help companies benchmark their performances and mitigate their negative impacts. Lastly, discussion and conclusion complete the chapter.

2 Why Does Sustainability Matter?

2.1 The Real Cost of Fashion

2.1.1 Environmental Impact of Fashion Supply Chains

Carbon

Carbon emissions reached 35.7 billion tonnes (Gt) with an increase of 0.5% in 2014 (Olivier et al. 2015). Further, 2014 was recorded as the warmest year globally since the beginning of the records in 1880. The annually averaged temperature was 0.7 °C above the 20th century average of 13.9 °C (Olivier et al. 2015). The global textile and apparel industry plays a significant role in aforementioned numbers. The purchase and use of clothing results in the release of over 850 million tonnes CO_2 per year, accounting for around 3% of global production CO_2 emissions (Carbon Trust 2011a). Use phase emissions account for around 50% of a typical t-shirt's life cycle emissions (Carbon Trust 2011a). Reducing carbon-emissions requires the efforts of both retailers and consumers, from day-to-day actions to design and manufacturing processes. The United Nations Environment Program estimates that collaborative emissions reduction initiatives including companies, cities and regions could deliver an equivalent of 3 billion tonnes (Gt) of CO_2 reductions by 2020 (CDP 2015a). Transportation activities, high energy and inefficient production processes, textile disposal techniques must be optimized. To this end, FSCs must act as a whole to take effective steps to reduce footprints. Therefore, integration is needed.

Chemicals, Forestation and Water

In FSCs, there are many critical facets generating a tremendous impact on nature. For example, approximately 8000 synthetic chemicals are utilized to turn raw materials into textile among which many are released into freshwater sources (Aiama et al. 2016). Furthermore, land use and farming become vital topics to consider during the raw material sourcing stage. About 10–15% of the world's greenhouse gas emissions are generated by forest degradation and deforestation. Addressing deforestation is critical to meet international goals to fight against climate change. Up to 33% of the carbon mitigation needed to keep temperature rises in control could be achieved by addressing forest degradation and deforestation (CDP 2015b). A stable supply of good quality fresh water, on the other hand, might no longer be ensured in many regions. A 40% global shortfall in supply is expected by 2030 (CDP 2015c). Nevertheless, fashion is a thirsty business. One kg of cotton production requires 20,000 L of water (WWF 2014) and it must be stressed that the Aral Sea reduced 10% of its former volume consequent to the conventional cotton farming. The actual volume of water required to wash clothing is equivalent to around 10% of the global footprint (WRAP 2012). Additionally, cotton requires a very intense pesticide-intensive crop use. It is estimated that cotton consumes 16% of all the insecticides and 6.8% of all herbicides used worldwide (Organic Cotton 2015). Besides, approximately 4% of the world's nitrogen-based fertilizers for cotton production (Carbon Trust 2011b).

Waste

FSCs generate a considerable amount of waste. From raw material sourcing and textile production to garment manufacturing and distribution to retail stores, each stage constitutes a signature impact in terms of waste. As depicted in Fig. 1, subsequent to the raw material sourcing, material and fibre production stages occur. Subsequently, fabric production process takes place in which several stages, such as wet processing, contribute to waste generation. Following the stages of raw material sourcing, textile production, and garment manufacturing, fashion products are distributed to warehouses and/or distribution centres to be delivered to retail stores to reach end-consumers. Approximately 65% of the initial input is delivered to end-consumers as new clothing. In other terms, 35% of the initial input becomes waste in the SC stages till fashion products reach the final consumer. Reports show that an estimated 1.14 million tonnes of clothes are supplied onto the UK market each year and in order to produce these clothes, some 1.76 million tonnes of raw materials are utilized while around one-third of this becomes waste in the SC; furthermore during in-use stage, an estimated 10,000 tons of waste is generated (WRAP 2012). This mostly occurs when clothes are damaged while getting cleaned, such as shrinking or discoloration. Thus it could be stressed that the majority of clothing waste footprint is generated as post-consumer waste. An estimated 350,000 tonnes of used clothing goes to landfill in the UK every year (WRAP 2012). Only in Turkey, 287,000 tonnes of textile waste goes to landfill and it constitutes 2.62% of municipal solid waste (Buyukaslan et al. 2015). American consumers, on the other

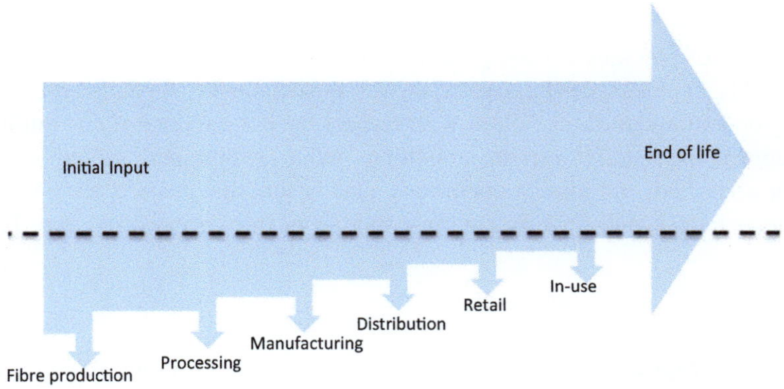

Fig. 1 Waste flow in the fashion industry

hand, send 10.5 million tons of clothing to landfills every year (Cline 2014). Nevertheless, only 20% of textiles are recycled each year (Ditty 2015). Considering that the average active lifetime of clothing is approximately 2 years, more post-consumer clothing waste must be collected for re-use and recycling purposes. As reported by the Fashion Revolution (2015), every tone of discarded textiles reused could save 20 tons of CO_2 from entering the atmosphere. If 5% more clothing at the end of its first life is reused the global waste footprint could be reduced by 0.7% (WRAP 2012). Figure 1 presents a depiction of waste footprint for clothing.

Animal Welfare

Animals are used for a variety of luxury goods. Animal treatment is a widely discussed topic due to a number of intimidating facts. Fur, widely used for luxury goods, is not a by-product of the meat industry. The European Union is the world's largest producer of factory-farmed fur. Around 30 million mink, 2 million fox and 100,000 raccoon dogs are killed each year in the European Union solely for their fur, as 60% of worldwide mink-fur production (30 million pelts) and 56% of worldwide fox-fur production (2.1 million pelts) comes from European farms (EFBA 2010). Angora rabbit fibre production is the largest animal fibre industry after wool and mohair, with an estimate output of 10,000 tonnes per year, according to Rural Industries Research and Development Corporation's report by McGregor (2012). Still, luxury fashion does not have to be a cruel business. There are many alternative solutions to deliver fashion with respect for the animal welfare. Animal skins and feathers could be sourced and supplied from the meat industry as a by-product. Additionally, fish skin could be further utilized to make leather rather than being dumped back into the sea. This could be a great alternative to the usage of exotic skins of crocodile and snake.

2.1.2 Social Impact of Fashion Supply Chains

Mining Operations

The mining of precious stones, gold and other valuable metals utilized in luxury goods holds significant environmental as well as social risks. As outlined by WWF, Mining operations must take place with respect for basic human rights outlined in international laws. Safe working conditions must be provided, labour standards must be respected, and mining operations must ensure that projects are not located in protected areas, and mine waste must stay away from oceans, rivers or lakes.

Child Labour

In its report of 'Marking Progress against Child Labour', ILO (2013) indicates that 168 million children are child labourers, accounting for almost 11% of the child population as a whole. Children in hazardous work make up almost half of all child labourers, totalling 85 million in absolute terms. However, the International Labour Organization shows that a progress against child labour has been marked over the years and there were about 78 million fewer child labourers at the end of the 12-year period that began in 2000. Still, child labourers are mainly employed by factories producing textile and garments to satisfy demands of consumers in developed countries (Moulds 2015).

Hidden Subcontractors

Purchasing practices of buying firms must respect human and labour rights at all stages of FSCs. Corporate accountability must be considered while defining purchasing policies. On the other hand, long-term and more stable buyer-supplier relationships must be built and maintained. Good production planning is a key criterion. Buying firms and manufacturers must agree on more equal and more realistic contractual agreements that would not unequally penalize any party for delays that occur outside the latter's responsibility (Theuws 2015). To meet strict deadlines manufacturers might subcontract certain production processes to other factories, even without informing the buying firm. To this end, the contractual supplier might shift the orders to such hidden suppliers. As addressed by The Centre for Research on Multinational Corporations (SOMO), such subcontracted factories are not considered in the boundaries of the buying firm's SC, thus audits do not take place there. Yet, there are many risks associated with hidden subcontractors. Such workplaces could be unregistered, informal companies where no taxes are paid, and where governmental labour inspections do not take place. There is a big risk of workplaces with insufficient conditions being incorporated into a fashion SC. The growing pressure on low prices, fast delivery, high unpredictability, along with volatile relationships in the fashion industry, has resulted in increasing levels of unauthorized subcontracting in production countries. Therefore, fashion retailers must be aware of the places where textiles and garments are produced. Without transparency, a strict monitoring and accountability secured and preventing human right violations become difficult.

Safety

Mass-production and uncertain working conditions have been a big problem since the mid-nineteenth century. The Triangle Shirtwaist Factory fire in Manhattan, New York City on March 25, 1911 was the deadliest disaster is known to be the deadliest disaster of its time, causing the deaths of 146 garment workers—123 women and 23 men. The disaster led to legislation requiring improved health and safety working conditions. A century later, another important progress was marked in the fashion history. On May 15th 2013, the Accord on Fire and Building Safety in Bangladesh was signed by many global fashion brands following the Rana Plaza disaster.

The accord is a five-year independent, legally binding agreement between global brands, retailers and trade unions that aims to build a safe and healthy Bangladeshi Ready Made Garment (RMG) Industry. Yet, it is reported that Rana Plaza disaster was less of a turning point than was expected. A backlash was reported against unions, increasing violence has been informed against unionized workers and unions were reportedly banned in Bangladesh's special export processing zones. However, as conveyed in Article 23 of The Universal Declaration of Human Rights (United Nations 1948), everyone has the right to form and to join trade unions for the protection of his interests.

Working Conditions

In spite of a growing number of international standards, certifications and regulations to improve working conditions in fashion SCs, textile and garment production is not always secure. In September 2015, Mulberry was reported to fail in protection of the worker rights in its supplier located in Izmir, Turkey. Despite a list of global sourcing principles set by Mulberry, workers in Mulberry's İzmir supplier were fired after they had joined a trade union (Clean Clothes Campaign 2015a). However, Article 23 of The Universal Declaration of Human Rights (UDHR) addresses that everyone has the right to form and to join trade unions for the protection of his interests. The Turkish clothing industry, on the other hand, was the third largest exporter to the European Union and the sixth largest globally in 2013 (Fibre2Fashion 2016). Given this importance, Business and Human Rights Resource Centre conducted a research and approached 28 major garment brands on how they deal with Syrian refugees in Turkish garment supply chains. Reportedly, 2.2 million refugees are living in Turkey, yet 250,000–400,000 refugees are estimated to be working illegally (McMullan 2016). Responses obtained from major fashion retailers, including C&A, H&M, NEXT, and Primark reveal that Syrian refugees in supplier factories were identified in 2015. NEXT, having 22 first tier suppliers collectively operating in 40 factories identified Syrian refugees in 10 factories and Syrian child labour in 2 factories. Reportedly, the refugees receive payment of a social security fee as part of their wages. H&M, on the other hand, identified Syrian refuges in 4 factories and Syrian child labour in 1 factory out of 210 factories in its first tier. These incidents occurring in FSCs underline the importance of performance assessment and strict monitoring of upstream suppliers.

Wages

A living wage is a human right as stated in the Universal Declaration of Human Rights by the United Nations. Article 23 further stresses that everybody has the right to work, to free choice of employment and to protection against unemployment. Everybody, without any discrimination, has the right to equal pay for equal work. Though, the legal minimum wage in most textile and garment producing countries is not enough for a decent life. Living wage is a challenging issue not only in Asia but also in Europe. It is estimated that the current minimum wage in Bangladesh only covers 60% of the cost living in a slum. Furthermore, in China and Cambodia the minimum wage would need to be almost double to cover the basic cost of living. It was reported that garment workers in the Ukraine, Romania, Bulgaria and Turkey also face low wages, which are even below those in China (Clean Clothes Campaign 2015b). In Italy, immigrant workers were reportedly working in clandestine workshops producing for high-end fashion brands. A recent research conducted by the Clean Clothes Campaign into working conditions shows the net minimum wage in Poland and in the Czech Republic was €312 and €390 respectively (Clean Clothes Campaign 2016). Another study conducted into Italian shoe and garment industry in 2013 reveals that the competition with Eastern Europe and Asia drives wages down in Italy. It is shown that factory workers' wages at the entry level do not exceed 1200 € per month. In the Veneto region, trainees were reportedly paid 730 € a month, while home-based workers got only 850 €—far below the amount needed to survive with dignity (Clean Clothes Campaign 2015c). The illegal industry was thus found to be growing as a response to price competition. Some subcontracting firms were reportedly hiring workers at low prices on contracts where the workers were subject to work excessive hours to stay employed. To eliminate such problems, purchasing practices must be revised and stricter controls must be ensured.

3 How Could Sustainability Truly Meet Fashion?

3.1 Toward Sustainability Integration in Luxury Fashion Supply Chains

Luxury goods companies unveil their collections and set the trends in iconic runway shows. Subsequent to fashion shows, at least six months are needed to arrange planning, ordering, production, and distribution. Yet, pace is getting quicker, and today's luxury goods industry seeks ways to reduce process lead times from months to weeks to meet end-consumers' demands. Consequently, luxury fashion buying firms outsource and restructure their logistics and SC designs (Brun et al. 2008; Brun and Castelli 2008). Nonetheless, this growth jeopardizes numerous patterns. Growingly scarce natural resources and rising commodity prices create critical variables for luxury goods companies to remain profitable. During the favourable economic situation of the 1990s, companies were relatively reluctant to respond to concerns over social and environmental needs. Nevertheless, with today's global economic uncertainty, globalization and growing business complexity, a renewed focus to sustainability is required. Caniato et al. (2009) address a set of critical success factors to compete in the luxury market, including premium quality, craftsmanship, exclusivity, brand building, style and design, emotional appeal, country of origin, uniqueness, performance and innovation and lifestyle creation. However, Brun and Castelli (2013) suggest that the set of critical success factors must be revised and extended by taking recent market trends into considerations such as ethical aspects and sustainability. Hence, luxury fashion companies must develop and deploy sustainability management capabilities to manage complex relationships with each actor in their SCs. This raises the growing importance and relevance of upstream SCs for sustainability management in the luxury goods industry.

3.2 The Strategic Importance of Upstream Supply Chain in Luxury Fashion

Social and environmental management need to be enhanced in globally dispersed FSCs. However, this could be demanding since other competitive pressures, such as cost, quality, unpredictability of market demand and product variety continue to escalate. The analysis of Parmigiani et al. (2011) shows that the nature of stakeholder exposure determines how social and environmental capabilities impact social and environmental outcomes. Fostering sustainable behaviours, effectively responding environmental and social pressures, and reducing the impact are getting more and more challenging. Many small and medium sized suppliers fail in responding to environmental pressures due to limited capabilities and available resources (Lee and Klassen 2008), which could ultimately damage big buying firms. Similarly, as stated by Tachizawa and Wong (2014), the most critical environmental and social issues in SCs are generated by suppliers located in the second tier or further upstream. This thus increases the pressure to monitor sustainability in industries—such as luxury goods—where environmental and social impacts are serious due to material sourcing, energy consumption, and labour intensity. Social and environmental performance of suppliers is an area of mounting concern, yet an integrative method to assess sustainability at upstream levels could be considered a missing link. High-end fashion companies are exposed to scrutiny to improve their sustainability performance in the context of social and environmental behaviours. Smaller suppliers or sub-contractors face few in contrast. Luxury goods companies' purchasing power within the SC and technical capabilities are critical to diffuse the sustainable behavior throughout the chain. Collectively, sufficient coordination between SC partners is needed. Despite the attempt coming from high-end buying firms, suppliers often fail to appropriately implement sustainability initiatives. A requirement for supplier flexibility negatively affects the suppliers' ability to plan the demands on business resources. As an example, Laura Ashley requested a 10% discount on cost price from upstream suppliers with immediate affect, and similarly Monsoon Accessorize demanded 4% discount from all suppliers as well as an increase in payment terms from 60 to 90 days (Perry et al. 2014). Given these demands, compliance with code of conduct has become more challenging for suppliers throughout FSCs. The rise of complex relationships results in reduced visibility and control of ethical issues. Therefore, the risk of sustainability failures addresses the significance of upstream and downstream integration. Environmental Protection Agency (EPA) lists tanneries as top polluters on its "Superfund" list. SCM therefore requires a closer involvement with suppliers, employees, customers, communities and environment by establishing long-term relationships, enduring sustainability management capabilities and information sharing systems through a more structured approach.

Growing pressure on companies to integrate sustainability in operations management results in challenges including integrating environmental, health and safety concerns, green-product design, lean and green operations, and closed-loop supply chains (Kleindorfer et al. 2005). Thus, it is becoming vital to understand how to address environmental and social facets while maintaining competitiveness. Integrating environmentally and socially sound behaviours, transforming the entire SC into becoming more sustainable, monitoring the performance and further creating value require significant resources. Once companies move beyond the 'low hanging fruit' philosophy, such as reducing energy use, and begin inspecting more central issues such as sustainability integration into business models, reengineering of current SCs could bring performance improvements. Operations management literature often illustrates the positive connection between environmental practices and financial outcomes. Schoenherr (2012) demonstrates that environmental management practices could lead to a positive performance on a plant's quality, delivery, flexibility and cost performance. Golicic and Smith (2013) relatedly show a positive relationship between environmental management and focal company performance. Further, Beske and Seuring (2014) address that win-win situations could be achieved, including minimizing waste for the environment, less pollution burdens for the society, and cost savings for the company. From the inbound point of view, making SCs more sustainable could bring numerous benefits to a company such as cost and risk reduction. Total quality management perspective provides a strong association between quality management and environmental management systems. The processes, which improve quality, reduce waste, and cut costs could be used to advance environmental outcomes, implying that multiple stakeholders could be satisfied simultaneously (Wu and Pagell 2011). It is worth stressing that stakeholders are instrumental in integrating sustainability and enriching the overall value proposition. In the 1980s, stakeholder theory was introduced and expressed to acknowledge the individuals who could affect and/or be affected by the achievement of corporate goals. Stakeholder theory has become the most cited and discussed theory in the sustainable SCM (SSCM) literature (Varsei et al. 2014). Freeman et al. (2006) further stress the importance of companies' responsibilities for stakeholder. Accordingly, stakeholders could influence companies to improve sustainability practices and the integration of sustainability throughout the entire value chain. Internal sustainability practices could become antecedents for external collaboration. Thus, an integrated framework encapsulating various stakeholders and dynamic sustainability capabilities as constituents of a corporate behavior is needed to advance sustainability performance in luxury FSCs.

3.3 Integrating and Monitoring Sustainability

Not only upstream suppliers but also buying firms, customers and communities have a great authority in the translation of sustainability management capabilities into value creating action programs. Subsequently, upstream and downstream integration could be a bridge between external stakeholder assessments and internal operational changes. In this sense, all stakeholders play multifunctional roles, such as monitoring progress, seeking advancements and scanning for external opportunities. The attitude of stakeholders toward sustainability and the commitment of buying firms to social and environmental issues could create successful corporate strategies. To this end, a coherent model to create and monitor sustainability in fashion SCs is required. In this domain, Ageron et al. (2012) establish a framework to manifest sustainability in upstream SCM. However, the model, incorporating drivers, barriers and sustainability approaches, integrates all blocks independently. Nevertheless, elements required to fully embark sustainability in upstream SCs must be coordinated holistically. Furthermore, reliable and mutually beneficial relationships among all stakeholders in the network must be established. This continuity considers the overall SC performance and not only the performance of individual SC members. In order to further a more complete theory, Pagell and Wu (2009) build a coherent and testable model to show elements necessary to create SSCM. While providing useful insights, what is not yet tested is how these social and environmental management practices could translate to actors, such as second movers, commodity suppliers or an entire industry. Afterwards, Wu and Pagell (2011) further investigate how organizations could deal with short-term pressures to remain economically viable while implementing SSCM models. Subsequently, they develop and propose configurations of environmental behaviours as well as business strategies. Nevertheless, social aspect of sustainability is not covered within the study. Yet, an inclusive analysis of sustainable operations must consider all three dimensions simultaneously. Additionally, Lee and Klassen (2008) examine drivers and enablers that could foster environmental management capabilities (EMCs) in small and medium sized suppliers. Despite the relevancy and accuracy of the findings, the consequences of the improved EMCs of suppliers and the ways that both buying firms and their suppliers could benefit from buying firms' sustainability management are yet to be investigated. Gualandris et al. (2015) further synthesize a model to address accountability for sustainability issues in SCs. Sustainable evaluation and verification (SEV) model, encompassing all activities that identify key metrics, process data, verify reliability and disclose the information of sustainability performance to all stakeholders, brings esteemed evidence, yet it does not address how local firms could leverage inclusivity to enhance SSCM development. Respectively, it becomes important to present a holistic view to monitor environmental and social performance through stakeholder unification.

4 Where Does Sustainability Stand?

4.1 Interrogating Sustainability Disclosure in the Fashion Industry

Sustainability reporting to disclose sustainability performance is receiving a prominence among global fashion companies. Recently, there is an increase in the number of fashion companies sharing details of their sustainability performance in publicly available reports. There is no universally agreed definition of a sustainability report, however there are many definitions available in the literature. The World Business Council for Sustainable Development states that sustainability reports are public reports to provide external as well as internal stakeholders with a depiction of corporate position and activities on environment, society, and economy. Providing accurate and fully transparent information on topics such as governance, intangible assets, non-financial issues, and sustainability performance is vital, as this information is publicly available. Nevertheless, despite the proliferation of these sustainability reports, a general consensus is yet to be reached on what these reports channel, what they must encompass, and how accountability must be transferred along the value chain. The global reporting initiative (GRI) is the best known and the most frequently used framework for voluntary reporting with the overall purpose of standardizing various reporting systems implemented simultaneously. Being launched in 1997 by the United Nations Environment Program (UNEP), GRI highlights that companies that want to commit to business sustainability must report on the social, environmental, and economic aspects. Academic debate on sustainability reporting is growing, however concerns and doubts exist regarding the accuracy and accountability of these reports. This occurs due to the high degree of qualitative information disclosed in such reports. Thus, incorporating quantitative information plays a fundamental role in corporate transparency. Despite the proliferation of such reports, questions exist on what information sustainability reports should contain and how the reports should be structured (Roca and Searcy 2012). Additionally, it is important to know that the level of commitment, dissemination, and performance disclosures vary from industry to industry. Given inconclusive results of the academic debate and diversity of sustainability reports, an industry-level analysis through an assessment of the comprehensiveness of sustainability reports published by luxury goods companies could deliver novel and rigorous outcomes.

4.2 Sustainability Assessment of Luxury Fashion Companies

According to marketing gurus, consumers everywhere, regardless of their economic level, desire more luxury (Brun et al. 2008). Luxury encompasses several industries including personal goods, footwear, leather, jewellery and cosmetics. In this cross-sector setting, many luxury goods companies have been diversified into shoes, bags even cosmetic and jewellery. Having this enriched industrial depiction in mind, the selection of luxury brands becomes pivotal to deeply examine sustainability in the luxury goods industry. To this end, Financial Times Global 500 (Dullforce 2015) was initially visited to see the biggest global luxury goods companies. Subsequent to the preliminary scan on the economical component, social and environmental sustainability performance was deeply examined. The United Nations Global Compact (UNGC) is a strategic policy initiative for businesses to ensure that markets, commerce, technology and finance advance in ways that benefit economies and societies everywhere. To this end, UNGC database was visited to unveil which luxury goods companies joined the initiative. Furthermore, Carbon Disclosure Project (CDP) was considered. CDP is a global initiative working to transform the way the world does business so as to prevent climate change and protect natural resources. Similarly, luxury fashion companies disclosing their carbon performance were deliberated. Having said that, three main luxury goods holding companies, taking together owning the majority of the entire luxury goods industry, were finally selected. **Kering**, *owning luxury brands of Gucci, Bottega Veneta, Saint Laurent, Alexander McQueen, Balenciaga, Brioni, Christopher Kane, McQ, Stella McCartney, Tomas Maier, Boucheron, Dodo, Girard-Perregaux, JeanRichard, Pomellato, Qeelin and Ulysse Nardin*; **LVMH**, *owning high-end fashion and leather goods brands, such as Berluti, Céline, Christian Dior, Donna Karan, Edun, Emilio Pucci, Fendi, Givenchy, Kenzo, Loewe, Loro Piana, Louis Vuitton, Marc Jacobs, Nicholas Kirkwood, Thomas Pink*; and **Richemont**, *owning key luxury goods mansions, such as Alaia, Chloé, Lancel, Montblanc, Peter Millar, The Neta-Porter Group* were designated. Kering joined UNGC on 2008, LVMH on 2003 and Richemont on 2013. Additionally, these three are active on their carbon performance disclosure. Our sample luxury holding companies disclose their environmental and social performance at the group level. This implies that reporting is issued under the name of the holding company while incorporating all practices and strategies implemented at the brand level. For example, Kering launches its corresponding reports in which all brands owning by Kering are excessively elaborated in terms of their sustainability records. Subsequent to the sample selection, content analysis was employed. Sustainability reports for the last five years (2010–2015) were obtained from corporate websites for further analyses. All reports were assessed, key performance indicators (KPIs) were highlighted, all quantitative information was documented then a database encompassing KPIs, corresponding aspects, and companies' environmental sustainability practices, was further developed.

4.3 Environmental Sustainability Aspects of the Luxury Fashion Industry

KPIs were classified in terms of aspects to reflect organizations' significant environmental impact. GRI (2015) catalogues fundamental environmental indicators in following aspects.

- Material,
- Energy,
- Water,
- Biodiversity,
- Emissions,
- Effluents and waste,
- Product and services,
- Compliance, and
- Transport.

The breakdown of KPIs into environmental aspects is rigorous, yet it is important to highlight that some performance indicators could belong to more than one aspect. Furthermore, GRI provides a list of topics identified as relevant for apparel, accessories and luxury goods manufacturing (GRI 2013). Belonging to the environmental category, the most fundamental topics to the textile, apparel and footwear industry are:

- Material sourcing
- Material sourcing and use
- Material use
- Energy consumption
- Water management systems
- Water use efficiency
- Water use efficiency and quality
- Waste water treatment
- Emissions to air
- Emissions to air—GHG emissions
- Chemicals use and discharges
- Textile waste management
- Cotton sourcing strategy and policies
- Packaging
- Life cycle assessment of products
- Industrial process innovation.

Environmental sustainability practices of luxury goods companies were identified based on their publicly available reports. This study turned these environmental sustainability practices into numerical scores, with each measure given equal weight in making up the overall score. In the absence of any practice to determine the relative importance of the KPI, the weighting was not applied. A ratio was

Fig. 2 The breakdown of environmental sustainability aspects in luxury FSCs

estimated between the number of environmental practices employed by the company, and overall number of practices deployed within the industry. Industry average was assigned and each company was positioned according to their score. Based on the weights, the following breakdown is provided to display the most addressed aspects in the luxury goods industry, as displayed in Fig. 2.

- **Energy Consumption**: This category groups all indicators related to direct energy consumption, indirect energy consumption, energy savings, renewable energy based products, energy efficiency, initiatives for energy reduction, and energy improvements.
- **Process and Packaging**: Indicators in this category deal with the initiatives to mitigate environmental impact, plastic use and management, resource use efficiency, chemical use and discharge, heavy metal and chemical use, and industrial processes.
- **Waste and Emissions**: These indicators detail total direct and indirect greenhouse gases emissions, other relevant indirect and significant air emissions, waste by type and disposal method, any kind of spills, and percentage of transported waste.
- **Biodiversity**: This category deals with areas of high biodiversity value, description of significant impacts of activities, and services on biodiversity, and activities to mitigate the impact on numerous species and biodiversity.
- **Materials Sourcing and Use**: These category regroups all indicators related to materials used, percentage of materials used that are recycled, cotton sourcing, recycled materials and synthetic fibres, wood-based products from responsibly managed forests.
- **Water**: Indicators linked to water consumption, water treatment, water sources affected by withdrawal of water, percentage and total volume of water recycled and reused.

The aspects do not represent a balanced breakdown within the environmental category. The majority of focus was devoted to energy and packaging. Nevertheless, water management, biodiversity, and material sourcing require more attention due to their profound importance for luxury. With respect to **Energy Consumption**, Richemont fostered energy efficiency through solar panel installation. Furthermore, the company is currently producing its own hot water. Gucci, and similarly Louis Vuitton, were actively engaged in installation of automatic dimmer switch, and in the development of heat recovery. Other companies in the sample were partially certified with LEED for their warehouses, and energy efficient store lighting emerge as a hot trend. Loewe largely installed energy saving equipment at its plants. Nevertheless, energy efficiency should be further improved across the industry. Creation of e-learning modules on sustainability, sustainable building constructions, efficient use of heating and cooling systems could be promoted.

Process and packaging related practices accounted for a total of substantial 22% of all practices deployed in the environmental category. Bottega Veneta and Stella McCartney profoundly dominated the industry through the use of natural materials, recycled plastic, and alternative materials. Loewe utilizes Forest Stewardship Council (FSC) certified packaging materials in addition to using reusable boxes. Gucci similarly focused on the issue by modifying box sizes to reduce the packaging material required. Fashion companies commonly use recycled and certified raw materials; however it was observed that incentives are required to incorporate these actions at upstream supplier level.

When it comes to **Emissions and Waste**, a total of 20% of environmental practices were implemented to deal with residues, climate change, and waste. Richemont principally utilizes environmental friendly HVAC (heating, ventilating and air conditioning) systems. Loewe invests in natural leather, rubber and crepe, metals, cardboard and machine oils recycling. Kering brands, on the other hand, aim to increase the performance by going towards zero waste and full emission control. Cooling and heating systems could be replaced to reduce greenhouse gas emissions to this end. Waste recycling could be further promoted within the industry.

Biodiversity, on the other hand, is a vital topic, and therefore suppliers must employ accepted practices and humane treatment in sourcing. Buying firms and/or fashion groups must regularly control suppliers to ensure high standards. The procurement of precious skin should follow international regulations and procedures. Within the industry, there is a need of taking a number of initiatives to enhance traceable, and sustainable sourcing. Leather must be sourced from verified sources, which do not cause damage in sensitive ecosystems. Thus, improving sustainability in the leather trade, and tracking the leather through a traceability system could be ensured with a very important priority. Gold, diamonds, and precious stones are also important raw materials for the industry. In the pursuit of sustainable raw materials, mining operations should also be traced not to have a negative impact on biodiversity and local communities. Sourcing from the verified and certified mines possessing high social and environmental standards must become a business strategy. Stella McCartney has dedicated her company vision to improve animal welfare and to protect precious species. Similarly, Gucci, Chloé and Bottega Veneta deployed practices to source materials from certified tanneries. Nonetheless, practices deployed for biodiversity should be urgently improved as luxury fashion heavily depends on precious materials.

A total of 11% of the practices implemented were classified in **Material Sourcing and Use**. Bottega Veneta and Gucci were observed dominating this category through their well-established material sourcing strategies, and their collaborative actions with upstream suppliers. Nevertheless, environmental impact is largely generated during raw material sourcing. Thus, fashion companies must implement strategic collaborations.

Water management is also an extremely important topic that should be carefully taken into account. Guidelines including recommendation on management systems, certification and supplier engagement must be provided to ensure water efficiency across supply chains. In this vein, Richemont emerges as a significant example with a rainwater recycling system while Bottega Veneta opts to recycle water. Hence, in this fashion water must be managed in an efficient fashion through awareness raising programs and management guidelines. Inefficient use of materials causing waste generation leads to more production and extraction of raw materials. Efficient material usage, responsible sourcing, and waste minimisation could lead to reductions in energy and water use. Awareness raising campaigns, supplier engagement, and smart utilization and recycling based programs must be enhanced within the industry.

It is important to highlight that each company reported their progression on **Supplier Management**. However, sustainability integration must be further passed upstream and stricter monitoring throughout the chain must be ensured. During various stages of FSCs a number of stakeholders are exposed to critical environmental and social risks. To this end, an integrated approach is needed to further develop and deploy sustainability integration in luxury FSCs in order to mitigate negative environmental and social impacts and create new opportunities through value destroyed. The next section provides a novel approach to appraise environmental and social sustainability throughout the entire chain of luxury fashion companies.

In addition to the environmental practices developed and deployed within the industry, it is equally important to channel which practices employed in the social context. In terms of social sustainability, luxury brands seek ways to foster social development. In this vein, Richemont deploys a 3-year CSR implementation plan. CSR culture with a goal of increasing employee engagement is embarked at the group level. For example, Richemont Dubai has established a cross functional CSR committee and employees are encouraged to volunteer for charitable causes. Community investment is another growing topic in which RICHeros program was initiated in 2015 to support charitable activities. Kering, on the other hand, has been executing some significant projects to advance social sustainability across globe. The Kering Corporate Foundation, since its launch in 2009, has been dedicated to battling with violence against women. In an alignment with Kering's identity, the Kering Foundation focuses its activities on three geographical areas by combatting sexual violence in the Americas, domestic violence in Asia and harmful traditional practices in Western Europe. Since 2009 more than 140,000 women have been benefitted from this project and 47 partnerships with local as well as global NGOs have been established. Furthermore, LVMH grounds its social responsibility on four essentials, namely workplace well-bring, and quality working conditions, talent development, discrimination prevention and local community support. In this sense, some signature initiatives take place. The group is committed to gender equality and the United Nations Women's Empowerment Principles are further promoted. LVMH also founded the Institut des Metiers d'Excellence to preserve and convey artisanship and promote the attractiveness of careers in traditional craftsmanship. Additionally, LVMH supports social entrepreneurs that create economically viable businesses addressing environmental as well as social needs.

5 Sustainability Resolution

5.1 Impact Mapping Throughout Luxury Fashion Supply Chains

Concerns are mounting over environmental and social impacts and sustainable development is becoming vital. For example, the 2030 Agenda for Sustainable Development at the United Nations Sustainable Development Summit was adopted on 25 September 2015. The agenda includes a set of 17 global Sustainable Development Goals (SDGs) to end poverty, fight inequality and injustice, and tackle climate change by 2030. The negative impact could be mitigated if corporate vision is driven toward innovative solutions. Sustainability vision could lead to value generation when embedded in the development of new technologies, products and processes. As stressed, a wide range of stakeholders is widely affected and therefore looking at various stages of life cycle is required to advance environmental efficiency and social improvement. To this end, an integrated approach is needed to further develop and deploy sustainability in luxury FSCs. This study thus pledges a significant step towards a testable model based on environmental and social performance assessment of luxury FSCs. A rigorous methodology is developed to pilot with global luxury fashion companies from a 360-degree perspective.

5.2 360 Degrees Performance Appraisal

360 Degrees Performance Appraisal indicators have been created based on the UN Global Compact's Ten Principles, the Universal Declaration of Human Rights, Global Reporting Initiative (GRI) G4 Sustainability Reporting Guidelines, UN Guiding Principles and Corporate Human Right Benchmark. The model provides a concise set of indicators to which fashion companies must have answers to meet their responsibilities to respect UN Global Compact Areas. On the whole, 360 Degrees Performance Appraisal aims to:

- Set UN Global Compact areas, namely human rights, labour, environment and anti corruption, in the business context of luxury FSCs,
- Meet a reliable threshold of information,

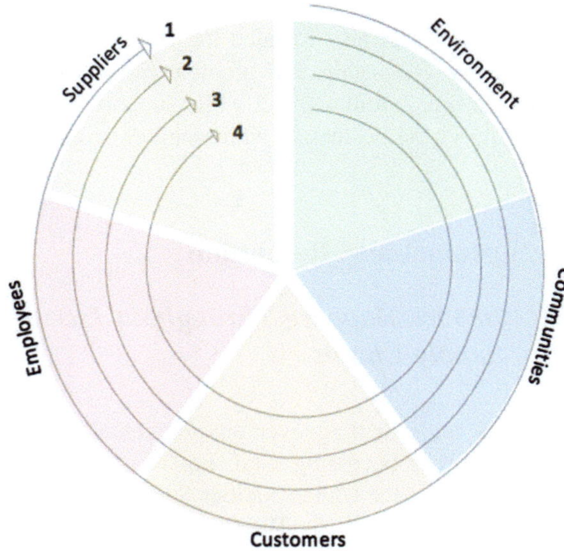

Fig. 3 360 degrees performance appraisal model

- Demonstrate a continuous improvement,
- Address the most critical impacts of fashion operations,
- Transform value destroyed into new opportunities in luxury FSCs.

As depicted in Fig. 3, a 360 Degrees Performance Appraisal model is presented to assess sustainability throughout the entire chain with a final goal of reducing negative impacts and transforming value destroyed and value wasted into new resources.

5.3 360 Degrees Performance Appraisal: Steps to Pursue

1st Step: Review Sourcing, Purchasing, Manufacturing and Distribution Strategies

- Understand current design, sourcing, purchasing, manufacturing and distribution structure
- Identify who is involved in SC processes

2nd Step: Map the critical path

- Determine the current value delivered to each stakeholder
- Define the negative social and environmental impact each stakeholder faces Identify value destroyed, e.g. emissions, and value wasted, underused aspects

3rd Step: Develop an improvement plan

- Develop strategies to eliminate negative environmental impact, e.g. pollution and waste prevention

- Improve techniques to transform negative impact into new opportunities, e.g. creating value from waste
- Explore new methods to reduce negative social impact, e.g. resolving conflicts among stakeholders

4th Step: Continuous Performance Appraisal

- Monitor performance regularly
- Evaluate the progress and improve appraisal procedure
- Disclose the results and enhance transparency throughout the chain

Luxury fashion companies must initially review their strategies on design, sourcing, purchasing, manufacturing and distribution. All parties and individuals involved in SC processes must be clearly identified e.g. communities, customers, employees, first and second tier suppliers, raw material suppliers. Defining stakeholders strategically is important, as stakeholder engagement is pivotal for business efficiency as well as value generation. Operational policies must be further comprehended by which a series of practices and processes in the system could be identified. The positive impact generated, in the form of value delivered to stakeholders, must be acknowledged. Once positive impacts are outlined, negative impacts on environment and society must be critically determined. Assessing waste generation and measuring carbon footprint is indispensible to detect unnecessary activities to be avoided at the source. The identification of value destroyed, e.g. emissions and value missed, underused capabilities could create new value opportunities.

In order to transform negative environmental and social impact into value, such as creating value from waste to be utilized in new product development, an improvement plan must be developed. A number of improvement techniques could be industrialized and conceptualized throughout the SC. New methods must be explored for environmental mitigation. For example, during the design phase, which has effects on all other process steps from raw materials to end of life, environmental impact could be drastically reduced through a wide range of sustainable design options. Emissions generated by cotton production could be reduced through the use of a number of measures such as fertilizer application rates, tilling methods, fertilizer and pesticide optimization, and increasing energy efficiency as well as uptake of low carbon sources of electricity for cotton ginning. During the manufacturing phase, utilizing improved and alternative fibres and investment in new equipment of product reconfigurations could bring significant reductions. Throughout the consumption phase, the modification of consumer behaviour, including the increase of product lifetime and the increase of reuse and recycling, result in impact mitigation.

Prior studies show that (WRAP 2012), washing clothes at a slightly lower temperature results in 1.4% reduction in carbon footprint, 0.03% in water footprint, and 0.1% in waste footprint. Extending the useful life of clothes through design, use and re-use by 10%, equivalent to 3 months longer use, could massively result in 7.7% in carbon, 10% in water, and 8.8% in waste reduction (WRAP 2012).

A closed loop approach must be endorsed through a circular manner for apparel development, consumer use and end of life. Similarly, from physical to virtual, production on demand and dematerialization could be endorsed as new methodologies to further embed environmental stewardship. Still, to this end, new techniques must be explored and corporate strategies must be embarked through a sustainable vision.

360 degrees performance appraisal is decisive to increase transparency and provide continuous improvements. Nevertheless, monitoring and evaluation must be performed through quantitative data and performance must be regularly evaluated. Fashion retailers must operationalize an ethical code of conduct. To this end, environmental and social performance must be monitored, evaluated and reported across FSCs and make available to public. Furthermore, SC information including factory lists, supplier and vendor list across further tiers must be published publicly. Compliance, corrective action plans, and remediation activities must be shared along with wages information to further support a living wage principle. In order to ensure a rigorous value creation through SCs, fashion companies must move from reactive approaches to innovative business models.

5.4 360 Degrees Performance Appraisal: Measuring Sustainability

Performance indicators assigned for each sustainability dimension aim to derive both qualitative and quantitative information. Luxury fashion companies will be asked to provide information on below mentioned aspects. The main components of sustainability performance appraisal model encompass (i) governance, (ii) embedding sustainability and (iii) performance measurement.

A. Governance

A.1. Policy Commitments

A.1.1. Commitment to human rights
A.1.2. Commitment to labour practices
A.1.3. Commitment to local communities
A.1.4. Commitment to environment
A.1.5. Commitment to product responsibility
A.1.6. Commitment to anti-corruption
A.1.7. Commitment to stakeholder engagement

B. Embedding Sustainability

B.1. Sustainability Management

B.1.1. Performance management
B.1.2. Responsibilities assigned and resources provided for sustainability implementation
B.1.3. Training on sustainability issues

B.2. Sustainability Communication

B.2.1. Communication of policy commitment within the company
B.2.2. Communication of policy commitment across the SC
B.2.3. Communication of sustainability performance within the company
B.2.3. Communication of sustainability performance across the industry

B.3. Sustainability Integration

B.3.1. Sustainability integration within risk management
B.3.2. Ethics and integrity
B.3.3. Stakeholder engagement
B.3.4. Sustainability in design, sourcing, manufacturing, distribution, and retail processes

C. Performance Measurement

C.1. Environmental Management

C.1.1. Materials
C.1.2. Energy
C.1.3. Water
C.1.4. Biodiversity
C.1.5. Emissions
C.1.6. Waste
C.1.7. Products and Services
C.1.8. Transport
C.1.9. Supplier Environmental Assessment

C.2. Social Management

C.2.1. Labour practices and decent work

- Employment
- Labour/management relations
- Occupational health and safety
- Training and education
- Diversity and equal opportunity
- Equal remuneration for women and men
- Supplier assessment for labour practices

C.2.2. Human Rights

- Non-discrimination
- Freedom of association and collective bargaining
- Child labour
- Forced or compulsory labour
- Security practices
- Indigenous rights
- Assessment
- Supplier human rights assessment

C.2.3. Society

- Local communities
- Anti-corruption
- Supplier assessment for impacts on society
- Grievance mechanisms for impacts on society

C.2.4. Product Responsibility

- Customer health and safety
- Product and service labelling

5.5 360 Degrees Performance Appraisal: Scoring the Sustainability Maturity

Subsequent to the implementation of the 360 Degrees Appraisal Model, this scoring is further suggested to designate where companies operating in a FSC is located in terms of their environmental and social sustainability performance. In this sense, it is conceptualized in a way that performance indicators follow a set of structure granting points to the companies based on demonstrating example of their practices. For each indicator a company could score zero points if there is no sufficient evidence. To gain higher scores, the company must fulfil the requirements outlined for each indicator. For example, a raw material supplier could be awarded with five if a water recycling system takes place in its plant. Further, a garment manufacturer could be given zero if there is no social sustainability assessment is performed to monitor its sub-contractors' social performance. As for the consumers, if a fashion company completes post sales activities concerning waste disposal and/or product responsibility, the governance performance is likely to be awarded with a higher score.

The number of points awarded for each category in each dimension will be added and divided by the maximum number of points available for that dimension.

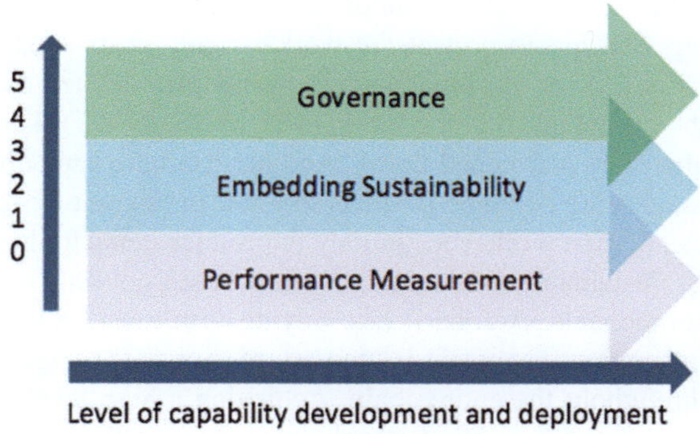

Fig. 4 Sustainability maturity level

That is, this score will bring an overall score for each measurement aspect. Not only could a company see its overall performance, but also critical spots and improvements could be highlighted through a benchmark on a longitudinal ground. Furthermore, fashion companies could further implement this technique to assess their suppliers' sustainability performance as well. Given that social and environmental risks often occur at further levels of FSCs, luxury fashion companies are strongly suggested to formulate sustainability performance assessment for their suppliers as well as for their sub-contractors. Figure 4 depicts how system maturity would be addressed based on the scores obtained through performance appraisal.

Companies with no sustainability awareness or processes are to be assigned with *Level 0* while *Level* 1 is given to companies with little development of sustainability management. *Level 2, on the other hand*, constitutes limited development of sustainability management with sporadic implementation. *Level 3* refers to the companies in which sustainability management adopted but development and deployment is inconsistent whilst *Level 4* companies develop and deploy sustainability management well and implemented internally. *Level 5*, as the highest score, goes to the companies that perform mature sustainability management internally and externally, and embed continual improvement in operations. Subsequent to this section's methodology justification, the following part provides a discussion followed by the chapter's conclusion.

6 Discussion

Luxury goods industry poses numerous environmental and social challenges due to complex and heterogeneous nature of FSCs. As a consequence of rapidly changing market dynamics, globalization and more and more vertically network configurations impact assessment is getting more demanding. Prior studies stress that environmental and social risks mostly occur outside the buying firm's physical boundaries, implying that suppliers play a vital role in company performance. However, buying luxury fashion firms are not always fully aware of their network members' operational impacts. For example, in September 2015, Mulberry was reported not to protect the worker rights in its supplier located in Izmir, Turkey. Despite a set of global sourcing principles defined by Mulberry, its supplier failed the compliance as the workers in Mulberry's Turkish supplier were fired shortly after they had joined union. Another example depicting the importance of an entire monitoring could be given as follows. Immigrant workers are reportedly working in clandestine workshops in Italy to produce items for high-end luxury fashion brands.

Sustainability integration is thus based not only on technical, but also relational components. A control of a buying firm must be strictly ensured. To this end, an integrative approach is needed to provide a rigorous appraisal to be applied throughout the entire chain. Achieving a high level of sustainability integration is critical and it requires development of a consolidated information flow, optimization, know-how transfer, and reliability among members. Negative environmental and social impacts could be reduced only when corporate vision is directed through a rigorous analysis to be conducted at the network level. Sustainability integration must be passed upstream and as expressed, and the entire value chain must be strictly monitored to achieve substantial results. To this end, this study initiates a novel step towards a testable model based on environmental and social performance assessment of luxury FSCs. A rigorous methodology is developed to pilot with global fashion companies from a 360-degree perspective. The 360 Degrees Performance Appraisal model is presented to assess sustainability throughout the entire chain in order to reduce negative impacts and to transform value destroyed into new resources. Performance indicators to measure social and environmental sustainability have been generated in accordance with the UN Global Compact's Ten Principles, the Universal Declaration of Human Rights, Global Reporting Initiative (GRI) G4 Sustainability Reporting Guidelines, UN Guiding Principles and Corporate Human Right Benchmark. Indicators follow a set of structure awarding points to the companies based on demonstrating example of their practices. For each indicator a company could score zero points if there is no sufficient evidence. To gain higher scores the company must fulfil the requirements outlined for each indicator.

Drawing a wide range of stakeholders and relating them to social and environmental sustainability performance appraisal could unveil a number of revolutions that could have significant academic and managerial implications. The collaborative approach, although it presents many challenges, appears to be the most suitable to achieve a better sustainability integration. Buying firms must act like a change agent to innovate and provide the coordination throughout the value chain. A legislative perspective plays a vital role in the dissemination of sustainability. The growing importance attached to sustainable development has generated a proliferation of standards, regulations and certifications over the years. However, a lack of coordination and monitoring has led to confusion and inconsistencies. Accordingly, fashion companies must go beyond regulatory compliance and hence performance appraisal must be formulated to holistically embrace a longitudinal perspective.

7 Conclusion

This study develops a testable model, which employs capabilities literature by considering how technical and relational components of sustainability management must be assessed. Luxury goods industry, which has widely been investigated from marketing point of view, was examined from the SCM perspective. Personal luxury goods are designed, manufactured, distributed and sold throughout a number of stakeholders. Product components, coming from a number of suppliers, are controlled through a number of routines to ensure no hazardous substance is attached to products. Nonetheless, some other crucial issues in the upstream SCs are associated with human rights. Hence, monitoring a multitude of upstream suppliers' environmental and social behaviour is fundamental despite its cost and management challenges. A stricter assessment and monitoring of further suppliers must be secured to improve social facets. Due to the length of global FSCs, the vulnerability must arise from both technical and relational aspects, e.g. how products were produced and delivered, and to what extent social necessities were fulfilled. In this sense, transactions among members should not be taken as a unit of analysis; instead a broader set of provisions to protect the entire value chain must be considered simultaneously. This expanded foundation requires a further integration. Such accounting through this model could enable companies to have a comprehensive and longitudinal picture of where they currently stand and to what extent progress could be advanced.

This study aims for an alignment with international standards and universal principles. In this vein, it takes its biggest inspiration from the United Nation Global Compact's ten principles. Further, the valuable work that Nordic Fashion Association and Danish Fashion Institute are accomplishing (including the NICE Code of Conduct and Manual: For the Fashion and Textile Industry, striving for the UNGC's principles while providing additional specificity from an industrial perspective, and Copenhagen Fashion Summit) have become pivotal wake up calls to realize the urgency for this context. In order to come up with quantifiable measures and to provide a set of indicators, the UN Global Compact's Ten Principles, the Universal Declaration of Human Rights, Global Reporting Initiative (GRI) G4 Sustainability Reporting Guidelines, UN Guiding Principles and Corporate Human Right Benchmark have been applied. The model aims to provide a concise set of indicators to which fashion companies ought to have answers to meet their responsibilities.

The broad scale impact of social and environmental issues within public and private domains implies that an effective coordination is required to create efficient, accountable and sustainable SCs. The major managerial implication of this study is that fashion companies could assess sustainability performance effectively and comprehensively. Companies could contemplate on their performance scores, learn from the best practices and evaluate how overall performance could be improved. Future research can explore many avenues. As a methodological note, it can be seen that the conceptual framework as a step in theory building, but is not inevitably a final product. Thus, the model must go through the various stages of validation and tested before being considered a theoretical framework. In this vein, to increase reliability, case study-based research is further needed. In order to improve the model, the very next step is going to be to implement the 360 Degrees Performance Appraisal in a luxury FSC, preferably leather and cashmere. Lastly, studying fashion companies from different countries as well as segments could enhance the novelty in terms of the extent to which regulations affect sustainability performance.

Acknowledgements This research was conducted within the framework of the European Doctorate in Industrial Management (EDIM). This doctorate program is funded by European Commission's the Education, Audio-visual and Culture Executive Agency (EACEA), under Erasmus Mundus Action 1 programs.

References

Ageron B, Gunasekaran A, Spalanzani A (2012) Sustainable supply management: an empirical study. Int J Prod Econ 140:168–182. doi:10.1016/j.ijpe.2011.04.007

Aiama D, Carbone G, Cator D, Challender D (2016) Biodiversity risks and opportunities in the apparel sector. Gland

Beske P, Seuring S (2014) Putting sustainability into supply chain management. Supply Chain Manage Int J 19:322–331. doi:10.1108/SCM-12-2013-0432

Brun A, Castelli C (2008) Supply chain strategy in the fashion industry: developing a portfolio model depending on product, retail channel and brand. Int J Prod Econ 116:169–181. doi:10.1016/j.ijpe.2008.09.011

Brun A, Castelli C (2013) The nature of luxury: a consumer perspective. Int J Retail Distrib Manage 41:823–847. doi:10.1108/IJRDM-01-2013-0006

Brun A, Caniato F, Caridi M et al (2008) Logistics and supply chain management in luxury fashion retail: empirical investigation of Italian firms. Int J Prod Econ 114:554–570. doi:10.1016/j.ijpe.2008.02.003

Buyukaslan E, Jevsnik S, Kalaoglu F (2015) A sustainable approach to collect post-consumer textile waste in developing countries. Marmara J Pure Appl Sci: 107–111. doi:10.7240/mufbed.43773

Caniato F, Caridi M, Castelli CM, Golini R (2009) A contingency approach for SC strategy in the Italian luxury industry: do consolidated models fit? Int J Prod Econ 120:176–189. doi:10.1016/j.ijpe.2008.07.027

Carbon Trust (2011a) International carbon flows: clothing. In: The carbon trust. Available via https://www.carbontrust.com/media/38358/ctc793-international-carbon-flows-clothing.pdf. Accessed 15 Oct 2015

Carbon Trust (2011b) International carbon flows: cotton. In: The carbon trust. Available via http://www.carbontrust.co.uk/policy-legislation/international-carbon-flows/global-flows/pages/uk.aspx. Accessed 24 Oct 2015

CDP (2015a) CDP global 500 climate change report 2015. London

CDP (2015b) Realizing zero-deforestation transforming supply chains for the future. London

CDP (2015c) Accelerating action: CDP global water report 2015. London

Clean Clothes Campaign (2015a) Mulberry slammed for worker exploitation at Turkish factory. In: Clean Clothes Campaign. http://www.cleanclothes.org/news/press-releases/2015/09/15/mulberry-slammed-for-worker-exploitation-at-turkish-factory. Accessed 20 Nov 2015

Clean Clothes Campaign (2015b) Global wage forum confirms living wage as core human right. http://www.cleanclothes.org/news/2015/10/12/global-wage-forum-confirms-living-wage-as-core-human-right. Accessed 12 Dec 2015

Clean Clothes Campaign (2015c) Italian factory conditions deteriorate. http://www.cleanclothes.org/news/press-releases/2015/01/23/italian-factory-conditions-deteriorate. Accessed 23 Dec 2015

Clean Clothes Campaign (2016) European garment workers face forced overtime and poverty wages. http://www.cleanclothes.org/news/press-releases/2016/02/02/european-garment-workers-face-forced-overtime-and-poverty-wages. Accessed 3 Feb 2016

Cline E (2014) Where does discarded clothing go? In: The Atlantic. http://www.theatlantic.com/business/archive/2014/07/where-does-discarded-clothing-go/374613/. Accessed 25 Feb 2016

D'Arpizio C, Levato F, Zito D, de Montgolfier J (2015) Luxury goods worldwide market study. Milan

Ditty S (2015) It's time for a fashion revolution. The United Kingdom

Dullforce A (2015) FT global 500

EFBA (2010) Annual report 2010. Brussels

EURATEX (2015) The EU-28 textile and clothing industry. In: EU-28 textile and clothing industry year 2014. http://euratex.eu/library/statistics/key-data/key-data-details/?tx_ttnews%5Btt_news%5D=4846&cHash=a43f98d261ba88233db94846a8f7e08e

Fibre2Fashion (2016) Turkey textile industry overview. http://www.fibre2fashion.com/market-intelligence/countryprofile/turkey-textile-industry-overview/index.asp. Accessed 10 Feb 2016

Freeman E, Velamuri R, Moriarty B (2006) Company stakeholder responsibility: a new approach to CSR. Virginia

Golicic SL, Smith CD (2013) A meta-analysis of environmentally sustainable supply chain management practices and firm performance. J Supply Chain Manage 49:78–95

GRI (2013) Textiles, apparel, footwear and luxury goods: 59 topics. Amsterdam

GRI (2015) G4 sustainability reporting guidelines. Amsterdam

Gualandris J, Klassen RD, Vachon S, Kalchschmidt M (2015) Sustainable evaluation and verification in supply chains: aligning and leveraging accountability to stakeholders. J Oper Manage 38:1–13. doi:10.1016/j.jom.2015.06.002

ILO (2013) Making progress against child labour—global estimates and trends 2000–2012. Geneva

Kleindorfer P, Singhal K, Van Wassenhove L (2005) Sustainable operations management. Prod Oper Manage 14:482–492. doi:10.1111/j.1937-5956.2005.tb00235.x

Lee S-Y, Klassen RD (2008) Drivers and enablers that foster environmental management capabilities in small- and medium-sized suppliers in supply chains. Prod Oper Manage 17:573–586. doi:10.3401/poms.1080.0063

McGregor BA (2012) Properties, processing and performance of rare and natural fibres: a review and interpretation of existing research results. Barton

McMullan D (2016) Syrian refugees in Turkish garment supply chains

Moulds J (2015) Child labour in the fashion supply chain. In: Guard. http://labs.theguardian.com/unicef-child-labour/. Accessed 19 Jan 2016

Olivier JGJ, Janssens-Maenhout G, Muntean M, Peters JAHW (2015) Trends in global CO_2 emissions: 2015 report. The Hague

Organic Cotton (2015) Environmental risks. In: Risks of cotton farming. http://www.organiccotton.org/oc/Cotton-general/Impact-of-cotton/Risk-of-cotton-farming.php

Pagell M, Wu ZH (2009) Building a more complete theory of sustainable supply chain management using case studies of 10 exemplars. J Supply Chain Manage 45:37–56. doi:10.1111/j.1745-493X.2009.03162.x

Parmigiani A, Klassen RD, Russo MV (2011) Efficiency meets accountability: performance implications of supply chain configuration, control, and capabilities. J Oper Manage 29:212–223. doi:10.1016/j.jom.2011.01.001

Perry P, Fernie J, Wood S (2014) The international fashion supply chain and corporate social responsibility. In: Ferrie J, Sparks L (eds) Logistics and retail management: emerging issues and new challenges, 4th edn. Kogan Page Limited, London, pp 77–100

Research and Markets (2015) Apparel retail: global industry guide. Dublin

Roca LC, Searcy C (2012) An analysis of indicators disclosed in corporate sustainability reports. J Clean Prod 20:103–118. doi:10.1016/j.jclepro.2011.08.002

Schoenherr T (2012) The role of environmental management in sustainable business development: a multi-country investigation. Int J Prod Econ 140:116–128. doi:10.1016/j.ijpe.2011.04.009

Tachizawa EM, Wong CY (2014) Towards a theory of multi-tier sustainable supply chains: a systematic literature review. Supply Chain Manage Int J 19:643–663. doi:10.1108/SCM-02-2014-0070

Theuws M (2015) Hidden subcontracting in the garment industry. Amsterdam

United Nations (1948) The universal declaration of human rights. New York

Varsei M, Soosay C, Fahimnia B, Sarkis J (2014) Framing sustainability performance of supply chains with multidimensional indicators. Supply Chain Manage Int J 19:242–257. doi:10.1108/SCM-12-2013-0436

WRAP (2012) Valuing our clothes. In: The true cost of how we design, use and dispose of clothing in the UK. Available via http://www.wrap.org.uk/sites/files/wrap/VoC FINAL online 2012 07 11.pdf. Accessed 14 Nov 2015

Wu Z, Pagell M (2011) Balancing priorities: decision-making in sustainable supply chain management. J Oper Manage 29:577–590. doi:10.1016/j.jom.2010.10.001

WWF (2014) Cotton. In: Sustain. Agric. http://www.worldwildlife.org/industries/cotton. Accessed 10 Jan 2016

Application of Digital Enterprise Technology (DET) for Green Made-to-Measure in Korean Luxury Fashion Industry

Kyung Eun Lee

Abstract Due to the significant market size and growth rate of the luxury fashion industry, environmental impacts of the luxury products become more and more serious. Currently, many luxury companies continuously adopt Information Technologies, especially Digital Enterprise Systems (DET) that support strategically designing product life cycle in environmentally sustainable approaches. Made-to-Measure (M2M), customized luxury products that is one of impactful product differentiation strategies of global luxury companies, has some issues of M2M related to limited business scalability and environmental impact. To overcome these issues, some Korean luxury fashion companies use DET technology systems in M2M product life cycle management, since the DET systems assist to reduce waste materials as well as to protect the environment. This study will help the product designers and executives, in charge of M2M in the luxury fashion industry, recognize that the use of innovative technologies such as the DET systems, is essential for luxury companies' green M2M capability enhancement and for longevity of the firms in the luxury fashion industry.

Keywords Digital Enterprise Technology (DET) · Made-To-Measure (M2M) · Sustainability · Luxury market

1 Introduction

The luxury industry is one of the most harmful environmental impact generators from the excessive use of chemical materials and processes as well as energy sources (Gardetti and Muthu 2015). These harmful effects occur throughout the luxury product life cycle, from raw materials manufacturing (e.g., fibers, yarns, textiles) to garment production processes (e.g., assembly, packaging, logistics),

K.E. Lee (✉)
Department of Apparel, Events and Hospitality Management,
Iowa State University, 28 Mackay Hall, Ames, IA 50011, USA
e-mail: Kyungeun@iastate.edu

M.A. Gardetti (ed.), *Sustainable Management of Luxury*,
Environmental Footprints and Eco-design of Products and Processes,
DOI 10.1007/978-981-10-2917-2_15

through post-production activities (e.g., end-use, recycling, disposal) (Khan and Islam 2015). Due to the significant market size and growth rate of the luxury fashion industry, environmental impacts of the luxury products become more and more serious. Thus, currently, many luxury companies continuously adopt Information Technologies, especially Digital Enterprise Systems (DET), that support strategically designing product life cycle, especially product design and development processes in environmentally sustainable approaches. Among various DET systems, 3D CAD software and 3D imaging technology allow multiple benefits to luxury companies: For designers, virtual 3D simulation of a product allows luxury designers to craft design ideas and reduce physical sample making while enhancing product qualities; for consumers, these technologies create a new e-commerce platform for luxury brands with the capabilities in providing complete virtual consumer-product interactions before manufacturing or purchasing the product (Delamore 2014).

Made-to-Measure (M2M), customized luxury products to fit consumer's particular requirement are one of the most impactful strategies in the global luxury fashion industry (Carl 2012). Due to issues of M2M related to limited business scalability and environmental impacts, some Korean luxury fashion companies use DET technology systems in M2M product design and development as well as environmentally sustainable material waste disposal. This study was an attempt to examine the effects of DET system usage in green M2M business in Company A (pseudonym), the luxury fashion company of Korea. More specifically, the objectives of this research were as follows: (a) identifying DET systems being used for green M2M in the global luxury fashion industry, (b) analyzing the effectiveness of DET system application to green M2M to the environment and society, (c) discussing opportunities and challenges of green M2M, and (d) suggesting future research direction of DET system application for M2M in the global luxury industry. Specifically the context under examination was a case of Korean luxury company's M2M business.

2 Sustainable Luxury

In current fashion industry, luxury goods are considered as premium priced with finest product quality providing exceptional product experience to consumers (Tang and Tang 2012). From 2010 and 2015, a market size of global luxury industry increased approximately by 65% while creating both positive (e.g., economic profits) and negative impacts (e.g., deforestation, chemical material usage). Luxury fashion industry creates harmful effects to the environment through an excessive amount of pesticide usage for cotton manufacturing and waste material creation (Lochard and Murat 2011).

In light of negative environmental issues, in today's luxury fashion industry, more and more luxury brands are engaging in corporate social responsibility (CSR) that is concentrated on socially and environmentally responsible development and consumption (Tang and Tang 2012). For instance, Gucci gives 25% of profits obtained from producing a special UNICEF accessory line back to UNICEF, since 2005 (UNICEF, N.D.). Hermès produces a special accessory line called Petit H by using leftover materials (Leitch 2013). Gucci, Kering, and LVMH now use organic leather tanning method using vegetable material instead of using toxic chromium (Givhan 2015). These sustainable practices promote luxury companies' business performance, as an increasing number of luxury consumers are associated with CSR (Tang and Tang 2012). Furthermore, sustainable development aid luxury companies to differentiate their brand and company image from competitors based on consumers' growing social and environmental awareness (Kim et al. 2012).

3 Digital Enterprise Technologies and the Luxury Fashion Industry

Today's luxury fashion industry heavily depends on information technology (IT) systems to respond to global challenges associated with the environmental impacts occurring throughout the life cycle of luxury products and reduced product lifecycles and frequently changing consumer demands and market conditions. One of the most green M2M leaders in the luxury fashion industry Company A is a large, global, luxury apparel design, and retail company with annual sales that exceed two billion dollars (USD) in Korea. In Company A, digital Enterprise Technologies (DET), the automated digital product modeling systems is one of the major technologies being used for the company's luxury M2M product development process management (Mourtzis et al. 2015). DET systems enable effectively tracking material usage and manufacturing processes throughout product design and development stages in a consolidated virtual system environment. Hence, through the use of DET systems, Company A receives major benefits for its green M2M business practices: (a) eco-friendly material and manufacturing process selection, (b) material waste reduction, (c) eco-friendly luxury business process optimization, and (d) product up-cycling facilitation by tracking left-over inventories (e.g., fabrics, produced garments, consumer returns, defected merchandises). To maximize the system efficiencies, DET systems being used by Company A integrate multiple IT systems (e.g., Enterprise Resourcing Planning, Knowledge Management, Supply Chain Management, Consumer Relationship Management, E-commerce, Product lifecycle Management) into one platform (Mourtzis et al. 2015; Myung 2003).

Among these systems, Product Lifecycle Management (PLM) refers to systems managing the entire lifecycle of a product through engineering processes from design to production and from usage to disposal of manufactured products (Cantamessa et al. 2012; Gecevska et al. 2011). For global luxury companies operating product design offices in different geographic locations, PLM is one of the most beneficial DET system technologies remotely assisting designers for inter-department collaborations (Cantamessa et al. 2012; Gecevska et al. 2011). There are three key PLM system capabilities for Company A and other luxury fashion companies: short Time to Market (duration of time taken in product design development processes), low cost, and high product quality (Ha-Brookshire and Hawley 2013; Park and DeLong 2009).

Within the integrated DET system platform, one of the most important PLM system components for the company A's M2M is the integrated design systems consisting of Computer-Aided Design (CAD), Computer-Aided Manufacturing (CAM), and Computer-Aided Engineering (CAE) for four major reasons: (a) to expedite work process simulation, (b) to reduce errors by the use of consistent software environment, (c) to simplify data configuration processes of system users (e.g., designers, system engineers) among each product, part, and machine/tool design application, and (d) to save software costs and simulation time (Siemens PLM Software, N.D.a; Waurzyniak 2012). See Fig. 1.

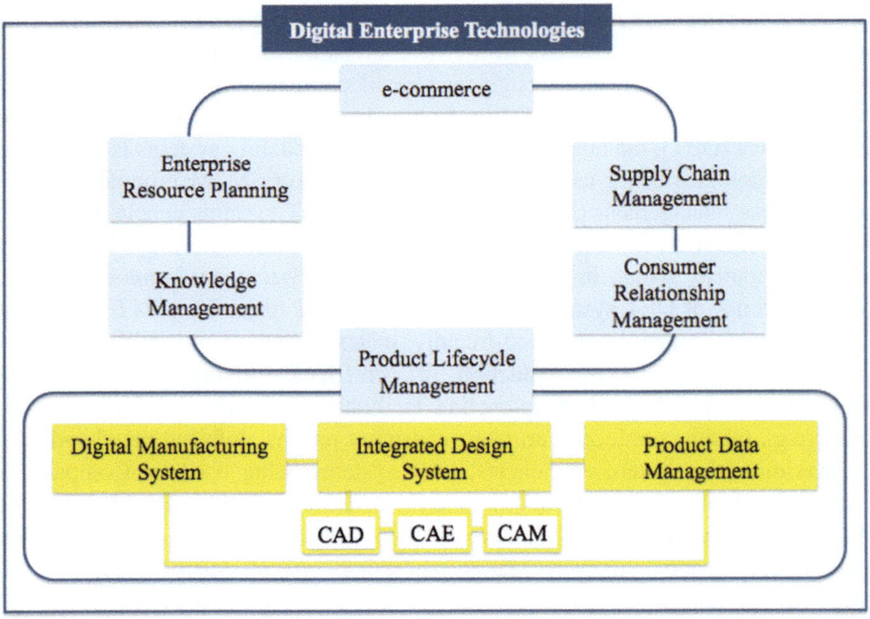

Fig. 1 Current digital enterprise technology structure adapted from "3D PLM in digital enterprise" (Myung 2003)

3.1 Computer-Aided Design (CAD)

Computer-Aided Design (CAD) is one of the core PLM functions used for creating virtual illustration of physical design components in a two or three-dimensional solid or surface model (Siemens PLM Software, N.D.a). For the company A's M2M, key CAD features include conceptual design, digital sampling, assembly simulation, and design elaboration, which can be utilized throughout virtual product design and development processes (Bodein et al. 2013). In a traditional design approach, sampling is one of the most time and cost demanding stages in luxury product development processes, because at least three to four rounds of sample makings are required to finalize the right style and fit of the garment (Businesswire 2004). With the use of digital sampling technology however, as the initial step, designers make sketch styles and patterns in 2D CAD models and then, draw a construction mock-up of the garment with seam-line connection. After that, digitally construct 3D CAD models before making physical samples or initiating main production (Park and DeLong 2009). Hence, with the use of digital sampling technology, the number of physical samples made is reduced through 3D virtual simulation techniques that allow designers to control product design processes (e.g., pattern-making, draping, fitting) in a consolidated seamless digital process (Businesswire 2004).

Critical roles of CAD in the company A's luxury product design and development include: (a) identifying an optimal shape and size of a physical product to generate planned performances as well as to minimize waste materials (Siemens PLM Software, N.D.a), (b) reducing the number of physical prototypes made through visualization of the final product design and assembly (Bodein et al. 2013), (c) reducing Time to Market, (d) reinforcing more accurate and retrievable design data documentation (Siemens PLM Software, N.D.a), (e) facilitating design data reuse from a digital repository, which enhances designers' knowledge integration and collaboration at each product development stage, and (f) enhancing overall qualities of luxury products (Bodein et al. 2013). The usability of CAD performing these roles was noted by the industry specialists in pattern development departments from prestigious global luxury apparel brands such as Christian Dior Homme and Christian Dior Homme (Reuters 2013).

In addition to product design and development, 3D imaging technology is often used by luxury e-commerce retailers to enhance the consumer shopping experience (Park and DeLong 2009; Transparency Market Research 2015). The use of 3D imaging technology creates a new e-commerce platform for luxury brands with the capabilities in providing complete virtual consumer-product interactions before manufacturing or purchasing products (Delamore 2014). For example, in 2011, the luxury fine jewelry brand De Beers presented the mobile app allowing virtual try-on experiences to consumers using 3D augmented reality (AR) technologies to promote the new Forevermark Millemoi collection. Through this mobile app, consumers can virtually wear different rings, earrings, and pendants from the new Forevermark Millemoi collection in a live 3D environment, so they can create the connection to the brand and their products without making a purchase (DeMarco 2011).

3.2 Computer-Aided Manufacturing (CAM)

Computer-Aided Manufacturing (CAM) assigns the relationships between manufacturing plans and available automated machines using the Computerized Numerical Control (CNC) function (Siemens PLM Software, N.D.c; Waurzyniak 2012). CAM provides four major benefits to the company A: (a) it allows expanded machine usage based on accurate and standardized resource data in CAD for high quality part manufacturing (Waurzyniak 2012); (b) Computerized Numerical Control (CNC) applications connected to CAM accordingly match standardized machine tools with shop floor jobs (Siemens PLM Software, N.D.c); (c) shop documents are generated automatically whenever CAD information is updated (Siemens PLM Software, N.D.c); and (d) system engineers can create efficient tool paths to reduce engineering errors among related manufacturing software applications, which increase the amount of time wasted on the shop floor. In addition to the company A, in the luxury fashion industry, Gucci is one of CAM users to manufacture premium quality watch. Since 2000, in Gucci watch's production plant located in La Chaux-de-Fonds, Swiss, complete product design and development processes, such as sampling, validating production sample qualities, part-manufacturing, and assembly, are performed by using CAM and CNC (Gucci 2012). In summary, the latest CAM features allow luxury fashion designers easier and more efficient luxury product design of complicated and multi-functioned manufacturing processes (Gucci 2012; Waurzyniak 2012).

In 2014, the French software company Lectra became the world largest CAD/CAM system provider after acquisition of Investronica Sistemas, the Spanish CAD/CAM systems provider that had over 6000 customers across industries. These customers included major premium fashion brands, such as Armani, Christian Dior, Dolce & Gabanna, Ermenegildo Zegna, Gucci, and Ralph Lauren, that operate M2M business in the luxury fashion industry (Businesswire 2004).

3.3 Computer-Aided Engineering (CAE)

Computer-aided engineering (CAE) simulates product design performance to prevent potential engineering problems and enhance design quality (Siemens PLM Systems, N.D.b). CAE provides practical system functions, such as simulation, validation, and optimization of product, process, and machine/tool design, involved in manufacturing (Siemens PLM Systems, N.D.b). To provide these system functions, CAE is integrated with CAD to develop optimized design and engineering processes based upon geometric analysis of the relationship between these two systems (Shapiro et al. 2011; Siemens PLM Systems, N.D.b). There are two ways to integrate CAD and CAE: forward and inverse integration. In forward integration, CAE analyzes demands of a CAD model and generates solutions automatically. By contrast, inverse integration requires revision of a CAD model in accordance with

CAE analysis results (Shapiro et al. 2011). Normally, in inverse integration, manual revision of a CAD model is executed by a product designer who makes decisions on design changes (Shapiro et al. 2011).

For the company A's M2M, key benefits of using CAE include features such as: (a) product, process, and machine/tool designs can be previewed for their performance to resolve potential problems prior to starting actual engineering processes, (b) it reduces costs related to design and manufacturing throughout product lifecycle, and (c) accumulated design simulation data enhances designs and their performances (Siemens PLM Systems, N.D.b; Waurzyniak 2015). Due to these benefits, CAE became one of the largest contributors to the 6.8% growth of the PLM market in 2014 from the previous year (Waurzyniak 2015). See Fig. 1 for current DET system structure.

4 Luxury Fashion and Environmental Sustainability

In 2014, the market size of the luxury fashion industry was increased approximately to €224 billion, 3% from the previous year, that accounts for nearly 7.5% of total global fashion industry market size (€2.7 trillion) (Consultancy.UK 2015; Gardetti and Muthu 2015). Among luxury products composing this market, apparel and accessories occupied around 54%. By 2017, the luxury industry is expected to grow to €265 billion with 6% growth rate (Consultancy.UK 2015). See Fig. 2. Due to the significant market size and growth rate, environmental impacts of the luxury industry become bigger issues.

Thus, currently, many luxury companies continuously adopt Information Technologies, especially Digital Enterprise Systems (DET) that support strategically designing product life cycle, especially product design and development processes in environmentally sustainable approaches. Among various DET systems, 3D CAD software and 3D imaging technology allow multiple benefits to luxury companies: For designers, virtual 3D simulation of product allows luxury

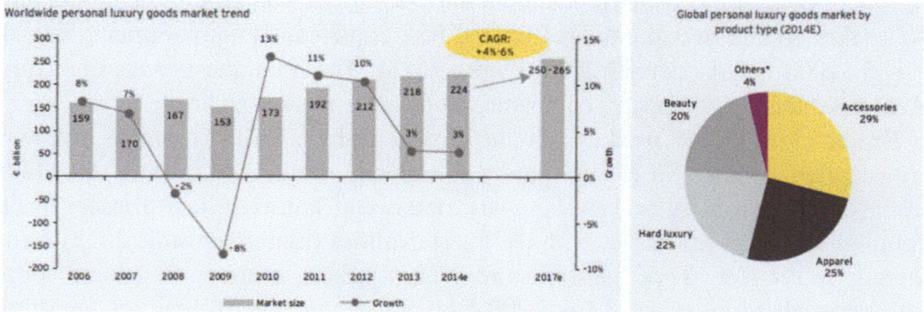

Fig. 2 Worldwide personal luxury goods market trend and global personal luxury goods market by product type in 2014 (Unit: Euro) (Consultancy.UK 2015)

designers to craft design ideas more efficiently while reducing physical sample making while enhancing product qualities; for consumers, these technologies create a new e-commerce platform for luxury brands with the capabilities in providing complete virtual consumer-product interactions before manufacturing or purchasing the product (Delamore 2014).

5 Green Made-to-Measure

5.1 What Is Made-to-Measure?

Made to Measure (M2M) refers to specially made clothing in accordance with a particular specification required by consumers to fit their individual purposes (e.g., sizes, tastes, customs, usage) (Carl 2012). M2M is one of major product and brand differentiation strategies that allow production flexibility in multiple colors, designs, styles, and other forms of customization (Okonkwo 2007). Currently, there are growing luxury consumers seeking customization fulfilling their individual fashion tastes and providing exceptional and exclusive brand experiences (Okonkwo 2007). Thus, with strong product exclusivity, M2M has been one of essential elements of luxury branding to maintain prestige stance in the fashion industry for a long time. Luxury companies continuously carry M2M to provide unique and personalized brand experience to consumers with premium product qualities and ensure long-term growth of the companies with differentiated brand images (Carl 2012).

Luxury brand M2M include suits, jackets, coat, outerwear, knitwear, trousers, shirts, and bags and one of the most important components of these M2M products is fitting perfectly crafted to an individual consumer' body shape and wearing comfort (Brooke 2007). There are several characteristics of luxury M2M products: (a) base patterns are available to be modified to fit individual consumers, (b) measurement processes are expedited and standardized, (c) basic fabric and sundry options are available, (d) price increase is minimized due to the use of standardized patterns and materials, (e) combination of hand-made and machine construction steps, (f) stitch tensions are controllable, (g) limited fitting sessions are required (normally once prior to the production and one or two more before completion), (h) it takes around two month lead-time to be completed, (i) normal pricing begins at U.S.$1000 (Mydapperself 2015; Quora 2013). These unique aspects of luxury M2M products allow apparel companies to create a market niche.

Examples of M2M products in the luxury fashion industry are as follows: Ermenegildo Zegna's Su Misura produces annually the limited quantity 60 M2M products per year in suits, jackets, coats, outerwear, knitwear, and trousers made from lightweight wool fabrics with the finest qualities (Longroadmedia 2015). Dior Homme in the New York flagship store, makes jackets, trousers, vests, and shirts with materials to be selected from 200 fabrics for suits and 160 fabrics for shirts using hand-finishing techniques. Ralph Lauren's purple label offers to-order suits,

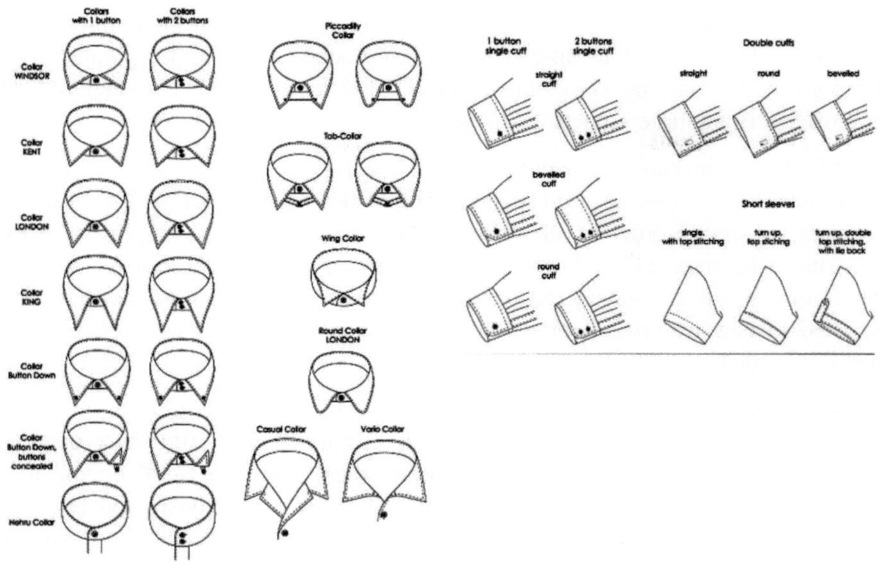

Fig. 3 The options available with our made-to-measure shirts (Themastertailor, N.D.)

dress shirts, formal wear, accessories, shoes, and leather goods in selected worldwide locations (Reena 2013).

In luxury fashion industry, there are both opportunities and challenges in the M2M business. For opportunities, M2M is an effective approach to build brand exclusivity to consumers. Due to the nature of customization per order made by a consumer, M2M doesn't create left-over produced garments; this is a positive aspect of M2M business to the environmental sustainability (Carl 2012). Despite these opportunities, M2M is difficult to become a scalable business due to high costs for the use of luxury materials and hand-made manufacturing processes (Murdock 2013). Hence, luxury companies are currently seeking new green M2M product design and development approaches, promoting the future economy and the environment. See Fig. 3.

6 Green Made-to-Measure (M2M)

To be responsive to the challenges identified in traditional M2M process associated with the business scalability and environmental issues, the emergence of green M2M approaches were led by large luxury fashion companies in Korea. For green M2M practices, the Company A currently retains highly stabilized and advanced DET systems integrating ERP, PLM, SCM, and Customer Relationship Management (CRM). These systems serve for over the company's 30 luxury brand businesses. According to the Company A's sustainability initiative report, there are two distinctive aspects of Green M2M: The DET system applications throughout the product lifecycle and the up-cycling of used products for CSR (Corporate Social Responsibility) retailing. The Company A's DET systems contribute to green M2M approaches in multiple aspects, as listed below.

Standardized material preparation
In standardized materials and pattern preparation, traditional M2M business requires placing an order of all standardized materials and making physical base patterns to be selected by the consumer before an actual M2M order is placed. Traditionally, these standardized materials were seasonal and they needed to be updated to new materials every season; unused materials after one season were often discarded.

Pattern making
A lot of physical materials were traditionally consumed every time base patterns were customized for an individual consumer order. In green M2M approaches, digital material catalogs linked to the PLM systems to continuously update quantities, pricing, and delivery dates of available materials on-line and allow placing orders of selected materials by the consumer instantly. Hence, unused materials are preserved after the season.

Body measurement
Instead of taking time for measurement sessions, 3D body scanning technology allows the precise measurement of the body and minimizes errors on measured data. Digital pattern making software connected to 3D CAD customizes the selected pattern type by the consumer instead of making a physical pattern. Digitally created patterns are easy to be modified, saved and reused by pattern makers for future pattern works.

3D simulation
Before physical sample making, the digitized pattern is transformed to a 3D virtual simulation of garment assembly and specifications by using 3D imaging technology. Thus, designers can make corrections on components related to overall construction of garments such as patterns, specifications, assembly and design details.

Up-cycling and CSR retailing
The PLM systems are capable of alerting campaigns and events for M2M consumers to encourage to donate used garments, so these garments are up-cycled to be sold in Corporate Social Responsibility (CSR) stores that are socially responsible retailing approaches returning the profits back to the society. Since 2014, Company A operates the CSR selling green apparel products, consisting of up-cycled used M2M items in collaboration with new contemporary fashion designers, natural material apparel and accessories, and donated luxury labels. The company returns half of the profits gained from this store back to the society in the form of donations, campaigns, and events, which promote environmental sustainability. See Fig. 4 for comparison of traditional and green M2M production process.

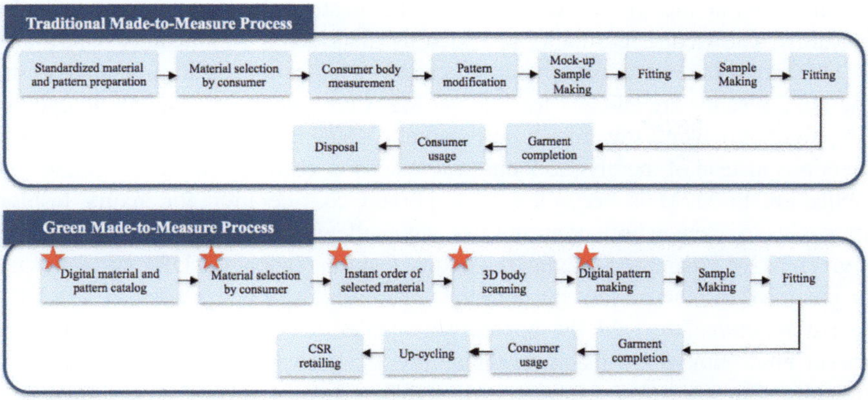

Fig. 4 Comparison of traditional Made-to-Measure and the company A's green Made-to-Measure process (*star marks* indicate the areas applied the DET technology)

Besides the Company A, several other global luxury fashion companies adopt the DET technology for their green M2M business. For example, according to Lectra (N.D.), Façonnable, French luxury M2M brand adopts Lectra's PLM and 3D design solutions throughout product design and development process in a company wide systems including the ones used by supply chain. As the results, these system usage allowed Façonnable to grow its M2M business to a mass scale by resolving limitations in a traditional handmade tailoring approach. Furthermore, the company became capable of saving a large amount of annual cost and material consumption and waste saving, as well as streamlining production line to be able to generating the maximized efficiency. Façonnable's DET system helps to keep tracking on each production stage and material usage in the entire supply chain, thus, the company can control their development and manufacturing processes toward environmentally sustainable direction.

Canadian M2M brand IndoChino uses 3D CAD solution of Optitex that allows virtually to simulate produced styles before making a physical sample. By using 3D CAD, IndoChino can offer affordable price with luxury product quality based on reduced production costs. Especially, IndoChino as on-line retailor provides 3D production views for their both supply chain and consumers.

7 Conclusion

For M2M business of the luxury fashion companies, the DET systems assist to not only reduce waste materials but also to allow luxury companies to contribute to protect the environment. Thanks to the use of DET technologies, many unnecessary steps and errors can be prevented when a physical sample is made. Additionally, all information and records are saved in the PLM systems as historical data for future reference. Even though M2M requires customization for every order made by consumer, knowledge reusability provided by the systems significantly contributes to M2M design and development processes with shorter lead-time and costs. The DET systems are efficient tools to get M2M consumers involved in up-cycling campaigns and events towards environmentally sustainable and socially responsible directions instead of merely producing wastes. Thus, currently, the DET technology application to M2M is becoming an important consideration for luxury fashion companies to manage both opportunities and challenges related to the future growth of green M2M business. Studying the luxury fashion company's DET system usage for green M2M is important for both the apparel industry and academia. The DET system functionalities aid the apparel companies to stay exclusively at present market environment by using sustainability as innovative product development strategies. For academia, examining the DET system usage of luxury brand customers contributes to developing quantitative scales to evaluate the effects of the systems for green M2M business practices.

The current study will help the product designers and executives, in charge of M2M in the luxury fashion industry, recognize that the use of innovative technologies such as the DET systems, is essential for luxury companies' green M2M capability enhancement and for longevity of the firms in the luxury fashion industry. Sustainable development should be continuously promoted to luxury brands not only to protect the environment by reducing harmful material usage and waste but also improve company image to growing consumers with social and environmental awareness. Therefore, assisting tools to sustainable development like the DET system must be considered by luxury companies to contribute to environmental sustainability in more efficient approach.

8 Future of Green M2M

In the future luxury industry, M2M will stay as an important parts of product strategies for luxury companies for enhancing exclusive and prestigious brand images. Despite multiple benefits of the DET systems, there are limitations to implement such systems by the luxury fashion companies due to the requirements in the monetary investment and work process changes. For the future M2M, 3D CAD software should be capable of providing more interactive gaming experience during interaction with consumers. To provide more engaged virtual consumer-product interaction, 3D imaging technologies must be capable of functions as followed: (a) illustrating complex clothing physics such as clothing and body interaction in multiple layers, (b) integrating games engines for creating real-time interaction, and (c) rendering in high quality textures, and allowing to view real-time video streams (Delamore 2014). Further research is needed to focus on the evolution of technologies aiming at a luxury green M2M business not making waste material beyond recycling/up-cycling before or after the use of M2M products.

References

Bodein Y, Bertrand R, Emmanuel C (2013) A roadmap for parametric CAD efficiency in the automotive industry. Comput Aided Des 45(10):1198–1214

Brooke S (2007) Just how personal can you get. Financial Times, 31 Mar 2007

Businesswire (2004) Lectra acquires Investronica Sistemas; the acquisition of Spanish-based world number three player significantly strengthens Lectra's world leadership. http://www.businesswire.com/news/home/20040105005539/en/Lectra-Acquires-Investronica-Sistemas-Acquisition-Spanish-Based-World. Accessed 3 Feb 2016

Cantamessa M, Montagna F, Neirotti P (2012) Understanding the organizational impact of PLM systems: evidence from an aerospace company. Int J Oper Prod Manag 32(2):191–215

Carl M (2012) Custom cobbling. Vanity Fair. vanityfair.com. Retrieved 16 Nov 2015

Consultancy.UK (2015) Global luxury market worth 224 billion, top 20 brands. http://www.consultancy.uk/news/2803/global-luxury-market-worth-224-billion-top-20-brands. Accessed 5 Feb 2016

Delamore P (2014) Change of paradigm: delivering realtime 3D luxury fashion. http://www.gold.ac.uk/calendar/?id=7585. Accessed 13 Feb 2016

DeMarco A (2011) De Beers turns to augmented reality to market its branded diamond. http://www.forbes.com/sites/anthonydemarco/2011/10/12/de-beers-turns-to-augmented-reality-to-market-its-branded-diamond/#3ca3fdde2cdc. Accessed 13 Feb 2016

Gardetti M, Muthu S (2015) Sustainable apparel? Is the innovation in the business model?—The case of IOU project. Text Clothing Sustain 1(1):1–9

Gecevska V, Cus F, Polenakovic R, Chabert P (2011) Process of innovation in product lifecycle management business strategy. Perspect Innovations Econ Bus 9(3):53–56

Givhan R (2015) Luxury fashion brands are going green. But why are they keeping it a secret? https://www.washingtonpost.com/lifestyle/style/luxury-fashion-brands-are-going-green-but-why-are-they-keeping-it-a-secret/2015/12/08/d3d93678–8c8a-11e5-acff-673ae92ddd2b_story.html. Accessed 9 Feb 2016

Gucci (2012) GUCCI Watches: Italian fashion meets Swiss-made watchmaking. http://www.guccitimeless.com/en/news/2012/05/02/gucci-watches-italian-fashion-meets-swiss-made-watchmaking. Accessed 9 Feb 2016

Ha-Brookshire J-E, Hawley J-M (2013) Envisioning the clothing and textile-related discipline for the 21st century its scientific nature and domain from the global supply chain perspective. Clothing Text Res J 31(1):17–31

Khan M-R, Islam M (2015) Materials and manufacturing environmental sustainability evaluation of apparel product: knitted T-shirt case study. Text Clothing Sustain 1(1):1–12

Kim K-H, Ko E, Xu B, Han Y (2012) Increasing customer equity of luxury fashion brands through nurturing consumer attitude. J Bus Res 65(10):1495–1499

Lectra (N.D.) Lectra counts the largest tailored-clothing brands as its customers. http://www.lectra.com/en/fashion-apparel/tailored-clothing-menswear-solutions

Leitch L (2013) Hermès embraces recycling with petit h. http://www.telegraph.co.uk/luxury/womens-style/15811/hermès-embraces-recycling-with-petit-h.html. Accessed 9 Feb 2016

Lochard C, Murat A (2011) Luxe et développement durable: La nouvelle alliance. Eyrolles, Paris

Longroadmedia (2015) Advantages and disadvantages of bespoke or Made-to-Measure. http://www.longroadmedia.com/2015/12/advantages-and-disadvantages-of-bespoke-or-made-to-measure/. Accessed 7 Feb 2016

Mourtzis D, Maropoulos P, Chryssolouris G (2015) Digital enterprise technology: systems and methods for the digital modelling and analysis of the global product development and realisation process. Int J Comput Integr Manuf 28:1–2

Murdock (2013) What is the difference between Made To Measure and bespoke? http://www.forbes.com/sites/quora/2013/01/16/what-is-the-difference-between-made-to-measure-and-bespoke/#1aa72ae15d85. Accessed 13 Feb 2016

Mydapperself (2015) What is the difference between bespoke and Made To Measure suits? http://mydapperself.com/what-is-the-difference-between-bespoke-and-made-to-measure-suits/. Accessed 9 Feb 2016

Myung S (2003) Implementation of the digital manufacturing in automotive industry. Dassault Systems Korea Annual Conference

Okonkwo U (2007) Luxury fashion branding: trends, tactics, techniques. Palgrave Macmillan, Basingstoke

Park J, DeLong M (2009) User perceptions of technology adoption and implementation: a case study of virtual sampling technology for footwear production in a global market. Fashion Pract (The Journal of Design, Creative Process and the Fashion Industry) 1(1):82–108

Quora (2013) What is the difference between made to measure and bespoke? http://www.forbes.com/sites/quora/2013/01/16/what-is-the-difference-between-made-to-measure-and-bespoke/#1aa72ae15d85. Accessed 9 Feb 2016

Reena (2013) Dior made-to-measure service offered at New York's flagship store. http://luxurylaunches.com/fashion/dior-made-to-measure-service-offered-at-new-yorks-flagship-store.php. Accessed 5 Feb 2016

Reuters (2013) LECTRA: Lectra brings the 3D revolution to the fashion industry. http://www.reuters.com/article/idUSnHUGd6gw+70+ONE20130523. Accessed 2 Feb 2016

Shapiro V, Tsukanov I, Grishin A (2011) Geometric issues in computer aided design/computer aided engineering integration. http://computingengineering.asmedigitalcollection.asme.org.proxy.lib.iastate.edu/article.aspx?articleid=1402343. Accessed 2 Feb 2016

Siemens PLM Software (N.D.a) CAD/Computer-Aided Design. http://www.plm.automation.siemens.com/en_us/plm/cad.shtml. Accessed 2 Feb 2016

Siemens PLM Systems (N.D.b) CAE/Computer-Aided Engineering. http://www.plm.automation.siemens.com/en_us/plm/cae.shtml. Accessed 2 Feb 2016

Siemens PLM Software (N.D.c) CAM/Computer-Aided Manufacturing. http://www.plm.automation.siemens.com/en_us/plm/cam.shtml

Tang Z, Tang J (2012) Stakeholder–firm power difference, stakeholders' CSR orientation, and SMEs' environmental performance in China. J Bus Ventur 27:436–455

Transparency Market Research (2015) Product Lifecycle Management (PLM) market expected to reach US$75.87 Bn by 2022 globally. http://www.transparencymarketresearch.com/pressrelease/product-lifecycle-management-market.htm. Accessed 5 Feb 2016

Themastertailor (N.D.) Look and feel your best in a made-to-measure shirt. http://www.mastertailoruk.com/made-to-measure-shirts/bespoke-shirt-styles.htm. Accessed 5 Feb 2016

UNICEF (N.D.) UNICEF's corporate partnerships. http://www.unicef.org/corporate_partners/index_gucci.html. Accessed 13 Feb 2016

Waurzyniak P (2012) CAD/CAM software advances. Manuf Eng 148(2):63–69

Waurzyniak P (2015) Why manufacturing needs real-time data collection. http://www.sme.org/MEMagazine/Article.aspx?id=8589936436. Accessed 9 Feb 2016

Part V
Sustainable Luxury, Entrepreneurship (Including Family Business) and Craftsmanship

Sustainable Luxury Fashion: The Entrepreneurs' Vision

Miguel Angel Gardetti and Rachida Justo

Abstract Luxury depends on cultural, economic or regional contexts. This transforms luxury into an ambiguous concept (Low, undated). Godart and Seong (Sustainability luxury: managing social and environmental performance in iconic brands. Greenleaf Publishing, Sheffield, p 12, 2014) show us that the relationship between luxury and fashion is quite an ambiguous one, as fashion does not fully belong to the luxury world, but overlaps with luxury in its most expensive and exclusive segments. On one hand, sustainability within fashion means that the development and use of something or process are not harmful to people or the planet, and once put into action, such thing or process can enhance the well-being of those people who interact with it, and the environment it is developed and used within. One other hand, sustainable luxury would not only be the vehicle for more respect for the environment and social development, but it will also be synonym of culture, art and innovation of different nationalities, maintaining the legacy of local craftsmanship (Gardetti in Conference dictated at the seminar sustainable luxury and design within the framework of the MBA of IE. Instituto de Empresa, Madrid, 2011). This chapter shows the results and analysis of a study conducted with fifteen entrepreneurs of different nationalities with sustainable projects in the luxury fashion and fashion accessory sector. All these entrepreneurs participated in the IE Award for Sustainability in the Premium and Luxury Sectors. Some of them won in the category "fashion and accessories" and others reached the final stages of the award. The purpose of this study is not only to get to know what entrepreneurs understand by sustainable luxury and, particularly sustainable luxury fashion, but also to learn their vision about the future of this sector. The methodology used was a survey with closed and open questions. These surveys also included semi-structured interviews.

M.A. Gardetti (✉)
Center for Studies on Sustainable Luxury, Av. San Isidro
4166 PB "A", C1429 Buenos Aires, Argentina
e-mail: mag@lujosustentable.org

R. Justo
IE Business School, Calle Alvarez de Baena 4, 28006 Madrid, Spain
e-mail: rachida.justo@ie.edu

M.A. Gardetti (ed.), *Sustainable Management of Luxury*,
Environmental Footprints and Eco-design of Products and Processes,
DOI 10.1007/978-981-10-2917-2_16

Keywords Fashion · Luxury · Sustainability · Sustainable luxury fashion · Entrepreneurship

1 Introduction

Coco Chanel once defined luxury as "*(...) a necessity that begins when necessity ends*" [Coco Chanel quoted in Okonkwo (2007)]. In this same line Heine (2011) defines luxury as something desirable and more than a necessity. However luxury has been a sign of prosperity, power and social status since ancient times (Kapferer and Basten 2010). Also, according to Ricca and Robins (2012) luxury is a source of inspiration, controversy, admiration and considerable economic success. Sustainable luxury is the return to the ancestral essence of luxury, to the thoughtful purchase, to the artisan manufacture, to the beauty of materials in its broadest sense, and to the respect for social and environmental issues (Girón 2009). The foundations of sustainable luxury are social entrepreneurship and craftsmanship. Social entrepreneurs are well motivated to break the rules and promote disruptive solutions to the environmental and social issues. They are leaders who transform, inspire, and care deeply about people and the environment (Short et al. 2009). They reappraise native culture and craftsmanship.

This paper shows the analysis and results of an exploratory study into the high-end luxury fashion market drawn from surveys conducted to fifteen entrepreneurs of different nationalities with sustainable premium and luxury projects. These entrepreneurs have participated in the above mentioned award and have been nominated or even awardees. These findings have been combined with those extracted from in-depth analysis of several case studies of entrepreneurship in the sustainable luxury sector, which have been published in Gardetti and Giron's (2014) book. The purpose of this study is not only to get to know what entrepreneurs understand by sustainable luxury and, particularly sustainable luxury fashion, but also to learn from their vision about the future of this sector.

We begin this paper with an overview of the concept of luxury, fashion, sustainable development. Then we describe the relationship between sustainable luxury and entrepreneurship. We then describe our methodology and analyse the main findings emerging from our study.

2 Luxury, Fashion, and Sustainable Development

Luxury depends on cultural, economic or regional contexts. This transforms luxury into an ambiguous concept (Low, undated) and shows for some researchers such as, Scheibel (unknown year), the absurd aspect of this industry. Thus, luxury is a matter of seeing and being seen. 'Seeing' as a search of the latest distinctive signifiers that can be used to 'be seen' in different distinctive group processes

(Mortelmans 2005). Christopher L. Berry in his work "The idea of Luxury" from 1994—one of the most interesting and comprehensive pieces on the concept of luxury particularly its intellectual history—establishes that luxury has changed throughout time and that it reflects social norms and aspirations.

True elements of (authentic) luxury rely on the search for beauty, refinement, innovation, purity, the well-made, what remains, the essence of things, the ultimate best (Girón 2012). However this luxury has given way to the *new luxury* through its democratisation (massification?) that occurred when family and artisanal luxury companies sagged against the large conglomerates which had a strong focus on economic aspects. It means that the image—*neither reputation nor legitimacy*—was the way, and marketing was the function (Gardetti and Torres 2014).

Already in 1999 Robert H. Frank stated in his book 'Luxury Fever—Weighing the Cost of Excess' the need to minimise the culture of "excess" to restore the true values of life. And this is in line with the World Commission on Environment and Development (WCED 1987) report, *Our Common Future*, also known as the Brundtland Report, which defines sustainable development as the development model that allows us to meet present needs, without compromising the ability of future generations to meet their own needs. Luxury—according to Kleanthous (2011)—is becoming less exclusive and less wasteful and more about helping people to express their deepest values.

In the introduction of the book titled "Sustainable Luxury: Managing Social and Environmental Performance in Iconic Brands," Gardetti and Torres (2014) describe the evolution of the sustainability-luxury relationship between 2003 and 2011—a relationship that they first noticed in the book "Deluxe—How Luxury Lost its Luster," by Thomas (2007). Some of the aspects of this evolution are as follows:

> The two most important conferences of this sector held in 2009 addressed these issues and focused their discussions on the assessment of these changes in the consumer and the new concept of success for achieving a "sustainable" luxury. One of them, organised by the International *Herald Tribune* in New Delhi (India) was called 'Sustainable Luxury Conference'. The other, promoted by *Financial Times* in Monaco with the attendance of Prince Albert was titled "Business and Luxury Summit - Beyond Green: economics, ethics and enticement".
>
> The book "Inside Luxury" written by Girón (2009) that presents a documented study of luxury and its future and how "sustainability" has influence on it.
>
> The conduction by UNCTAD -United Nations Conference on Trade and Development- together with Green2Greener of the Conference 'Redefining Sustainability in the International Agenda - Inspiring Greater Engagement in Biodiversity Issues' with the participation of several luxury brands and the subsequent creation of the RESP (Responsible Ecosystem Source Platform) initiative.
>
> The creation -in early 2010- of the "Centre for Studies on Sustainable Luxury" which mission is to assist companies in this sector in the transition towards sustainability, thus encouraging sustainable business practices across all areas of the organisation and their supply chain. To this end, academic learning and research will become vital and current for future 'sustainable' leaders.

Sustainable luxury would not only be the vehicle for more respect for the environment and social development, but it will also be synonym of culture, art and innovation of different nationalities, maintaining the legacy of local craftsmanship (Gardetti 2011). The foundations of Sustainable Luxury are social entrepreneurship and craftsmanship.

According to Giacosa (2014) in order to set the context of luxury fashion and, then, sustainable luxury fashion, we need to differentiate the terms *fashion* and *luxury*. While the term fashion refers to products, such as clothing and accessories, according to Fletcher (2008, 2014) fashion is the way in which our clothes reflect and communicate our individual vision within society, linking us to time and space. Clothing is the material thing that gives fashion a contextual vision in society (Cataldi et al. 2010). Luxury represents items perceived as a symbol of status. Luxury is also a symbol of elegance and sophistication with emphasis on the intrinsic value of many categories.

Godart and Seong (2014) show us that the relationship between luxury and fashion is quite an ambiguous one, as fashion does not fully belong to the luxury world, but overlaps with luxury in its most expensive and exclusive segments. These authors also show us that the relationship between luxury and fashion is quite an ambiguous one, because fashion does not fully belong to the luxury world as it once did up until the turn of the 19th Century, but overlaps with luxury in its most expensive and exclusive segments. Luxury fashion is a recurrent change at its highest level, and it is distinguished from other luxury segments by its constant pressure to change. In other words, we can say that when we talk about fashion we are referring to clothes and accessories, while the term luxury means status, elegance and sophistication given by a brand, quality, price, originality and creative content.

According to Kapferer (2012) both luxury and fashion share the common need for social differentiation, but they differ in two major aspects: first, whereas luxury is timeless, fashion is ephemeral, and whereas luxury is for self-reward, fashion is not. Thus, luxury fashion seems to be a contradiction in terms: as luxury, it is supposed to last, although as fashion it is supposed to change frequently. However, since the essence of fashion is change, luxury fashion gives exclusive access to enforced change. Luxury fashion is recurrent change at its highest level, and it is distinguished from other luxury segments by its constant pressure for change. This is shared by Pinkhasov and Nair (2014) who add that in a celebrity-driven culture, fashion has come to dominate the image and attitude of luxury.

The term "sustainable development" dates back to the United Nations Conference on the Human Environment in 1972 where it was first coined. Sustainable development is a problematic expression on which meaning few people agree. Each person can take the term and "reinvent" it considering his/her own needs. This is a concept that continuously leads us to change objectives and priorities since it is an open process and as such, it cannot be reached definitely. However, one of the most widely accepted definitions of sustainable development

—though diffuse and non-operating—is the one proposed by the World Commission on Environment and Development (WCED) report—Our Common Future, which defines sustainable development as "the development model that allows us to meet the present needs, without compromising the ability of future generations to meet their own needs" (WCED 1987: 43). At its core is the notion that all natural systems have limits and that human well-being requires living within those limits. The essential objective of this development model is to raise the quality of life by long-term maximisation of the productive potential of ecosystems with the appropriate and relevant technologies (Gardetti 2005).

Some authors, such as Walker (2006) and Koefoed and Skov (unknown year), have studied the contradictions between fashion and sustainability: sustainability requires a drastic reduction in our ecological footprint, and increasing a product's useful life; fashion, on the other hand suggest a passing trend or fad—something transient, superficial and often rather wasteful. But, beyond these contradictions, fashion should not necessarily come into conflict with sustainable principles. Indeed, fashion plays a role in the promotion and achievement of sustainability, and it may even be a key to more sustainable ways of living. According to Hethorn and Ulasewicz (2008) fashion is a process, which is expressed and worn by people, and as a material object, it has a direct link to the environment. It is embedded in everyday life. So, sustainability within fashion means that the development and use of something or process are not harmful to people or the planet, and once put into action, such thing or process can enhance the well-being of those people who interact with it, and the environment it is developed and used within. Also Sustainable fashion is an approach to the fashion system intended to minimise negative environmental impacts, and, in turn, maximise positive impacts (benefits) for workers and their families all along the value chain, hence playing a decisive role in poverty reduction. For this reason, Kate Fletcher (2012) in the preface of the book "Sustainability in Fashion and Textiles: Values, Design, Production and Consumption" wrote, "*For me the fostering of alternatives to the status quo in fashion and textiles is essential if we are to deeply engage with the process of sustainability...*" So, our challenge is to adjust sustainability paradigms and develop new ones that would fit the nature of the luxury fashion industry, one of the sectors called "core luxury sectors" within the luxury framework.

According to Godart and Seong (2014) luxury can offer a unique opportunity for creating sustainable business environments due to its two core features that set it apart from other market segments or industries. First, luxury is (often times) based on unique skills. This allows luxury to provide high-quality and rewarding business conditions. And, second, luxury is characterised by its peculiar relation with time, for its value is inscribed in the long-term. This allows luxury to offer a sustainable business model for resource management and high-quality product development, just to name a few relevant elements of sustainable luxury.

3 Sustainable Luxury, Entrepreneurship

The development of the luxury industry has gone in parallel to extensive economic growth and increases in the quality of life over the last century. Despite these benefits, there remains significant concerns that this growth has also had substantial negative effects on the natural environment and on some of the most vulnerable populations around the world. It is believed that this, in turn, will diminish the vitality and sustainability of our economies and communities (Schmidheiny 1992; United Nations 1987; World Resources Institute 2004).

Most studies based on traditional views on environmental and welfare economics tend indeed to consider business activities and the preservation of the environment or social welfare are inherently contradictory. Some indeed consider that market failures generally motivate environmentally degrading entrepreneurial activities (Tietenberg 2000; Cropper and Oates 1992; Balor 1958). However, more recently experts have stated to argue that entrepreneurship can in fact help resolve environmental problems through the exploitation of opportunities inherent in environmentally relevant market failures and thereby help move global economic systems toward sustainability (Anderson and Leal 1997; Cohen and Winn 2007; Dean and McMullen 2007).

There are indeed entrepreneurs with a profound perspective towards environmental and social issues and who are well motivated to "break" the rules and promote disruptive solutions to these issues. They are less constrained by existing realities than larger organisations, have less vested interest in the status quo, and have less to lose and more to gain from innovation (O'Malley 2011), including the type of innovation that addresses social or environmental issues (Cohen and Winn 2007; Senge and Carstedt 2001). They are transformational leaders that inspire, provide intellectual stimulus, and care deeply about people and the environment. And leadership and inspiration are crucial to move towards sustainability in the luxury sector (Gardetti and Girón 2014). Personal values are very important: idealistic values regarding environmental and social goals can be translated into valuable economic assets (Dixon and Clifford 2007). These entrepreneurs inspire and guide the fundamental transformation that sustainability requires (Egri and Herman 2000). Some of them were equally able to impact the decision-making of the giants. For example, Tesla Motors, a Silicon Valley car manufacturer founded during the last decade, was able to prove that eco-friendly electric cars can be appealing and efficient.[1] Their success is credited for pushing BMW to develop a green offering, despite it being centred on performance and polluting cars.[2]

[1]Several attempts have been made to create an environment friendly car. However, most of them were sacrificing aesthetics, and suffered of range and practicality issues (i.e. Chevrolet Volt, Nissan Leaf, etc.) discouraging customers. Tesla was able to demonstrate for the first time that electric cars are both viable and attractive.

[2]For example, http://www.greencarreports.com/news/1096051_audi-bmw-mercedes-porsche-electric-cars-to-target-tesla-model-s-plug-in-hybrids-too.

And, in practice, the characteristics prevailed in the process of the first four editions of the IE Award for Sustainability in the Premium and Luxury Sectors at a global level.[3]

Yet, despite the increasing interest for sustainable entrepreneurship, and in particular in the sector of luxury, there remains considerable uncertainty regarding the nature of entrepreneur's role in the area of sustainability and how it may unfold (Hall et al. 2010). In particular, we still have little understanding of how entrepreneurs in the luxury industry discover and develop those opportunities that lie beyond the pull of existing markets and why they decide to embark in such an unusual and uncertain journey. The aim of our study is to be a first step towards answering these questions.

4 Objectives and Work Methodology

The purpose of this study is not only to get to know what entrepreneurs understand by sustainable luxury and, particularly sustainable luxury fashion, but also to learn their vision about the future of this sector.

To develop this paper, the authors conducted an open question survey to fifteen entrepreneurs of different nationalities with sustainable fashion projects in the premium and luxury sectors. See the questions in Appendix. All these entrepreneurs participated in the IE Award for Sustainability in the Premium and Luxury Sectors. Some of them won in the category "fashion and accessories" and others reached the final stages of the award. In addition to the results stemming from the survey, we have also relied on insights emerging from in-depth analyses of several case studies of entrepreneurship in the sustainable luxury sector, which have been published in Gardetti and Giron's (2014) book. The book recognises the exemplary work done by several organizations like the London based Elvis and Kresse that reuses firefighters' hoses to create handbags and belts; Aïny Savoirs des Peuples of France that focuses on selling organic cosmetics and contribute in the development of indigenous communities in South America or We Are Knitters of Spain that surfed on the knitting and do it yourself trends to sell natural Peruvian wool.

5 Findings

One of the main findings derived from the surveys as well as the in-depth interviews realized with the sustainable entrepreneurs subject of our study, is that they both have common features and stark differences. In what follows we will

[3]The main purpose of this Award is to acknowledge the culture and practice of sustainability in the premium and luxury sectors, and hence of their communication, in order to encourage "more sustainable" and, therefore, "more authentic" sectors.

details the communalities and differences that span across a wide range of interviewees.

5.1 Two Different Understandings of "Sustainable Luxury Fashion"

As can be seen from the appendix, our survey includes four sets of questions. The first question inquires about the entrepreneur's personal understanding of the concept of sustainable luxury fashion. The third one focuses instead on explaining they key factors that make the studied enterprises sustainable. Given that all the entrepreneurs' interviewed belonged to the Sustainable Luxury Fashion industry, our initial assumption was that entrepreneurs would give similar answers to questions 1 and 3, or that we would find at least some of the components of question 1 reflected in their answer to question 3. Our findings revealed instead that this was not always the case. Specifically responses revealed the existence of two categories of sustainable luxury entrepreneurs (see Fig. 1):

- The first category of entrepreneurs, which we have called "Integrators", share a holistic view on sustainable luxury. Indeed, eight of our respondents directly associated 'sustainable luxury' as both outstanding products and sustainable practices at the same time. In that sense, their answer to question 3 (keys to sustainability) was similar or englobed part of their answer to question 1 (understanding of sustainable luxury fashion).
- Contrary to our expectations, however, there exists a second category of entrepreneurs whom we have labeled as "Decouplers" because, unlike the previous set of entrepreneurs, they decouple the concepts of Sustainability and luxury and seem to regard "sustainable luxury" as two components: luxury on one hand, sustainability on the other. Indeed, 6 entrepreneurs out of the 14 gave different answers between question 1 and 3. Their answers answer to question 1 seemed to regard the product itself, and words like "Quality", "Durability", "Materials" and "Design" were frequently used. However, the entrepreneurs did

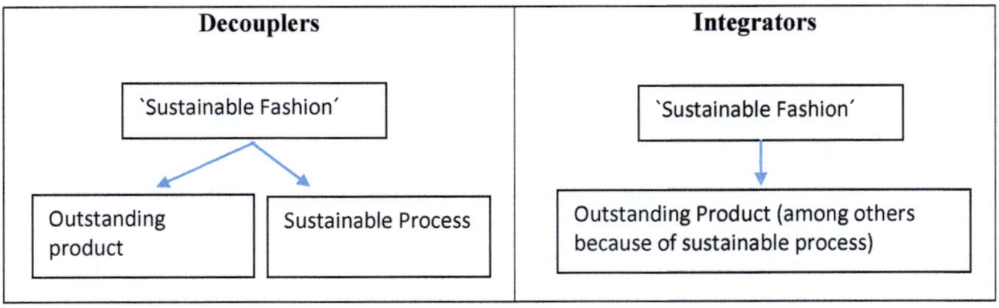

Fig. 1 Two types of sustainable luxury entrepreneurs. *Source* Made by authors

not mention the sustainable aspect/production of the product. Their answer to question 3, on the other hand, introduced the sustainable aspect of the products. For example, entrepreneurs referred to aspects such as using "upcycled materials" or "a sustainable approach in each of the life cycle phases of our products" and "giving authorship and pride back to the artisan".

Thus, it is only when specifically asked about sustainability that those entrepreneurs include it in their answer. The sustainable aspect of a luxury brand, according to those entrepreneurs, usually comes from traceability, sustainable materials and respect of the artisans. Therefore, it seems like luxury is often assigned to the object/product itself while sustainability is assigned to the process. This observation comes in contradiction with the answers given by the second set of entrepreneurs who have sustainability aspects as a prominent aspect of their notion of sustainable luxury, as exemplified by the following responses to answer 1: "… incorporates sustainable strategy across environmental, social and economic aspects" and "will contribute to, rather than detract from, the lives of all current and future potential stakeholders and citizens."

In many ways we can say that the *Decouplers* share the same approach to sustainable luxury as traditional luxury brands which have recently adopted strategies that might help improve the sustainability of their business, while preserving their financial interests. Memorable examples are those of Loro Piana and Ermenegildo Zegna, two Italian high-end brands that developed a local culture of vicuñas in the Peruvian Andes, respecting the environment, empowering local societies and benefitting to their business. Even famous French brand Saint Laurent Paris has created capsule collection made out of vintage fabrics.[4] This increased attention underlines a higher awareness where the future of the planet is valued much more than a simple Corporate Social Responsibility program. The *Integrators* instead, stand out from traditional luxury brands in the sense that they have created their businesses in the first place in order to give a new more authentic meaning to luxury.

5.2 Personal Motivation and the Identification of a Need Versus Its Creation

A clear pattern in motivations emerges with most of the entrepreneurs falling into two categories. Most of the *Decouplers* have started their social business after having identified a need in the market, which was the core motive for the creation of the venture. On the other hands, many of the *Integrators* first founded their business idea/product out of passion and then created the need for their products in the

[4]An article by the French version of Vogue relates the information (retrieved on March 11th, 2015): http://www.vogue.fr/mode/news-mode/articles/la-collection-new-vintage-iii-dyves-saint-laurent/18970.

markets in which they were involved. For the latter, the motive that pushed them to create their social business was the will to do something good for society of the passion for a material. Market opportunity was not what triggered their businesses. However, it may be interesting to note that the company We Are Knitters followed both behaviors described: They first felt passionate about the product and then succeeded in identifying in the market prior to launching their venture. In any case, available information does not allow us to assess whether one pattern is more successful than the other.

5.3 Different Generations and Views of the Future

While the entrepreneurs interviewed varied in their personal background and profile, a closer look at the age distribution across the two categories identified above reveal some interesting insights that might deserve a deeper inquiry in future studies. Indeed, the average age of the entrepreneurs, as a whole, is 35.31 years old. However, on average, the set 1 of entrepreneurs is higher than set 2: the first is 38 years old while the latter is 32, which suggests that some age factor could be at play in the differing approaches taken by entrepreneurs with respect to sustainable luxury.

In addition, another trend appears in relation to the views our entrepreneurs hold about the future. To the question whether they think that think brands will continue integrating sustainability to their business, or whether, in contract, the number of those doing it will decline, Decouplers are of the view that the integration of sustainability will happen, but remain sceptic that its success will depend on external forces such as market, entrepreneurial commitment or consumer knowledge. Integrators, on the other hand, are quite affirmative about the successful integration of sustainability and used phrases such as 'for sure', 'will increase', 'will become a norm'. Underlying this more optimistic view of the future of sustainable luxury seem to be a belief that there exists a broader and growing consensus among all stakeholders of the organization (industry members but also consumers regulators and the general public) about the importance of having sustainability as a key aspect of the very definition of luxury.

5.4 Additional Insights

Beyond the insights derived from the survey, interviews and desk research on the sustainable luxury entrepreneurs featured in Gardetti and Giron's (2014) book reveal some interesting patterns described below.

An entrepreneurial mindset: just like conventional entrepreneurs, the ones examined in this study exhibit "typical" entrepreneurial attitudes and mindsets. For example, either from the creation of the business or throughout the venture's journey, the entrepreneurs in this book have paired-up or have heavily relied on their team. The second pattern identified regards their attitudes toward failure. Many of them have experienced previous setbacks but see failure as a powerful learning tool and insist on the importance of overcoming the difficulty and growing from it. Last patterns regards the feeling of never being satisfied that some of the entrepreneurs share. They always want to keep on improving every aspect of the business, especially with respect to the sustainability aspect of their business.

Environmental and Social Impact: Most of the social ventures presented analyzed in this study have a social impact on a developing community or supports causes from which their product is derived. However, the impact those businesses have on their community is very wide. Among the social businesses, the entrepreneurs could either focus on environment, society or environment and society. Yet even for the ventures started out of the passion for nature such as Elvis and Kresse[5] or Estancia Peuma Hue,[6] the business has a positive impact on the society. Thus, no ventures presented in this book are engage only in sustainable environment and many are active both for the society and nature, and some of them, such as AÏny savoir des peuples[7] have focused their effort on promoting native communities in developing countries.

Supply Chain: As explained earlier, sustainable luxury often involves a complete reevaluation of the supply chain and production processes. However, our research indicates that this is easier said than done as working out successfully through the production chain was a challenge for many of the entrepreneurs examined. Specifically, half of these entrepreneurs faced a same difficulty: finding a producer that accepts to work with their unusual material or production pattern. This difficulty illustrates probably the fact that we are dealing with visionaries who are in many way ahead of their time. In that sense, they often need to create a completely new commercial/entrepreneurial set up to design, manufacture and distribute their products or services.

6 Discussion and Conclusions

The field of luxury is witnessing today a strong shift that realigns its priorities and resets its values, morals and finalities. This change, largely due to a strong shift in consumers' perceptions and expectations has forced several luxury powerhouses to question their methods and practices and rethink the image and lifestyle they are promoting. The materialistic approach, largely promoted during the last decades has been ousted by more experience oriented conscious and responsible patterns that value authenticity and respect for communities and human nature.

[5]http://elvisandkresse.com/.

[6]http://www.peuma-hue.com/.

[7]http://www.ainy.fr/.

Formerly greatly disconnected from the increasing awareness about salient social issues, luxury had to get out of its artificial bubble and to address growing customer requirements in terms of sustainability, an issue attracting continuously attention in the last years and affecting business practices in the world. If the luxury sector has been only a little affected by the change until recently, it is now forced to completely rethink its methods, valuing sustainability and its three pillars: people, planet, and prosperity.

But most importantly, the new mindset lead to the creation of a multitude of start-ups in different countries that are eager to give a new more authentic meaning to luxury. These innovative businesses follow an approach that diverges significantly from the more static and conservative well-established powerhouses and promote new diverse ideas in various fields. They are extremely entrepreneurial and profoundly reactive, thus profiting of the larger firms' inability cope rapidly with change.

Our exploratory survey of 15 sustainable luxury entrepreneurs operating in the fashion industry reveals two differentiated patterns. Older sceptic entrepreneurs often assign luxury to the object/product itself while sustainability is assigned to the process. On the other hand, the younger affirmative group of entrepreneurs sees 'sustainable luxury' as a combination of an outstanding product and a sustainable process together.

Appendix: Survey to Entrepreneurs

1. Your Data

Name and Surname:
Gender:
Age:
Nationality:

2. Project/Brand Data

Brand or Company Name:
Country of origin of the project/brand:

3. About Luxury Fashion and Sustainability

Notes for respondents:
For this research work, *sustainability means the integration of environmental, social and economic aspects.*

(1) What do you understand for sustainable luxury fashion?
(2) About sustainable luxury future: Do you think brands will continue integrating sustainability to their business? Or, in contract, do you think the number of brands integrating this concept will decline? In both cases, why?
(3) In your opinion, what factors make your enterprise or business sustainable?
(4) Why has sustainability become your business core?

References

Anderson TL, Leal D (1997) Enviro-capitalists: doing good while doing well. Rowman and Littlefield Publishers, Lanham

Balor FM (1958) The anatomy of market failure. Q J Econ 72:351–379

Berry CJ (1994) The idea of luxury—a conceptual and historical investigation. Cambridge University Press, New York

Cataldi C, Dickson M, Grover C (2010) Slow fashion: tailoring a strategic approach towards sustainability. Thesis submitted for completion of Master in Strategic Leadership towards Sustainability, School of Engineering, Blekinge Institute of Technology, Karlskrona, Sweden

Cohen B, Winn MI (2007) Market imperfections, opportunity and sustainable entrepreneurship. J Bus Ventur 22(1):29–49

Cropper ML, Oates WE (1992) Environmental economics: a survey. J Econ Lit: 675–740

Dean TJ, McMullen JS (2007) Toward a theory of sustainable entrepreneurship: reducing environmental degradation through entrepreneurial action. J Bus Ventur 22(1):50–76

Dixon SEA, Clifford A (2007) Ecopreneurship: a new approach to managing the triple bottom line. J Organ Change Manage 20(3):326–345

Egri CP, Herman S (2000) Leadership in North American environmental sector: values, leadership styles and contexts of environmental leaders and their organizations. Acad Manage J 43 (4):523–553

Fletcher K (2008) Sustainable fashion and textiles—design journeys. Earthscan, London

Fletcher K (2012) Preface. In: Gardetti MA, Torres AL (eds) Sustainability in fashion and textiles: values, design, production and consumption. Greenleaf Publishing, Sheffiel, p ix

Fletcher K (2014) Sustainable fashion and textiles—design journeys, 2nd edn. Earthscan, London

Frank RH (1999) Luxury fever—weighing the cost of excess. Princeton University Press, New Jersey

Gardetti MA (2005) Sustainable development, sustainability and corporate sustainability. In: Gardetti MA (ed) Texts in corporate sustainability—integrating social, environmental and economic considerations with short and long term. LA-BELL, Buenos Aires

Gardetti MA (2011) Sustainable luxury in Latin America. In: Conference dictated at the seminar sustainable luxury and design within the framework of the MBA of IE—Instituto de Empresa —Business School, Madrid, Spain

Gardetti MA, Girón ME (2014) Sustainable luxury and social entrepreneurship: stories from the pioneers. Greenleaf Publishing, Sheffield

Gardetti MA, Torres AL (2014) Introduction. In: Gardetti MA, Torres AL (eds) Sustainability luxury: managing social and environmental performance in iconic brands. Greenleaf Publishing, Sheffield, p 1

Giacosa E (2014) Innovation in luxury fashion family business: processes and products innovation as a means of growth. Palgrave Macmillan, New York

Girón ME (2009) Secretos de Lujo. LID Editorial Empresarial SL, Madrid

Girón ME (2012) Diccionario LID sobre Lujo y Responsabilidad. Editorial LID, Madrid

Godart F, Seong S (2014) Is sustainable fashion possible? In: Gardetti MA, Torres AL (eds) Sustainability luxury: managing social and environmental performance in iconic brands. Greenleaf Publishing, Sheffield, p 12
Hall JK, Daneke GA, Lenox MJ (2010) Sustainable development and entrepreneurship: past contributions and future directions. J Bus Ventur 25(5):439–448
Heine K (2011) The concept of luxury brands. Technische Universität Berlin, Department of Marketing, Berlin
Hethorn J, Ulasewicz C (eds) (2008) Sustainable fashion, why now?—a conversation about issues, practices, and possibilities. Fairchild Books Inc., New York
Kapferer JN (2012) The luxury strategy: break the rules of marketing to build luxury brands. Kogan Page, London
Kapferer JN, Basten V (2010) The luxury strategy: break the rules of marketing to build luxury brands. Kogan Page, London
Kleanthous A (2011) Simple the best is no longer simple—*Raconteur on* sustainable luxury, July 2011, 3. Source: http://theraconteur.co.uk/category/sustainability/sustainable-luxury/. Accessed Dec 2012
Koefoed O, Skov L (unknown year) Sustainability and fashion. In: Openwear: sustainability, openness and P2P production in the world of fashion, research report of the EDU fashion project. http://openwear.org/data/files/Openwear%20e-book%20final.pdf. Accessed 9 May 2012
Low T (unknown year) Sustainable luxury: a case of strange bedfellows. University of Bedfordshire, Institute for Tourism Research Bedfordshire, Bedfordshire
Mortelmans D (2005) Sign values in processes of distinction: the concept of luxury. Semiotica 157:497–520
Okonkwo U (2007) Luxury fashion branding: trends, tactics, and techniques. Palgrave Macmillan, New York
O'Malley Ch (2011) Lessons from the entrepreneurial path. In: Marshall J, Colemna G, Reason P (eds) Leadership for sustainability—an action research approach. Greenleaf Publishing, Sheffield, p 138
Pinkhasov M, Nair RJ (2014) Real luxury: how luxury brands can create value for the long term. Palgrave Macmillan, New York
Ricca M, Robins R (2012) Meta-luxury—brands and the culture of excellence. Palgrave Macmillan, New York
Scheibel S (unknown year) Ethical luxury—myth or trend? Essay. The London School of Economics and Political Science, London
Schmidheiny S (1992) Changing course: a global perspective on development and the environment. MIT Press, Cambridge
Senge P, Carstedt G (2001) Innovating our way to the next industrial revolution. Sloan Manage Rev 42(2):24–38
Short JC, Moss TW, Lumpkin GT (2009) Research in social entrepreneurship: past contributions and future opportunities. Strateg Entrepreneurship J 3(2):161–194
Tietenberg T (2000) Environmental and natural resource economics. Addison Wesley, New York
Thomas D (2007) How luxury lost its luster. Penguin Books, New York
Walker S (2006) Sustainable by design—explorations in theory and practice. Earthscan, London
World Commission on Environment and Development—WCED (1987) Our common future. Oxford University Press, Oxford
World Resources Institute (2004) World resources 2002–2004. World Resources Institute, Washington, DC

The Family Business Phenomenon as a Means for a Sustainable Business in the Clothing Luxury Business

Elisa Giacosa

Abstract The family business phenomenon represents the most widespread business model in the world, representing both a relevant source of growth, development and economic stability, and a social determinant. The family business phenomenon has an important role in the clothing fashion business, as a large number of the medium and large companies operating are family businesses. The chapter focuses on the family business model in the clothing luxury business. The main goal is to analyse how the innovation strategy implemented by family businesses operating in the clothing context could represent a means of sustainable development in the society, in terms of economic, social and cultural aspects. Indeed, thanks to its history and reputation, the family could increase the integrity of its business, generating a positive impact both on its turnover, on the employment and, in general, on the society. In addition, the clothing luxury democratization makes a choice in a scale of desires. As the luxury is a social value manifestation within the community, the possession of a clothing luxury product leads to certain individual benefits, expressing the extent to which a person belongs to a certain social class. Consequently, the luxury could be a synonym of culture, art and innovation, instead of a simple vehicle for companies to make money. The theoretical framework is based on the study of Miller and Le Breton-Miller model (Managing for the long run: Lessons in competitive advantage from great family businesses. Harvard Business School Press, Boston, 2005), which identified a series of family business success drivers which, in a strong combination, impact on both strategic decisions and daily management. We created a qualitative model, identifying a series of specific features of family businesses which constitute a competitive advantage in the context of clothing luxury fashion, in order to make the generate a "sustainable luxury business". This model was created by means of a qualitative method: despite criticism of its trustworthiness, the qualitative method was used due to the fact that it allows for an accurate understanding of the dynamics of the family business, which is a complex context, while avoiding a purely

E. Giacosa (✉)
Department of Management, University of Turin, Corso Unione Sovietica, 218 bis, 10134 Turin, Italy
e-mail: elisa.giacosa@unito.it

M.A. Gardetti (ed.), *Sustainable Management of Luxury*,
Environmental Footprints and Eco-design of Products and Processes,
DOI 10.1007/978-981-10-2917-2_17

descriptive approach which would be equally limiting. In addition, the effectiveness of the framework theories can be directly contextualised in the analysed sector. Lastly, the qualitative method is frequently used in family business research, both to understand the family decision-making process and to compare family businesses and non-family businesses. The choice of this topic is due to a lack of studies in the literature on the impact of the family variable on the innovation strategy in the clothing luxury business. In fact, analysis of the contents of the major family business journals and books revealed that no research has been published on this subject.

Keywords Family business phenomenon • Luxury business • Clothing luxury business • Democratization • Sustainable luxury business

1 Introduction

The family business phenomenon is considered as the most widespread business model in the world (Acquaah 2013; Contò et al. 2014; Mediobanca 2013), representing both a dominant source of growth, development and economic stability for several countries (Culasso et al. 2012a, b; IFERA 2003). In addition, family firms (FFs) have a positive impact on regional competitiveness and its drivers (Viassone 2009; Vrontis and Viassone 2013).

In the European context, 60% of the companies are FFs, compared to 80% in the United States. In Italy, FFs are especially small and medium-sized companies, representing a distinctive factor in the economic context and producing around 70% of the total Italian GDP (Bertoldi 2015).

For this reason, many researchers have focused on the phenomenon (Astrachan and Shanker 2003; Claessens et al. 2000; Colli 2013; Culasso et al. 2013; Faccio and Lang 2002; La Porta et al. 1999; Morck and Yeung 2004; Garcìa-Ramos and Garcìa-Olalla 2011).

Family business phenomenon is involved in several economic sectors, such as luxury fashion business. In Italy, where the luxury fashion context is so strong (Giacosa 2011, 2012; Vecchi 2008), 80% of the medium and large companies operating in the luxury context in Italy are family businesses (Osservatorio AIDAF-Unicredit, Bocconi (AUB) 2011). Therefore, the family business phenomenon has an important role in the clothing fashion business, becoming a driving force despite the world consumption crisis. Thanks to a joint presence of historic family businesses and large multi-brand groups, the FFs in the clothing fashion business continue to attract domestic and foreign customers, reaching interesting performances.

The luxury context is affected by a strong competitiveness and new paradigms—which impact on the companies' business models—try to face this problem: a slow process of democratisation, and a market globalisation. Firstly, a luxury democratization is coming (Silverstein and Fiske 2003a, b): many luxury companies have been reducing the exclusivity of their items, permitting a wider range of target customers and the company to look for profitable opportunities outside their usual market niche. Luxury has expanded into affordable luxury segments, generating a less marked social stratification and permitting a wider range of customers to purchase luxury products. Therefore, luxury democratization creates a choice in a sense of desires. As the possession of a luxurious product is a kind of social manifestation (person expresses belonging to a certain social class), it leads to certain individual benefits. As a result, luxury doesn't have to be understand as simple instrument used to make money but as a synonym of art, culture and innovation, and as a manifestation of a certain social status. As a consequence of the democratisation process, the luxury industry has sometimes sacrificed its integrity (Thomas 2007), in terms of historical corporate values and familiness factor. This causes damaging the corporate image, in terms of reputation and legitimacy (Giacosa 2012).

Secondly, the market globalisation negatively impacts on luxury pureness by moving production processes to such countries. The transfer of manufacturing process might cause a decline in the quality status of the luxury products and, consequently, in the brand image: for instance, French haute couture started its decline when production processes was outsourced in foreign countries. Therefore, any company that decides to outsource in certain countries should consider the risk of losing its luxury status, and have to strengthen its country of origin factor (Kapferer and Bastien 2009; Corbellini and Saviolo 2009).

Consequently, literature has stressed on a *sustainable luxury*: a sustainable luxury development is a new paradigm, and requires looking at company's business model in a different way. While luxury business maintains its role as a social determinant, despite the democratization process (Kleanthous 2011) and the market globalisation, it needs to be a vehicle for more respect both the reaching of good companies performance, environment and social development, cultural art and innovation of several nationalities, and conserving local craftsmanship (Gardetti 2011).

The chapter focuses on the family business model in the clothing luxury business. The main goal is to analyse how the investments strategy implemented by a family firm could represent a means of sustainable development, in terms of economic, social and cultural aspects. Indeed, thanks to its history and reputation, the family firm could increase the integrity of its business, generating a positive impact on its turnover, on the employment and, in general, on the society. Consequently, the involved companies needs to follow an internal change approach, thus encouraging sustainable business practices across all functional areas.

The chapter is structured as follows. A short background on the luxury business and on the innovativeness of family businesses has been made in the second section. The research method is outlined in the third section. The description of solutions and recommendations is conducted in the fourth paragraph. Finally, the conclusions and implications of the study are set out, along with the limitations of the research.

2 Background

The luxury context is not easy to define, as the literature gave several definitions for different reasons (Chevalier and Mazzalovo 2008; Corbellini and Saviolo 2009; Giacosa 2012; Jackson 2004):

- firstly, due to different interpretations and contextualization, meaning that individuals consider in a subjective way if a product is luxurious for them;
- secondly, social value of a luxury product depends on the context and a sphere of action;
- lastly, doesn't exist a single sector for luxury items but several sectors are involved, such as core luxury sectors (they are traditional luxury sectors, such as clothing, accessories, jewelleries and watches, cars, perfumes and cosmetics) and new luxury sectors (which have began to enter the luxury market, such as tourism and catering, wines, spirits and other gourmet products, and furniture and household items). Consequently, many values and sectors permit the customers to demonstrate their membership of a certain social class.

Focusing on the clothing luxury business and for making a complete contextual framework, it was necessary to qualify the type of offer to meet the needs of the customers. For this reason, we could distinguish different typologies of clothing luxury fashion, according to affordability of a brand (Chevalier and Mazzalovo 2008; Giacosa 2011; Okonkwo 2007):

- non-affordable clothing luxury fashion: products are characterized by outstanding quality and creativity, and high price is affordable for a select few customers. Products are created in limited editions or one-off pieces, are handmade or semi-handcrafted, are customised or tailored, and give the customers a high social status. De Grisogono is an example of non-affordable clothing luxury fashion in the field of jewellery;
- affordable clothing luxury fashion: price is more affordable than in case of non-affordable clothing luxury, despite a satisfactory quality and strong stylistic content. Customers could perceive the social status of the brand, which generates a sense of belonging to a certain social group (Kapferer and Bastien 2009, 2012). Missoni is an example of non-affordable clothing luxury fashion in the field of clothing luxury.

One of the topic analyzed in the luxury context is represented by the innovation strategy. Innovation strategy constitutes a set of factors such as human resources, goals and perspective, having as an objective ensuring the company's long-term survival (Giacosa et al. 2014a, b; Re 2013) and in the luxury context the innovation strategy has specific features (Giacosa 2012).

We said that family business phenomenon is widespread in the luxury fashion business. For this reason, we focus on this phenomenon.

One of the dominant research topics in family business is represented by the management of innovation, as the innovation strategy represents a critical issue for family businesses (Aronoff 1998; Craig and Moores 2006).

Many researchers categorized internal, external, and a mixed innovation solutions (Cohen and Levinthal 1990; Chesbrough 2003; Schilling 2009) in the family firms business model. In addition, the products and processes innovation has been distinguished and has relevant theoretical and practical consequences, even if this distinction is not always clear cut (Adner and Levinthal 2001; Bresciani 2009, 2010; Bresciani et al. 2015; Sood and Tellis 2009).

Innovation strategy in family businesses stimulated different branches of thinking (Craig and Moores 2006; Culasso et al. 2012a, b, 2013):

- family businesses have been considered less attracted by innovative policies than non-family businesses, due to a more risk aversion. Indeed, innovative investments are generally risky and uncertain (Block 2012; Bresciani et al. 2013a, b; Chen and Hsu 2009) and these features impact on their decision-making process. In addition, there is a concentration of ownership in a few persons (such as family members or family members and non-family members) that increases aversion risk. Researchers stated that family businesses are more conservative in relation to innovative investment and less interesting in performing new risky opportunities (Chrisman and Patel 2012; Czarnitzki and Kraft 2004; Dunn 1996; Matzler et al. 2014; Munoz-Bullon and Sanchez-Bueno 2011;
- conversely, family businesses have been considered more attracted by innovative policies than non-family businesses, as their sense of long-term investment as well as vision is much stronger (Chua et al. 2009; Le Breton-Miller and Miller 2006; Zellweger 2007; Zellweger et al. 2012). Family firms are aware that thanks to innovation they can improve their competitiveness in long-term period (Astrachan 2010; Cefis and Marsili 2005; Zahra et al. 2008).
- more recent studies focused on the combination between traditional and innovative strategies (Re 2013), thanks to which the innovative policy increases its competitive advantage over generations (Kellermanns et al. 2008).

Behaviour of family businesses in terms of innovative strategy has been delineated. The "familiness" factor (Arregle et al. 2007; Frank et al. 2010; Irava and Moores 2010; Minichilli et al. 2010; Sirmon and Hitt 2003) impacts on the family businesses behaviour and, in particular, on the attitude for a long-term vision in their investments strategy. The familiness factor represents a key factor in their competitive advantage, as it consists of a series of special, characteristic competencies, abilities and resources which result from the interaction between family members and their involvement in the business, which is higher than in non-family businesses.

The familiness factor impacts on the capital invested, defined as "patient capital" (Dobrzynski 1993; Teece 1992). Investments are focused on creativity and innovation with a long-term vision, respecting in the same time tradition and unity of the family business; if it comes about money invested—they don't have a predicable return. In relation to the values which luxury brand should hold, only so-called patient capital is considered as perfect for long-term creation of them.

Alternately, instead of considering the familiness as a single variable, the family business has been considered as a set of heterogeneous ideas, objectives, family spirit and attitude (Block 2012; Carcano et al. 2011; Hülsbeck and Wirsching 2014; Sirmon et al. 2008). This values impact on the innovation strategy and, in particular, on the strengths and weaknesses of the company due to several factors, such as (Astrachan and Jaskiewicz 2008; Naldi et al. 2007; Zellweger and Astrachan 2008), family unity, relationship between company's goals and a family and its emotional attitude to the company, dimension and age of a company, family risk propensity, business features; as well as capital structure in the context of frequency and types of leverage.

3 Methodology

The main goal of the chapter is to analyse how the investments strategy implemented by family businesses operating in the clothing luxury context could represent a means for sustainable development, in terms of economic, social and cultural terms.

Our framework is represented by Miller and Le Breton-Miller model (2005), where a range of family business key success drivers have been identified, having a significant impact on strategic decisions and daily management as well, and representing a competitive advantage in terms of luxury fashion.

The methodology of present research has been conducted in the following phases:

- the first phase: we analyzed existing literature concerning the luxury business and the innovativeness of family businesses, as it enables to understand the role which investments strategy has in developing the company competitive advantage and development of the context;
- the second phase: as the objective was qualitative in nature, the treatise was conducted using qualitative approach. Thanks to the observation of multiple internationally-recognized clothing luxury medium-sized companies, which are characterized by a highly innovative business approach that has been handed down from generation to generation, our analysis has been possible.

The second phase requires deepen illustration. We created a qualitative model which identifies a series of specific features of clothing luxury family firms innovation strategy, which reflects a competitive advantage in terms of clothing luxury fashion to ensure a sustainable luxury business.

Despite criticism of the qualitative method, we decided to apply it. Our choice permits an accurate understanding family firms dynamism as it constitutes a complex issue, and averting a descriptive approach which would be equally limiting. Moreover, effectiveness of the qualitative methods applied can be directly adapted in the luxury context. Lastly, this kind of methods are commonly used in family business research because of the possibility of understanding better the family decision-making process and because the qualitative methods permit a comparison between family and non-family businesses (George and Bennett 2005; McKeown 2004; Yin 2009).

To create our model, schematic view (Lau and Woodman 1995) has been used, as it enables identification of a series of key issues and critical factors which simplify and support the decision-making process carried out by management and/or ownership, according to the considered issue. Selection of the schematic view is justified by the following reasons:

- it facilitates the challenges which the management has to face, considering any signal deriving from the external environment and from the analysis of the company's circumstances;
- the recipients are enabled to develop several valuable insights into phenomenon discussed, which allow to take a look at the business' state of affair and simplify the decision-making process;
- the final model enables to define a number of key issues and simplifies a further reasoning.

This topic has been chosen due to a lack on the issue of implementing a *sustainable business* into the clothing luxury context, especially thanks to investments strategy. The most important research databases like Google Scholars, EBSCO as well as international journal on family business (Family Business Review, Journal of Family Business Strategy) don't pay attention this problem, overall by using a qualitative method, which is adequate because of the reasons highlighted before.

This study is interested in solving the problem of understanding and identifying the benefits of investments for the companies in the clothing luxury sector, in terms of sustainable luxury.

4 Solutions and Recommendations

The luxury business is characterized by a strong competitiveness. To gain the competitive challenge, several number of companies moves production processes to the countries where the labour costs are lower. This action can lead to deterioration of the brand image and can create a risk of losing the luxury status; therefore the company should take into consideration this risk and reinforce its country of origin factor.

In addition, a few luxurious companies have started to reduce the exclusivity of their products in order to extend a range of potential customers, allowing the company to gain new opportunities beyond an usual market niche. As a result, the luxury, in the context of familiness factor and historical corporate values, sometimes suffers on its integrity (Thomas 2007), what has a negative impact on the company's image (its reputation and legitimacy) (Giacosa 2012).

For the reason mentioned above, companies has to pursue a *sustainable luxury*: the luxury phenomenon should be an instrumentality of reaching good companies' performance, respecting the environment and social development, cultural art and innovations of several nationalities, and protecting in the same time local craftsmanship (Gardetti 2011; Giacosa 2011). To reach this purpose, the company may introduce an effective and efficient investments strategy and the choice of this strategy business model depends on several variables, reflecting the features of each company.

Thanks to the schematic view, and to reach the purpose of the research, we referred to our framework (Miller and Breton-Miller 2005), which identifies a series of key issues and critical factors whose role is to support and simplify the management and/or ownership in the decision-making process, considering the investments strategy. The purpose is then to approach a sustainable development in economic, social and cultural context. According to the framework, the critical factors thanks to which our business model is properly matched are presented below:

- Continuity: one of the relevant features of the family firms is a certain continuity in reaching the company mission, which requires the ownership of several core competencies necessary for reaching the goals in a long-term vision and achieving or maintaining the competitive advantage;
- Community: family firms need to create a strong sense of belonging to the company, which is due to a strong unity and collaboration between the human resources;
- Connections: family firms need to establish a potent sustained relationships with several external parties, to enable developing a productive network favourable in the long-term competitiveness of the company;
- Command: family firms, especially small and medium-sized ones, have a freedom in the context of actions and are able to make a fast decisions in reference to management and innovation.

Taking consideration to our framework, we created a new model, identifying a series of specific features of family firms thanks to which the investments strategy could represent a means of sustainable development, in terms of economic, social and cultural aspects. The key drivers are the following (Fig. 1):

- Company heritage;
- Long-term vision in the investment attitude;
- Specialization;
- Relationships;
- Innovativeness.

Fig. 1 Success drivers of our model. *Source* personal elaboration

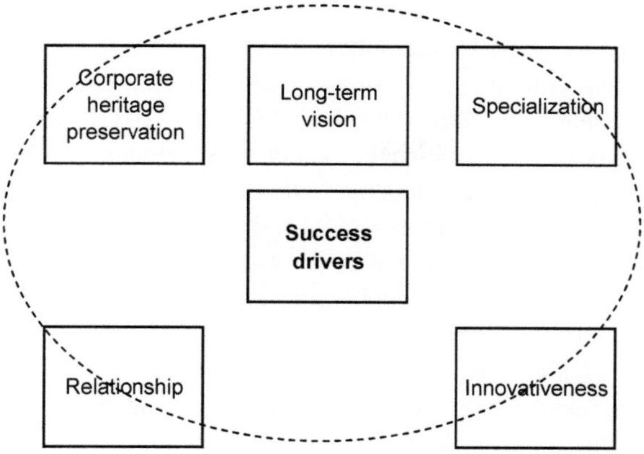

Recognition of the key drivers mentioned above was necessary to understand how the investments strategy of clothing luxury fashion family firms could help to promote a sustainable luxury.

(A) The company's heritage

Firstly, the innovation strategy has to consider corporate heritage values (Bordieu 1977; Davis 1983) (Fig. 2), reflecting a history heritage, which constitutes an effect of accumulation over time of the features handed down through the generations (such as trust, reputation, values and prestige), in order to give a company's brand a long-lasting authenticity (Davidow and Narva 2000; Eckrich and Loughead 1996; Lambrech 2005; Narva 2001, 2010; Perricone et al. 2011). For example, in Missoni the corporate heritage values reflect the family's philosophy, which is the same across generations.

The company's heritage is is related to the company's reputation. For this reason, innovation has to consider the elements of the corporate heritage having as an aim the protection of the company's brand perception (Lipovetsky and Roux 2003).

Fig. 2 Corporate's heritage. *Source* personal elaboration

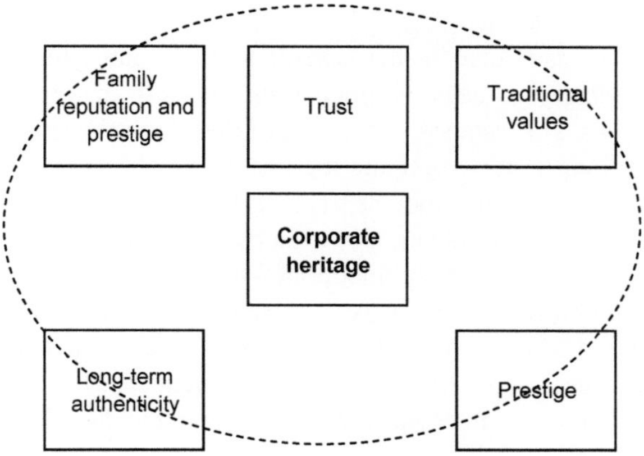

Perception of the heritage factor is so relevant, since the patrimony of the firm's history helps to strengthen a prestige of the brand. That is why the brand's reputation constitutes so important element in a family business: it reflects not only the company, but also the family (Attanzio 2011; Ferda 2010).

In its products range, the company should offer an internal quality of perfection, meaning that a company's attention about details in terms of production and customer relations management has to be coherent with the traditional corporate values. Strong attachment to the corporate heritage is the only feature which helps to keep the consumer's loyalty, reducing the customers tendency to abandon the brand.

In addition, the prestige of the brand is also due to the quality level of the products: when the quality level is not in line with the brand value, the quality factor impacts on the customers perception (Bresciani et al. 2013a, b).

Also the policy price impact on the company's heritage. Indeed, variation of a price depends on the availability of the luxurious product. For instance, non-affordable luxury goods companies practise a policy of Everyday High Pricing, i.e. price is not reduced during the sales season, and in this way the companies can record long-lasting clientele level, increasing products' attractiveness for the consumer. De Grisogono, for instance, doesn't make reduction price. Process of product's price decrease could be considered by the customer as non-luxurious, and in result could lead to losing the consumer's loyalty (Aaker 1997; Dubois and Czellar 2002; Okonkwo 2007; Silverstein and Fiske 2003a, b). On the contrary, affordable luxury goods companies apply the policy of High/Low Pricing, what means that during a season high price is set out; then, during the sales period a company reduces a price of non-continuative products. In this situation, the level of legitimacy threshold has to be raised up in order to improve business opportunities.

If the innovation strategy doesn't respect the brand identity, the products are not compatible with brand identity; in addition, the products are not in line with the permanent stylistic identity of the company (Kapferer 2008). Therefore, a wrong innovation strategy damages brand's prestige reflects a symbolic issue differentiating the product, and strengthens social status of individual who is wearing it (Ko and Megehee 2012; Miller and Mills 2012; Shipman 2004). A wrong innovation strategy might cause a non sustainable business in terms of the respect of company's heritage.

In particular, in the case of non-affordable luxury, damage of the corporate's heritage has negative impact on the sense of rarity, reducing the sense of prestigious. For instance, in the case of the Channel or Kelly products, limited production or long waiting time are the factors which lead to increase the sense of rarity and luxury of their products, making the brand even more exclusive.

In order to obtain a sustainable business, protection of the corporate's heritage should be extremely important issue, even if the fashion collections reflect a strong variability in every season. Indeed, mentioned collections have to be linked also to brand identity (De Chernatony 2001; Fabris and Minestroni 2004; Kapferer and Bastien 2009, 2012; Ross and Harradine 2011) reflecting the heritage factor, to which a customer showed his loyalty in the past. The identity of a luxury brand has a relevant impact on the attractiveness and prestige of the product and highlights its internal quality, rarity and creative content. Even if the fashion trends need to vary each collection, Max Mara collection tends to be aligned to the corporate's heritage.

The maintenance of the corporate heritage has to be related to the degree of the brand's exclusivity and to be aligned to the customer perception in terms of social status. In this context, different strategies depend on the luxury typologies:

- in the case of affordable luxury, the aim is to enhance the brand's attractiveness, strengthening in the same time the customer's loyalty: increase of the company's marketability and gaining the competitive advantage is possible thanks to strong creativity and innovation activity;
- in case of non-affordable luxury, the aim is to improve the intrinsic value of the products, leading to the situation where some products are affordable not for each customer but for a few selected.

In both cases, the company should take into consideration needs of all customers (Moody et al. 2010): therefore, the innovation strategy constitutes the instrument, thanks to which a consumer is able to satisfy his own needs, also in terms of social class achievement. Reaching this purpose, the business becomes sustainable also in relation to the consumers' satisfaction.

(B) The long-term vision in the investments attitude

Family firms are characterized by a longer-term vision in their investments strategy (Fig. 3) than non-family businesses (Block 2012; Chua et al. 2009; Le Breton-Miller and Miller 2006; Miller and Le Breton-Miller 2003; Sirmon and Hitt 2003; Uotila et al. 2009; Zellweger and Sieger 2012).

This is due to the purpose of the family firms, where the transfer of the company to the next generation is typical; consequently, a long-term vision is coherent with the transfer of the company to the future generations (Basco and Rodríguez 2009; Neumann 1997; Ward 1997; Zellweger 2007).

Fig. 3 Long-term vision. *Source* personal elaboration

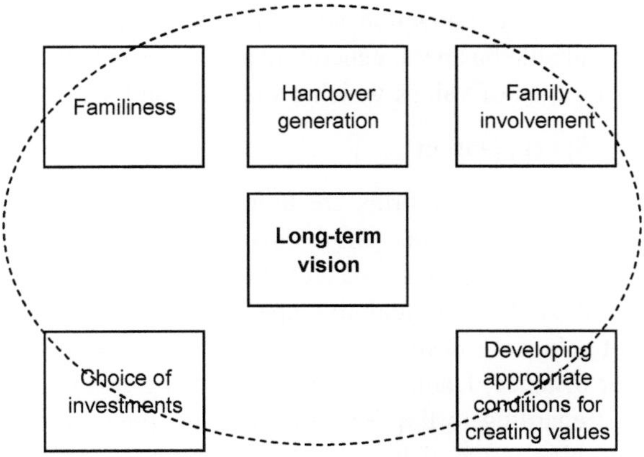

The long-term vision indicates only a choice of investments and developing the most appropriate conditions for generating value over time (Davis 1983; Habbershon and Pistrui 2002; Naldi et al. 2007; Zahra 2005; Zahra et al. 2004).

If the entrepreneur or the management has not the prospect of a company's activity continuation (James 1999), their attitude to the company is less strong (James 1999). In this situation, shareholders' priority is no longer the return on equity in a short period of time, but the company's survival and taking care of maximizing the return on the company's social and symbolic assets.

The attitude for a long-term vision of a family business in their investments strategy is also due to a "familiness" factor. Human capital of the company has relevant impact on the company's familiness and improves a sense of community and closeness between its members (Sirmon and Hitt 2003). For instance, Missoni is characterized by a strong familiness factor, which has been strong also after the disappearance of some family members last years.

In case of the absence of the company's long-term vision the competition is very hard. This is due to the fact that the brand's creation and extending the marketability is a long-lasting process; that's why only this kind of capital which characterizes family firms is considered as perfect capital for creation all the values which luxury brand should have.

A strategy with a short-term vision is not able to create and valorise a brand, and leads to loss of its prestige in the long term period. Consequently, the company should improve the brand's exclusivity because only a strong luxury brand is able to attract consumer's attention, as the possessing this certain brand can increase his social status.

Due to an ineffective investment strategy with long-term vision, the damage in the prestige of the brand leads to the damage of the company's image. This issue is all the more marked, the larger a level of family engagement in management process (Aldrich and Cliff 2003; Anderson and Reeb 2003), and the greater capacity of adapting the company's business is, and as a result, a positive features are linked not only to the family, but also to the company (Micelotta and Raynard 2011; Miller and Le Breton-Miller 2005).

Incorrect innovation strategy in terms of long-term vision might cause a non sustainable business, generating a damage of both brand and long-term construction of a series of values which a luxury brand should conserve.

(C) **Specialization**

Typically, family firms are limited in their investments strategy. In this context, several reasons emerged: the investment sectors reflects an extension of the family business operational area and evolution of its tradition (Morris 1998; Donckels and Frohlich 1991); in addition, specialization (Fig. 4) constitutes a solution in case of not achieving desired level of company performance (Chrisman and Patel 2012). For instance, Damiani maintains a strong specialization into the jewellery sector.

Regarding to the family businesses, they are characterized by the tendency of investing in the niche markets as opposed to non-family businesses: they focus on

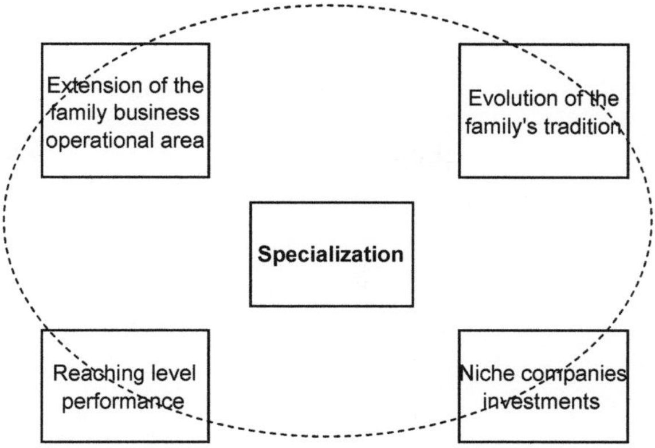

Fig. 4 Specialization. *Source* personal elaboration

differentiating some product ranges operating in certain production phases, without vertical integration in the value chain.

In case of small-sized family firms, lack of the capital obtained from the external (limited bargaining power with banks caused by financial world crisis) and internal factors (company's age, dimension and growth rate) as well as family and non-family ownership have a relevant impact on the specialization strategy (Coleman and Carsky 1999; Harvey and Evans 1995; Mahérault 2000).

When the company's strategy is not coherent with the company's brand portfolio, the company concentrates all its efforts and investments on a wrong object. Additionally, brand's attractiveness is likely to suffer, if the specialization is less marked: we observe decrease of the effectiveness caused by the synergies between the differentiated products which belong to different categories (Terzani 2007), as well as decrease of the customer's loyalty.

In addition, when the company's strategy is not in line with the company's attitudes and know-how, the investments are not focused on the right products range. This fact can cause both a dispersion of competencies and focus on the products according to which the company is not competitive. It might cause a reduction in the product range appeal and the increase of cost production.

The combination of these factors might lead to a non sustainable business, as it generates both brand damage, and a wrong long-term construction of products range appeal.

(D) **Relationships**

The creation of networks between the company and its external parties (Fig. 5), such as suppliers of goods and services, is recommended (Dicken 1994; Flatten et al. 2011; Giacosa 2012; Gulati and Martin 1999; Human and Provan 1997; Gulati et al. 2000; Miles et al. 1999; Nieto and Santamaría 2010; Walker et al. 1997).

Typically, networks produce a cluster (Crestanello and Tattara 2011; Dunford 2006; Molina-Morales and Martínez-Fernández 2010; Vecchi 2008), in which the companies reach a competitive advantage thanks to the sharing of resources, skills and attitudes.

Fig. 5 Relationships. *Source* personal elaboration

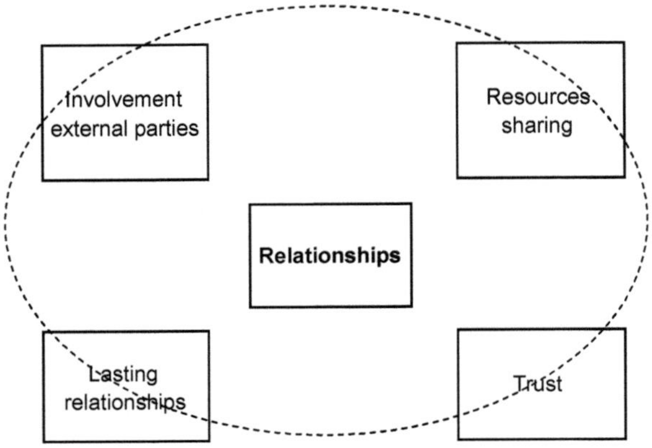

Family firms are generally characterized by long-term relationships with other parties based on trust (Walters 1982), which are necessary for increasing trust and altruism. In this context, sharing the resources enables the network to generate innovation.

The company has a wide range of relationships with numerous entities involved in supply of products and services in both upstream production process (suppliers and subcontractor) and downstream (distributors, the middlemen and the retailers). Small family business manages this kind of relationships at personal level, increasing the entities' loyalty to the family business (Bresciani et al. 2013a, b) and enhancing a sense of trust, what favours supply of the production factors. In general, family business relies on long-lasting relationships with suppliers, based on the trust and altruism (Anderson et al. 2005; Carney 2005).

The absence or the inefficacy in the relationships constitutes a negative factor for the companies' development and competitiveness. In particular, the absence or the ineffectiveness of formal or informal relations becomes a barrier in exchanging of resources and doesn't permit the company to operate beyond its boundaries (Ghoshal and Bartlett 1990; Garzella 2000; Lorange and Roos 1993; Liebeskind et al. 1996; Rosenkopf and Almeida 2003; Zander and Kogut 1995).

When the company doesn't participate in a cluster and the its skills and resources sharing, the company reduces the possibility of innovation. It leads to decrease of the productivity level and causes both an absence of a virtuous cluster-innovation circle (Granovetter 1992; Hansen 1999; Schilling and Izzo 2013) and a regional development.

In addition, the absence of the relationship between the company and its territory generates the lose in customers attracting. It is caused by strong attitude to its origin country, especially when it presents an important attachment to luxury fashion [this trend is observed for example in case of Italy or France, which are considered in the world as the centre of quality, glamour and exclusivity (Giacosa, 2012)]. The absence of the respect for the origins of the company, its family and community could impact on the possibility to maintain a strong attachment to the corporate values, and also on the creation of a pleasant workplace. Think about Brunello

Cucinelli: "the beauty of the area and attachment of his employees to their community of origin, combined with the tranquillity of the workplace, promotes the creativity of the human mind. All our employees, regardless of their role, participate in the life of the firm: each one knows that his/her work makes an essential contribution to common growth; our 'integral quality' is the result of the inner quality of each one of them" (www.brunellocucinelli.com/filosofia).

Concluding, the absence or the inefficacy in these relationships might cause the reduction of the sustainable business, in terms of brand appeal (especially when the brand is linked to a country of origin) and the possibilities to share resources and skills within a cluster.

(E) **Innovativeness**

The company has to choose between an incremental or radical innovation, depending on the purpose to reach, the perception of the customers and the company risk aversion. An incremental innovation might be adapted when a marginal or slight adaptation of existing products is preferred; on the contrary, a radical innovation consists of new products, different from that offered by the competitors.

The innovativeness (Fig. 6) has to be a combination between productive, organizational and commercial aspects (Giacosa 2012):

- in productive aspects: stringent quality control has to be applied in order to increase the product's quality (O'Donnell et al. 2002). In this aspect is observed a possibility of reducing the unacceptable manufacturing faults and improvement of internal product characteristics, permitting the reduction of production costs;
- in commercial aspects: the innovation strategy has to attract old and new consumers. In particular, it takes into consideration the type, characteristic and

Fig. 6 Innovativeness

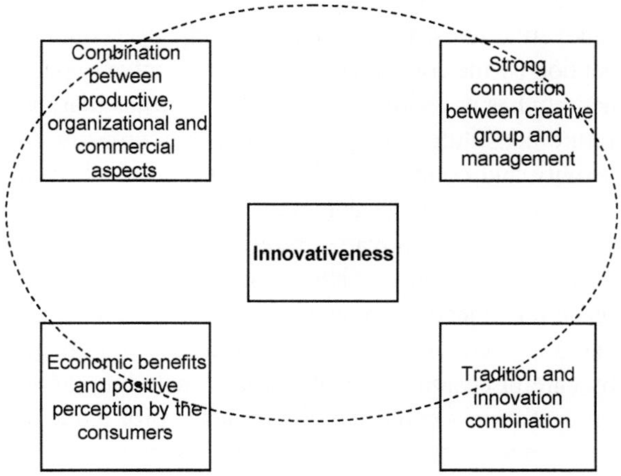

location of the shops, as well as communication and marketing channels. Multi-sensorial instruments are useful for attracting the consumers and distinguishing them from the competitors (Binz et al. 2013; Kapferer 2002; Poza et al. 1997; Ward 1997). New digital and innovative instruments (Donaldson 2011; Geerts and Veg-Sala 2011; Ha and Perks 2005; Kim and Ko 2012; Okonkwo 2009) should evaluate the products features, creating new desires: consequently, the brand myth invigorates;
- in organizational aspects: organizational structure of the innovation strategy need to be improved, with the purpose to suit to the innovation strategy needs. For this reason, a strong involvement of the human resources at every level is necessary. Every functional area has to be involved, both in separately way and considering the company as a whole system.

Brunello Cucinelli is a perfect example of innovativeness, as the company is usual to maintain its core business, but using innovative processes to increase the quality and the attractiveness of its product range.

When the combinations mentioned above are not present, it emerged the risk for a limited vision of innovation. The innovation strategy creates a risk to be considered as a limited investments, not involving the company considered as an integral system (Ferrero 1987; Cantino 2007).

In addition, each aspect discussed before, requires a strong connection between creative group and management (Giacosa 2011): on one side, the creative group must have an entrepreneur's individual market's vision, with the purpose to reach attractive products range; on the other, the management group should reach to company's targets in economic-financial context, respecting in the same time consumers' needs and their preferences. When this combination is not present, the company risks its survival in a long-term vision (Bresciani et al. 2013a, b; Tardivo et al. 2011).

Lastly, the innovativeness should not be imposed at all costs, but only at the innovative ones, to provide the company economic benefits and its positive perception by the consumer, giving an incentive to purchase products. If the innovative activity has not commercial results, it means that the survival of the company in the future constitutes the priority, and the company uses the combination of creative activity and commercial venture only to reached that aim (Giacosa 2012). In this situation, lack of the commercial perspective leads to very low or even negative performance of innovative products. Moreover, excessive innovation can results in producing the items which imitate those one developed by the competition, and as a result it can lead to commercial immediate damage (meaning reduction in turnover) and a damage in long period of time (meaning damage on corporate image) caused by careful customers' evaluation. Concluding, an ineffectiveness or the absence of innovativeness could represent a reason for reducing the company competitiveness and, consequently, the sustainable degree of its business.

5 Conclusions, Implications and Limitations

The luxury business is one of the most relevant industry in some countries, such as Italy and France, representing a significant source of growth, development and economics sustainability. The luxury business is characterized by a strong presence of the family business phenomenon.

New paradigms affecting the luxury business (such as the process of democratisation and a market globalisation) require a deepened company's attention in defining its business model. On the one hand, the luxury democratization reduces the exclusivity of the products. Even if the democratization permits a choice in a sense of desires, the company risks to sacrifice its integrity, in the context of historical corporate values and the familiness factor, what leads to deterioration of the company image, its reputation and heritage. On the other hand, the market globalisation has a negative influence on luxury pureness as it causes the transfer of production process to the countries, where the production costs are lower, but frequently the quality level is not so good. It results in deterioration of both, the company's brand image, and quality status of its luxurious products: therefore, each company has to take into consideration the risks of losing its luxury status, and strengthen its country of origin factor.

To face the risks mentioned above, the company needs to apply a sustainable business model. In this research, we elaborated a business model which identifies a series of specific features of innovation strategy of clothing luxury family firms. It represents competitive advantage in terms of clothing luxury fashion and its aim is to ensure sustainable luxury business in terms of economic, social and cultural aspects. A sustainable luxury business respects very important issues, such as environment and social development, cultural art and innovation, and protection of local craftsmanship.

Our model identifies a series of specific features of family firms combination of which could represent the meaning of sustainable development, in terms of economic, social and cultural aspects. The following key drivers have been distinguished:

- Company heritage;
- Long-term vision in the investment attitude;
- Specialization;
- Relationships;
- Innovativeness.

According to the company heritage, the maintenance of its values constitutes the most important factor necessary for developing the sustainable business, even if the fashion collections are characterized by strong variability every season. In fact, every single collection should be coherent with brand identity, reflecting the heritage factor. On the contrary, when the company strategy doesn't respect the brand identity, the products are not consistent with permanent stylistic identity of the company. This strategy generates a deterioration of the brand's prestige, reduces social status of possessing it, and causes non sustainable business in terms of the respect the company's heritage. In particular, in the case of non-affordable luxury, damage of the corporate heritage impacts in negative way on the sense of rarity, reducing the prestige. It means that corporate heritage is a factor having extremely high importance, without which developing of sustainable business is impossible (even despite the fact, that fashion collections shows strong evolution every season).

Secondly, the absence of a long-term vision in their investments strategy creates a hard competition. Indeed, certain productive and commercial investments require several years to be effective. Note the brands, which creation and recognition in the market requires a long time. When the company has not elaborated a long-term vision, it causes difficulties in obtaining effective investment, in both economic and commercial terms. In addition, the absence of a prospect of continuation of the company's activity reduces the family members attachment to the company, weakening the familiness factor. For this reason, a short-term vision of the business generates harms in the brand's valorisation, causing a loss of its prestige. A wrong investments strategy can lead to developing non-sustainable business, meaning damage of both the company's brand as well as the construction of the set of values which luxurious brand should have.

Thirdly, when a company is not in line both with the brand's portfolio and its attitudes and know-how, the actions and investments are focused on inappropriate object. In addition, a low specialization might reduce the brand's appeal. This can lead to dispersion of competencies and concentration on the products in relation to which the company is not competitive.

Fourthly, the absence or the inefficacy in the relationships with external parties constitutes a negative factor for the companies' development and competitiveness. Indeed, exchange of different typologies of resources is not possible, and the company can't operate beyond its boundaries. It reduces the innovation opportunities: consequently, it impacts on each productivity level of the company, on regional development, and on the attractiveness of the products range. Concluding, inefficient relationships with external parties (or their absence) result in deterioration of the sustainable business in the context of brand's attractiveness (notably in case when the brand is strongly related to a country of origin) as well as in terms of sharing of competencies and resources within the cluster.

Lastly, the innovation strategy means opportunity for the companies to improve their competitive advantage. Importance of the innovation strategy is approved in economic and social network and, of course in literature related to business research. In recent years, the innovation strategy has been considered as a way to strengthen the business model, despite the consumption crisis. In other words, ability of attracting foreign resources to invest in innovation strategy should be analysed by researches in order to overcome the crisis situations. The absence or the ineffective innovation strategy doesn't permit to the company and its territory to gain economic benefits and referring to innovative products—suggests positive sensitivity in the perception of the consumers. In addition, when the innovative activity's results are not good, it means that performance of undertaken investments can be really low or even negative. This issue has a negative impact on the sustainability of the business.

Only a combination between key factors previously discussed enables to understand the issue of promoting a sustainable luxury by clothing luxury fashion family firms. Importance of each key driver is equal. In addition, there is a joint force of those factors which permits to strengthen the company's business model, making it more competitive in the international context. It derives from the unity of the company's system: company is defined as a system, while components are interrelated in united manner. Only a marked combination of these components allows the performance of the company's activity. The system components are connected together by intense relationships of interrelations and complementarities: these relations make the company's system is inseparable.

In addition, the company's system functions in time only if the relationships between the components of the company system are maintained. Indeed, the strong relationships between its components favour the continuity of the system, despite the significant internal and external changes of the company. In the luxury context, the seasonal factor is very strong: nevertheless, according to the alternation of the various collections, maintaining the relationships between components of the system is a fundamental condition, without which the company's survival would be threatened.

The study has developed the following theoretical and practical implications:

- The theoretical implications are related to the potential of the proposed model in the clothing luxury fashion business. The main benefits of the business model proposed is by adopting it the strategies of investments can be managed in more efficient end effective way, according to *sustainable luxury*. In particular, the research analyses the role and the connection between several key factors, and their impact on the business model in the analyzed context. Only a combination between the mentioned key factors enable to understand the impact of clothing luxury fashion family business can on promoting a sustainable luxury;
- The practical implications are linked to the luxury context, which represents an interesting area of opportunities for companies in many typologies of sub-context, such as the core luxury sectors and the new luxury sectors. In addition, certain luxury business features reflect a competitive advantage because they identify the optimal investment strategy in a better way, in terms of the management of several factors (such as the company heritage, a long-term vision in the investment attitude, the specialization; the external relationships, and the innovativeness).

In current presentation a qualitative method has been chosen and we focused on innovative approach of luxury companies. It would be interesting to extend the research about the comparison of potential dynamics of the luxury sectors with dynamics of other sectors, which have a special meaning for many countries like for example Italy and France. In addition, the usage of suitable econometric models would be better to make a simulation of the impact of key factors developed by the model on the investment strategy benefits.

References

Aaker DA (1997) Brand equity. La gestione del valore della marca. FrancoAngeli, Milano

Acquaah M (2013) Management control systems, business strategy and performance: a comparative analysis of family and non-family business in a transition economy in sub-Saharan Africa. J Fam Bus Strategy 4:131–146

Adner R, Levinthal D (2001) Technology evolution and demand heterogeneity: implications for product and process innovation. Manage Sci 47(5):611–628

Aldrich HE, Cliff JE (2003) The pervasive effects of family on entrepreneurship: toward a family embeddedness perspective. J Bus Ventur 18(5):573–596

Anderson RC, Reeb DM (2003) Founding family ownership and firm performance: evidence from the S&P 500. J Finan 58(3):1301–1327

Anderson AR, Jack SL, Dodd SD (2005) The role of family members in entrepreneurial networks: beyond the boundaries of the family firm. Fam Bus Rev 18(2):135–154

Aronoff CE (1998) Megatrends in family business. Fam Bus Rev 11(3):181–186

Arregle JL, Hitt MA, Sirmon DG, Very P (2007) The development of organizational social capital: attributes of family firms. J Manage Stud 44(1):72–95

Astrachan JH (2010) Strategy in family business: toward a multidimensional research agenda. J Fam Bus Strategy 1(1):6–14

Astrachan JH, Shanker MC (2003) Family businesses' contribution to the US economy: a closet look. Fam Bus Rev 16(3):211–219

Astrachan JH, Jaskiewicz P (2008) Emotional returns and emotional costs in privately held family businesses: advancing traditional business valuation. Fam Bus Rev 21(2):139–149

Attanzio G (2011) Le imprese familiari resistono meglio alla crisi economica. http://www.impresanews.it (home page). Accessed 4 Aug 2014

Basco R, Rodríguez MJP (2009) Studying the family enterprise holistically evidence for integrated family and business systems. Fam Bus Rev 22(1):82–95

Bertoldi B (2015) Manager di famiglia: come i manager vengono scelti e hanno successo nel capitalismo familiare. Gruppo 24 ore, Milano

Binz C, Hair JF Jr, Pieper TM, Baldauf A (2013) Exploring the effect of distinct family firm reputation on consumers' preferences. J Fam Bus Strategy 4(1):3–11

Block JH (2012) R&D investments in family and founder firms: an agency perspective. J Bus Ventur 27(2):248–265

Bordieu P (1977) Outline of a theory of practice. Cambridge University Press, Cambridge

Bresciani S (2009) Innovation and creativity within firms. An empirical demonstration in the Piedmont area. Int J Bus Innov Res 3(4):427–443

Bresciani S (2010) Innovation within firms. A survey in the Piedmont area. Int J Qual Innov 1 (2):138–152

Bresciani S, Bertoldi B, Giachino C, Ferraris A (2013a) The approach of family businesses in the luxury industry. In: Conference readings book proceedings, in 6th Euromed conference of the Euromed academy of business. Confronting Contemporary Business Challenges through Management Innovation, Estoril, Portugal, 3–4 Oct, pp 423–433

Bresciani S, Vrontis D, Thrassou A (2013b) Change through innovation in family businesses: evidence from an Italian sample. World Rev Entrepreneurship Manag Sustain Dev 9(2): 195–215

Bresciani S, Vrontis D, Thrassou A (2015) The internationalization of R&D activities to Asian developing countries. World Rev Entrepreneurship, Manag Sustain Dev 11(2/3):200–216
Cantino V (2007) Corporate governance, misurazione della performance e compliance del sistema di controllo interno. Giuffrè, Milano
Carcano L, Corbetta G, Minichilli A (2011) Why luxury firms are often family firms? Family identity, symbolic capital and value creation in luxury-related industries. Universia Bus Rev 32:40–52
Carney M (2005) Corporate governance and competitive advantage in family-controlled firms. Entrepreneurship Theory Pract 29(3):249–265
Cefis SE, Marsili O (2005) A matter of life and death: innovation and firm survival. Ind Corp Change 14(6):1167–1192
Chen HL, Hsu WT (2009) Family ownership, board independence, and R&D. Investment Fam Bus Rev 22(4):347–362
Chesbrough H (2003) Open innovation, the new imperative for creating and profiting from technology. Harvard Business School Press, Boston
Chevalier M, Mazzalovo G (2008) Luxury brand management. Franco Angeli, Milano
Chrisman J, Patel PJ (2012) Variations in R&D investments of family and non-family firms: behavioral agency and myopic loss aversion perspectives. Acad Manag J 55(4):976–997
Chua H, Chrisman JJ, Bergiel EB (2009) An agency theoretic analysis of the professionalized family firm. Entrepreneurship Theory Pract 33(2):355–372
Claessens S, Djankov S, Lang LHP (2000) The separation of ownership and control in East Asian corporations. J Financ Econ 58(1):81–112
Cohen WM, Levinthal DA (1990) Absorptive capacity: a new perspective on learning and innovation. Adm Sci Q 35(1):140–152
Coleman S, Carsky M (1999) Sources of capital for small family-owned businesses: evidence from the national survey of small business finances. Fam Bus Rev 12(1):73–85
Colli A (2013) State of the art—family firms: risks and opportunities: the state of the art of the debate. Socio-Econ Rev 11:1–23
Contò F, Vrontis D, Fiore M, Thrassou A (2014) Strengthening regional identities and culture through wine industry cross border collaboration. Br Food J 116(11):1788–1807
Corbellini E, Saviolo S (2009) Managing fashion and luxury companies. Etas, Milano
Craig JBL, Moores K (2006) A 10-year longitudinal investigation of strategy, systems, and environment on innovation in family firms. Fam Bus Rev 19(1):1–10
Crestanello P, Tattara G (2011) Industrial clusters and the governance of the global value chain: the romania-veneto network in footwear and clothing. Reg Stud 45(2):187–203
Culasso F, Broccardo L, Giacosa E, Mazzoleni A (2012a) Corporate governance in listed Italian family firms: impact on performance and comparison with not family firms. J Manag Change 29:67–88
Culasso F, Giacosa E, Mazzoleni A (2012b) Fashion system: the case of Italy. World J Soc Sci 2 (3):209–226
Culasso F, Manzi LM, Giacosa E, Broccardo L (2013) Family Italian listed firms: comparison in performances and identification of two main configurations. In: 6th Euromed conference of the EuroMed academy of business. Confronting contemporary business challenges through management innovation. Conference readings book proceedings, Euromed Press, Nicosia, pp 719–748
Czarnitzki D, Kraft K (2004) Firm leadership and innovative performance: evidence from seven EU countries. Small Bus Econ 22(5):325–332
Davidow T, Narva R (2000) How multi-generational family firms transfer management successful. www.genusresources.com (home page). Accessed 4 Aug 2014
Davis JH (1983) Realizing the potential of the family business. Org Dyn 12(1):47–56
De Chernatony L (2001) From brand vision to brand evaluation strategically building and sustaining brands. Butterworth Heinemann, Oxford
Dicken P (1994) Global-local tensions: firms and states in the global space-economy. Adv Strateg Manag 10:217–247

Dobrzynski JH (1993) Relationship investing. Bus Week 3309:68–75
Donaldson MS (2011) Promoting luxury goods in china through social media. MultiLingual (October/November) 29–31
Donckels R, Frohlich E (1991) Are family businesses really different? European experiences from STRATOS. Fam Bus Rev 4(2):149–160
Dubois B, Czellar S (2002) Prestige brands or luxury brands? An exploratory inquiry on consumer perceptions. In: Conference readings book proceedings, 31st European Marketing Academy, Braga, Portugal, 28–31 May 2002
Dunford M (2006) Industrial districts, magic circles and the restructuring of the Italian textiles and clothing chain. Econ Geogr 82(1):27–59
Dunn B (1996) Family enterprises in the UK: a special sector? Fam Bus Rev 9(2):139–155
Eckrich CJ, Loughead TA (1996) Effects of family business membership and psychological separation on the career development of late adolescents. Fam Bus Rev 9(4):369–386
Fabris G, Minestroni L (2004) Valore e valori della marca. FrancoAngeli, Milano
Faccio M, Lang L (2002) The ultimate ownership of western corporations. J Financ Econ 65 (3):365–395
Ferda E (2010) Family business reputation: a literature review and some research questions electronic. J Fam Bus Stud 4(2):133–146
Ferrero G (1987) Impresa e management. Giuffré, Milano
Flatten TC, Greve GI, Brettel M (2011) Absorptive capacity and firm performance in SMEs: the mediating influence of strategic alliances. Eur Manag Rev 8(3):137–152
Frank H, Lueger M, Nosé L, Suchy D (2010) The concept of 'familiness'. Literature review and systems theory-based reflections. J Fam Bus Strategy 1(3):119–130
Garcìa-Ramos M, Garcìa-Olalla R (2011) Board characteristics and firm performance in public founder and nonfounder-led family business. J Fam Bus Strategy 2(4):220–231
Gardetti MA (2011) Sustainable luxury in Latin America. In: Conference dictated at the seminar sustainable luxury & design within the framework of the MBA of IE Business School, Madrid, Spain
Garzella S (2000) I confini dell'azienda. Giuffrè, Milano
Geerts A, Veg-Sala N (2011) Evidence on internet communication—management strategies for luxury brands. Glob J Bus Res 5(5):81–94
George A, Bennett A (2005) Case studies and theory development in the social sciences. MIT Press, Cambridge
Ghoshal S, Bartlett AC (1990) The multinational corporation as an interorganizational network. Acad Manag Rev 15(4):603–625
Giacosa E (2011) L'economia della aziende di abbigliamento. Giappichelli, Torino
Giacosa E (2012) Mergers and acquisitions (M&As) in the luxury business. McGraw-Hill, Milano
Giacosa E, Giachino C, Bertoldi B, Stupino M (2014a) Innovativeness of Ceretto Aziende Vitivinicole: a first investigation into a wine company. Int Food Agribusiness Manag Rev 17 (4):223–236
Giacosa E, Giovando G, Mazzoleni A (2014b) Wine sector as a driver of growth for the Italian economy. In: 2nd international symposium systems thinking for a sustainable economy advancements in economic and managerial theory and practice conference proceedings, 23–24 Jan, Roma, Italy, pp 1–17
Granovetter M (1992) Problems of explanation in economic sociology. In: Nohria N, Eccles R (eds) Networks and organizations. Harvard Business School Press, Boston
Gulati R, Martin G (1999) Where do networks come from? Am J Sociol 104(5):1439–1493
Gulati R, Nitin N, Akbar Z (2000) Strategic networks. Strateg Manag J 21:203–215
Ha HY, Perks H (2005) Effects of consumer perceptions of brand experience on the web: brand familiarity, satisfaction, and trust. J Consum Behav 4(6):438–452
Habbershon TG, Pistrui J (2002) Enterprising families domain: family-influenced ownership groups in pursuit of trans generational wealth. Fam Bus Rev 15(3):223–238
Hansen MT (1999) The search-transfer problem: the role of weak ties in sharing knowledge across organization subunits. Adm Sci Q 44:82–111

Harvey M, Evans R (1995) Forgotten sources of capital for the family-owned business. Fam Bus Rev 8(3):159–176
Hülsbeck M, Wirsching K (2014) Types of family firm governance and innovation in family firms. In: Conference proceedings, 10th workshop on family firm management research innovation, family firms and economic development, 22–24 May 2014, Bergamo, Italy, pp 1–3
Human SE, Provan KG (1997) An emergent theory of structure and outcomes in small-firm strategic manufacturing networks. Acad Manag J 40(2):368–403
IFERA (2003) Family businesses dominate. Fam Bus Rev 16(4):235–240
Irava WJ, Moores K (2010) Clarifying the strategic advantage of familiness: unbundling its dimensions and highlighting its paradoxes. J Fam Bus Strategy 1(3):131–144
Jackson TB (2004) International retail marketing. Elsevier Butterworth-Heinemann, Oxford
James H (1999) Owner as manager, extended horizons and the family firm. Int J Econ Bus 6 (1):41–55
Kapferer JN (2002) Ce qui va changer les marques. Editions d'Organisation, Paris
Kapferer JN, Bastien V (2009) Luxury strategy. FrancoAngeli, Milano
Kapferer JN, Bastien V (2012) The luxury strategy: break the rules of marketing to build luxury brands. Kogan Page Limited, London
Kellermanns FW, Eddleston KA, Barnett T, Pearson A (2008) An exploratory study of family member characteristics and involvement: effects on entrepreneurial behavior in the family firm. Fam Bus Rev 21(1):1–14
Kim K, Ko E (2012) Do social media marketing activities enhance customer equity? An empirical study of luxury fashion brand. J Bus Res 65(10):1480–1486
Kleanthous A (2011) Simple the best is no longer simple raconteur on sustainable luxury. theraconteurcouk/category/sustainability/sustainable-luxury. Accessed Dec 2012
Ko E, Megehee CM (2012) Fashion marketing of luxury brands: recent research issues and contributions. J Bus Res 65(10):1395–1398
La Porta R, Lopez-de-Silanes L, Shleifer A (1999) Corporate ownership around the world. J Finan 54(2):471–517
Lambrech J (2005) Multigenerational transition in family businesses: a new explanatory model. Fam Bus Rev 18(4):267–365
Lau C, Woodman RC (1995) Understanding organizational change: a schematic perspective. Acad Manag J 38(2):537–554
Le Breton-Miller I, Miller D (2006) Why do some family businesses out-compete? Entrepreneurship Theory Pract 30(6):731–746
Liebeskind JP, Oliver AL, Zucker L, Brewer M (1996) Social networks, learning and flexibility: sourcing scientific knowledge in new biotechnology firms. Organ Sci 7(4):428–443
Lipovetsky G, Roux E (2003) Le luxe éternel De l'âge du sacré au temps des marques. Gallimard, Paris
Lorange P, Roos J (1993) Strategic alliances: formation, implementation, and evolution. Blackwell Business, Cambridge
Mahérault L (2000) The Influence of going public on investment policy: an empirical study of French family-owned businesses. Fam Bus Rev 13(1):71–79
Matzler K, Veider V, Hautz J, Stadler C (2014) The impact of family ownership, management and governance on innovation journal of product innovation management, Special issue on management research on governance, ownership and innovation: opening up the agenda to family firms forthcoming. Available at SSRN: http://ssrncom/abstract=2414351
McKeown T (2004) Case studies and the limits of the quantitative worldview. In: Brady H, Collier D (eds) Rethinking social inquiry. Lanham, Rowman and Littlefield
Mediobanca (2013) Dati cumulativi di 2035 società italiane. Mediobanca, Milano
Micelotta ER, Raynard M (2011) Concealing or revealing the family? Corporate brand identity strategies in family firms. Fam Bus Rev 20(10):1–20
Miles G, Preece SB, Baetz MC (1999) Dangers of dependence: the impact of strategic alliance use by small technology-based firms. J Small Bus Manage 37(2):20–29

Miller D, Le Breton-Miller I (2003) Challenge versus advantage in family business. Strateg Organ 1(1):127–134
Miller D, Le Breton-Miller I (2005) Managing for the long run: lessons in competitive advantage from great family businesses. Harvard Business Press, Boston
Miller KW, Mills MK (2012) Contributing clarity by examining brand luxury in the fashion market. J Bus Res 65(10):1471–1479
Minichilli A, Corbetta G, MacMillan IC (2010) Top management teams in family controlled companies: familiness, faultlines, and their impact on financial performance. J Manage Stud 47 (2):205–222
Molina-Morales FX, Martínez-Fernández MT (2010) Social networks: effects of social capital on firm innovation. J Small Bus Manage 48(2):258–279
Moody W, Kinderman P, Sinha P (2010) An exploratory study: relationships between trying on clothing, mood, emotion, personality and clothing preference. J Fashion Mark Manag 14 (1):161–179
Morck R, Yeung B (2004) Family control and the rent seeking society. Entrepreneurship Theor Pract 28(4):391–409
Morris MH (1998) Entrepreneurial intensity. Quorum Books, Westport
Munoz-Bullon F, Sanchez-Bueno MJ (2011) The impact of family involvement on the R&D intensity of publicly traded firms. Fam Bus Rev 24(1):62–70
Naldi L, Nordqvist M, Sjoberg K, Wiklund J (2007) Entrepreneurial orientation, risk taking, and performance in family firms. Fam Bus Rev 20(1):33–47
Narva R (2001) Heritage and tradition in family business: how family-conuolled enterprises connect the experience of their past to the promise of their future. In: McCann G, Upton N (eds) Demoying myths and creating value infamily business. Stetson University, Deland
Narva L (2010) Heritage and tradition in family business: how family-controlled enterprises connect the experience of their past to the promise of their future. Narva & Company, Newtonville
Neumann M (1997) The rise and fall of the wealth of nations: long waves in economics and international politics. Edward Elgar, Cheltenham and Lyme
Nieto MJ, Santamaría L (2010) Technological collaboration: bridging the innovation gap between small and large firms. J Small Bus Manage 48(1):44–69
O'Donnell A, Gilmore A, Carson D, Cummins D (2002) Competitive advantage in small to medium sized enterprises. J Strateg Mark 10(3):205–223
Okonkwo U (2007) Luxury fashion branding. Palgrave Macmillan, New York
Okonkwo U (2009) The luxury brand strategy challenge. J Brand Manag 16(5/6):287–289
Osservatorio AIDAF-Unicredit, Bocconi (AUB) (2011) Luxury and family business in Italia. http://www.aidafit/wp-content/uploads/2014/08/3%25C2%25B0-Osservatorio-AUB-2011pdf. Accessed 2 Sept 2014
Perricone PJ, Earle JR, Taplin IM (2011) Patterns of succession and continuity in family-owned businesses: study of an ethnic community. Fam Bus Rev 14(2):105–121
Poza EJ, Alfred T, Maheshwari A (1997) Stakeholder perceptions of culture and management practices in family and family firms: a preliminary report. Fam Bus Rev 10(2):135–156
Re P (2013) La gestione dell'innovazione nelle aziende familiari. Giappichelli, Torino
Rosenkopf L, Almeida P (2003) Overcoming local search through alliances and mobility. Manage Sci 49(6):751–766
Ross J, Harradine R (2011) Fashion value brands: the relationship between identity and image. J Fashion Mark Manag 15(3):306–325
Schilling M (2009) Strategic management of technological innovation. McGraw Hill Higher Education, Boston
Schilling MA, Izzo F (2013) Gestione dell'innovazione. McGraw-Hill, Milano
Shipman A (2004) Lauding the leisure class: symbolic content and conspicuous consumption. Rev Soc Econ 62(3):277–289
Silverstein MJ, Fiske N (2003a) Luxury for the masses. Harvard Bus Rev 81(4):48–57

Silverstein MJ, Fiske N (2003b) Trading up: why consumers want new luxury goods and how companies create them. Portfolio Penguin, Group New York
Sirmon DG, Hitt MA (2003) Managing resources: linking unique resources, management, and wealth creation in family firms. Entrepreneurship: Theor Pract 27(4):339–358
Sirmon DG, Arregle JL, Hitt MA, Webb JW (2008) The role of family influence in firms' strategic responses to threat of imitation. Entrepreneurship Theor Pract 32(6):979–998
Sood A, Tellis GJ (2009) Do innovations really payoff? Total stock market returns to innovation. Mark Sci 28(3):442–456
Tardivo G, Bresciani S, Fabris F (2011) Internal dealing and insider trading: focus on financial market in Italy research findings. J Fin Manag Anal 24(1):24–38
Teece DJ (1992) Competition, cooperation, and innovation: organizational arrangements for regimes of rapid technological progress. J Econ Behav Organ 18(1):1–25
Terzani S (2007) Controllo di gestione nelle imprese di alta moda. FrancoAngeli, Milano
Thomas D (2007) Deluxe—how luxury lost its luster. Penguin Books, New York
Uotila J, Maula M, Keil T, Zahra SA (2009) Exploration, exploitation, and financial performance: analysis of S&P 500 corporations. Strateg Manag J 30(2):221–231
Vecchi A (2008) Globalisation and the viability of industrial districts. Verlag Dr Muller, Saarbrucken
Viassone M (2009) What makes Piedmont catching up. In: 2nd annual EuroMed conference of the EuroMed academy of business conference readings book proceedings, EuroMed Press, Nicosia, pp 1825–1837
Vrontis D, Viassone M (2013) When wine meets territory: the Italian scenario. In: 6th Euromed conference of the EuroMed academy of business. Confronting contemporary business challenges through management innovation. Conference readings book proceedings, EuroMed Press, Nicosia, pp 2457–2468
Walker G, Kogut B, Shan W (1997) Social capital, structural holes and the formation of an industry network. Organ Sci 8(2):109–125
Walters LH (1982) Are families different from other groups? J Marriage Fam 44(4):841–850
Ward JL (1997) Growing the family business: special challenges and best practices. Fam Bus Rev 10(4):323–337
Yin RK (2009) Study research: design and methods, 4th edn. Sage Publications, Thousand Oaks
Zahra SA (2005) Entrepreneurial risk taking in family firms. Fam Bus Rev 18(1):23–40
Zahra SA, Hayton JC, Salvato C (2004) Entrepreneurship in family vs non-family firms: a resource based analysis of the effect of organizational culture. Entrepreneurship Theor Pract 28(4):363–381
Zahra SA, Hayton JC, Neubaum DO, Dibrell C, Craig JBL (2008) Culture of family commitment and strategic flexibility: the moderating effect of stewardship. Entrepreneurship Theor Pract 32 (6):1035–1054
Zander U, Kogut B (1995) Knowledge and the speed of transfer and imitation of organisational capabilities: an empirical test. Organ Sci 6(1):76–92
Zellweger T (2007) Time horizon, costs of equity capital, and generic investment strategies of firms. Fam Bus Rev 20(1):1–15
Zellweger TM, Astrachan JH (2008) On the emotional value of owning a firm. Fam Bus Rev 21 (4):347–363
Zellweger TM, Sieger P (2012) Entrepreneurial orientation in long-lived family firms. Small Bus Econ 38(1):67–84
Zellweger T, Nason R, Nordqvist M (2012) From longevity of firms to transgenerational entrepreneurship of families: introducing family entrepreneurial orientation. Fam Bus Rev 25 (2):136–155

Craft's Path to the Luxury Market: Sustaining Cultures and Communities Along the Way

Jana M. Hawley and Judy Frater

Abstract This chapter focuses on the value and place of traditional craft in the luxury market. We argue that craft embodies a unique combination of skill with a deeper expression of human values that transcend culture, time, and space. While couture designers have often used craft artisans to produce their lines, we argue that the craftsmen themselves can also have a pathway to the luxury market, thus contributing to a shorter and more sustainable footprint. If luxury is defined as something that is of high quality, limited quantity, creative, and rater—then indeed, traditional craft qualifies.

Keywords Sustainable luxury · Artisans · Craftsman · Global markets

1 Introduction

Traditional craft embodies a unique blending of skill with a deeper expression of human values that transcend culture, time, and space. Each work of traditional craft is by its nature and heritage unique, evoking the bespoke. If luxury is defined as something that is of high quality, limited quantity, creative and rare—then indeed, traditional craft qualifies. While couture designers have often used traditional craftsmen to develop their lines, we argue that the artisans themselves could also have a pathway to the luxury market, thus contributing to sustainability. We explore the value and place of traditional craft and what is required for traditional craft and artisans to find their pathway to sustainable high-end markets and perhaps even luxury markets.

J.M. Hawley (✉)
University of Arizona, Tucson, USA
e-mail: hawleyj@email.arizona.edu

J. Frater
Somaiya Kala Vidya, Kutch, India
e-mail: judyfrater@hotmail.com

M.A. Gardetti (ed.), *Sustainable Management of Luxury*,
Environmental Footprints and Eco-design of Products and Processes,
DOI 10.1007/978-981-10-2917-2_18

The definition of luxury is ever-changing, evolving now to encompass tenets of sustainability. In Professor Montgomery's essay on *Craft the new luxury* (2016) he outlines four indicators that predict that craft and luxury will converge. When the luxury market became more and more democratized, it began to lose its exclusivity, an essential aspect of luxury. Crafts, those items made by human hands, are: (1) unique (2) authentic to the maker, (3) sustainable to the maker and community, and (4) tactile thus providing intimacy of touch. Exceptional craftsmanship for the luxury consumer is a compilation of time, skill, passion, and ideas. Luxury consumers want to distinguish themselves as people of taste and style and can accomplish this with craft products that are instilled with the human traits of personality, knowledge, traditions, and memory. Traditional crafts can capitalize the bespoke element and bring exclusivity in a new form.

Consumers of luxury products have often been seen as connoisseurs, that is someone with an educated eye who can pass critical judgment. Connoisseurs typically know a lot about the object and are able to pass judgment in the matter of taste or relevancy. Connoisseurship, then, is a way of thinking or an attitude toward an object. Those items that are handcrafted, small-batch, or artisanal are often celebrated examples of connoisseurship. Just as wine connoisseurs pass judgment on tasting notes, depth of color, provenance, aroma, or varietal, connoisseurs of craft can recognize the nuances of handcrafted quality, the region from where it derives, and its authenticity to the culture. In addition, connoisseurs of craft products develop a relationship with the person who made the item, much like the relationship between a bespoke tailor and the client. The result is a changed landscape in the luxury market, moving the consumption of luxury goods away from the relationship between the *consumer* and the *brand* toward a relationship between the *consumer* and the *maker*. The authentication of the product becomes a primary indicator of the relationship. As this occurs, master artisans will begin to establish their place in the luxury market.

To make our point, we use the International Folk Art Market (IFAM) of Santa Fe New Mexico, USA as an excellent example of the process by which traditional artisans can reach higher end markets. Through the IFAM processes and events, connoisseurs of traditional craft are being cultivated. We further illustrate our point by using a case study of traditional artisans of Kutch, India, which provides an illustration of how this path to a sustainable luxury market can be developed.

2 Background

2.1 The Idea of Luxury

For centuries, people worldwide have satisfied themselves with the possession of beautiful things. Traditionally, luxury goods have been defined as "goods for which the mere use or display of a particular branded product brings prestige to the owner" (Husic and Cicic 2007). In 1994, Dubois and Laurent (1994) defined luxury as a specific tier of offer in almost any product or service category. Okonkwo (2007) notes that luxury products are not just over-priced goods, but rather offer a complete package of benefits to consumers that appeals to their social and psychological well-being (p. 2). Kapferer (2010) described luxury as extra pleasure derived from sensory satisfaction. Webster's New World dictionary defined it as the use and enjoyment of the best and the most costly things that offer the most physical comfort and satisfaction (2013). Spiegel (2015) said that it is a personal experience of pleasure outside of daily routine.

But we argue that luxury always requires a context. Luxury is in the eyes of the beholder; today, it is more about the consumer's experience or the personal dimensions of the luxury. Luxury is about aspiration, and luxury is niche. Luxury products are distinguished from premium products in their balance of material value, price, quality, symbolic value, and scarcity (Sinha and Som 2016). Two key components of luxury are thus craftsmanship and exclusivity. Singh (2015, April 28) claims that luxury can be "defined by craftsmanship and design...skill and investment of time" and shows how craftsmanship is an intrinsic part of modern luxury. No doubt, luxury is also a thriving industry that has global influence. While concepts such as prestige, premium, exclusivity, or high quality are all hallmarks of luxury, Singh states that luxury is a socially oriented practice of consumption and can be demonstrated in products that require an extraordinary investment and skill in craftsmanship.

Rarity illuminates the notion of desire; it validates the idea that something is special and somehow worthy of the term luxury (Sinha and Som 2016). *Bespoke luxury* conjures up ideas of patience, one-of-a-kind, or custom-made. In other words, making products that the customer wants—not what the company wants to sell. The term stems from creating something truly unique for an individual, something that will not be duplicated. Bespoke is handmade by highly skilled craftsmen using traditional methods and tools. As a result, the products can command high prices that make them unattainable by many consumers. As the master artisan pool becomes scarce, the products become even more rare, commanding even higher prices.

Almond (2011) notes that Vivienne Westwood and Alexander McQueen pushed craft traditions of bespoke to the forefront of their luxury collections, which "in turn provoked a resurgence of interest in the craft" (p. 2). Consider *Quoddy*, a company based in Maine, USA that makes one-of-a-kind shoes. The company takes pride in crafting handmade shoes that are customized throughout. Or consider *London Connoisseur,* a company that strives to bridge the gap between superb artisanship and customers who seek extraordinary products. Shih and Agraflotis (2016) write about the craft revival that is occurring in the luxury market where apprentices are clamoring to learn skills and trade. They also point out that some luxury brands have showcased their craft production to the public in order to raise awareness of the highly skilled artisans. As an example, in 2009, Oscar de la Renta produced a video that showcased his collection using ikat from Uzbekistan artisans (https://www.youtube.com/watch?v=xEiWPlgEnkE).

2.2 Sustainability and Luxury

In this paper, we are concerned not only with understanding who the luxury customer is, but more importantly, what exemplifies sustainable luxury products and how we can assure that the production of these products contributes positively to the lives of the producers and their communities.

As the meaning of luxury continues to evolve, Maria Girón claims that "sustainability [is] one of the new meanings of luxury for savvy, well-educated and demanding luxury clients" (2015, p vii). Furthermore, in the *Deeper Luxury Report*, the World Wildlife Fund (WWF-UK, 2007) noted that the luxury consumer is a part of the "affluent and global elite that is increasingly well educated and concerned about social and environmental issues" (p. 2). The report goes on to say that the trustworthiness of luxury brands will be closely connected to their ability to generate wellbeing for not only the customer but also those involved with the production, use, reuse, and disposal, and that customers are ready to pay more for responsibly produced goods.

Research indicates that there are significant shifts in consumer demand towards products that are value-centric, honoring the maker and environmentally friendly processes in which the products are made (Hawley 2015). Since the rise of consciousness toward sustainability, the luxury sector has come under enormous scrutiny (Bendell and Kleanthouse 2007). It is even declared that sustainability and luxury are antagonistic concepts. But after the recession of 2008, customer awareness of and concern for social, ethical, and environmental issues began to rise and reinforced their desire to purchase lasting and meaningful products as a new form of luxury consumption. Thus consumption patterns, production methods, and social consciousness find themselves in a disruptive state. TIME Magazine called this the "responsibility revolution' and reported that customers purchased products because they liked the social or political values that underpinned the product (Stengel 2009).

Today, the most oft-cited definition of sustainability stems from what is commonly referred to as the "Brundtland Report" which actually defines sustainable development, not sustainability. The report states that sustainable development "meets the needs of the present without compromising the ability of future generations to meet their own needs" (United Nations 1987). This report emphasizes the continuing need for economic growth, but indicates limits must be set and advantage must be taken of technological innovation to ensure a minimization of damage (Jones 2014). Elkington is considered the major contributor to the concept of Triple Bottom Line (Elkington 1998). He places emphasis on social, environmental, and economic impacts of industries and businesses.

As consumers are giving more attention to social and environmental impacts of their purchase decisions, the luxury sector has begun to understand their role in sustainability and strategically uses it as a way to maintain their niche in the marketplace. Jenkins (2012) defined sustainability as the capacity to endure, reduced impact on the earth, stewardship, ethical consumption, and innovations to reduce consumption. She cited several small luxury brands such as Viridus Luxe, Pachacuti, Etrala Londa, and Theiry Mugler who are aligning more closely with the concept of sustainability as they become pro-active in reducing their environmental impact. Gardetti and Muthu (2015) point out that the luxury icons are turning to sustainability using strategies such as artisan manufacturing, quality materials, and attention to the environment.

A young Los Angeles designer, Christina Kim, attracts A-list admirers such as Nicole Kidman. Kim uses Jamdani saris from Bengal in her designs. *Jamdani*, a fine handwoven muslin features thicker cotton threads to create discontinuous supplementary weft patterning that produces the illusion of a suspended pattern on the surface of an almost transparent fabric. In describing her jamdani line, Kim fully capitalizes the romantic mystique of the fabric to increase its value and allure. For Kim, every piece is precious. She uses fair-labor practices and preference for long-lasting garments made with recycled scraps and natural dyes with a result that resembles a sustainable version of Westwood and McQueen's bespoke creations, and have placed her in demand not only as a designer but also as a thinker. Time magazine named her one of its innovators of the year in 2003, and in 2009 she was a panelist on a UCLA Hammer program titled "Conscientious Consumption: The Future of Luxury in an Age of Sustainability" (Johnson 2009). For Kim, and other designers that embrace sustainability, the process of bringing ethics and environment into the luxury market is a conscious one.

Luxury brands have the power to influence consumer aspiration and behaviour by editing consumer choices through product design, distribution and marketing; and by influencing how, when and for how long consumers use their products. They have both the opportunity and the responsibility to promote sustainable consumption. Bendell and Thomas (2013) called this "elegant disruption" and pointed out that luxury companies are well-positioned to use disruptive strategies to reframe how we define luxury calling on heritage and longevity as key factors that can make social change in the industry. Lehman (2015) said that this "elegant disruption" comes from design thinking, customer experience, and then displacement of the status quo. He went on to say that small steps toward change can win new customers while at the same not alienate the current customer.

2.3 Craft and Cultural Consumption

The word *craft* means to make or fashion with skill, especially by hand (Hanks 1979) and includes a wide variety of processes including weaving, block printing, jewelry making, basketry, and so on. Craft has also been defined as a product made and designed by the same person (Harrod 1995) with the craft producer being someone who has control over the processes involved in the manufacture of the good. Therefore, the artisan becomes the one who invests his or her self into the product so that the end-product is imbued with authentic aspects of humanness (Campbell 2005).

In McIntyre's (2010) survey of the craft consumer, he found that the craft consumer associates craft with other categories such as design, luxury, or art, but that s/he purchases craft because it suggests handmade, authenticity, workmanship, and genuineness. Compared to the other associative categories, craft was least likely to be mass-produced. Craft consumers felt that luxury items were high quality, sophisticated, expensive, and connoting status, but not necessarily more desirable than craft. Crafts are often seen as something that will have increased value over time. Consumers purchased craft because it makes a unique gift, they admire the skill involved, it fulfills a need for aesthetic beauty, and it provides an opportunity to own a unique object (McIntyre 2010, p. 26). Craft consumers report that they purchase so they "have something nobody else owns" or because they are "drawn to the story behind the object or the ideas it represents" (p. 26). McIntyre concludes that the craft market would be well served to target consumers of luxury goods who are looking for new ways to signal connoisseurship. Market opportunities for craft lie in trends that indicate interest in personalization, demand for authenticity, or the desire to support ethical consumption.

2.4 Traditional Craftsmanship

Mcintyre's study is based on contemporary studio craft in England. Traditional craft is another genre of hand made. While Risatti (2007) comments that craft objects remain strikingly similar worldwide in terms of function, form, technique, and material, traditional craft is a distinct expression of cultural heritage and identity, created within a specific cultural context, often in rural, remote regions of the world. It is part of a long-held shared way of life.

The result is that traditional craft serves to meet the social, cultural and increasingly economic needs of emerging economies, but it can also serve to bridge cultures. Traditional crafts are often collected by those who have had the luxury of traveling the world, as examples of unique, extraordinary art and also as remembrances of experiences of other cultures. Examples include carpets from Nepal, ebony inlaid silver jewelry from Niger, or intricate bandhani from India. For example, a lifelong collector of textiles from Colorado, USA has acquired a collection of museum quality textiles from her global travels that honors the textile traditions of other cultures, links to history, showcases technical ingenuity, and provides evidence of connection to the person who masterminded the impressive object. Her collection ranges from rich colors to intricate patterns and reflects both her travels and exquisite taste.

Traditional craft has not been thought of as luxury. But as ideas of luxury are changing and luxury brands have become more accessible, unique objects of cultural heritage can provide an ethical, ecological version of the exclusivity that is a key attribute of luxury. In his article on Elegant Disruption, Bendell (2012) says "disruption may not arise from a direct transfer of market share, but by influencing the way a whole category of products is understood, and thereby challenging existing stories of value" (p 22). We ask, then, can traditional craft position itself as sustainable luxury?

To make this possible, challenges must be identified and met. Traditional craft is created for a local market. To reach the luxury market, it must have excellence in quality, and it must appeal to a sophisticated taste. It must evolve from home made to hand made. It must capitalize its aspect of exclusivity and experience for the consumer. It must also be given a pathway or access to the luxury market.

2.5 The International Folk Art Alliance: The Creation of an Elegant Disruptor

The International Folk Art Alliance (IFAA—sometimes referred to as IFAM) was established in 2004 as a way to share and celebrate traditional crafts of the world. Sponsors and partners include, among others, UNESCO, the Museum of New Mexico Foundation, and the State of New Mexico Department of Cultural Affairs. A primary goal of IFAA is to provide opportunities for folk artists to have access to the world marketplace and at the same time address the economic and social needs of their communities. IFAA "envisions a world that values the dignity and humanity of the handmade, honors timeless cultural traditions, and supports the work of artists serving as entrepreneurs and catalysts for positive social change" (https://www.folkartalliance.org/about/who-we-are/).

In 2015, the International Folk Art Market (IFAM) was named the best art festival in the United States (http://www.10best.com/awards/travel/best-art-festival/). In its thirteenth year, IFAM has proven to make economic impact in both the local Santa Fe economy, and in the lives of traditional artisans from around the globe. With more than 600 applicants, approximately 170 finalists represent master crafters and artisans who are selected based on the quality of their work by a careful jury process that focuses on quality, authenticity, and variety. Because of this strict jury process, the market serves as a gatekeeper to some of the finest crafts from developing countries. Unlike other global markets of craft, the International Folk Art Market gives close attention to the human impact factor, including relationships that develop between the artisans, the volunteers, and the customers. A primary goal of the market organizers is to assure artisans' success so that they can become a "catalyst for positive social change in their communities by improving employment, healthcare, education, and well-being." (https://www.folkartalliance.org/events-programs/international-folk-art-market-santa-fe/, Accessed March 15, 2016).

When the market first started, the intent was to preserve tradition, make impact, and support social change. But since its inception, the market has bolstered its selection process so that only carefully juried products reach buyers. While some of the artists are time-honored traditional masters, others are young creators reinterpreting the traditional crafts. The result is traditional folk art that ranges from the "highly affordable to museum-quality masterworks" (Hertel 2016). Today's market has slowly evolved into a conscious consideration of craft for the luxury consumer while at the same time preserving tradition and making social impact in the lives and communities of the artists (Schrandt, April 17, 2015).

The artist selection committee evaluates global applications based on quality, traditionalism, aesthetic, production process, artistic knowledge, and the role or use of the art form in the artist's community. Each year, special attention is given to applications from new artists. The craft must be an expression of traditional cultures, rooted in tradition, and produced by individuals whose creative skills convey cultural identity. Each year the number of applications increases while at the same time the jury process becomes more selective. This increasingly rigorous process assures that the entries can meet expectations that the marketplace demands and ultimately so that the artisans are truly on their pathway to the luxury market.

Artists come from more than fifty-seven countries including Bangladesh, China, India, Uzbekistan, or Niger. In 2015, over 18,000 people attended the market with sales of over $2.9 M. Ninety percent of the proceeds travel back to the home countries with the artists. Because many of the artists represent cooperatives in their home countries, the economic impact is spread to more than 25,000 artists and their extended families (IFAA Impact Report 2015).

While craft is the second largest income in the developing world (Snell 2015), most artisans lack sophisticated business skills to interact in a global marketplace. IFAM has developed *Mentor-to-Market,* a workshop held immediately prior to the two and a half day Market to help artisans develop business and marketing skills. The two-day Mentor to Market (M2M) program initiates understanding of global markets and prepares artisans to maximize their experience of IFAM. The ultimate goal of M2M is to provide economic security for the artisans, their families, and communities. Artists learn how to price their products, merchandise their selling space, build assets, and improve selling techniques. Artisans are encouraged to "tell their story" to the customers. The stories told by many of the artists are often heartfelt stories about struggles in the developing world. These extraordinary stories help build the connection between the artisan and their customer.

Friday night of the market is limited to special ticket holders, or VIP guests. This event was developed to instigate connoisseurship where VIP guests have an exclusive preview of products in a party-like atmosphere that includes wine, food, and music. These consumers have first access to the best of the market and often purchase showcase pieces. Local, national, and international celebrities are among the shoppers, but there are also many shoppers who simply seek some of the best master crafts in the world. Observing the Friday night consumers affirms McIntyre's (2010) description of the craft consumer: primarily female, educated, culturally active, curious, have a general knowledge of the objects, and believe that arts and culture improve quality of life for all. It could be argued that these consumers are collectors of high-end objects who by purchasing the showcase pieces provide a pathway for traditional artisans to the luxury market.

2.6 Changing Times, Changing Consumers

At the other end of the world, in developing countries people of simple, subsistence economies—people like the artisans who produce Kim's jamdani—have for generations enjoyed a surprisingly similar luxury in their traditional craft. The creators and users of traditional craft are intimately connected. Work is thus created in a "bespoke" environment. Traditional craft had both quality and a personalized element. Alimohamed Isha, a bandhani artist from Bhuj, Kutch, explains the connection:

> We give life to fabric, using a gift of God. The maker has to understand the people for whom s/he is making. Traditionally there was a relationship between maker and user. What is made in such a relationship will be excellent because of that relationship. The sensibility was, 'I'll make you something that no one else can make.' Traditional work was for earning love (From an interview with J. Frater 2008).

Also Interviewed in 2008, Vishramji Valji, a master weaver from Bhujodi, Kutch, agreed, "Traditionally, the feeling was that we knew that as the blanket we wove slowly wore away, the user would remember us." In 2016, his son Shyamji Vishramji echoed, "Then, we knew who would wear what we wove, and the wearer knew the weaver."

From the early 1960s, with inflation and the influx of cheaper industrially produced products, traditional clients of rural India began to prefer plastics, synthetics and mill made fabrics to hand craft. At the same time, the concept of design as an entity came to India. There is no word in Indian languages for design as separate from the creating process. In the 1950s, as India began nation building, balancing traditions with modern technology and ideas, it focused on rapid industrialization. Under the advice of Pupul Jayakar, a noted writer on Indian craft traditions and the founder of the Indian Handlooms and Handicrafts Export Council (HHEC), and others, the Indian Government considered establishing an institute of design (Jayakar 2011).

Jayakar met the renowned American designer Charles Eames. In 1957 the Indian Government asked the Ford Foundation to invite Charles and Ray Eames to visit India, to meet writers, craftspeople, architects, scientists, industrialists, educators and philosophers. On April 7, 1958, they presented the India Report to the Indian Government. The Report defined the beginning of design education in India. It recommended linking learning with experience, and suggested that the designer could be a bridge between tradition and modernity. In September 1961, based on the India Report, the Indian Government established the National Institute of Design in Ahmedabad (Eames and Eames 1958).

Designers came to the fore, and within the decade they and buyers began to approach artisans and give them designs to produce for commercial markets. Artisans did not know the new commercial markets and they welcomed direction as well as the cash income. Having worked in barter economies, they did not know how to value their work, and usually equated it with other forms of manual labor, such as agricultural labor or construction.

Vishramji Valji Siju, a master weaver with one of the most well known weaving businesses in Kutch recalls the beginning of design in Bhujodi.

In 1965 Prabha Shah and Malti Jhaveri, two sisters from Mumbai, came to Kutch. They ran an organization called Sohan in Mumbai. God sent Prabhaben to us. The time was such that we had no sales. Traditional clients didn't want our work anymore, so when we would take it to their villages to sell, we would bring a lot of it back.

Prabhaben was intelligent. She had the idea to take new designs to new places. She sat with us. She stayed with us for a week at a time. She had a feeling for Kutch. Prabhaben gave us new designs. She took inspiration from our traditions, and remixed them. The first thing she did was a Rabari ludi shawl. We enjoyed making these new products. Her designs were simple and good, and sold instantly. They are designs that we still aren't tired of looking at today.

Sohan worked with a limited number of artisans in Kutch from 1965 until 1975. In 1975, just after Vishramji won his President's award for craftsmanship, Brij Bhasin launched the Gujarat government emporium, Gurjari. Bhasin gave work to all artisans of Kutch, anyone who wanted work. There was a drought and the idea was that artisans should not have to dig ditches. They did craft work and Gurjari purchased it all (personal communication with Siju, March 25, 2016). The majority of businesses and agencies working with artisans aimed for the middle class market, and used production goals of scaling up: making products faster, cheaper and more standard, ironically emulating the industries that had displaced hand craft. Inexpensive craft became widely available.

The Indian designer Ritu Kumar was one of the first to understand the greater value of craft. "The artisans' skills were as fine as ever," she observed. "But they no longer had access to commensurate materials." (personal communication with Judy Frater 1994). In 1969, re-introducing quality silk and metallic yarns, she launched high quality formal wear for the Indian elite. Her company was built on the patronage of craftspeople, and a gloriously rich Indian aesthetic. Kumar's understanding of ancient designs and the innovative use of traditional crafts created a new classicism. She demonstrated that hand-made products can be as profitable and even more glamorous than those made by machine. The Ritu Kumar brand is patronized by Indian stars including Aishwarya Rai, Priyanka Chopra, Lara Dutta, Dia Mirza and international celebrities such as the late Princess Diana, Mischa Barton, Anoushka Shankar.

Today many other Indian designers including Sabyasachi Mukherjee, Abu Jani and Sandeep Khosla, Abraham and Thakore, Rajesh Pratap Singh, Muzaffar Ali, and Sanjay Garg use hand work for their luxury garments. However, neither businesses and agencies aiming for the middle class market, nor haute couture designers engage the full potential of artisans while working with craft. Steeped in the new, western concept of design as a discrete entity, and wanting a quick, insured route to sophisticated taste, businesses and designers employed artisans as technicians to execute their designs and paid them minimum wages. In commercializing craft, the creative role of the artisan and the personal quality that were integral to the identity and value of traditional craft were overlooked. Art became commodity; craftsmanship became labor.

Many good artisans left their traditional livelihood simply because it could not provide an equitable, sustainable income. After the massive earthquake in Kutch in 2001, industry was aggressively encouraged in the region as a means of development. Artisans of ordinary economic means found that industry paid equal or better wages than craft could. In Bhujodi, a village that enjoys an exceptionally robust market as a weaving center, hand looms have declined from 250 prior to 2001, to 70 today. In villages which do not have developed markets, such as Rampar Vekra, hand looms have dwindled from 30 to 5.

2.7 Re-valuing Craft Traditions

Having worked with textile artisans of Kutch since 1970, Judy Frater, an American anthropologist, sought to reactivate the creative potential of artisans, with the hope of increasing the value of hand craft and sustaining traditions more genuinely and holistically. In 2003 she received an Ashoka Fellowship to develop a curriculum of design education for traditional artisans. The curriculum focuses on acquiring knowledge and skills that will enable artisans to *use* design and business effectively in their work, in order to successfully reach appropriate new markets, while at the same time strengthening traditional identity. It aims to develop a new approach to education based on existing traditions. The core concept is that tradition is more than technique; it comprises concept and knowledge as well. The method of the course is to work within traditions, by understanding and drawing from their salient features.

The curriculum's stated objectives are:

- to enable artisans to significantly improve their standard of living—socio—culturally as well as economically;
- to increase artisans' confidence in knowing what they know;
- to strengthen the vitality and raise the value of crafts.

Frater's design education program was initiated at Kala Raksha Vidhyalaya in 2005. Currently it operates as Somaiya Kala Vidya, in Adipur, Kutch.

The year-long course is open to working traditional artisans of Kutch, the premise being that the participants have prerequisite knowledge and skills and can learn design in a year. Its strength has rested in great part on sustained input over a year, and local orientation: teaching in the vernacular language and drawing from local cultural concepts. With focus on effectively reaching appropriate new markets., a post-graduate course in business and management for artisans was launched at Somaiya Kala Vidya in 2014. Both courses employ professional educators as visiting faculty, and are taught as practically as possible, with immediate application of theory in hands-on activities including the crafts themselves. Between intensive residential modules, artisans return to their homes to implement their learning in their crafts. Local faculty visit them and guide in execution of homework.

An important objective of both design and business education is strengthening traditional identity, which is understood as evolving rather than static, and is recognized as having important potential as an edge in the contemporary market. The institute is advised by a board of master artisans. Courses begin by teaching artisans to understand and value their traditions, and then to draw inspiration from and innovate within those traditions. Each artisan is encouraged to find his or her unique individual response to the contemporary world. This approach intended to mitigate the counterproductive competition of artisans producing the same products for the same markets. By diversifying the market, it was hoped that a larger number of artisans could earn more equitable income.

A decade of design education has clearly demonstrated success in connecting graduate artisans to new markets and increasing their incomes. About 75% of male graduates say they have increased their income 10–300%. Women increased their income from 10% to six times. All graduates say they have increased their creative capacity. Graduates have traveled to the USA, Europe, Indonesia and China, taught workshops and courses, and spoken in conferences. Recently, three graduates' work was exhibited in the contemporary design section—along with that of Sabyasachi Mukherjee and others—of the major exhibition "The Fabric of India" at the Victoria and Albert Museum in London. Graduates have become a new genre: *Artisan Designers*, and their work has helped to expand the market for craft. As hoped, education has increased diversity. In a decade, there has been virtually no duplication in collections the artisans develop. Distinctive stylistic interpretations have in fact led to more artisans earning more equitably. As one small-scale artisan noted, "My income has increased ten times, while the long-time major producer's income has not suffered at all. It is a win-win situation!" Graduates have formed lasting networks of communication and support. Children of traditional artisans who once wanted to do anything but craft are now motivated to become artisan designers. The number of looms in Bhujodi is reported to be increasing again as of 2015.

The impact of education is evident in the thinking as well as products of graduates. As artisan designers report:

> I have woven all my life. What I have done with my weaving is what is important.
>
> Pachan Premji Siju, weaver 2015.
>
> I learned to bring what is inside outside. I don't have to see what everyone is doing.
>
> Poonam Arjun Vankar, weaver 2015.
>
> I have confidence that I can do something new.
>
> Pravin Premji Siju, weaver 2015.
>
> I learned the value of my craft. It mattered to me to learn.
>
> Ravji Lakhmshi Meriya, weaver 2015.
>
> I realized the value of one good piece vs. production.
>
> Sohel Anwar Khatri, ajrakh printer 2015.
>
> Tradition was a livelihood for me. But I realized it is more important. It is our heritage.
>
> Aslam Abdulkarim Khatri, ajrakh printer 2015.

Frater's work with artisans in Kutch provides strong evidence that education helps invigorate and sustain traditions. But artisan designers must find markets commensurate to their work. The design and business courses are open-ended, intending to provide artisans with tools to make creative decisions about their livelihoods. However, with the central goal of increasing value and income, they recognize the importance of understanding higher end market taste.

Somaiya Kala Vidya consciously introduces artisans to higher end markets that they would not otherwise directly experience. Students visit higher end shops, and the homes of connoisseurs in Ahmedabad. Homework after the exposure trip is to create products for individuals they met. This exercise has consistently resulted in excellent creative work, verifying that the personal orientation of craft and the role of personal connection in inspiration are still vital, and effective in design. Artisans are further encouraged to aim high by final jury members, who are consciously drawn from among people who are well versed in craft traditions and higher end design markets. Further, as part of the course they exhibit and sell their collections at high end galleries and analyze the experience afterwards.

Given access to a range of markets, most artisan designers would prefer this type of smaller, higher end segment. As examples, Kutch artisan designers report:

> Whatever production I do should have something of me in it, myself. So I do limited quantity, but it is good. When people see it they should feel this is good work and realize it should be worth it. I have never been told to make it cheaper.
>
> Alimahamad Isha Khatri, SKV Advisor and master bandhani artist,
>
> With increase in scale, there is a compromise in quality. For us, it is most important to keep the quality.
>
> Abdulaziz Alimamad Khatri, bandhani artist,
>
> I don't believe in large scale for craft. It is then not craft. The way for us is clear. We have the tools. We must vary design constantly and work for niche markets, always evolving.
>
> Dahyalal Atmaram Kudecha, weaver 2016.

Artisan designers are also clear that creativity is important to their work. In a survey on aspirations, they ranked recognition as equal to increased income. Further, in a discussion on hand work in luxury brands, they expressed frustration in working with designers. "Designers often give designs that are not producible," they cited. They stated that a chance to speak and explain their work is important, and that they are more interested in customers who appreciate their craft than in fashion per se.

While the design and business education courses at Somaiya Kala Vidya encourage artisans to aim for higher end markets, they have not specifically targeted luxury markets. Market orientation and marketing are included in the curricula. However, positioning has not been addressed. While artisans are encouraged to develop diversity, exclusivity and consumer experience have not been a focus. And while *artisan designer* has evolved as a genre, *design craft* has not yet been branded as a distinctive niche of hand work.

2.8 Developing a Luxury Craft Market

Klamer (2012) and other cultural economists at Erasmus University write of nurturing a Creative Craft Culture. In such a culture, the profile of craft production would encompass what the design and business courses have achieved:

- Young people viewing the creative crafts as a career worth striving for
- Strong traditions of apprenticeship
- A strong sense of tradition
- Recognition of the masters; fair and effective
- A strong sense of collegiality among creative craftspeople
- A spirit of creativity and innovativeness
- A strong appreciation of entrepreneurship
- Core values and a clear sense of mission (promoting and sustaining quality, contributing to a joyful and inspiring life)

Klamer further states that essential to sustainability of this creative craft culture is an appropriate market, which could arguably describe a luxury market:

- An active and interested community of stakeholders
- A well developed distinctive taste for and appreciation of creative quality (usually implying the existence of an elite)
- Willingness to pay
- Significant local demand as well as international interest (where international buyers do not crowd out local buyers)

Considering their inherent capabilities along with their values and orientation, artisan designers are well placed to target small, personalized high-end production.

However, while artisan designers may aspire to high-end markets, direct access to them is very limited. In the last ten years, the IFAM has become well known in Kutch as the apex of accessible high quality craft markets. In terms of the economies of artisans of developing countries, the stakes are enormous. The average booth sales are equivalent to more than two years' sales for a small-scale producer. Acceptance into the IFAM is competitive and coveted. The encouragement of excellent and unique work through design education gives artisan designers an edge. To date eight artisan design graduates have had the honor and privilege to be juried into the IFAM.

Direct contact with a new market is the major benefit for artisans who attend IFAM. They experience the potential for excellent work aimed at the luxury taste the IFAM consciously promotes. More important, they meet individual representatives of that niche. Artisans understand how to innovate for clients as an essential aspect of their traditional crafts. Design education prepares them to recognize and use this original unique strength of traditional craft: personal connection. With exposure and opportunity, artisan designers can absorb as much as possible from an IFAM experience, and tap this potential effectively. After his first experience at IFAM in 2014, Dahyalal Kudecha noted, "I learned about customers of the USA.

I learned what people like, what is marketable- colors, finishing, and the importance of tradition. I realized in a real way that we need to retain our identity. We need to present our culture." With his earnings, Dahyalal sourced new, better quality materials, and tried experiments that he was confident would sell. Abdulaziz Khatri, at his first IFAM in 2015 was providing live coverage to his brother Suleman, also a design graduate, by What's App. They developed next year's designs before the 2015 Market [IFAM] was over. All three artisans were juried into the 2016 IFAM.

2.9 Sustainability

If IFAM is considered among artisans as the apex of accessible high quality craft markets, it has also taken on the role of an emerging luxury brand, in its shaping of tastes of consumers, influencing their aspirations and behavior by controlling the products available through the strong jurying process.

The direct contact of IFAM opens the possibility of sustainability of this new luxury craft market, in terms of insuring continual, evolving production. When artisans meet customers, they can effectively create for them, and when they, themselves are creating, their traditions remain vital. Familiarity brings appreciation, making creating for the market more conceivable and more organic. Dahyalal and Abdulaziz work with increasing sophistication, attending to drape, texture, wearability, subtlety and overall impact. They have honed the ability to narrate the stories of their work to clients. Abdulaziz has imaginatively named pieces "Kutch Monsoon," and "Galaxy." Figure 1 illustrates the *Galaxy* design that has made its pathway to the high-end market. This becomes possible because these artisan designers now prefer the quality materials, sophisticated colors and restrained patterns of luxury taste. Lachhuben Raja Rabari, embroidery graduate, after two IFAM experiences said, "We learned to like these colors." The artisans that Dahyalal and Abdulaziz employ will in turn absorb understanding of materials and aesthetics of a luxury market. See Fig. 1.

Luxury oriented products bring markets to their creators. Ramu Devraj, a quilt maker from Kutch, enjoyed several experiences at IFAM. With the exposure and earnings, he upgraded his product, organized a unit of 300 women and 10 tailors, and now creates for the higher end Indian market. This year he is doing an order of RS 3,000,000 for Fabindia, India's largest private platform for products from traditional techniques, skills, and hand-based processes. Local demand, as Klamer (2012) points out, is essential to sustainability. Domestic markets are within reach for artisans, in terms of the cost of travel and transportation, and in terms of interaction that will insure appropriate innovation. The validation of having been at IFAM gave Ramu an edge to the domestic market. He no longer needs to attend IFAM. See Fig. 2.

Fig. 1 The galaxy design for the high-end market. Photo courtesy of K Brooke

Dahyalal and Abdulaziz now sell to a range of clients, largely in the Indian domestic market. Dahyalal has expanded his capacity from two to eleven looms in three years. Abdulaziz has expanded to employ 250 women and 10 men. In 2015, while Abdulaziz was experiencing IFAM in Santa Fe, his brother Suleman was selling saris in the Paramparik Karigar exhibition in Mumbai. All told, Suleman had the higher sales. Equally important, he contributed to the increasing value of his own work by selling at higher prices without apology. When an organizer commented that his saris were expensive, he observed simply that they were selling very well. See Fig. 3.

IFAM has encouraged artisan designers to focus innovation on luxury markets. If the experience can become a springboard rather than an end point, sustainability for artisans in the luxury market is possible. What is needed for real sustainability is more Ramus, Dahyalals and Abdulazizes. Sustainable luxury craft markets need more unique interpretations, not more production. See Figs. 4 and 5.

Fig. 2 Ramu Devraj shows quilts to clients in his village

Fig. 3 Dahyalal with client at IFAM. Photo courtesy of J. Frater

Fig. 4 Dahyalal has an opportunity to expand sales in a private sale to high-end customers. Photo courtesy of J Hawley

Fig. 5 Abdulaziz explains the bandhini process to young consumer at IFAM. Photo courtesy of J. Frater

3 Conclusion

Luxury is about aspiration. A significant number of customers define luxury by iconic brand, as it is associated with status. But we contend that luxury is about consumer experience. And luxury is niche. It is sophisticated, exclusive in its positioning, and while it has impressive reputation, it has a selective system of distribution, limited distribution and low promotional activity. The secret ingredient to marketing to luxury consumers is how strong desirability is to influence purchase, how great a share of the consumer's income a product can command. Luxury starts well before the purchase, and ends only when memory fails (Sinha 2011).

In closing, we would like to offer a new way to think of traditional craft as luxury. We believe that it requires attention to marketing that addresses aspiration, experience, exclusivity and desire. We also believe that traditional craft offers something that mass manufactured goods can never attain—*uniqueness, an embodied relationship,* and *sustainability* because it reaffirms the connection between quality and price. Traditional craft has also been formed by the human hand—giving it distinction from mass manufactured goods. This is why craft skills and crafted products are going to become even more important to what is meant by luxury. Unique traditional crafts allow us to value something as luxury, even when it is not necessarily expensive. Exquisitely made traditional craft speaks to one-of-a-kind authenticity and originality. It embraces the human relationship with the product, the result of which is the consumer's relationship with authenticity. As Montgomery says, "the product will change to become an emblem of a mutually appreciative relationship" (2016) where such products will no longer be "totems of soul-less consumerism, but authentic expressions of creativity." When the relationship between the consumer and the crafter becomes mutually appreciative, the role of the product is elevated to the category of luxury in the marketplace.

Celebrity endorsement can play an important role as well. Dahyalal's success at his first experience of IFAM was considerably bolstered by the enthusiasm of American film start Ali McGraw. Ms. McGraw purchased his signature piece early on the first day of the Market and wore it through the day, directing people to his booth. In Klamer's Creative Crafts Culture (2012) the third critical element is the presence of strong intermediaries and intensive discussions of the works of creative crafts people. Intermediaries include modern guilds, special shops, experts, journalists, newspapers, websites, scholarship, festivals, labs, fairs and so on.

Creative crafts do not just come about because people make creative products but also because others are able to appreciate those products. Creative crafts are not just products but contributions to the conversation that constitutes creative crafts. Success is first determined by the people who know, by the experts who can judge the quality, and then by others who pay attention to such judgment and are able and willing to pay the price. (Klamer 2012).

The current social climate is conducive for artisans to consider the luxury market. The luxury market is in a dynamic phase where artisans have an opportunity to learn, develop, and ultimately find their path. Montgomery (2016) and

Mcintyre (2010) have specifically identified opportunities for craft to appear as a new luxury and a new signal of connoisseurship. IFAM has been a vehicle for shaping the taste and behavior of an emerging luxury craft market. As it moves consciously toward developing a luxury craft market, traditional artisans from around the world can learn to define and make inroads to this important niche market.

Through special events, like the IFAM, consumers are able to interact with the maker and the relationship is strengthened. In other words, consumers and the maker build special bonds that are imbued with knowledge, memory, and personality. Through its Mentor-to-Market workshops but more importantly through its direct experience, IFAM educates artisans about the contemporary high end market. As more artisans "graduate" and new ones enjoy the experience, IFAM and traditional artisans together can contribute to a phenomenon, an elegant disruption, which can change the perception of traditional craft to a luxury that is sustainable and in turn serves to sustain traditional cultures.

References

Almond K (2011) Bespoke tailoring: the luxury and heritage we can afford. Proceedings of the the international conference of technology and society, University of Basque Country, Spain, pp 25–27, March 2011

Bendell J (2012) Elegant disruption: how luxury and society can change each other for good. Asia Pacific Centre for Sustainable Enterprise Queensland

Bendell J, Kleanthous A (2007) Deeper luxury report. WWF, London

Bendell J, Thomas L (2013) The appearance of elegant disruption. J Corp Citizensh 52:9–24

Campbell C (2005) The craft consumer: culture, craft and consumption in postmodern society. J Consum Cult 5(1):23–24

Dubois B, Laurent G (1994) Attitudes toward the concept of luxury: an exploratory analysis. Asia Pac Adv Consum Res 1:272–278

Eames C, Eames R (1958) The India report. National Institute of Design, Ahmedabad

Elkington J (1998) Cannibals with forks: the triple bottom line of 21st century business. New Society Publishing, Stony Creek

Frater J (2008) Interviews for the film series 'The Masters' Voices, Directed by Nilosree Biswas and produced by Kala Raksha

Gardetti M, Muthu SS (2015) Handbook of sustainable luxury textiles and fashion

Giron ME (2015) Forward. In: Gardetti MA, Torres AL (eds) Sustainable luxury: managing social and environmental performance in iconic brands. Greenleaf Publishing, Sheffield

Hanks (1979) Collins dictionary of the English language. Collins, London

Harrod T (1995) The crafts in Britain in the 20th century. Yale University Press, New Haven and London

Hawley JM (2015) Economic impact of textile and apparel recycling. In: Hethorn J, Ulasewicz C (eds) Sustainable fashion: What's next?, 2nd edn. Bloomsbury, New York

Hertel C (2016) Press release: The 13th Annual International Folk Art Market. https://www.folkartalliance.org/press-release-market-2016/. Accessed 15 Mar 2016

Husic M, Cicic M (2007) Luxury consumption factors. J Fashion Mark Manage 13(2):231–245

IFAA (2015) International folk art alliance impact report. https://www.folkartalliance.org/about/who-we-are/. Accessed 15 March 2016

Jayakar P (2011) Craft traditions of India: past, present, and future. National Council of Educational Research and Training

Jenkins H (2012) CSR in sport: investigating environmental sustainability in UK premier league football clubs. Paper presented at the corporate responsibility research conference. University of Leeds

Johnson R (2009) Dosa's Christina Kim: every piece is precious. Available via Los Angeles Times online. http://articles.latimes.com/2009/apr/19/image/ig-greendosa19. Accessed 06 Mar 2016

Jones KB (2014) Looking at fashion through green-colored glasses: a multimodal critical discourse analysis of Vogue's sustainable fashion editorials. Dissertation. University of Missouri, Columbia

Kapferer JN (2010) All that glitters is not green: the challenge of sustainable luxury. The European Business Review. www.eurpeanbusinessreview.com. Accessed 25 Jan 2016

Klamer A (2012) Crafting culture: the importance of craftsmanship for the world of the arts and the economy at large. Paper presented at Erasmus University of Rotterdam

Lehman M (2015) Elegant disruption—innovation refined. Handbook of sustainable luxury textiles and fashion. pp 137–161. Accessed 18 Mar 2016

McIntyre MH (2010) Consuming craft: the contemporary craft market in a changing economy. UK Crafts Council

Montgomery B (2016) Craft the new luxury. Craft & Design. https://www.craftanddesign.net/features/214/craft-the-new-luxury. Accessed 2 Mar 2016

Okonkwo U (2007) Luxury fashion branding: trends, tactics, techniques. Palgrave-MacMillan, New York

Risatti H (2007) A theory of craft: function and aesthetic expression. The University of North Carolina Press, Chapel Hill

Schrandt L (2015). American festivals: 10 best readers' choice winners. USA Today, 17 Apr 2015

Shih WC, Agrafiotis K (2016) The sustainable luxury craft of bespoke tailoring and its enduring competitive advantage In: Gardetti MA, Muthu SS (eds). Handbook of sustainable luxury textiles and fashion. pp 137–161

Singh R (2015) Redefining luxury the V&A way in 2015. Forbes/Asia. Retrieved 12 Feb 2016 from http://www.forbes.com/sites/ranisingh/2015/04/28/what-is-luxury/#6ca420686cce. Accessed 26 Jan 2016

Sinha PK (2011) The luxury roundtable: does the luxury business need a rethink? J Decis Makers 36(1):73–74

Sinha P, Som A (2016) Luxury as we do not know it—part 1. Crafting Luxury and Lifestyle Businesses. Indian Institute of Management, Ahmedabad, 1 Feb 2016

Snell J (2015) Social innovation and entrepreneurship; teaching students to create solutions to social problems through innovation. Keynote delivered at the 2015 international textile and apparel association annual conference, Santa Fe, New Mexico, USA

Spiegel F (2015). What is luxury? Exhibition at the Victoria & Albert Museum. http://decodedarts.com/luxury-exhibition-victoria-albert-museum/2466 Accessed 18 Mar 2016

Stengel R (2009). For American consumers, a responsibility revolution. TIME Magazine, 10 Sept 2009

United Nations (1987) Report of the World Commission on Environment and Development: our common future. Document a42-427. 42nd Session

Websters New World Dictionary (2013) Simon and Schuster

World Wildlife Fund (2007) Deeper luxury report: quality and style when the world matters. http://www.wwf.org.uk/deeperluxury/. Accessed 18 Mar 2016

Part VI
Sustainable Luxury and Consumption

(Not) Made in Italy: Can Sustainability and Luxury Co-exist?

Vertica Bhardwaj and Sergio C. Bedford

Abstract In the past, *luxury* often resonated with a tradition, superior quality, and a pampered buying experience, something that only very privileged few could afford. In recent years, the meaning of luxury has become profoundly different and transitioned into a phenomenon that primarily focuses on selling expensive designer brand names with a mere illusion of exclusiveness and superior quality to the masses. As a result of the mass luxury phenomena, a designer handbag or watch no longer serves the same purpose as before. Today owning a luxury brands does not transmit reliable information about its wearer's socioeconomic status or background. With such dramatic changes in luxury industry, a large body of consumer research has looked into understanding why individuals seek luxury brands and high-end logo-ed merchandise across the globe. However, until recently, a rather unexplored area remains in the domain of 'sustainable' consumption of luxury brands. The overall question arises of the role of consumers with regards to sustainable practices in luxury industry and how these practices influence decisions or intentions to indulge in a behavior. For instance, does it matter to the consumer if a 'Made in Italy' handbag was actually made (in-part) off-shore in a factory? Does lack of transparency in manufacturing process and information play any role in forcing consumers to rethink about their perception of luxury brand and, of course, *luxury* itself? This chapter outlines a conceptual framework that explores factors that influence perception on consumption of sustainable luxury. It is built upon the idea that luxury brands acts as a psychological meaning of self-expression and symbolic association that individuals aspire to create or maintain. The ultimate goal of this chapter is to explore whether consumers are increasingly becoming conscious of the adverse effects of the social, moral, and environmental imbalances

V. Bhardwaj (✉)
Merchandising and Consumer Sciences, School of Family and Consumer Sciences,
Texas State University-San Marcos, San Marcos, TX 78666, USA
e-mail: vertica@txstate.edu

S.C. Bedford
Retail and Consumer Sciences, The University of Tennessee,
244A Jessie Harris Building, 1215 W. Cumberland Ave., Knoxville,
TN 37996-1911, USA

M.A. Gardetti (ed.), *Sustainable Management of Luxury*,
Environmental Footprints and Eco-design of Products and Processes,
DOI 10.1007/978-981-10-2917-2_19

created by luxury brands. Based on existing theoretical and empirical insights from *Theory of Impression Management* and *Value-Attitude-Behavior* paradigm, it becomes evident that psychological determinants, such as symbolic value perception (i.e., self-directed and others-directed perception of the symbolic association with luxury bands), tendency to conform, *moral judgment* (i.e., ability to rationally engage one's knowledge in performing actions that are responsible to the society), and hedonism, as well as individuals' attitude toward and involvement with sustainable luxury brands can represent antecedents of intentions to purchase sustainable luxury. This exploratory work will be first of its kind and will contribute to the limited literature on consumers' perception toward sustainable luxury.

1 Introduction

Luxury industry has flourished for the past 10 years, but the good times have started to stall

Fflur Roberts, Global Head of Luxury Goods Research

In the past, *luxury* often resonated with a tradition, superior quality, and a pampered buying experience, something that only very privileged few could afford. Luxury brands are often associated with prestige because of their superior craftsmanship, premium quality, and iconic and exclusive designs. These have been ostensibly emphasized to be authentic, and reflect concomitant respect for artisans and the environment, that fosters values of both quality and sustainability. The appetite for luxury has increased manifold in recent years. Consumers' have shown a steep increase in the demands for luxury brands such as Louis Vuitton, Burberry, Gucci, and Prada, to name a few. Unlike previous decades when the market for luxury brands didn't exhibit high growth patterns—when such brands were mainly for the privileged few, the growth of current luxury market has been phenomenal, increasing from US $317 billion in sales in 2015 to an expected sales of US$328 billion by 2016 and projected growth of 9% annually by 2010 (Bain & Co. 2014).

Buying 'luxury (brands)' has become a mass-phenomena. Among many reasons, the vast reach of the contemporary advertising, digital advances, product placement via media and marketing channels, brand awareness, and constantly increasing social acceptance among peers has even boosted democratization of luxury. As a result, the craftsmanship has waned and (once) an exclusive designer handbag, jewelry, or a watch no longer serves the same purpose as before. Today owning a luxury brands does not transmit reliable information about its wearer's socioeconomic status or background. Unfortunately, the widespread notion that a luxury brand still symbolizes the same essence as it did years ago has only exacerbated its effects.

Among other changes in luxury industry in the past decade, the most prominent one is routine manufacturing of luxury fashion labels in China (Passariello 2011). European brands such as Burberry, Armani, Prada, and Dolce & Gabbana have shifted their manufacturing to factories in China in the last few years. These factories do not necessarily classify as factories like Rana Plaza in Bangladesh (known for factory collapse); rather they are often equipped with technology based machinery to manufacture goods in order to achieve the craftsmanship that luxury brands often claim to be associated with (Bain 2015). Undoubtedly, the quality of luxury goods manufactured in (some) of these outsourced factories has remarkably improved and is getting better over time. Besides quality, overseas manufacturing process also gives an advantage to quickly respond to the exponential demands for luxury fashion products such as bags, scarves, sunglasses, watches etc. across the globe. What remains uncertain is whether consumers are aware of such changes, and whether it really matters to them or not? For instance, it is expected that consumers associate a luxury watch to be Swiss, expensive perfumes and cosmetics to be French, their cars to be German, and their designer clothing, handbags, and shoes to be from Italy or France. Paradoxically, if their handbag is not French, it may not hold the same reputation as it did before. So far, the literature on consumer behavior that looks into luxury consumption has explored various motivations and antecedents that explain purchase consumption toward luxury brands.

As consumers become more sensitive to their eco-consciousness about various products they buy and consumer on a regular basis, they have high expectations from the fashion companies, especially in the luxury industry, to use practices in running their business that are ‘actually’ eco-conscious. Brands that merely claim to be sustainable or eco-conscious are scrutinized by their followers and have received severe backlash. For instance, brands such as Burberry, Hugo Boss, and Ralph Lauren have been criticized enough recently for not being ‘true’ to be a sustainable brand as they claim to be in their social responsibility reports (Passariello 2011). Figure 1 shows some examples that were made public as a result of the above mentioned scrutiny.

With this background, this chapter is an attempt to initiate a discussion on sustainable luxury and see how it can change perceptions towards sustainable consumption among individuals. This chapter outlines a conceptual framework (see Fig. 2) that explores factors that influence perception on consumption of sustainable luxury. It is built upon the idea that luxury brands acts as a psychological meaning of self-expression and symbolic association that individuals aspire to create or maintain. The ultimate goal of this chapter is to explore whether consumers are increasingly becoming conscious of the adverse effects of the social, moral, and environmental imbalances created by luxury brands. The first portion of this chapter attempts to look into what we understand by ‘sustainable luxury’ and discussed various dimensions that help readers to expand their understanding of the same. Next, the theoretical background builds the foundation of this conceptual research. Following, the literature review examines previous research on motivations to buy luxury in general and specific to sustainable consumption of luxury brands. Finally, the chapter closes with a discussion of possible future research approaches and managerial implications as opportunities to develop marketing campaigns that adequately respond to consumers’ values and motivations.

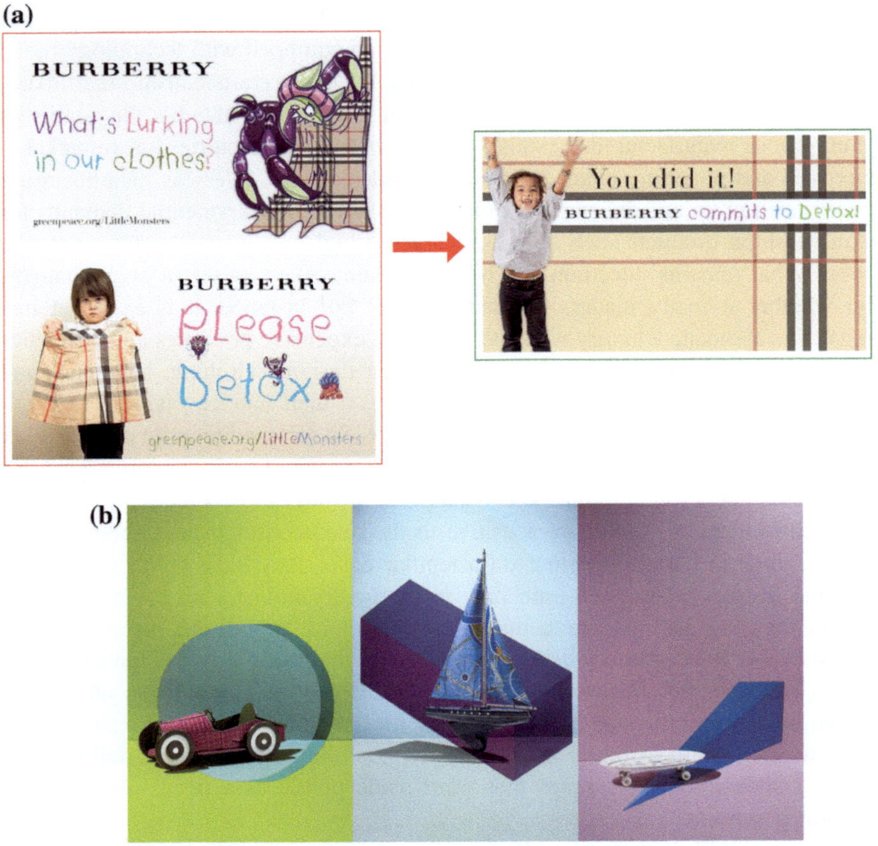

Fig. 1 **a** Burberry commits to toxic free clothing for kids by 2020. *Source* Greenpeace.org. **b** The petit h collection by Hermes: showcasing their eco-friendly upscale up-cycling efforts

1.1 Defining 'Sustainable Luxury'

Sustainable consumption, in general is complex to understand simply because there are more than 50 definitions and circumscriptions of sustainability that exist (Faber et al. 2005). In line with the authors, each of these attempts to define sustainability from a perspective of either a marketer, an environmentalist, a sociologist, or an economist, to name a few—essentially pointing to the fact that it is discipline oriented. Keeping in mind the most accepted and versatile definition that addresses a need to "meet the needs of the present without compromising the ability of future generations to meet their needs" (WCED 1987, p. 43), we argue that it is important to take a holistic approach to integrate our understanding of 'sustainability' as a concept. Therefore, instead of proposing another definition, we treat it as a conceptual foundation that may have different interpretations depending on the

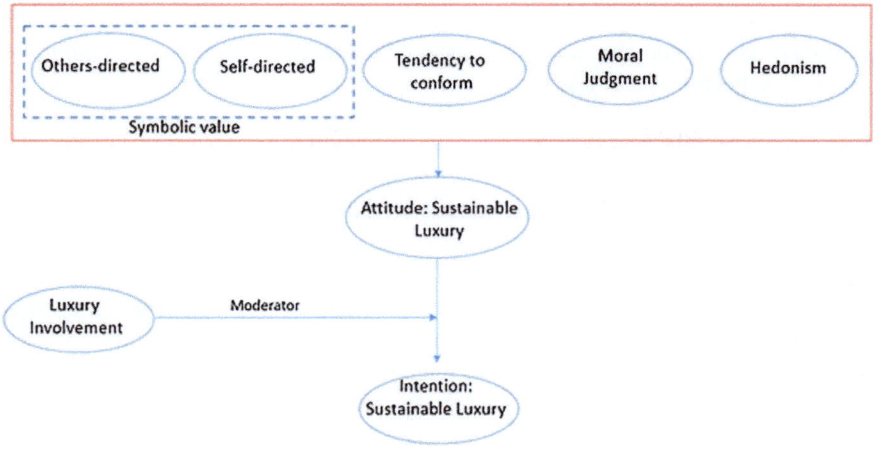

Fig. 2 Conceptual model

product/brand in question. Moreover, situational factors such as lack of information, lack of availability of sustainable options, high price points, simple and basic designs etc. may force individuals not to be inconsistent when making a purchase decision for a sustainable product (Dickson 2001; Shaw et al. 2006).

In an attempt to broaden the scope of '*Sustainable*' luxury, there is a need to follow multidimensional approach encompassing three broad areas, often referred to as Triple Bottom Line concept (TBL) that underpins and reviews how luxury brands contribute to economic, environmental, and social impact to the society, not necessarily in that order (Gardetti 2011; Goworek et al. 2012). These three dimensions can be further broken down into: product transparency, packaging, raw material sourcing, production and manufacturing, reusability/recyclability, animal-friendliness, philanthropy and fair labor conditions—essentially pointing to an entire value chain in the luxury industry. Each of these dimensions should be considered equally critical to qualify a luxury brand as 'sustainable'. For instance, luxury brands that scrutinize to ensure that they are using ethically sourced leathers and furs, providing a living wage to factory workers, and minimizing waste in the packaging process when sending the final products to the retail stores can be qualified to be considered as 'sustainable'. On the other hand, a luxury brand that has been known for in-house manufacturing that recently moved majority of production overseas to meet high demand in the market may not qualify to be classified as 'sustainable'. Adding to the lack of a clear understanding of sustainability, luxury and sustainable consumption are often looked as opposite concepts. For instance, sustainable consumption inherent ethically-grounded values of altruism, restraint, and moderation in contrast with luxury's inherent hedonism, aestheticism, rarity, affluence, and at times superfluity (Carrier and Luetchford 2012). It is highly likely that consumers often fail to understand or simply ignore to address the relation between these two juxtaposed concepts. The potential confusion for consumers to differentiate luxury and sustainable luxury rekindles interest in a key questions: how should '*Sustainable* luxury' be defined? 'Luxury' and '*Sustainable*' luxury has been looked upon as different concepts; more specifically as two extremes of a continuum. This is evidenced in recent study where individuals showed unfavorable responses for luxury products that were made from recycled materials (Achabou and Dekhili 2013). If these two concepts are looked as a continuum, then it is up to the individuals to judge whether they perceive luxury brands/products as either 'luxury' or 'sustainable'. For instance, it is worth exploring whether they perceive a luxury automobile, expensive watches, clothing, perfumes, and accessories as 'luxury' or as 'sustainable luxury'? Taking all these different perspectives in account, this chapters attempts to open the dialog to explore multi-disciplinary approach in understanding sustainable luxury from the mindset of a consumer. As consumers become increasingly concerned about products they buy and where they come from, an argument can be made that they should have same concerns about luxury products as well.

1.2 Sustainability in the Luxury Industry

Recent years have found that consumers are becoming increasingly aware of sustainable development and seek sustainable luxury products. With this change, luxury brands have also made structural decisions to include sustainable development as a theme in running their businesses. However a majority of the retailers have been criticized for not being sustainable. Studies find that the luxury industry is still behind other industries, as far as some stating that sustainable and luxury is incompatible (Kapferer 2010). For instance, diamonds mined from areas of conflict, has created a change in the jewelry industry (DeBeers 2009). Media has also reported instances of racial discrimination and Prada was exploited for using Chinese immigrants in Italian sweatshops (Davies et al. 2012). Although there have been numerous cases that identify unethical and unsustainable issues in luxury brands, very little laws and enforcement has been done to regulate the industry.

Studies find that luxury brands have both the opportunity and the responsibility to promote sustainable consumption. The trends in sustainable development will naturally evolve luxury brands and create a market for sustainable luxury products. While consumers are showing more interest in sustainable products and have altered their behaviors to be more sustainable, the industry has not seen change happening as radically, however, the luxury industry has already seen changes. For instance, Cartier is a member of the Responsible Jewelry Council, an organization provides certification that diamonds and gold used to make their jewelry comes from socially and environmentally stable areas (Achabou and Dekhili 2013). Ethical fashion shows have become a norm since 2004 and continue to be a big part of Fashion Week (Davies et al. 2012). In addition, a number of luxury brands have adopted recycling processes, even including a collection of accessories, origami, and keepsakes in their collection *Petit h* by Hermes using fabric scraps and seconds from their workshops (Achabou and Dekhili 2013). As shown in Fig. 1, *Greenpeace's detox* campaign has driven luxury retailers such as including Burberry, Dolce and Gabbana, and Valentino to reduce and eventually eliminate the use of hazardous chemicals by 2020 (Chua 2014a, b). While some designers have incorporated a more sustainable approach in their manufacturing and sourcing, it has still not been an industry norm, as many do not consider it an obligation or duty to their customers. Overall, the key to successfully achieve 'sustainability' in luxury products is to make informed decision at the design stage, share proper information specifying the use of materials which are not toxic, proper labelling to name a few.

2 Theoretical Background

2.1 Theory of Impression Management

Modern luxury has become a medium of self-expression and often a way to impress others, especially for aspirational masses (Chandon et al. 2016). With this in mind, current research is grounded under the *Theory of Impression Management*, pointing to the fact that people are highly affected by their internal drive to create a favorable social image from their purchase behavior outcomes (Eagly and Chaiken 1993; Mandrik 1996). The notion 'to impress others' has been often the basic motivation behind consumption of luxury products as a way to pursue ideals and aspirational lifestyle. Numerous conceptual models and empirical studies on luxury brand consumption have pointed to the tendency of consumers to be associated with high social status, and identified as being fashionable, stylish (Ko and Megehee 2012).

2.2 Value-Attitude-Behavior Paradigm

This study is underpinned on the conceptual framework of value-attitude-behavior (V-A-B) model proposed by Homer and Kahle. Past research in sociology has frequently employed V-A-B model which implies that "the influence should theoretically flow from abstract values to mid-range attitudes to a specific behavior" (p. 638) indicating that values influence attitude, and attitude in turn influences behavior. Values are enduring beliefs that may affect individual's views and have been characterized as determinants of specific attitudes and outcome behavior. Ultimately, consumers buying behavior has been found to include values and motivations (Parsons and Shillis 1967).

Values lead a consumer to specific behaviors based on their beliefs; motivations lead a consumer to purchase based on the products ability to fulfill certain needs (Nwankwo et al. 2014). In recent years, the luxury fashion industry has seen a shift from an overindulgent, wasteful system to one that has included sustainability, overcoming counterfeits products, poor labor standards and illegally sourced materials Hennigs et al. 2013). As consumers concern for sustainability has shifted, luxury brands have had to shift as well in order to maintain relevant (Hennigs et al. 2013). This study argues that as consumers become more conscientious about the environment by being more sustainable (recycling, fair trade, etc.) they expect luxury brands to do the same. Consequently, luxury brands awareness of the shift in consumer values should lead to a natural progression of sustainable luxury.

3 Conceptual Model

A sustainable consumer may reflect wider range of motivations that influence decision making; thus creating complexity in identifying various motivations with respect to sustainable luxury. This can be supported by the hierarchical structure of the decision making process—individuals may be driven either by functional (utilitarian) value or attributes of the products as a way to problem-solving, or by psychological, emotional, and social motivations (Reynolds and Olson 2001). Adding to the list of motivations, Shaw et al. (2006) refer to self-expression, aesthetic satisfaction, and group conformity as other potential motivations toward sustainable fashion products. In line with this, we propose a set of motivations including: symbolic value, tendency to conform, moral judgment, and hedonism that may be potential antecedents to attitudes and intentions for consumption of sustainable luxury. In addition, product involvement has often been cited to have an influence on purchase intentions. Therefore, we argue that luxury involvement will play a key role to understand intentions in the context of sustainable luxury. The proposed conceptual model is shown in Fig. 2.

3.1 Determinants of Sustainable Luxury Consumption

Numerous conceptual and empirical studies have examined the interaction of values, attitudes, goals, and motivations on behavior toward luxury brands and sustainability separately. Recent literature has started to highlight the need to focus on consumption of sustainable luxury (Achabou and Dekhili 2013; Steinhart et al. 2013; Lundblad and Davies 2015). Among handful of studies that have examined luxury with respect to sustainability, a recent research explored the extent to which consumers associate luxury brands with recycled material. In their examination, Achabou and Dekhili (2013) discovered the propensity of consumers to consider recycled material in luxury purchase. Unlike expected, the respondents indicated negative perceptions about luxury product made with recycled material. They further indicated that they didn't have any desire to buy 'sustainable' luxury as it is not perceived to have the exclusivity and prestige, irrespective of what brand it is. The findings from other studies are somewhat contradictory. For instance, Steinhart et al. (2013) found positive reactions from consumers about luxury products that claim to benefit the environment either by reducing deforestation or global warming. This study proposes to examine the motivations that depict consumers' intentions to purchase 'sustainable luxury', which is an area that remain primarily unexplored, both conceptually and empirically.

3.2 Symbolic Value

It is well established that people make inferences about others based on product selection and usage, especially in public (Belk 1988). This becomes extremely relevant for luxury brands as they have strong symbolic associations to social status, power, prestige, affluence, and recognition to within a social context (Dubois and Laurent 1994). Symbolic value, the extent to which individuals attach psychological connections, in the context of sustainable luxury can be either 'self-directed'—as a way to reward themselves; or 'others-directed'—as a way to signal wealth, power, and social class of individuals (Shukla and Purani 2012). Based on findings from previous research on luxury consumption (e.g., Nwankwo et al. 2014; Shukla and Purani 2012; Wong and Ahuvia 1998), it can be argued that individuals that rely on luxury brands to create an impression on others are likely to violate sustainable consumption of luxury brands rather than actively engaging with them. This is supported by the findings from Xianchi and Fries (2008) that suggest that powerful individuals believe they have greater capacity to be resistant toward normative influence that makes them more likely to behave in socially-unacceptable ways. On the contrary, individuals who aspire for luxury brands as a way to reward themselves are likely to be open to have favorable attitudes and intentions to purchase sustainable luxury.

3.3 Tendency to Conform

In the modern society, individuals tend to adopt a mentality to conform to the social groups that share a common culture and opinions with a desire for public compliance or to gain social acceptance (Bearden and Rose 1990). People try to "keep up with the Joneses" with the social groups that they belong to and with those of a similar socio-economic stature (Seth et al. 2011). Though conformity may be somewhat synonymous to symbolic value, theoretically it is a related yet different concept in the context of luxury brands. Specifically, individuals may purchase luxury brands as a way to express their uniqueness instead of showing off their wealth and status or desire for public compliance (Amaldoss and Jain 2005). Previous studies indicate the need to maintain one's appearance and reputation in public creates an ever-escalating desire to purchase luxury brands. Riff (1989) has described conformity as a negative trait and suggested individuals need to have an internal locus of evaluation, whereby one does not look to others for approval but evaluates oneself by personal standards. On the other hand, Smith (1985) considered the ability to conform to role expectations, keeping aside personal preferences, as a commitment to the society that signifies strength and maturity of the individual. Previous research suggests that individuals who have a high tendency to conform value luxury products more as they are publically consumed and have high visibility among ones social class; versus commodity or necessity products that are often used in private. In line with Achabou and Dekhili (2013), it can be argued that individuals who have high tendency to conform will likely to have favorable attitudes toward luxury, instead of 'sustainable' luxury.

3.4 Moral Judgment

The most basic human value or motivation is believed to aim at consumption optimal for ones' and society's well-being, in that order. In our discussion, we include moral judgment, i.e., an ability to judge what's right and what's not, in the context of sustainable luxury consumption to be a critical factor in understanding the relationship between attitudes and intentions to behave. This is in line with previous studies that have looked into moral judgment in contexts other than luxury brands. These studies suggest that moral judgments indicate one's ecological and social awareness, as well as one's ability to rationally engage in performing actions that are responsible to the society (Furnham and Valgeirsson 2007; Green and Smith 2002; Nill and Shultz 1996). Tan (2002) suggested that moral judgment allows individuals to evaluate how far certain actions might be perceived as ethically acceptable. Furnham and Valgeirsson (2007) examined ethical judgments toward consumption of counterfeit luxury brands. In doing so they suggest that consumers with a high standard of moral judgment may perceive a higher risk associated with counterfeit consumption. Due to minimal research on ethics or moral judgment toward sustainable luxury, it is unclear whether existing literature on moral judgment would be applicable in the sustainable luxury context. We argue that this is an area that needs further exploration in the form of mixed methods to get a deeper understanding of the construct.

3.5 Hedonism

It has been well established that consumers are predisposed to consuming luxury products as they are associated with the belief that products bring a sense of joy, happiness. Luxury consumption is associated to carry an emotional value that provides an intangible benefit and intrinsic enjoyment (Vigneron and Johnson 2004). Moreover, since luxury products offer an excellent quality and deliver pleasure and sensory gratification to individuals, it is apparent that ones' perceived level of hedonism is related to positive perception of consumption of luxury brands. A large amount of literature on hedonism or hedonic value has mainly focused on consumption of luxury brands. However, it is yet to be explored whether availability of sustainable luxury provides similar experiential value. Due to scarce literature in this domain, a big question remain unanswered: can sustainable luxury provide its user the same level of experiential value?

3.6 Attitudes and Intentions Toward Sustainable Luxury

In the interest of exploring 'Sustainable luxury', it becomes important to understand what consumers think of it. By focusing on their relevant attitudes and related behavior such as disposal activities, purchasing choices, and transfer of ownership of products, useful insights can be helpful to identify policies and strategies to circulate sustainable luxury. Studies suggest that consumers are moving from 'conspicuous' consumption to a more 'meaningful' consumption model (Carr 2013). This so called "ethics era," where consumers are looking for sustainable luxury products demonstrates that consumers have adapted their buying habits and have become aware of the effects that their purchase may have on the rest of the supply chain (Crane and Matten 2007; Davies et al. 2012; McGoldrick and Freestone 2008). Luxury consumers have also purchased more timeless styles that are of higher quality instead of purchasing trendier pieces as an attempt to purchase something that may address social and environmental concerns they may have (Kendall 2010).

In general, there has been a controversy in recent decades as marketers and consumers alike do not see the relationship between these seemingly different constructs. On one hand, luxury brands have acted upon consumers' demands for products to be more sustainable finding that consumers reacted positively to sustainable luxury products. For instance, Gucci developed an eco-friendly handbag collection made from leathers that came from Rainforest Alliance Certified ranches (Ki and Kim 2016). In addition, Stella McCartney developed biodegradable materials that allowed them the ability to refrain using leather and fur. On the other hand, consumers have also reported negative attitudes toward luxury brands that are made of recycled materials (Achabou and Dekhili 2013). While there have been significant strides towards sustainable luxury on the manufacturers' side, consumers ultimately have to make a decision to seek out sustainable luxury products.

3.7 Luxury Involvement

(Product) involvement has been studied since 1947 by psychologists, who identified involvement as the relationship between a stimulus either consciously or subconsciously being elated to the ego (Sherif and Cantril 1947). Product involvement indicates how consumers assign importance to product categories based on their inherent needs, values and interests (Bian and Moutinho 2011). Existing research suggests that product involvement plays an important role in shaping intentions to purchase in the decision making process. Subsequently, involvement is attributed to explain a large percentage of consumer purchasing behavior (Quester and Lim 2003). It is imperative to understand consumer product involvement, particularly with sustainable luxury products, to better understand consumers' intention to purchase these.

Previous research indicates that for consumers who have high involvement with brands, the decision process undergoes extensive stages of information search and criteria evaluation (Bian and Moutinho 2011). Owning luxury products has been identified as a manifestation of one's social self as consumers who own luxury products use them to communicate status, wealth, and in turn, both social and economic powers to those who see them (Beckham and Voyer 2014). As a result, consumers who display higher levels of (luxury) product involvement are more familiar with the brand image, characteristics of the products, and have an internal reference of the price (Lin 2013).

The study further suggests that consumers who purchase luxury products spend more time and effort to ensure that the products quality, features, and benefits align with their interests and expectations. As involvement with luxury products increase, consumers expect to have personal and symbolic gains to compensate for the high product price (Dubois and Duquesne 1993). Moreover, consumers who experience high involvement with luxury products establish close relationships with the brands including: brand satisfaction, trust, affective commitment, and loyalty (Choo et al. 2012). Quester and Lim (2003) also reported that high involvement with brands is a precondition to building loyalty for the brand. In a similar vein, consumers with high involvement with luxury products leads to customer equity (Kim et al. 2012). As a result, this equity benefits the luxury company; creating lifetime customers (those with high involvement) become the most profitable for the brand, increasing WOM, willingness to pay premiums, and repeat purchasing (Kim et al. 2012).

Considering the results from empirical studies on involvement construct mentioned above, it is our contention that luxury involvement will play a critical role in moderating the relationship between attitudes towards and intentions to purchase sustainable luxury. As consumers purchase luxury products regularly, they display positive attitudes towards luxury goods. Consumers who buy from luxury brands often expect to pay a premium for a high quality product; and as suggested throughout this chapter, there has been a shift in luxury consumers to purchase sustainable products. This attitude shift has created a new era in the luxury consumer. This so called "ethics era," where consumers are looking for sustainable luxury products demonstrates that consumers have adapted their buying habits and have become aware of the effects that their purchase may have on the rest of the supply chain (Crane and Matten 2007; Davies et al. 2012; McGoldrick and Freestone 2008).

4 Conclusion

Sustainability in regards to luxury has gained traction recently in press and fashion magazines; however, it is still a fairly new topic in the field of academia. It is evident that sustainability, along with innovation, are considered to be the biggest and most recent trends in the industry (Tutty 2016). As the awareness grows in the industry and among individuals, luxury industry cannot ignore the concerns in the TBL concept—including environmental impacts, human rights and wellness, and other social practices associated with their products. In view of identifying and understanding consumption of sustainable luxury on a broad level, the primary goal of this chapter was to establish a multidimensional framework to further the dialog regarding consumption behavior for sustainable luxury.

Although there have been moves to create a co-existence of luxury and "sustainability," the industry needs to continue developing a sustainable luxury market. As consumers gain more interest in sustainable luxury, the industry needs to be ready provide a clear message about its benefits to individuals and society; and increase availability of sustainable luxury (Davies et al. 2012). For those luxury brands that employ more sustainable tactics will allow them to sustain brand differentiation and establish a lasting relationship with their customers (Achabou and Dekhili 2013). As indicated by the authors, consumers gaining interest in sustainable products coupled with high luxury involvement are likely to have a lasting relationship with these brands.

Luxury brands that become more sustainable will find that consumers can take part in the co-creation of a sustainable luxury brand; allowing them to have a sense of responsibility toward the environment and humankind in general. Luxury brands should shift their product offerings to encompass sustainable practices. Specifically, offering timeless cuts, durable products, and incorporating natural materials can entice both consumers seeking sustainable products as well as those who have not been exposed to sustainable luxury previously (Lundblad and Davies 2015). In addition, shifting luxury brand offerings to include repairing, recycling, and repurposing products will allow the consumers to see an extended usable life of the product and ultimately increase awareness of sustainable behavior (Lundblad and Davies 2015). Ultimately, consumers are willing to pay more for a product when they know that it was made under conditions that support the TBL concept, which will manifest into continued brand value and loyalty for the luxury brands (Kapferer 2010; Hennigs et al. 2013).

As with most consumer products, promoting and marketing consumption of sustainable luxury requires identifying key market segments, understanding the motivations and needs of each segment. Strategic efforts in spreading awareness could boost the speed with which people embrace sustainable luxury practices to become a new norm; however, it is up to luxury brands to act first and show their involvement in sustainable development for consumers to understand its merits. In brief, the industry needs to take an active stance in realizing its responsibility to lead sustainable efforts, as being ethical, and being seen to take concrete actions by bringing institutional changes to engage its customer in accepting the co-existence of sustainable luxury.

> There is no beauty in the finest cloth if it makes hunger and unhappiness
>
> —Mahatma Gandhi

References

Achabou MA, Dekhili S (2013) Luxury and sustainable development: is there a match? J Bus Res 66(10):1896–1903

Amaldoss W, Jain S (2005) Pricing of conspicuous goods: a competitive analysis of social effects. J Mark Res 42:30–42

Bain & Co. (2014) World luxury market report. Bain & Co, Paris

Bain M (2015) “Made in China” really doesn’t mean what it used to. Quartz. Retrieved 29 Mar 2016 from http://qz.com/417052/made-in-china-really-doesnt-mean-what-it-used-to/

Bearden WO, Rose RL (1990) Attention to social comparison information: An individual factor affecting consumer conformity. J Consum Res 16(4):461–471

Beckham D, Voyer BG (2014) Can sustainability be luxurious? A mixed-method investigation of implicit and explicit attitudes towards sustainable luxury consumption. Adv Consum Res 42:245–250

Belk R (1988) Possessions and self. Wiley, New York

Bian X, Moutinho L (2011) The role of brand image, product involvement, and knowledge in explaining consumer purchase behaviour of counterfeits: direct and indirect effects. Eur J Mark 45(1/2):191–216

Carr T (2013) BCG: four trends driving the new age of luxury. Luxury Daily. Retrieved from http://www.luxurydaily.com/bcg-four-trends-tend-to-drive-retargeting-of-luxury-consumers/

Carrier JG, Luetchford PG (eds) (2012) Ethical consumption: social value and economic practice. Berghahn Books, New York City

Chandon J-L, Laurent G, Valette-Florence P (2016) Pursuing the concept of luxury: introduction to the JBR special issue on “Luxury marketing from tradition to innovation”. J Bus Res 69:299–303

Choo HJ, Moon H, Kim H, Yoon N (2012) Luxury customer value. J Fash Mark Manage 16(1):81–101

Chua JM (2014a) Burberry commits to eliminating hazardous chemicals by 2020. Ecouterre. Retrieved from http://www.ecouterre.com/burberry-commits-to-eliminating-hazardous-chemicals-by-2020/

Chua JM (2014b) Italy’s largest textile supplies pledge to detox their supply chains. Ecouterre. Retrieved from http://www.ecouterre.com/italys-largest-textile-suppliers-pledge-to-detox-their-supply-chains/

Crane A, Matten D (2007) Business ethics. Oxford University Press, Oxford

Davies IA, Lee Z, Ahonkhai I (2012) Do consumers care about ethical-luxury? J Bus Ethics 106:37–51

DeBeers (2009) Luxury: considered. Ledbury Research, London

Dickson MA (2001) Utility of no sweat labels for apparel consumers: profiling label users and predicting their purchases. J Consum Aff 35(1):96–119

Dubois B, Duquesne P (1993) The market for luxury goods: income versus culture. Eur J Mark 27(1):35–44

Dubois B, Laurent G (1994) Attitudes toward the concept of luxury: an exploratory analysis. In: Leong S, Cote J (eds) Asia Pacific advances in consumer research. Association for Consumer Research, Provo, pp 273–278

Eagly AH, Chaiken S (1993) The psychology of attitudes. Harcourt Brace Jovanovich College Publishers, San Diego

Faber N, Jorna R, Van Engelen JO (2005) The sustainability of "sustainability"—a study into the conceptual foundations of the notion of "sustainability". J Environ Assess Policy Manage 7(1):1–33

Furnham A, Valgeirsson H (2007) The effect of life values and materialism on buying counterfeit products. J Soc Econ 36(5):677–685

Gardetti MA (2011) Sustainable luxury in Latin America. In: Lecture delivered at the seminar sustainable luxury & design within the framework of the MBA of IE. Instituto de Empresa—Business School, Madrid, vol 18

Goworek H, Fisher T, Cooper T, Woodward S, Hiller A (2012) The sustainable clothing market: an evaluation of potential strategies for UK retailers. Int J Retail Distrib Manage 40(12):935–955

Green RT, Smith T (2002) Countering brand counterfeiters. J Int Mark 10(4):89–106

Hennigs N, Weidmann K-P, Klarmann C, Behrens S (2013) Sustainability as part of the luxury essence: delivering value through social and environmental excellence. J Corp Citizensh 52:25–35

Kapferer JN (2010). All that glitters is not green: the challenge of sustainable luxury. Eur Bus Rev 40–45

Kendall J (2010) Responsible luxury: a report on the new opportunities for business to make a difference. CIBJO, The World Jewellery Confederation. Retrieved from http://www.cibjo.org/download/responsible_luxury.pdf

Ki CW, Kim YK (2016) Sustainable versus conspicuous luxury fashion purchase: applying self-determination theory. Family Consum Sci Res J 44(3):309–323

Kim KH, Ko E, Xu B, Han Y (2012) Increasing customer equity of luxury fashion brands through nurturing consumer attitude. J Bus Res 65:1495–1499

Ko E, Megehee CM (2012) Fashion marketing of luxury brands: recent research issues and contributions. J Bus Res 65:1395–1398

Lin WB (2013) Factors affecting high involvement product purchasing behavior. Qual Quant 47(6):3113–3133

Lundblad L, Davies IA (2015) The values and motivations behind sustainable fashion consumption. J Consum Behav

Mandrik CA (1996) Consumer heuristics: the tradeoff between processing effort and value in brand choice. Adv Consum Res 23(1):301–307

McGoldrick PJ, Freestone OM (2008) Ethical product premiums: antecedents and extent of consumers' willingness to pay. Int Rev Retail Distrib Consum Res 18(2):185–201

Nill A, Shultz CJ (1996) The scourge of global counterfeiting. Bus Horiz 39(6):37–42

Nwankwo S, Hamelin N, Khaled M (2014) Consumer values, motivation and purchase intention for luxury goods. J Retail Consum Sci 21:735–744

Parsons T, Shillis EA (1967) Towards a general theory of action. Harper & Row, New York

Passariello C (2011) Prada is making fashion in China. Wall Str J. Retrieved from http://www.wsj.com/articles/SB10001424052702304231204576403680967866692

Quester PG, Lim AL (2003) The product involvement/brand loyalty link: an empirical examination. J Prod Brand Manage 12(1):22–38

Reynolds TJ, Olson JC (eds) (2001) Understanding consumer decision making: the means-end approach to marketing and advertising strategy. Psychology Press, UK

Riff CD (1989) Happiness is everything, or is it? Explorations on the meaning of psychological well-being. J Pers Soc Psychol 57(6):1069–1081

Seth JN, Sethia NK, Srinivas S (2011) Mindful consumption: a customer-centric approach to sustainability. J Acad Mark Sci 39:21–39

Shaw D, Hogg G, Wilson E, Shiu E, Hassan L (2006) Fashion victim: the impact of fair trade concerns on clothing choice. J Strateg Mark 14(4):427–440

Sherif M, Cantril H (1947) The psychology of ego involvement. Wiley, New York

Shukla P, Purani K (2012) Comparing the importance of luxury value perceptions in cross-national contexts. J Bus Res 65(10):1417–1424

Steinhart Y, Ayalon O, Puterman H (2013) The effect of an environmental claim on consumers' perceptions about luxury and utilitarian products. J Cleaner Prod 53:277–286

Smith RJ (1985) Japanese society: Tradition, self, and the social order. CUP Archive, Cambridge

Tan B (2002) Understanding consumer ethical decision making with respect to purchase of pirated software. J Consum Mark 19(2):96–111

Tutty J (2016) 2016 predictions for the luxury industry: sustainability and innovation. Posit Luxury. Retrieved 24 Mar 2016 from http://blog.positiveluxury.com/2016/01/2016-predictions-luxury-world-sustainability-innovation/

Vigneron F, Johnson I (2004) Measuring perceptions of brand luxury. J Brand Manage 11(6):484–506

Wong NY, Ahuvia AC (1998) Personal taste and family face: luxury consumption in Confucian and Western societies. Psychol Mark 15(5):423–441

World Commission on Environment and Development (WCED) (1987) Our common future. Oxford University Press, Oxford

Xianchi D, Fries A (2008) Power and donation: does perceived power decrease or increase charitable contribution? Adv Consum Res 35:984

Luxury, Sustainability, and Corporate Social Responsibility: Insights from Fashion Luxury Case Studies and Consumers' Perceptions

Marco D'Anolfo, Cesare Amatulli, Matteo De Angelis and Giovanni Pino

Abstract Since the mid-twentieth century, rapid technological and scientific developments have produced continuous and significant improvements in people's daily lives. However, this growth has had considerable, and sometimes horrifying, consequences for the climate and environment (Egri and Ralston 2008). As a result, sustainability has become a central point in public discourse—not only for consumers, but also for the companies, and in particular those in the luxury industry, who generally operate at the international level and may serve as relevant actors in economic development (Fraj and Martinez 2007; Kerr et al. 2009). Our focus on luxury firms was born from the fact that such firms are still uncertain about how to combine luxury and sustainability; indeed, these two concepts are often seen as contradictory (Joy et al. 2012; Kapferer and Michaut-Denizeau 2014). The very term "Luxus" has a double meaning: It positively connotes such concepts as beauty, prestige, and power, but also implies excessive and ostentatious displays of wealth (Kapferer 1997, 2012). This duality helps to explain consumers' difficulty in reconciling luxury, sustainability, and Corporate Social Responsibility (CSR), and also why a company like Prada recently launched a website dedicated to highlighting initiatives that reflect the brand's commitment to sustainability. The present chapter thus delves into a currently neglected area of research—namely, consumers' perceptions of the relationship between luxury, sustainability, and the CSR strategies that major luxury companies adopt to counter said perceptions. To analyze luxury

M. D'Anolfo (✉) · M. De Angelis
LUISS Guido Carli University, Rome, Italy
e-mail: marco.danolfo@gmail.com

M. De Angelis
e-mail: mdeangelis@luiss.it

C. Amatulli
University of Bari, Bari, Italy
e-mail: cesare.amatulli@uniba.it

G. Pino
University of Salento, Lecce, Italy
e-mail: giovanni.pino@unisalento.it

M.A. Gardetti (ed.), *Sustainable Management of Luxury*,
Environmental Footprints and Eco-design of Products and Processes,
DOI 10.1007/978-981-10-2917-2_20

brands' initiatives, we conducted a qualitative study on three well-known brands—Gucci, Tod's, and Bulgari—and examined their eco-sustainable collections and approaches. We supplemented this research with a survey of consumers' perceptions regarding luxury, sustainability, and CSR specifically by interviewing 200 consumers via an online questionnaire. From these data sources, we draw theoretical and managerial implications.

1 Sustainable Development in Modern Society

In the decades immediately following World War II, societal development was defined primarily in economic terms and measured by the growth in countries' gross domestic product (GDP) and per capita incomes. However, many of the production activities that arose from such development have had a detrimental impact on the environment. As a result, the notion of sustainability has risen to the forefront of public discourse, reflecting the emerging green orientation of companies and their customers (Bansal and Roth 2000). Sustainability captures the idea of satisfying the current generation's needs without compromising those of future generations (Fletcher 2008). In the Referencing a report from the World Commission on Environment and Development (1987), Seidman (2007: 58) notes that "sustainability is about much more than our relationship with the environment; it is about our relationship with ourselves, our communities, and our institutions."

Sustainability comprises three dimensions that reflect the complex and fluctuating environmental dynamics that affect human livelihoods and well-being (Langenwater 2009). The first dimension of sustainability is a *social* one, which assumes that sustainable development leads to a general improvement of life quality (Fukuda-Parr 2003). Development should be judged not only by the so-called "basket of goods", or the GDP guaranteed to a person or country, but also by the level of environmental protection and trust in social relationships; the inability to guarantee these conditions engenders a low quality of life (Gasper 2000). The second dimension is *environmental*, which builds on the idea that natural resources form the basis for satisfying material needs. According to this idea, human activities reduce natural resources and, when coupled with production emissions and waste, can compromise humans' ability to satisfy their needs (Glassberg 1998).

> "Man cannot live more than two or three minutes without air, or more than one or two days without water. We completely depend on what our ancestors called the 'Creator's gifts' or 'Nature's goodness.' Although these natural goods, with the exception of the Earth, are essential to men's lives, they have not been considered from economists." (de Jouvenel 1957)

Contrary to earlier beliefs, which saw natural resources as inexhaustible, modern information clearly demonstrates that raw materials are limited—both in terms of the productive resources that humans transform into commodities or goods (e.g., minerals, water, energy reserves, fauna, forest capital), and these resources' ability to absorb residues, emissions, and waste. Thus, the principle goal of sustainable production approaches involves reducing natural resource usage while producing eco-friendly goods and/or services that respect eco-efficiency principles and dematerialization standards.

The third dimension of sustainability is *economic*, which refers to the need to regulate resources usage across generations. Excessive levels of consumption are detrimental for the environment because they accelerate the extraction rates of non-renewable resources and shorten the rotation times of renewable ones. Therefore, people must reduce their present consumption if they hope to save resources for future generations.

2 Sustainability and CSR in the Luxury Industry

The increasing relevance of sustainability issues has driven luxury firms to invest in Corporate Social Responsibility (CSR), which embodies an enterprise's commitment to behaving in an ethical way that supersedes the mere observation of regulatory standards (Aguilera et al. 2007; Bhattacharya and Sen 2004; Kerr et al. 2009). In particular, CSR represents the set of initiatives that a company undertakes in order to assess and control the potential and real effects of its activities on both the environment and society at large. CSR may also refer to company efforts that exceed the requirements imposed by regulators or environmental protection groups. In short, the concept can be delineated into two dimensions: (1) *internal*, which includes human resources management, health and occupational safety, business organization, natural resource management, and effects on the environment; and (2) *external*, which encompasses local communities, economic partners, suppliers, consumers, respect for human rights across the whole supply chain, and global environmental issues.

For businesses, sustainability generally reflects their concern for the long-term impact of their activities on future generations and the environment. Therefore, CSR is a means by which companies formalize their commitment to sustainable development. The CSR department typically strives to support the company's economic goals while also developing initiatives aimed at increasing societal wellbeing—for instance, by respecting regulations and ethical principles, or preserving natural resources for future generations. However, being socially responsible requires more than simply respecting laws; companies are expected to invest in human capital, the environment, and their relationships with interested groups. As the European Commission wrote in their 2001 Green Paper "Promoting a European Framework for Corporate Social Responsibility":

> "The concept of corporate social responsibility basically means that companies spontaneously decide to contribute to improving society and to make the environment cleaner. […] Companies integrate social and environmental concerns in their business operations and in their interaction with their stakeholders on a voluntary basis. […] Being compared, in the 'globalization' context and particularly the internal market, to the challenges set by a

transforming environment, the enterprises are even more aware that social responsibility can perform as a direct economic value. Even if their main responsibility is that of taking advantage, companies are at the same time able to contribute to social objectives and environmental safeguarding by completing their social responsibility as a strategic investment for their commercial strategy in their management tools and in their actions." (European Commission 2001, p. 5)

Moreover, the Green Paper specifies that:

"By affirming their social responsibility and assuming spontaneous commitments that go beyond the regular and conventional needs with which they have to conform, however, companies should strive to raise the rules connected to social development, environmental safeguarding, and the respect of fundamental rights by adopting an open government system that is able to appease the profits of the various parts involved within a global approach of quality and sustainable development." (European Commission 2001, p. 8)

Given this emphasis on voluntary self-regulation, enterprises' participation in CSR allows them to act not only as an economic player, invested in their own outcomes, but also as a social actor that contributes to society's well-being. This extends companies' accountability to the social, economic, and political contexts in which they are embedded. To solidify this relationship, it is necessary that companies' performance be measured in terms of their contributions to economic prosperity, environmental preservation, and the social capital of their host country (Spence et al. 2004). Thus, a firm is usually considered socially responsible if: (1) it invests in the training and management of human resources with respect to equal treatment and equal opportunity principles; (2) it concretely limits the impact of its own activities on the environment by investing in eco-sustainable technologies and productive processes, and (3) it guarantees the integration and respect of social and environmental problems.

Nowadays, CSR is seen as a fundamental component of the company vision and identity, capable of reconciling economic and collective interests. In sum, CSR represents a cultural shift in how businesses operate across different functional areas. As the European Commission (2001) explains:

"[...]it doesn't mean to abdicate the profit in order to make good deeds, but instead to make profit, even doing more of it, but in a correct manner, producing social welfare and not to cause damages to the collectivity, equilibrium and equity and not exploitation, the quality of the life and not pollution and diseases, wealth and culture diffusion and not rapacious destruction of human and environmental resources." (p. 16)

2.1 *The Link Between CSR and Luxury*

Sustainability has become extremely important for luxury companies and their customers, but as the root term "Luxus" suggests, the ideas of luxury and sustainability do not readily intertwine. For luxury consumers, luxury goods are primarily about the delivery of satisfaction and pleasure rather than addressing social or environmental needs. This stems from the fact that luxury goods cost a great deal relative to their utility or economic potential. As a result, several people consider such goods to be superfluous, detached from real needs and characterized by vain superfluity (Sombart 2003). For this reason, "luxury" is commonly associated with "excess," "intemperance," "debauchery," "splendor," and "magnificence," (Aiello and Donvito 2006), which can invite criticism with regard to social responsibility (Achabou and Dekhili 2013; Beckham and Voyer 2014). Thus, when luxury consumers hear about CSR in the luxury industry, they may suspect that these companies are simply trying to soften public opinion around luxury products.

While some consumers might interpret luxury companies' CSR as an optional—and therefore worthless—practice, consumers as a whole are increasingly interested in social and environmental policies. They thus expect companies to not only adhere to regulations, but also support charity programs, preserve the environment, and cooperate with local institutions. In response, many luxury companies now position ethics and responsibility as primary corporate values. Consequently, these brands are undertaking important challenges related to developing more sustainability-oriented production processes while delivering high-quality goods and services (Janssen et al. 2014).

The intent of most CSR plans is to mutually accomplish intrinsic and extrinsic aims. For many companies, the extrinsic (or egoistic) aim is to increase brand awareness. The intrinsic aim, by contrast, revolves around "doing good", that is, fulfilling a series of obligations either imposed on or undertaken by the company. Across both goals, the brand seeks to become a node in people's memory that can be "anchored" to other information, with the intent of forming a favorable "bundle" of associations in consumers' minds (Aaker 1991; Kotler 2011). To this end, it is important that companies position their CSR activities as a pillar of the larger brand's corporate values, rather than as an ancillary set of actions. Otherwise, consumers may perceive such activities as incoherent and form a negative brand impression.

In short, consumers respond best to a CSR activity when it demonstrates the company's commitment to being socially responsible rather than merely bolstering the brand image. By focusing on the intrinsic aim, firms can establish a sense of connection between their brand(s) and consumer perceptions, a phenomenon called "consumer-company identification" (Bhattacharya and Sen 2003; Bhattacharya et al. 2011; Sen and Bhattacharya 2001), which may foster a sense of attachment with the company (Fournier 1998). Companies that behave in a socially responsible manner are more attractive targets for identification. This process may foster a strong attachment among some consumers, who might then become "brand advocates" capable of positive word-of-mouth. Conversely, consumers may punish companies that create a false sense of social responsibility.

3 Case Studies: The CSR Approaches of Luxury Companies

The present research focuses on three leading luxury companies—Gucci, Bulgari and Tod's—that are making considerable efforts to act in a responsible way and be perceived as good corporate citizens. These companies are modifying many of their internal processes in order to increase the value of their offerings while simultaneously generating positive effects for society at large. We particularly selected these companies because they illustrate how the broader luxury industry could adopt new, eco-friendly management models as a prompt response to social, environmental and economic concerns. If successful, these three firms could become a benchmark for others in the luxury industry.

3.1 Gucci

3.1.1 Company Introduction

Gucci is a famous Italian brand and a symbol of style, class, high fashion, and luxury. It was created by Guccio Gucci in the 1920s in Florence (Gucci 2015a). The company specializes in leather products, which were originally designed for the English nobility. Because sophisticated persons indulging in horse hobbies comprised the chief commissions of that era, Gucci adopted two exclusive brand markings to appeal to those patrons: the clamp, which later became a fashion icon, and the green-red-green weft ribbon, which is reminiscent of a saddle girth. Nowadays, the company strongly invests in its local area and in each activity of the production chain, employing around 8000 people in its productive area and around 45,000 employees in the whole of Italy. In addition to local investment, Gucci's values include quality, exclusivity, a balance between modernity and tradition, and a "Made in Italy" commitment. Together, these values have made the firm one of the most popular Italian brands in the world (Gucci 2016).

3.1.2 Gucci's CSR Approach

Modern social and environmental problems have led Gucci to undertake a series of activities designed to sustain eco-friendly policies that guarantee high-level qualitative standards among their commodities while respecting labor rights (Gucci 2015b). In 2004, the company undertook a social liability certification process (SA 8000) involving their administrative offices, shops, and suppliers. Since 2008, the company has had a division exclusively designed for such purposes.

Gucci's CSR approach focuses on three main areas (Gucci 2015c): (1) human resource protection and promotion (e.g., labor rights, solidarity towards the most disadvantaged groups); (2) political and economic objectives and efforts to maintain art and culture, and (3) environmental impact management in all phases of the value chain by supporting the creation of sustainable commodities. In order to realize its program, the firm's CSR division coordinates with other internal departments.

Gucci wants to promote the idea that sustainable value, which has become an essential element of the company's mission statement, can and should be a cornerstone of its management policies and behaviors. The firm also emphasizes the importance of communicating with stakeholders, who hold the right to be informed of all processes that impact production. In a CSR perspective, consumers have the right to know where a product's raw materials originate from, so that they can feel confident that their consumption does not negatively affect the environment or other persons (Hennigs et al. 2013).

In 2009, the Italian Charter for Equal Opportunities and Equality at Work was introduced to guarantee equal dignity and working conditions to all employees via periodic inspections of firms' compliance with regulatory standards. In the same year, Gucci created the Committee for CSR and signed an agreement with the most important national labor unions. The purpose of these initiatives was to plan actions that would promote, valorize, and sustain Gucci as a unique icon of knowledge and value in terms of economic, environmental, and social sustainability. As part of this treaty, the company agreed "to valorize and to promote the production chain as a unique patrimony of knowledge and recognized value, in a logical of economic and social sustainability" (Italian Charter for Equal Opportunities and Equality at Work 2011).

In 2010, Gucci obtained the ISO 14001 environmental certification, an important outcome of several eco-friendly initiatives that reduced the environmental impact of its production activities. Moreover, Gucci implemented a moratorium against the raising of cattle livestock in the Amazon in response to an announcement by Greenpeace. These implementation measures called for an energy revolution to halt or at least slow climate change, protect marine reserves and prosecute unauthorized fishing, protect primary forests, namely those still "intact" or only slightly contaminated by human activities, eliminate toxic substances from production processes, and promote sustainable agriculture by protecting biodiversity and avoiding GMOs.

3.1.3 Eco-friendly Collections

Social responsibility, environmental protection, and economic sustainability form the basis of Gucci's commercial policies and marketing strategies. For this reason, the company designed a new green logo that embodies its commitment to these CSR goals.

> "A new logo is associated with our identity because we have thought that, from now, it was correct to identify our commitment in this area with a dedicated logo. These values are always part of the company DNA and have been carried out with great strength in the last years, through policies of social and environmental sustainability in the broad sense." (Patrizio di Marco, chairman and CEO of Gucci, Wise Society 2012)

CSR and Sustainability Director Rossella Ravagli described Gucci's green approach as:

> "…a very articulated and difficult path that asks for a detailed work and a necessary gradualness. Gucci's commitment has been forward-looking however, a sign, therefore, of a politics oriented towards the environmental and social sustainability undertaken by the group." (Rossella Ravagli, head of Corporate Sustainability & Responsibility Department, Wise Society 2013)

The new project is congruent with Gucci's responsibility mission of guaranteeing sustainable commodities to the modern consumer by balancing timeless style and love for the future (Gucci 2015b). The sections below describe these efforts in more detail.

Sunglasses: In 2013, Gucci and Safilo (the world leader in the luxury eyewear industry) announced the launch of four new models of glasses geared towards sustainability. The products in this collection were made with an innovative acetate that contained natural materials, contrary to those traditionally used in frames (Safilo 2013). To sensitize young consumers to sustainability, the brand also introduced the mini collection "Gucci Eyeweb," which included two bio-based sunglasses composed of a material extracted from castor oil seeds. Furthermore, the firms tailored the production of these items to reduce CO_2 emissions. These products were paired with a sunglasses frame made of liquid wood, a biodegradable and eco-friendly material derived from wood fiber (from forests with controlled reforestation), lignin (from paper) and natural waxes that replace traditional plastics.

Shoes: Gucci has launched a new line of eco-friendly men's and women's shoes designed by Creative Director Frida Giannini and featuring soles made of green material (Gucci 2015d). Named "Green Marola" (for women's ballerina shoes) and "Green California" (for men's sneakers), both shoes utilize bioplastic, a biodegradable material that substitutes for traditional plastic. It is an eco-sustainable material successfully tested in laboratories and certified in agreement with the UNI EN 13432 and ISO 17088 European and International standards. This material decomposes faster than traditionally manufactured plastics and does not produce any waste emission, thus limiting its environmental impact. The Green Marolas are entirely made of bioplastic, while the Green California feature a biodegradable rubber sole and an upper part in leather, laces in certified biological cotton, and the green Gucci logo impressed on a recycled polyester label.

Packaging: Gucci redesigned their packaging to be made with Forest Stewardship Council (FSC)-certified 100% recyclable paper. This new packaging is smaller and lighter, thus requiring fewer transportation loads and contributing to a 60% reduction in CO_2 emissions (Gucci 2015e). The new packaging features a modern and refined design emblazoned with the Möbius[1] symbol, an image of product recyclability employed and recognized all over the world. Buyers of Gucci products also receive a small informative leaflet and a pre-addressed envelope so that they can learn how to recycle the old object when buying a new article.

3.2 Tod's Group

3.2.1 Company Introduction

Tod's S.p.A ranks 38th among the top 100 luxury firms (Deloitte 2015), with 1.3 million USD in revenue. It is a holding company and a principal player in the production and commercialization of luxury footwear and leather goods, including the Tod's and Hogan brands; it also operates in the apparel sector with the Fay brand (Tod's Group 2015a). Each brand in the group is characterized by a balanced mix of tradition and modernity, high quality materials, creativity, and wide usability in every commodity. Because of specialization throughout the production chain, each item is handmade using craftsmen's techniques so that the end product is exclusive and highly recognizable. Each brand features iconic pieces, such as the Tod's *Driving Shoe* and *D Bag*, and Hogan's *Traditional* and *Interactive* footwear. Production occurs across both group factories (six for the footwear and two for the leather goods) and a small number of specialized laboratories, with which Tod's has long-established funding relationships. Material purchasing, production monitoring, and control stages are overseen by the group leader, even for products manufactured in external laboratories.

3.2.2 Tod's CSR Approach

Tod's embraces principles such as creativity, craftsmanship, tradition, "Made in Italy," and solidarity. Since 2008, the company has employed an Ethical Code and a Model of Organization, Management, and Control as a guideline for conducting training aimed at implementing such principles (Tod's Group 2015b). The group sees this support of human capital as an essential asset to its own activities, as well as a way to benefit the stakeholders in its countries of operation. Furthermore, Tod's devotes 1% of its net profits to supporting the poorest resident populations in its operating areas (Fondazione il Sole 2013). In this way, the company shares its own profits with the areas from which it draws resources, thus strengthening the group's bond with its local audience. Tod's Group is also committed to improving Italy's

[1]In mathematics, and more precisely in topology, the Möbius strip is an example of non-orientable surface and a grooved surface. It takes its name from the German mathematician August Ferdinand Möbius.

international reputation and preserving the nation's cultural heritage. It is a founding member of Teatro della Scala of Milan, and, in 2011, entirely financed a series of restoration works for the Colosseum (for a total of 25 million euros) in collaboration with the Ministry of Cultural Heritage and Activities and the Special Superintendence for the Archaeological Goods in Rome (Tod's Group 2016). This initiative aimed to protect and promote Italian culture, inspired by the company's conviction to invest in the "Made in Italy" concept, and hence in the abilities and traditions of local companies.

3.2.3 Eco-friendly Collections

Even though past criticisms were what led the company to behave more ethically and implement social responsibility management plans, Tod's has initiated numerous programs oriented toward environmental themes. One such priority is to decrease energy consumption in the company's production and retail establishments. With the help of specialized suppliers, the group can reduce electrical consumption by tailoring its buildings to be low energy. In its Italian headquarters, for instance, Tod's uses renewable energy resources (e.g., geothermal, photovoltaic). Meanwhile, a new building that has been under construction since 2014 abides by the newest green construction philosophies, featuring several floors composed of plant components and other natural materials. The group also embeds motor devices with *inverter* technology into its facilities, which reduce energy consumption by enabling operation modification in response to real needs. Additionally, Tod's utilizes LED[2] technology and recycled wood in both its Italian headquarters and the retail stores that compose the company's distribution network, further demonstrating the group's responsible use of resources.

Even though Tod's does not produce "green" products, the group engages in a series of activities aimed at preserving environmental resources. For instance, the firm prioritizes the environmental protection and sustainable development of the territories in which it operates, striving to minimize the adverse environmental impact of its business activities. To this end, the firm respects a series of European environmental regulations concerning emissions and land utilization. Furthermore, Tod's promotes activities and processes compatible with the environment, such as pursuing energy efficiency and the sustainable use of resources; assessing the environmental impact of all activities and business processes; collaborating with stakeholders, internal (e.g., employees) and external (e.g., public institutions), in order to optimally manage environmental problems, and contributing to environmental protection through the implementation of suitable management and monitoring systems.

[2]LED (Light Emitting Diode) is an optoelectronic device that uses the optical properties of some semiconductor materials to produce photons through a phenomenon of spontaneous emission.

3.3 *Bulgari*

3.3.1 Company Introduction

The Bulgari Company was founded in Naples in 1884 by Sotirios Voulgaris (in the original Greek; Bulgaris in Italian) and his business partner Demetrio Kremos (Bulgari 2015a). Bulgari drew inspiration from the Greek and Roman cultures for its constant brand renovation. The brand comprises six different Strategic Business Units (see Fig. 1): the percentages listed in Fig. 1 represent the incomes of each SBU (authors' elaboration of Bulgari's 2014 financial statement). With an overall turnover of 1.6 million euros, the brand is one of the most profitable subsidiaries of the Louis Vuitton Möet Hennessy (LVMH) group, a leading luxury goods conglomerate. Bulgari owns 180 retail stores around the world and plans to achieve further growth through its emphasis on such values as harmony and refinement.

3.3.2 Bulgari's CSR Approach

Bulgari's approach to CSR involves balancing the social, environmental, and economic values inherent to its strategies and activities while promoting long-term social and environmental sustainability (Bulgari 2015b). To this end, the firm has defined and continuously implemented an advanced ethical code and internal policy, as well as established reference values intended to promote diversity, prevent discrimination, and safeguard employee health. Internally, the group seeks to ensure safe job conditions, offer ongoing training to employees, and build a collaborative and conscious working environment (cf. Feltus, Petit and Dubois 2009) that values responsible behaviors, such as environmental protection. These principles are solidified in Bulgari's code of conduct (Bulgari 2015c), which relegates the provisioning of precious stones. Regarding community impact, Bulgari's strategy involves promoting youth education all over the world through the support of the World Scout Organization and other global or local associations devoted to the same objective. The firm also rewards improvements to its production chain by giving consideration to its nearest company partners. In 2014, for instance, Bulgari awarded its partner company Zordan with the Bulgari Design Award for sustainable construction.

3.3.3 Eco-friendly Collections

As said before, Bulgari's code of conduct regulates its supply of precious stones. Conflict diamonds (also called "blood diamonds") present a serious issue—for example, in those parts of the world where rebel movements use untreated precious stones to finance wars against governments. Therefore, companies cannot purchase

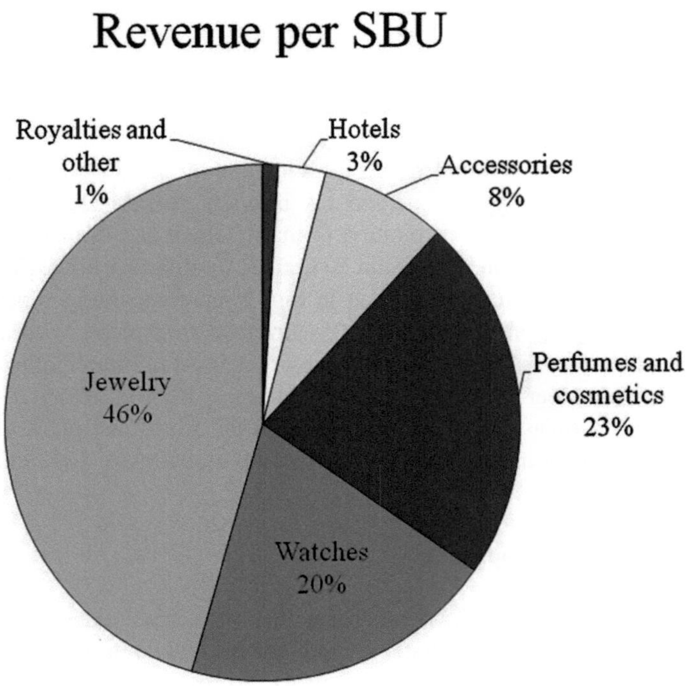

Fig. 1 Pie chart of Bulgari's SBUs

raw materials from such places without the risk of becoming complicit in atrocious felonies. As such, Bulgari exclusively purchases carved stones from qualified suppliers in the Kimberley Process (Bulgari 2015d). This is a combined initiative among governments, industries, the civil society, and companies that aims to stop the flow of conflict diamonds via an international control system of certified trade. In addition, Bulgari is a long-time member of the Responsible Jewelry Council (RJC), a non-profit organization that promotes ethically responsible business practices in order to safeguard human rights, in both social and working environments, across the whole jewelry production chain.

All companies directly and indirectly supervised by Bulgari have adopted policies and management systems to guarantee the respect of the following important principles: (1) *ethics in business*: to develop activities with respect to ethical norms, integrity, transparency, and conformity with enforced laws; (2) *human rights*: to respect the dignity of the individual according to the Universal Declaration of Human Rights enacted by the United Nations, and (3) *environment*: to suitably eliminate or minimize the adverse effects of production activities on the environment (Bulgari 2015c). In addition to earning the Green Certificate for the renewable energy of its centers, Bulgari also obtained LEED Certification

(Leadership in Energy and Environmental Design),[3] which attests to the minimal environmental impact of its buildings. Bulgari also aims to create optimal working environments where safety and prevention serve as cornerstone values. By advocating for the inestimable value of human life, the company is able to encourage environmentally conscious behaviors and delineate individual employees' ethical responsibilities.

Since 2012, the company has audited its own suppliers to confirm their respect for social and environmental responsibility in the distribution chain. In 2013, during the LVHM group's celebrations for 20 years of environment engagement, Bulgari launched *Green Week*: a week devoted to the environment (Bulgari 2015e). During this time, the company proposed activities meant to sensitize employees to environmental matters while also inaugurating a series of CSR projects linked to the green philosophy. For example, during the week's *Info Pills* campaign, which strove to spread information about new recycling measures, Bulgari's employees received gifts of ceramic cups for warm and cold drinks as substitutes for plastic glasses. The week ended with the sharing of good environmental practices across Bulgari's various centers in different countries.

4 Empirical Research

4.1 Objectives

To supplement the above analysis, we administered an online questionnaire with 19 questions to a sample of 200 Italian consumers (female 34.2%, male: 65.8%). We used this questionnaire (see Appendix) to understand the real impact that CSR policies have on consumers' behaviors and their perceptions of luxury brands and products.

4.2 Methodology

The survey was articulated into four parts that assessed respondents': (1) frequency of luxury purchases or reason for having never bought luxury goods; (2) knowledge of sustainability issues, CSR, and the luxury sector; (3) knowledge of the studied companies (Gucci, Tod's, Bulgari) and their green operations; and (4) demographic data.

[3]LEED is a voluntary certification program that can be applied to any type of building (both commercial and residential) and concerns the entire life cycle of the building, from design to construction.

4.3 Results

Regarding the frequency of luxury purchases, 41% of respondents considered themselves occasional luxury consumers, whereas 16% considered themselves usual luxury consumers (see Fig. 2). We found that consumption frequency began to increase among consumers aged between 26 and 35, who have more solid working positions and thus more opportunities to purchase luxury goods and/or services. The main obstacles to luxury purchases included economic impossibility (44%) and mismatch with the consumer personality (35%) (see Fig. 3).

In the second part of the survey, we introduced the topics of CSR, sustainability, and the relationship between these themes and the luxury sector, and assessed respondents' perceptions of these topics. More than half of the respondents (51.5%) had never heard of CSR. The most thorough definitions of sustainable development and CSR given by respondents were as follows:

> "All those actions intended for the use of energies derived from renewable and non-polluting sources, which are also cheap and eco-friendly for the company." (Female consumer, aged 18)

> "Sustainable development: a series of initiatives aimed at curbing the planet's decay through actions in respect of our environment; CSR: corporate policies that relate to compliance with social, environmental, and productive dimensions in the pursuit of sustainability." (Male consumer, aged 27)

> "Being able today to produce, without erasing the possibilities for future generations of living with dignity." (Female consumer, aged 35)

> "The implications of moral and ethical nature that may exist within the business strategy." (Male consumer, aged 51)

We asked respondents to indicate what dimensions of sustainability among the social, environmental, and economic ones, they considered most important and allowed them to provide more than one preference. Results revealed a significant preference for the environmental (140 responses) and social dimensions (116 responses). A lower number of respondents selected the economic dimension

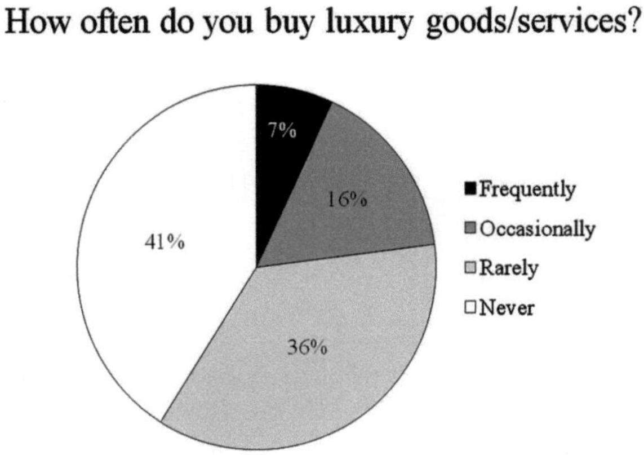

Fig. 2 Responses regarding the frequency of luxury consumption

Fig. 3 Reasons why people never bought luxury goods

(36 responses). These answers suggest that the environment and people's quality of life are closely linked. A subsequent question, however, revealed that only 27.3% of the respondents knew of luxury companies who care about sustainability and CSR.

Respondents were also asked to judge the production chain activities of luxury companies. Given a scale from one (negative judgment) to seven (positive judgment), the majority of respondents positioned the three companies in the range from one to three, thus indicating an overall more negative judgment. When asked to justify their responses, some of the respondents commented as follows:

> "Because I think they do not pose any problem." (Male consumer, aged 26)

> "These are processes that cause pollution, social and economic inequality, and conflict between employees and managers." (Female consumer, aged 28)

> "Often in their scale of values, sustainability is located at the lowest point of the pyramid." (Female consumer, aged 28)

> "Few companies today respect workforce rights or use green materials." (Male consumer, aged 34)

> "Exploitation of the poorest countries." (Male consumer, aged 57)

This part of the questionnaire also included multiple choice (Yes/No) questions that assessed respondents' perceptions about the ethicality of luxury companies and their own willingness to buy green luxury goods (Fig. 4). Over 70% of the respondents expressed skepticism toward the ethicality of luxury companies, asserting that CSR mainly represents a source of added value for these companies. Such perceptions reflect the belief that companies only adopt sustainable activities and CSR policies to distract and appease the public. Consumers who hold this point of view may push for luxury companies to rediscover their ethical and moral values (Abela and Murphy 2008) and thereby recover a positive image among consumers. Granted, it takes very little to destroy a company's brand image, but a very long time and considerable investment to recover its original position. Nevertheless, about 70% of respondents claimed that they were willing to buy green luxury goods and pay a premium price for them (Fig. 4).

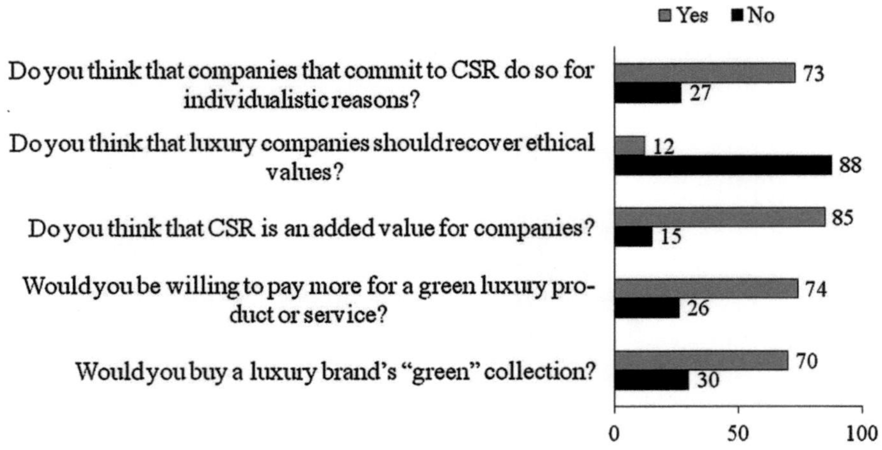

Fig. 4 Personal opinions of consumers about luxury companies and their CSR activities

In the third part of the survey, we asked respondents a series of questions related to Gucci, Tod's, Bulgari, and their green operations. First, we asked if participants knew of these companies; all of them did, as expected. We then asked respondents to share their opinions regarding whether and to what extent the studied companies are committed to CSR and sustainable development. Of the responses, 51.5% answered that the companies are "not committed to CSR and sustainability issues," while the remaining 48.5% claimed these brands have joined the ranks of the world's so-called *green companies*, but that their results are not satisfactory. We also asked respondents if they would buy the green collections produced by the three brands; the positive responses confirmed that sustainability may not scare consumers during the purchase phase. Finally, participants provided their gender, age, and occupation.

5 Discussion and Implications

The three luxury companies assessed in this chapter appear to be particularly sensitive to sustainable development. As a result, they have made considerable efforts to reduce the environmental impact of their activities, develop eco-compatible products and production processes, and contribute to societal wellbeing. In particular, Gucci focuses on the use of new materials for its products and packaging; Tod's emphasizes non-profit initiatives, and Bulgari invests in responsible production approaches and work environments. In addition, each company communicates its commitment to CSR in a singular way: Gucci uses green logos, Tod's develops cultural initiatives, and Bulgari formalizes its respect for ethical standards with certifications. Through those initiatives, these leading luxury companies demonstrate their sensitivity to relevant societal, economic, and environmental issues. They also hope to establish trusted relationships with their customers and foster a sense of consumer-company identification. However, consumers often lack awareness of these activities and thus do not identify with the companies that perform them; at best, consumers may take a company's CSR activities on good faith. To close this gap, luxury companies, such as those considered in this research, should cultivate their own ethical and moral values while increasing consumer awareness about their CSR commitment, emphasizing that CSR is a central to their corporate strategy rather than a mere secondary activity (Castaldo et al. 2009; Grappi et al. 2013).

Our survey ascertained that a significant number of consumers are still skeptical about the examined companies' CSR approaches. However, consumers seem willing to support these companies' efforts to act in a responsible way, so long as said companies are able to demonstrate a stronger commitment to social, economic, and environmental issues. Because modern consumers use alternative information sources, such as the Internet, to cultivate their understandings and opinions, they are less prone to accepting advertisements as truth. Further, they spend an increasing amount of their shopping time online—to the point that online purchases of personal luxury goods will represent almost 20% of the total amount by 2025 (McKinsey 2016). Companies with an online presence can use their websites to explain their production processes or material sources, or perhaps even devote entire pages to their CSR efforts, thus highlighting their contributions to public wellbeing. Using these communication platforms to spread awareness could be an important first step toward improving consumers' opinions about luxury brands.

As it happens, Gucci, Tod's and Bulgari are already very active in the digital sector (Contactlab 2015) and experiencing considerable growth. Nonetheless, these companies could enhance their positioning by broadcasting a green message to counteract any negative reputations. However, it is not enough to simply talk about these themes; consumers will likely be persuaded more by seeing the results of CSR activities. Moreover, companies must communicate that their initiatives constitute an integral part of a larger development policy aimed at benefiting all stakeholders, both internal and external.

Another way to create awareness is through green marketing activities, which comprise the set of initiatives aimed at planning and executing the marketing mix that is best able to facilitate consumption, production, distribution, promotion, packaging, and product reclamation in a manner that is sensitive or responsive to ecological concerns (Dahlstrom 2010). Through green marketing activities, luxury companies might propose and support new consumer lifestyles (Davies et al. 2012), thereby presenting customers with the opportunity to enhance their sense of identification with the company and the luxury brand. Those objectives could become particularly important for companies wanting to be perceived as ethical and responsible corporate citizens.

6 Conclusions

The results of this research suggest that leading luxury firms such as Gucci, Tod's and Bulgari invest considerable resources to ensure responsible operations in areas ranging from production and distribution processes, to working environments, to non-profit initiatives. In spite of this, consumers may suspect that these ethical initiatives mainly serve to increase these firms' profitability rather than meaningfully contribute to societal and environmental problems. To improve consumers' trust and support, luxury companies may need to better communicate their values, responsible actions, and non-profit goals. It is important, however, that consumers perceive sustainability initiatives as part of a structured production approach aimed at generating, re-investing, and re-distributing economic value in an ethical and responsible way, rather than as isolated actions design to improve the company's image and visibility. Given the importance of all CSR initiatives to a brand's health, luxury companies cannot afford to neglect one activity for another. Instead, they should seek to simultaneously benefit stakeholders, society, and the environment.

Appendix

Part I. Luxury purchases

1. *How often do you buy luxury goods/services?* (multiple choice question)
   - Frequently
   - Occasionally
   - Rarely
   - I never buy luxury products
2. *If you've never bought, why?* (multiple choice question)
   - I don't like the offers of luxury companies
   - I feel uncomfortable when I go in luxury stores
   - I believe that luxury goods offer nothing more than the common brand
   - Luxury goods don't match with my personality
   - Economic impossibility
   - Other

Part II. Sustainable development, Corporate Social Responsibility (CSR) and the luxury industry

3. *Have you ever heard about sustainable development and CSR?*
   - Yes
   - No
4. *If you heard about them, can you define these concepts with your own words?* (open-ended question)

5. *In your opinion, what dimensions of sustainability are most important?* (You can select more than one dimension)
   - Social: improving lifestyle quality
   - Enviromental: satisfying consumer needs without depleting natural resources
   - Economic: creating economic value for present and future generations
6. *Do you know any luxury company that has implemented sustainable production processes and/or CSR strategies?*
   - Yes
   - No
7. *How do you judge luxury companies' production chain* activities?* (1 = Negatively; 7 = Positively)
   *: all the stages of making a product, considered together.
8. *If you judged it negatively, why?* (open-ended question)
9. *Do you think that companies that implement sustainable choices and commit to CSR do so for individualistic reasons?*
   - Yes
   - No
10. *Do you think that luxury companies should recover ethical and moral values?*
    - Yes
    - No
11. *Do you think that CSR is an added value for companies?*
    - Yes
    - No
12. *Would you be willing to pay more for a green luxury product/service?*
    - Yes
    - No
13. *Would you buy a luxury brand's "green" collection?*
    - Yes
    - No

Part III. Three leading luxury companies: Gucci, Tod's, and Bulgari

14. *Do you know the luxury brands Gucci, Tod's, or Bulgari?*
    Mark the answer "other", if one of them is unknown to you.
    - Yes
    - No
    - Other

15. *Do you think that the three brands are committed to the topics of sustainable development and CSR?*
    - Yes
    - No
16. *Would you buy some of their "green" articles?*
    - Yes
    - No

Part IV. Socio-demographic data

17. *Gender*
    - Male
    - Female
18. *Age*
    18–25; 26–35; 36–50; 51–65; Over 65.
19. *Occupation* (open-ended question)

References

Aaker D (1991) Managing brand equity. The Free Press, New York

Abela AV, Murphy PE (2008) Marketing with integrity: ethics and the service-dominant logic for marketing. J Acad Mark Sci 36(1):39–53

Achabou MA, Dekhili S (2013) Luxury and sustainable development: is there a match? J Bus Res 66(10):1896–1903

Aguilera RV, Rupp DE, Williams CA, Ganapathi J (2007) Putting the S back in corporate social responsibility: a multilevel theory of social change in organizations. Acad Manag Rev 32 (3):836–863

Aiello G, Donvito R (2006) L'evoluzione del concetto di lusso e la gestione strategica della marca. Un'analisi qualitativa delle percezioni sul concetto, sulla marca e su un prodotto di lusso. In 5th international congress marketing trends, Venice, Italy, 20–21 Jan

Bansal P, Roth K (2000) Why companies go green: a model of ecological responsiveness. Acad Manag J 43(4):717–736

Bhattacharya CB, Sen S (2003) Consumer-company identification: A framework for understanding consumers' relationships with companies. J Mark 67(2):76–88

Bhattacharya CB, Sen S (2004) Doing better at doing good: when, why, and how consumers respond to corporate social initiatives. Calif Manag Rev 47(1):9–24

Bhattacharya CB, Sen S, Korschun D (2011) Leveraging corporate responsibility: the stakeholder route to maximizing business and social value. Cambridge University Press, Cambridge

Beckham D, Voyer BG (2014) Can sustainability be luxurious? A mixed-method investigation of implicit and explicit attitudes towards sustainable luxury consumption. Assoc Consum Res 42:245–250

Bulgari (2015a) Dalle Origini agli Anni '40. Retrieved on 25 May 2015 at: http://www.bulgari.com/it-it/from_the_origins

Bulgari (2015b) La Nostra Visione della CSR. Retrieved on 25 May 2015 at: http://www.bulgari.com/it-it/csr-in-bulgari-discover-explored

Bulgari (2015c) Bulgari code of conduct. Retrieved on 10 Jan 2016 at: http://www.bulgari.com/en-us/code_of_ethics

Bulgari (2015d) Bulgari and conflict diamonds. Retrieved on 12 Dec 2015 at: http://www.bulgari.com/en-us/csr_committment_resposible_sourcing

Bulgari (2015e) Green week. Retrieved on 3 Dec 2015 at: http://www.bulgari.com/en-us/special_initiatives

Castaldo S, Perrini F, Misani N, Tencati A (2009) The missing link between corporate social responsibility and consumer trust: the case of fair trade products. J Bus Ethics 84(1):1–15

Contactlab (2015) The online purchase experience ranking. http://www.contactlab.com/en/resources/the-online-purchase-experience-ranking. Accessed 20 Dec 2015

Dahlstrom R (2010) Green marketing management. Nelson Education, Toronto

Davies IA, Lee Z, Ahonkhai I (2012) Do consumers care about ethical-luxury? J Bus Ethics 106 (1):37–51

De Jouvenel B (1957) L'économie politique de la gratuité. Arcadie, Essais sur le Mieux-Vivre

Deloitte (2015) Global powers of luxury goods. http://www2.deloitte.com/it/it/pages/consumer-business/articles/gx-cb-global-powers-of-luxury-goods.html. Accessed 16 Sept 2015

Egri CP, Ralston DA (2008) Corporate responsibility: a review of international management research from 1998 to 2007. J Int Manag 14(4):319–339

European Commission (2001) Green paper. Promoting a European framework for corporate social responsibility. www.europa.eu/rapid/press-release_DOC-01-9_en.pdf. Accessed 2 Oct 2015

Feltus C, Petit M, Dubois E (2009) Strengthening employee's responsibility to enhance governance of IT: COBIT RACI chart case study. In: Proceedings of the first ACM workshop on information security governance, Chicago, 13 Nov 2009, pp 23–32

Fletcher K (2008) Sustainable fashion & textiles: design journeys. Earthscan, Oxford

Fondazione Il Sole (2013) Solidarietà/Appello di Diego Della Valle: « Le aziende diano l'1% degli utili a progetti di solidarietà » http://www.fondazioneilsole.it/solidarieta-appello-di-diego-della-valle-lle-aziende-diano-l1-degli-utili-a-progetti-di-solidarietar. Accessed 10 July 2015

Fournier S (1998) Consumers and their brands: developing relationship theory in consumer research. J Consum Res 24(4):343–353

Fraj E, Martinez E (2007) Ecological consumer behaviour: an empirical analysis. Int J Consum Stud 31(1):26–33

Fukuda-Parr S (2003) The human development paradigm: operationalizing Sen's ideas on capabilities. Feminist Econ 9(2–3):301–317

Gasper D (2000) Development as freedom: taking economics beyond commodities. The cautious boldness of Amartya Sen. J Int Dev 12(7):989–1001

Glassberg D (1998) Spectacular nature: corporate culture and the sea world experience. Environ Hist 3(4):534

Grappi S, Romani S, Bagozzi RP (2013) Consumer response to corporate irresponsible behavior: moral emotions and virtues. J Bus Res 66(10):1814–1821

Gucci (2015a) Gucci history—1920s. Retrieved on 3 Mar 2015 at: http://www.gucci.com/cn-en/worldofgucci/articles/history-1920

Gucci (2015b) People. Social responsibility across the value chain. Retrieved on 20 Apr 1015, at: https://www.gucci.com/us/en/st/sustainability-landing/people

Gucci (2015c) Corporate sustainability & responsibility. Retrieved on 3 Mar 2015 at: https://www.gucci.com/us/en/st/sustainability-landing

Gucci (2015d) Gucci sustainable soles. Retrieved on 26 Mar 2015 at: http://www.gucci.com/cn-en/worldofgucci/articles/the-sustainable-soles

Gucci (2015e) Operating efficiently in respect of the planet. Retrieved on 19 Oct 2015 at: https://www.gucci.com/us/en/st/sustainability-landing/environment

Gucci (2016) The house of Gucci. Made in Italy. Retrieved on 20 Oct 2015 at: http://www.gucci.com/cn-en/worldofgucci/articles/made-in-italy

Hennigs N, Wiedmann KP, Klarmann C, Behrens S (2013) Sustainability as part of the luxury essence: delivering value through social and environmental excellence. J Corp Citizensh 52:25–35

Italian Charter for Equal Opportunities and Equality at Work (2009) http://www.cartapariopportunita.it/contenuti/home.aspx. Accessed 11 May 2015

Janssen C, Vanhamme J, Lindgreen A, Lefebvre C (2014) The catch-22 of responsible luxury: effects of luxury product characteristics on consumers' perception of fit with corporate social responsibility. J Bus Ethics 119(1):45–57

Joy A, Sherry JF Jr, Venkatesh A, Wang J, Chan R (2012) Fast fashion, sustainability, and the ethical appeal of luxury brands. Fashion Theory 16(3):273–295

Kapferer JN (1997) Managing luxury brands. J Brand Manag 4(4):251–259

Kapferer JN (2012) The luxury strategy: break the rules of marketing to build luxury brands. Kogan Page Publishers, London

Kapferer JN, Michaut-Denizeau A (2014) Is luxury compatible with sustainability? Luxury consumers' viewpoint. J Brand Manag 21(1):1–22

Kerr M, Janda R, Pitts C (2009) Corporate social responsibility: a legal analysis. LexisNexis, Toronto

Kotler P (2011) Marketing insights from A to Z: 80 concepts every manager needs to know. Wiley, Hoboken

Langenwater G (2009) Planet first. Ind Manag 51(4):10–13

McKinsey (2016) Digital luxury experience 2015. https://www.mckinsey.it/idee/ecommerce-nel-2025-il-lusso-varra-il-18-delle-vendite. Accessed 20 Oct 2015

Safilo (2013) Gucci presents new version of sunglasses in liquid wood material. Retrieved on 6 Sept 2015 at: http://www.safilogroup.com/en/PR_2013-09-30_gucci-presents-new-version-of-sunglasses-in-liquid-wood-material

Seidman D (2007) How we do anything means everything. Wiley, Hoboken

Sen S, Bhattacharya CB (2001) Does doing good always lead to doing better? Consumer reactions to corporate social responsibility. J Mark Res 38(2):225–243

Sombart W (2003) Dal Lusso al Capitalismo, a cura di R. Armando Editore, Sassatelli, Roma

Spence L, Habisch A, Schmidpeter R (2004) Responsibility and social capital. The world of small and medium sized enterprises. East Econ J 32(3):567–569

Tod's Group (2015a) Storia. Retrieved on 6 June 2015 at: https://www.todsgroup.com/it/group/history

Tod's Group (2015b) Il codice etico. Retrieved on 30 Apr 2015 at: https://www.todsgroup.com/it/sostenibilita/codice-etico/

Tod's Group (2016) Collaboration agreements. https://www.todsgroup.com/en/sostenibilita/partnerships. Accessed 16 Mar 2015

Wise Society (2012) Gucci: un nuovo logo per le attività di sostenibilità economica. Retrieved on 7 July 2014. http://wisesociety.it/economia-e-impresa/gucci-un-nuovo-logo-per-le-attivita-di-sostenibilita-economica/

Wise Society (2013) Gucci: tracciabilità delle etichette e sicurezza delle persone. Ecco la nostra moda sostenibile. Retrieved on 7 July 2014 at: http://wisesociety.it/incontri/gucci-tracciabilita-delle-etichette-e-sicurezza-delle-persone-ecco-la-nostra-moda-sostenibile/

World Commission on Environment and Development (1987) Our common future (Chapter 2: Towards Sustainable Development; Paragraph 1). The Brundtland Commission to the United Nations. Oxford University Press, Oxford. http://www.un-documents.net/ocf-02.htm#I. Accessed 1 Feb 2010

Anything Worth Doing: The Ambiguity of Values in Sustainable Luxury

Nadine Hennigs, Evmorfia Karampournioti and Klaus-Peter Wiedmann

Abstract The 21st century is believed to be the rise of the ethical consumer who is concerned about a broad spectrum of issues ranging from the environment and animal welfare to societal concerns, including human rights. Ethical and environmental consumerism is regarded as a mainstream phenomenon in contemporary consumer culture related to all product categories and continues to build momentum around the world. Existing research finds evidence that even the average consumer prefers e.g., fair trade coffee and chocolate, organic cotton and cosmetics produced without animal testing. The ethical buyer, who is "*shopping for a better world*", is increasingly concerned about the consequences of consumption and "*intents to make certain consumption choices due to personal and moral beliefs and values*". Since the consumption of luxury provides the possibility to *express the deepest values*, there appears to be a close association to the concept of ethical consumerism. Nevertheless, a critical perspective on the economic reality supports the assumption that there exists an enormous gap between articulated individual values and behavioral intention and actual shopping behavior. Against this backdrop, the following key question has arisen: *Is the ethical consumer little more than a myth?* To shed light on this important question with special focus on the luxury market, in our paper, we investigate the relationship between (a) the set of core values and norms guiding consumer behavior defined as "*desirable goals, varying in importance, that serve as guiding principles in peoples' lives*" (Schwartz in J Soc Issues 50(4):19–45, 1994, p. 21) with special focus on environmental orientation on the one hand and (b) dimensions of customer perceived value understood as the "*consumer's overall assessment of the utility of a product (or service) based on perceptions of what is received and what is given*" (Zeithaml in J Mark 52(3):2–22, 1988, p. 14) on the other hand. We suggest that customers' general ethical and environmental orientation is translated into actual consumer behavior as represented by the demand for sustainable excellence in all business practices. Only if

N. Hennigs · E. Karampournioti (✉) · K.-P. Wiedmann
Institute of Marketing and Management, Leibniz University of Hannover,
Koenigsworther Platz 1, 30167 Hannover, Germany
e-mail: karampournioti@m2.uni-hannover.de

M.A. Gardetti (ed.), *Sustainable Management of Luxury*,
Environmental Footprints and Eco-design of Products and Processes,
DOI 10.1007/978-981-10-2917-2_21

consumers perceive superior value reflected in all respects of a certain brand or product, they are willing to bridge the gap between basic ethical value orientation and actual ethical consumption.

Keywords Environmental value orientation · Customer perceived value · Sustainability excellence · Luxury industry

1 Introduction

The issue of sustainability is one of the central challenges for all countries in the world and for companies in their production processes. The continuous population growth and rapid economic development demand innovative environmental and ethical solutions from both businesses and governments. With reference to the luxury industry, the concepts of luxury and sustainability, however, appear at first glance in contrasting relationship: While luxury is often associated with superficial pomp and extravagance, sustainability addresses responsible consumption, social justice and the protection of natural resources (Pascaud 2011; Brundtland 1987). In this context, luxury companies have been criticized for risky working conditions, local river pollution, deforestation, sourcing of blood diamonds and the use of hazardous chemicals what negatively impacts upon corporate reputation and brand image (Kapferer and Michaut 2015; Sarasin 2012; Konietzko et al. 2014). The consideration of sustainability is therefore an important aspect in order to ensure the survival of the brand and a long-term competitive advantage (Girón 2014; Kapferer 2015).

Reasoning that luxury consumers typically belong to a wealthy, cultured global elite that has increasing interest in the environment and society (Bendell and Kleanthous 2007), they "*want the brands they use to reflect their concerns and aspirations for a better world*" (Bendell and Kleanthous 2007, p. 5). As the consumption of luxury provides the possibility to express one's deepest values (Belk 1988; Jenkins 2004), there appears to be a close association of luxury to the concept of ethical and environmental consumerism. The ethical buyer demands products that meet his/her moral principles and boycotts companies involved in unethical practices (e.g., Barnett et al. 2005; Muncy and Vitell 1992). Expecting "*convincing answers to questions of environmental and social responsibility*" (Bendell and Kleanthous 2007, p. 8), consumers either reward or punish companies that stress or ignore the importance of social and environmental excellence (Grail Research 2010).

A critical perspective on the economic reality has led to the discussion if the ethical consumer is nothing more than a myth (e.g., Carrigan and Attalla 2001; Devinney et al. 2010). Global consumers tend to turn a blind eye to political and ethical malpractices of their favorite brands (BBC 2002) and believe the government, the market, companies or the overall system to be responsible for sustainability practices—not themselves (Devinney et al. 2010). Reflecting the enormous gap between articulated individual environmental values and actual shopping behavior, the present chapter investigates the relationship between

a. the set of core values and norms guiding consumer behavior defined as "*desirable goals, varying in importance, that serve as guiding principles in peoples' lives*" (Schwartz 1994, p. 21) with special focus on environmental orientation on the one hand and
b. dimensions of customer perceived value understood as the "*consumer's overall assessment of the utility of a product (or service) based on perceptions of what is received and what is given*" (Zeithaml 1988, p. 14) on the other hand.

We suggest that customers' general ethical and environmental orientation is translated into actual consumer behavior as represented by the demand for sustainable excellence in all business practices. Only if consumers perceive superior value reflected in all respects of a certain brand or product, they are willing to bridge the gap between basic ethical value orientation and actual ethical consumption.

The chapter is organized as follows: In the next paragraph, the theoretical background of the concept of environmental orientation and luxury value perception will be presented. After that, a taxonomy of value-based environmental orientation is discussed that links egoistic, altruistic and biospheric environmental orientation with the four dimensions of luxury value perception. This model is used for the evaluation of business practice in terms of sustainability excellence in the perspective of the customer. Finally, the chapter concludes with a discussion of next research steps and managerial implications to help managers in the luxury industry improve the efficiency and effectiveness of their sustainability practices.

2 Theoretical Background

2.1 Environmental Value Orientation

Usually defined as abstract standards and goals with the ability to shape human lives, value orientation has been reported to be one of the most influential factors for the description and prediction of human environmental beliefs and behavior (Howard and Woodside 1984; Pitts and Woodside 1984; Rokeach 1973; Richins and Dawson 1992; McCarty and Shrum 1994). Values are "*centrally held, enduring beliefs which guide actions and judgments across specific situations and beyond immediate goals to more ultimate end-states of existence*" (Rokeach 1968, p. 161). Similar to the definition of Schwartz (1994) reflecting on values as guiding principles in peoples live, both definitions clearly illustrate that values go beyond the evaluation of specific objects or situations. Rather, they embody value specific modes of conduct and end states of existence (enduring goals in life). Accordingly, consumers' specific value system serves as a basis of individual behavior, the emergence of needs as well as of the formation of attitudes toward certain objects (Rokeach 1968).

The "*[...] inevitability of "limits to growth," the necessity of achieving a "steady-state" economy, the importance of preserving the "balance of nature," and the need to reject the anthropocentric notion that nature exists solely for human use*" (Dunlap and Van Liere 1978, p. 10) clearly describes the emergence of a new world view, in which the necessity of environmental protection and the reduction of degradation is of great importance along with a higher environmental concern and awareness—The New Environmental Paradigm (Dunlap and Van Liere 1978). Furthermore, environmental problems "*in which self-interest choice is detrimental to shared resources in the environment and environmentally cooperative choice is beneficial to the environment*" (Cho 2012, p. 22) is often considered as a social dilemma (Dawes 1980; Parks 1994; Steg 2003). As social beings, humans are often required to make decision whose consequences are not confined to oneself, but—as an essence of social interdependence (McClintock and Allison 1989)—affects the welfare of those around them (Messick and Brewer 1983; Cho 2012). In challenging phases of social conflicts—and the continuum between acting to one's own interests or the best interest of a reference group (Dawes 1980; Messick and Brewer 1983; Geller 1995; Parks 1994; Steg 2003)—altruism serves as a basis to "*value outcomes that benefit others and can be motivated to act to prevent harm to other*" (Stern et al. 1993, p. 324). By treating environmentalism as a form of altruism, pro-environmental behavior can occur as a secondary effect due to arising concerns regarding the welfare of other human beings (Stern et al. 1993). In everyday life, humans face numerous conflicting situations and take decisions, which could possibly have positive consequences for themselves but negative consequences for the environment and the society at the same time. To encourage consumers' ethical and environmental behavior, with regard to luxury consumption in particular, it is important to gain a deeper insight into specific value orientations which may influence their perception about products and behaviors and their willingness to act in a pro-environmental and pro-social manner.

2.2 Customer Perceived Value and Sustainability Excellence in the Luxury Industry

Understood as the responsible and efficient use of resources as well as the consideration of ethical values in the form of social justice, intergenerational responsibility and the observance of human rights (Fien and Tilbury 2002; Girón 2014; Kapferer 2010), sustainability and ethics in the value chain preserve a company's legitimacy and prevents from reputation threats (Cervellon and Wernerfelt 2012; Bhaduri and Ha-Brookshire 2011; Sarasin 2012). Once associated with "*indulgence, extravagance, sheer look-at-me bling—the antithesis of responsibility*" (Simpson 2012), the luxury industry has started to addresses various dimensions of deeper value embodied in the companies' core business and reflected in the sourcing, manufacturing, marketing and distribution of luxury brands. With regard to consumption values that directly explain how consumers evaluate luxury brands and why they choose to buy or avoid them (Sheth et al. 1991), sustainability excellence is driven by the customer's subjective expectations and individual perceptions of luxury value. Based on the core elements of luxury value as proposed by Wiedmann et al. (2007, 2009) and the sustainability diamond of Hennigs et al. (2013), the key dimensions of value-based social and environmental excellence as perceived by the consumer can be specified as follows:

Financial Value: Addressing direct monetary aspects of a product such as the price and the resale price of the product, the discount or the investment made, luxury goods are generally associated with a premium price and a limited demand. The exclusivity and uniqueness of luxury products implies that fewer products are produced compared to mass market production and thus fewer resources are used (Kapferer 2010; Hennigs et al. 2013). Moreover, consumers often consider the purchase of luxury goods as a long term investment and since luxury goods have a long-lasting value, the possibility of resale at a secondhand (vintage) market exists (Bastien and Kapferer 2013; Kapferer 2010).

Functional Value: An important characteristic of luxury brands is the consistent premium quality and the long-term life of the products (Nueno and Quelch 1998; Blevis et al. 2007; Lim et al. 2012). Because durability is a core aspect of luxury and sustainability, the longevity reflects the important relationship of the luxury concept with sustainability excellence (Kapferer 2010; Hennigs et al. 2013). As stated above, the appreciation of a luxury product increases when it is perceived as rare and unique. Given that luxury products often rely on rare materials such as precious pearls or unique skins and leather, the luxury industry is highly dependent on preserving the earth's resources. Besides, craftsmanship, tailor-made and hand-made products differentiate luxury from mass production. Therefore, the entire supply chain has to be included in a promising sustainability strategy and refrain from the exploitation of workers in low-wage countries.

Individual Value: Focusing on the customer's identity, hedonistic and materialistic values, the ethical orientation of the consumer has a direct impact on the purchasing behavior (Tsai 2005; De Pelsmacker et al. 2005). In this sense, individuals who identify with the group of green consumers express their interest in environmental and social consumption. As consumers use brands to reflect their own concerns and aspirations (Bendell and Kleanthous 2007), they expect convincing information about the brand's environmental and social responsibility (Grail Research 2010).

Social Value: Considering that consumers use the symbolic meaning of brands to reveal their identity and their social status within society (Sarasin 2012; Husic and Cicic 2009), more ethical oriented consumers are expected to think not only about the effect a purchase has on themselves with reference to their social group(s), but also on the world around them (Harrison et al. 2005; Davies et al. 2012). Modern customers are more concerned about sourcing practices and working conditions in the country of manufacture and demand that luxury brand companies improve environmental and social standards (Davies et al. 2012; Cervellon and Wernerfeldt 2012; Sarasin 2012).

In an attempt to combine both research streams, customers' general ethical and environmental value orientation on the one hand and customer perceived value reflected in all respects of a certain brand or product on the other hand, the following paragraph presents a taxonomy that forms the basis of the assessment of value-based environmental orientation.

3 A Taxonomy of Value-Based Environmental Orientation

The current paper specifically addresses the gap within existing literature on the nature of individual environmental value orientation combined with customer perceived value of sustainable luxury. Having reviewed the literature on what constitutes a luxury brand, greater attention is now required on how consumers' perception of luxury differs based on their individual value orientation. Accordingly, there is a need for a taxonomy, which classifies different consumers with reference to their environmental value orientation, and equally considers their perception of luxury as an important classification scheme. To reach this goal, the theories used for the classification will be represented to subsequently interlink and describe their characteristics.

3.1 Dimensions of Environmental Value Orientation

The values hold by an individual are thought to be a motivating force for the perception of one's environment as well as for specific decisions to engage in pro-environmental behaviors (De Groot and Steg 2008; Dietz et al. 2005). Several theories have categorized environmental values in different ways. Thompson and Barton (1994) for example distinguish between two value systems that significantly influence environmentally responsible behavior. Ecocentrism has its focus on intrinsic values of the ecosystem, whereas anthropocentric individuals' belief that humans wellbeing is a major importance and represents a main reason for environmental protection. In the current paper, we rely on the theory of Stern et al. (1993) which is based on Schwartz's (1977) theory of altruism. Since human behavior is not exclusively explained and predicted by altruistic orientations and individuals consider the consequences of their behavior for themselves as well as for nature, Stern et al. (1993) distinguish between three different types of value orientation:

1. The *altruistic value orientation* is about "*concern for the welfare of other human beings*" (Stern et al. 1993, p. 326).
2. The *biospheric value orientation* focuses on "*concern with nonhuman species and the biosphere*" (Stern et al. 1993, p. 326).
3. The *egoistic value orientation* addresses environmental protection as humans' self-interest.

3.2 Dimensions of Sustainability Value Perception

As stated above, customer perceived value generally combines functional, economic, emotional value and social aspects of a product (or service) (Sheth et al. 1991; Smith and Colgate 2007; Sweeney and Soutar 2001). With special focus on the performance of luxury brands and products, the financial, functional, individual and social dimension of luxury value constitute the customer's evaluation (Wiedmann et al. 2007, 2009). While the *financial value* refers to direct monetary aspects expressed in dollars and cents that one is willing to spend to obtain a product (e.g., value for money, price, discount) (Ahtola 1984; Monroe and Krishnan 1985), the *functional value* addresses basic utilities and benefits such as quality, uniqueness, usability, reliability, and durability (Sheth et al. 1991). In contrast to that, the *individual value* focuses on the arousal of affective states such as enjoyment, amusement and happiness, whereas the *social value* represents the human desire for social recognition, status and prestige and the willingness to impress their social group and to meet reference group-related expectations. In line with Hennigs et al. (2013), the aforementioned dimensions of luxury value form the customer's subjective expectations and individual perceptions of sustainability excellence in luxury brand management.

The taxonomy as presented in Table 1 combines both types of values (individual value vs. product/brand-related value perception) and forms the basis for the empirical application to the context of sustainable luxury in the next paragraph.

Given the widespread use of values in marketing research to understand beliefs, attitudes, motives and behaviors, the current taxonomy examines the role of values as a determinant of customer perceived sustainability excellence of luxury products as well as internal driver of environmentally conscious behavior. We assume that, based on their different manifestation of values (environmental orientation), consumers differ in their perception of luxury in general and the sustainable augmentation of luxury products in particular.

Table 1 Taxonomy of values

	Egoistic value orientation	Altruistic value orientation	Biospheric value orientation
Financial value perception	• Rarity leads to rising values of luxury products	• High pollution costs for the whole society/humanity	• Nature as a cultural capital of a country
Functional value perception	• Uniqueness • Improvement of durability • Health benefits of fair trade and organic products • Enhanced quality of life	• Threats to health for society • Enhanced quality of life for societal members	• Climate change • Extinction of animal and plant species • Destruction of Nature • Quality of life
Individual value perception	• Nature provides pleasure and recreation	• Satisfaction/Happiness from making others happy/Sharing Happiness and by enhancing societal well-being • Philanthropy	• Maintaining diversity of animal and plant life • Promoting welfare of the planet • Autarky
Social value perception	• Good standing, when considered as sustainable • Being part of a movement (veganism, new religions) • Standing out from the crowd	• Avoiding exploitation • Preference for fair trade/no child labor • Thinking of future generations	• Protecting animals • Standing out from the crowd • Being part of a movement

4 Empirical Application to the Context of Sustainable Luxury

4.1 The Value of Egoistic Environmental Orientation

Consumers with a strong egoistic tendency are primarily focused on their own advantage and wellbeing. Nevertheless, even this group of consumers may express concern for sustainability related issues, particularly when their own lives are affected or high costs occur to themselves (Pereira and Forster 2015). Their personality is characterized by authority, social power, wealth and being influential (Roobottom 2004). The increased (monetary) value of sustainable luxury products represents one aspect of customers' perceived financial value. As "*luxury value is based on its objective rarity—rare skins, rare leathers, rare pearls, rare materials, rare craftsmanship*" (Kapferer 2010, p. 41), the success of luxury is dependent on the sustainability of its resources, showing that the concepts of luxury and sustainability go hand in hand. The high price of rare materials and luxury products have a positive role in determining high quality

(Wiedmann et al. 2007) and as high quality materials are perceived as long lasting and durable, sustainable luxury products increase in value over time. Thus, the rarity and scarcity of resources and luxury products enhance the product value (Lynn 1991), whereby serving as a long-term benefit for selfish individuals.

The egoistic orientation has even an impact on the perception of a products functional value. The aforementioned rarity and scarcity, which is related to the uniqueness of a product, as well as the quality and durability of luxury products and brands, contribute to an increased value creation of luxury goods over time.

Even though, organic and fair trade products share characteristics of luxury goods (Campbell and Li 2003; Schmidt et al. 2016), quite diverse reasons for their consumption exist. The personal benefits derived from the consumption of sustainable products are of great importance for egoistic consumers. Among these are the increased personal health benefits of organic products (Botonaki et al. 2006; Chryssohoidis and Krystallis 2005; Harper and Makatouni 2002), the perceived safety of organic farming (Kouba 2003; Sangkumchaliang and Huang 2012), the absence of harmful chemical fertilizer and the believe to have a higher nutritional value (Tregear et al. 1994; Baker et al. 2004; Pino et al. 2012; Squires et al. 2001; Lea and Worsley 2005). Hence, all this factors increase the personal well-being and quality of life of the consumer (Schifferstein and Ophuis 1998; Williams and Hammitt 2001; Zanoli and Naspetti 2002) and lead to the satisfaction of egoistic motives.

Similar aspects account for individual values ascribed to a product, since recreational activities such as tourism, contributes to perceived hedonistic values of the natural environment. Nevertheless, even if recreational behaviors (e.g., snowmobiles, jet skis) increase the pleasure-seeking perspective of egoistic consumers, there is a simultaneous destruction of the natural environment (Schultz 2002). As egoistic consumers would "*[…] oppose protection of the environment if the personal costs are perceived as high*" (Stern and Dietz 1994, p. 70), the loss óf nature and wildlife as a recreational area entails high costs as well and enhances its perceived value and individual conservation efforts.

Considering the social status and the need of admiration the social value plays a crucial role in shaping preference for a specific product. Hence, the usage of sustainable luxury products may have a positive influence on the societal status of its buyer and serves moreover as a communicator of altruism and pro-environmental behavior—as a strategy of forming good impressions (Griskevicius et al. 2010; Coste-Manière et al. 2016). Additionally, consumer demand for socially and environmentally friendly products is motivated by the need of being part of a movement (see Fig. 1) or a specific lifestyle (just for the sense of belonging), such as in the case of a veganism.

4.2 The Value of Altruistic Environmental Orientation

Consumers with strong social-altruistic values judge environmental issues on the basis of costs to or benefits for other people: specific individuals such as family and friends, a neighborhood or community, a social network, a country, future

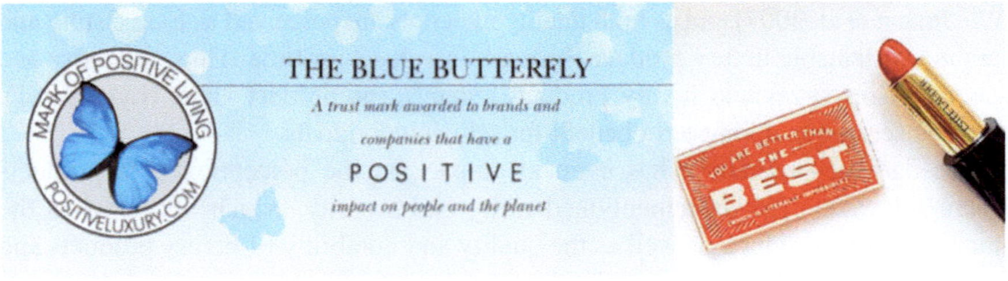

Fig. 1 Egoistic orientation in luxury product advertising: the desire to be appreciated. *Sources* http://38.media.tumblr.com/tumblr_m84xucpMDy1qc1vog.png; http://nadechedezwart.com/wp-content/up-oads/2015/05/11191616_428636800641370_2081907089_n.jpg

generations, or the humanity in general (Stern et al. 1993). As they focus on people other than self and perceive themselves as interconnected with others, they are driven by the desire to gain rewards or to avoid harmful consequence for others (Schultz 2000). Against this backdrop, altruism can be defined as a behavioral disposition that is aimed at the personal wellbeing of others, treating others fairly and maximizing others' benefits (Van Lange 2000)—regardless of demographic, biological, or personal characteristics (Corral-Verdugo et al. 2010). In the context of sustainability, altruism supports pro-environmental behaviors and promotes happiness (Corral-Verdugo et al. 2011). Driven by a "*warm glow of giving*" (Andreoni 1989, 1990; Boyce et al. 1992; Menges 2003; Nunes and Schokkaert 2003), social-altruistic consumers receive personal satisfaction and experience a feeling of well-being when acting environmentally conscious, protecting other individuals and thus contributing to the improvement of the common good environment (Ritov and Kahnemann 1997). Those actions aim at conditions that allow an equitable access to the use of natural resources (Ehrlich and Ehrlich 2004), moderate consumption of natural resources (De Young 1996; Iwata 2001), and assistance of others in need (Pol 2002).

In detail, altruistic environmental orientation focuses e.g., on effects of polluted air and polluted water on public health because it harms people all over the earth and reduces the quality of life for everyone, even for future generations. Moreover, social and ethical issues arising from poor labor standards, the use of sweatshops and child labor are in the focus of altruistic consumers' attention. Here, to preserve the rights and health of employees, codes of conduct for companies and their suppliers are of particular importance. In a luxury context, the mining of precious metals and stones and the use of blood diamonds to fund conflicts in Africa has been widely discussed. To reinforce consumer confidence in diamond, gold and platinum metals, the jewelry industry has set up an organization called the Responsible Jewellery Council (RJC). Certified membership aims to demonstrate the commitment to ethical, social, environmental, and human rights practices within the jewelry supply chain.

In sum, leading luxury brands (e.g., *Bulgari* supporting *Save the Children* or *Armani's Acqua for Life* charity campaign) have understood the importance of

Fig. 2 Altruistic orientation in luxury product advertising: the desire to benefit others. *Sources* http://www.patek.com/en/communication/news/product-advertising; http://www.luxurydaily.com/giorgio-armani-supports-unicef-tap-project-with-app/armani-3-420/

philanthropic or altruistic efforts as a powerful competitive differentiator and are more likely to align with the core values of increasingly critical customers (see Fig. 2). As a consequence, altruistic consumers are more likely to recommend and buy (even for a premium price) products or services from luxury brands that participate in philanthropic efforts, whereas brands that seem altruistically irresponsible are not trusted or even boycotted (American Express Publishing's Luxury Summit 2012). Even if corporate sustainability actions are often claimed to be greenwashing, it has to be stated that their ethical actions serve as a role model in the industry and in the customer perspective. Referring to the example of *Armani's Acqua for Life* program, over 880 million liters of clean water have been generated since the project started in 2011 (www.acquaforlife.org).

4.3 The Value of Biospheric Environmental Orientation

Despite egoistic and altruistic oriented consumers, whose aim for acting sustainably is the human welfare (either for themselves or other human beings), biospheric oriented individuals are concerned of the ecosystem and the biosphere as a whole (De Groot and Steg 2009). From a financial perspective, the cultural and natural heritage of a country is of high value and humans have the responsibility to protect them. The ideology of biospheric oriented consumers is often related with a frugal or simple lifestyle in which the conservation of natural resources is supported through the avoidance of unnecessary buying, expending and wasting of resources (De Young 1991). Functional benefits of their preferred products and brands result from the avoidance of environmental change and destruction of nature as well as the conservation of animal and plant species during the production and consumption process of products or even travel experiences (Radder and Han 2015). Hence, biospheric consumers pay high attention to the social and environmental augmentation of specific products and brands and are highly engaged in actions for environmental protection (Van der Werff et al. 2013).

Feeling as a part of the broader natural world fulfills the individual and emotional needs of biospheric consumers, which goes along with a high quality of life and is strongly related to ecological behavior and anti-consumerism (Mayer and Frantz 2004). Since individuals "*[…] view themselves as belonging to the natural world as much as it belongs to them*" (Schultz 2002, p. 67) empathic reactions towards nature as well as the willingness to help increases as a result of perceived relationship closeness (Cialdini et al. 1997). Through connectedness to nature "*the self is expanded to include the natural world, behavior leading to destruction of this world will be experienced as self-destruction*" (Roszak 1995, p. 504). Further, perceived individual values from nature connectedness and a biospheric oriented value system and ideology is the experienced well-being and overall life satisfaction (Mayer and Frantz 2004; Howell et al. 2011), among others characterized by autonomy, personal growth and positive relatedness (Capaldi et al. 2014).

By choosing sustainable product alternatives, consumers satisfy their social need for an entirely protected biospheric system and simultaneously fulfill their necessity to belong to the broader natural world and to feel like a valued member of a community (Kellert and Wilson 1993; Myers 2000). At the same time, biospheric consumers want to stand out from the crowd and demonstrate their exclusiveness by their autonomy and partially self-sufficient lifestyle. They feel as being part of a

Fig. 3 Biospheric orientation in luxury product advertising: the desire to protect natural beauty. *Source* http://www.iwc.com/en/help-protect-the-galapagos-islands/

movement where luxury means to free themselves from the system and to live a life in harmony and unity with nature.[1] Hence, biospheric consumers would choose products and brands that reflect their connectedness to the entire biospheric system even if they tend to avoid overconsumption. Consumers are more likely to buy brands that meet their expectations connected to their very specific lifestyle and worldview and enhance the welfare of all living organisms in the world. The watch manufacturer *IWC Schaffhausen* for example is part of the *Charles Darwin Membership Program* which scientifically explores the fragile Galapagos biotope and supports the conservation of the stunning world natural heritage. As a program partner, they created a luxurious chronograph called *Galapagos Island* (see Fig. 3).

5 Conclusion

Against the backdrop of serious challenges such as counterfeiting, fast fashion, the democratization of luxury as well as increasingly conscientious consumers, luxury marketers have to accept sustainability excellence as a fundamental market responsibility. The focus of this chapter was on the value-based assessment of sustainability in the luxury industry. To empirically assess the relationship between environmental orientation and customer value perception, a taxonomy has been presented and explained against the backdrop of luxury consumption. Consideration of sustainability is a value-driving aspect that impacts upon the customer buying decision: Consumers increasingly demand environmental-friendly and socially responsible products and perceive that particular luxury companies as pioneers and trendsetters hold the responsibility to use their high profits for implementing sustainability in all aspects of the value chain.

As discussed in this chapter, consumers evaluate the sustainability performance of luxury brands based on financial, functional, individual, and social components and in line with their individual environmental value orientation: egoistic, altruistic or biospheric. To verify that the luxury industry's commitment to sustainability is far more than hypocrisy and a sheer response to outside pressure, each management decision has to be reflected from the customer's perspective and the value that consumers attach to the multifaceted product attributes.

In this context, it has to be stated that the perception of luxury goods and products is situational contingent and depends on consumer specific needs and experiences (Wiedmann et al. 2007). Therefore, the assessment of the depth and sincerity of sustainability initiatives has to be regarded from an individual and cultural value perspective. It has been shown in accordance to post-materialism theory that environmental concern increases as societies grow more prosperous

[1]The *Free and Real* project serves as an example for this kind of lifestyle. Their «*Telaithrion Project*» hopes to put in actual perspective, that a self-sufficient sustainable society that is based on true incentives and selfless giving, can exist, and that can be applied to practice in everyday life, and that even if the entire culture of humanity adopted it could flourish. (Free and Real, n.d.).

(Inglehart 1995; Franzen and Meyer 2010). Consumers in wealthier nations are more willing to engage in global environment protection as compared to individuals that face pressing economic problems. In consequence, based on the taxonomy as presented in this chapter, cross-cultural studies focusing on the individual value orientation and the evaluation of the sustainability performance of luxury brands are a key challenge and useful direction for further research.

> We make a living by what we get, but we make a life by what we give.
> Winston Churchill

References

Ahtola OT (1984) Price as a 'give' component in an exchange theoretic multicomponent model. Adv Consum Res 11(1):623–636

American Express Publishing's Luxury Summit (2012) American express publishing luxury summit. Available via http://conzepts.net/wp-content/uploads/2012/10/American-Express-LUXURY-Summit-2012-White-Paper-Report.pdf. Accessed 20 Mar 2016

Andreoni J (1989) Giving with impure altruism: applications to charity and Ricardian equivalence. J Polit Econ 97(6):1447–1458

Andreoni J (1990) Impure altruism and donations to public goods: a theory of warm-glowing giving. ECON J 100(401):464–477

Baker S, Thompson KE, Engelken J et al (2004) Mapping the values driving organic food choice. Eur J Mark 38(8):995–1012

Barnett C, Cloke P, Clarke N et al (2005) Consuming ethics: articulating the subjects and spaces of ethical consumption. Antipode 37(1):23–45

Bastien V, Kapferer JN (2013) More on luxury anti-laws of marketing. In: Wiedmann KP, Hennigs N (eds) Luxury marketing. A challenge for theory and practice. Springer, Wiesbaden, pp 20–34

BBC (2002) Gap hit by 'sweatshop' protests. BBC NEWS. Available via http://news.bbc.co.uk/2/hi/business/2497957.stm. Accessed 20 Mar 2016

Belk RW (1988) Possessions and the extended self. J Consum Res 15(2):139–168

Bendell J, Kleanthous A (2007) Deeper luxury: quality and style when the world matters. Available via http://www.wwf.org.uk/deeperluxury/. Accessed 25 Feb 2016

Bhaduri G, Ha-Brookshire JE (2011) Do transparent business practices pay? Exploration of transparency and consumer purchase intention. Cloth Textiles Res J 29(2):135–149

Blevis E, Makice K, Odom W et al (2007) Luxury & new luxury, quality & equality. In: Proceedings of the 2007 conference on designing pleasurable products and interfaces, Helsinki

Botonaki A, Polymeros K, Tsakiridou E et al (2006) The role of food quality certification on consumers' food choices. Brit Food J 108(2):77–90

Boyce RR, Brown TC, McClelland GH et al (1992) An experimental examination of intrinsic values as a source of the WTA-WTP disparity. Am Econ Rev 82(5):1366–1373

Brundtland G (1987) Report of the world commission on environment and development: our common future. Available via http://www.un-documents.net/our-common-future.pdf. Accessed 25 Feb 2016

Campbell S, Li C (2003) Per capita consumption, luxury consumption and the presidential puzzle: a partial resolution. Brown University, Department of Economics

Capaldi C, Dopko RL, Zelenski JM (2014) The relationship between nature connectedness and happiness: a meta-analysis. Front Psychol 5(2014):1–15

Carrigan M, Attalla A (2001) The myth of the ethical consumer-do ethics matter in purchase behaviour? J Consum Mark 18(7):560–578
Cervellon MC, Wernerfeldt AS (2012) Knowledge sharing among green fashion communities online: lessons for the sustainable supply chain. JFMM 16(2):176–192
Cho Y (2012) Consumers value orientations and green advertising effectiveness: the moderating role of public self-awareness. Dissertation, The Graduate School of the University of Oregon
Chryssohoidis GM, Krystallis A (2005) Organic consumers' personal values research: testing and validating the list of values (LOV) scale and implementing a value-based segmentation task. Food Qual Prefer 16(7):585–599
Cialdini R, Brown S, Lewis B et al (1997) Reinterpreting the empathy–altruism relationship: when one into one equals oneness. J Pers Soc Psychol 73(3):481–494
Corral-Verdugo V, García C, Castro L et al (2010) Equity and sustainable lifestyles. In: Corral-Verdugo V, García C, Frías M (eds) Psychological approaches to sustainability. Nova Science Publishers, New York, pp 185–204
Corral-Verdugo V, Mireles-Acosta J, Tapia-Fonllem C et al (2011) Happiness as correlate of sustainable behavior: a study of pro-ecological, frugal, equitable and altruistic actions that promote subjective wellbeing. Hum Ecol Rev 18(2):95–104
Coste-Maniére I, Mukta R, Sudeep C et al (2016) Long-term sustainable sustainability in luxury. Where else? In: Gardetti MA, Muthu SS (eds) Handbook of sustainable luxury textiles and fashion. Springer Singapore, pp 17–34
Davies IA, Lee Z, Ahonkhai I (2012) Do consumers care about ethical luxury? J Bus Ethics 106 (1):37–51
Dawes RM (1980) Social dilemmas. Annu Rev Psychol 31(1):169–193
De Groot JI, Steg L (2008) Value orientations to explain beliefs related to environmental significant behavior how to measure egoistic, altruistic, and biospheric value orientations. Environ Behav 40(3):330–354
De Groot JI, Steg L (2009) Mean or green: which values can promote stable pro-environmental behavior? Conserv Lett 2(2):61–66
De Pelsmacker P, Driesen L, Rayp G (2005) Do consumer care abour ethics? Willingness to pay for fair-trade coffee. J Consum Aff 39(2):363–385
De Young R (1991) Some psychological aspects of living lightly: desired lifestyle patterns and conservation behavior. J Environ Syst 20(3):215–227
De Young R (1996) Some psychological aspects of a reduced consumption lifestyle: the role of intrinsic satisfaction and competence motivation. Environ Behav 28(3):358–409
Devinney T, Auger P, Eckhardt GM (2010) The myth of the ethical consumer. Cambridge University Press, New York
Dietz T, Fitzgerald A, Shwom R (2005) Environmental values. Annu Rev Env Resour 30:335–372
Dunlap R, Van Liere K (1978) The "new environmental paradigm": a proposed measuring instrument for environmental quality. Soc Sci Quart 65:1013–1028
Ehrlich P, Ehrlich A (2004) One with Niniveh. Politics, consumption and the human future. Shearwater Books, Washington, DC
Fien J, Tilbury D (2002) The global challenge of sustainability. In: Tilbury D, Stevenson RB, Fien J et al (eds) Education and sustainability: responding to the global challenge, Switzerland, pp 1–12
Franzen A, Meyer R (2010) Environmental attitudes in cross-national perspective: a multilevel analysis of the ISSP 1993 and 2000. Eur Sociol Rev 26(2):219–234
Free and real (n.d.) The Telaithrion project. Available via http://en.telaithrion.freeandreal.org/#!/project/. Accessed 18 Mar 2016
Geller ES (1995) Actively caring for the environment an integration of behaviorism and humanism. Environ Behav 27(2):184–195
Girón ME (2014) Sustainable luxury: stories from the pioneers. In: Gardetti MA, Girón ME (eds) Sustainable luxury and social entrepreneurship: stories from the pioneers. Greenleaf Publishing, Sheffield, pp 1–21

Grail Research (2010) Green—the new color of luxury: moving to a sustainable future. Available via http://www.grailresearch.com/pdf/Blog/2010-Dec-Grail-Research-Green-The-New-Color-of-Luxury_244.pdf. Accessed 25 Mar 2016

Griskevicius V, Tybur JM, Van den Bergh B (2010) Going green to be seen: status, reputation, and conspicuous conservation. J Pers Soc Psychol 98(3):392–404

Harper GC, Makatouni A (2002) Consumer perception of organic food production and farm animal welfare. Brit Food J 104(3/4/5):287–299

Harrison R, Newholm T, Shaw D (2005) The ethical consumer. Sage Publications, London

Hennigs N, Klarmann C, Behrens S et al (2013) Sustainability as part of the luxury essence: delivering value through social and environmental excellence. JCC 52:25–35

Howard JA, Woodside AG (1984) Personal values affecting consumer psychology. In: Pitts RE, Woodside AG (eds) Personal values and consumer psychology. Lexington Books, Lexington, MA, pp 3–12

Howell AJ, Dopko RL, Passmore HA et al (2011) Nature connectedness: associations with well-being and mindfulness. Pers Indiv Differ 51(2):166–171

Husic M, Cicic M (2009) Luxury consumption factors. JFMM 13(2):231–245

Inglehart R (1995) Public support for environmental protection: objective problems and subjective values in 43 societies. Polit Sci Polit 28(01):57–72

Iwata O (2001) Attitudinal determinants of environmentally responsible behavior. Soc Behav Personal 29(2):183–190

Jenkins R (2004) Social identity, 2nd edn. Routledge, London

Kapferer JN (2010) All that glitters is not green: the challenge of sustainable luxury. Eur Bus Rev 40–45

Kapferer JN (2015) Kapferer on luxury. How luxury brands can grow yet remain rare. Kogan Page, London

Kapferer JN, Michaut A (2015) Luxury and sustainability: a common future? The match depends on how consumers define luxury. LRJ 1(1):3–17

Kellert SR, Wilson EO (1993) The biophilia hypothesis. Island Press, Washington, DC

Konietzko J, van Woersem Y, Simpson J (2014) Feel good fashion, published by Rank a Brand. http://rankabrand.org/static/FeelGoodFashion_2014_Summary.pdf

Kouba M (2003) Quality of organic animal products. Livest Prod Sci 80(1):33–40

Lea E, Worsley T (2005) Australians' organic food beliefs, demographics and values. Brit Food J 107(11):855–869

Lim WM, Ting DH, Khoo PT et al (2012) Understanding consumer values and socialization—a case of luxury products. Manag Mark 7(2):209–220

Lynn M (1991) Scarcity effects on value: a quantitative review of the commodity theory literature. Psychol Market 8(1):43–57

Mayer FS, Frantz CM (2004) The connectedness to nature scale: a measure of individuals' feeling in community with nature. J Environ Psychol 24(4):503–515

McCarty JA, Shrum LJ (1994) The recycling of solid wastes: personal values, value orientations, and attitudes about recycling as antecedents of recycling behavior. J Bus Res 30(1):53–62

McClintock CG, Allison ST (1989) Social value orientation and helping Behavior. J Appl Soc Psychol 19(4):353–362

Menges R (2003) Supporting renewable energy on liberalized markets: green electricity between additionality and consumer sovereignty. Energ Policy 31(7):583–596

Messick DM, Brewer MB (1983) Solving social dilemmas. Pers Soc Psychol Rev 4(1):11–44

Monroe KB, Krishnan R (1985) The effect of price on subjective product evaluations. In: Jacoby J, Olson J (eds) The perception of merchandise and store quality. Lexington Books, Lexington, pp 209–232

Muncy JA, Vitell SJ (1992) Consumer ethics: an investigation of the ethical beliefs of the final consumer. J Bus Res 24(4):297–311

Myers D (2000) The funds, friends, and faith of happy people. Am Psychol 55(1):56–67

Nueno JL, Quelch JA (1998) The mass marketing of luxury. Bus Horizons 41(6):61–68

Nunes PA, Schokkaert E (2003) Identifying the warm glow effect in contingent valuation. J Environ Econ Manag 45(2):231–245

Parks CD (1994) The predictive ability of social values in resource dilemmas and public-goods games. Pers Soc Psychol B 20(4):431–438

Pascaud L (2011) Could sustainability be the future of luxury? Available via http://added-value.com/could-sustainability-be-the-future-of-luxury/. Accessed 25 Feb 2016

Pereira M, Forster PM (2015) The relationship between connectedness to nature, environmental values, and pro-environmental behaviours. Reinvention: Int J Undergraduate Res 8(2). Available via http://eprints.worc.ac.uk/4081/5/Nature_proofed_final-2.pdf. Accessed 26 Mar 2016

Pino G, Peluso AM, Gianluigi G (2012) Determinants of regular and occasional consumers' intentions to buy organic food. J Consum Aff 46(1):157–169

Pitts RE, Woodside AG (1984) Personal values and consumer psychology. Lexington Books, Lexington

Pol E (2002) The theoretical background of the city-identity-sustainability network. Environ Behav 34(1):8–25

Radder L, Han X (2015) An examination of the museum experience based on Pine and Gilmore's experience economy realms. J Appl Bus Res 31(2):455–469

Richins ML, Dawson S (1992) A consumer values orientation for materialism and its measurement: scale development and validation. J Consum Res 19(3):303–316

Ritov I, Kahnemann D (1997) How people value the environment. Attitudes versus economic values. In: Bazermann MHDM, Messick AE, Tenbrunsel KA et al (eds) Environment, ethics, and behavior. New Lexington Press, San Francisco, pp 33–51

Rokeach M (1968) Beliefs, attitudes and values: a theory of organization and change. Jossey Bass, San Francisco

Rokeach M (1973) The nature of human values. Free press, New York

Roobottom HE (2004) A proposed investigation into core value systems as predictors of ecotourist choice and behaviour, proceedings of Anzmac December 2004, Wellington, NZ. Available via http://www98.griffith.edu.au/dspace/bitstream/handle/10072/2356/25720_1.pdf; jsessionid=F20A249BA21B0716EFE46954129DFC67?sequence=1. Accessed 25 Mar 2016

Roszak T (1995) Where psyche meet Gaia. In: Roszak T, Gomes ME, Kanner AD (eds) Ecopsychology: restoring the earth, healing the mind. Sierra Club Books, San Francisco

Sangkumchaliang P, Huang WC (2012) Consumers' perceptions and attitudes of organic food products in Northern Thailand. Int Food Agribus Man 15(1):87–102

Sarasin (2012) The quest for authenticity—can luxury brands justify a premium price? Bank Sarasin & Co., Basel. Available via http://www.longfinance.net/images/reports/pdf/sarasin_luxurygoods_2012.pdf. Accessed 25 Mar 2016

Schifferstein HN, Ophuis PAMO (1998) Health-related determinants of organic food consumption in the Netherlands. Food Qual Prefer 9(3):119–133

Schmidt S, Hennigs N, Behrens S et al (2016) The luxury of sustainability: examining value-based drivers of fair trade consumption. In: Muthu SS, Gardetti MA (eds) Handbook of sustainable luxury textiles and fashion. Springer, Singapore, pp 121–136

Schultz PW (2000) Empathizing with nature: the effects of perspective taking on concern for environmental issues. J Soc Issues 56(3):391–406

Schultz PW (2002) Inclusion with nature: the psychology of human-nature relations. In: Schmuck P, Schultz WP (eds) Psychology of sustainable development. Kluwer Academic Publishers, Dordrecht, pp 61–78

Schwartz SH (1994) Are there universal aspects in the structure and contents of human values? J Soc Issues 50(4):19–45

Schwartz SH (1977) Normative influences on altruism. In: Berkowitz L (eds) Advances in experimental social psychology. Academic Press, New York, pp 221–279

Sheth JN, Newman BI, Gross BL (1991) Why we buy what we buy: a theory of consumption values. J Bus Res 22(2):159–170

Simpson A (2012) Luxury: the new epitome of green values? Available via green futures. http://www.forumforthefuture.org/greenfutures/articles/luxury-new-epitome-green-values. Accessed 25 Feb 2016

Smith JB, Colgate M (2007) Customer value creation: a practical framework. JMTP 15(1):7–23

Squires L, Biljana J, Cornwell TB (2001) Level of market development and intensity of organic food consumption: cross-cultural study of Danish and New Zealand consumers. J Consum Mark 18(5):392–409

Steg L (2003) Motives and behavior in social dilemmas relevant to the environment. In: Hendrickx L, Jager W, Steg L (eds) Human decision making and environmental perception: understanding and assisting human decision making in real-life settings. University of Groningen, Groningen, pp 83–102

Stern PC, Dietz T (1994) The value basis of environmental concern. J Soc Issues 50(3):65–84

Stern PC, Dietz T, Kalof L (1993) Value orientation, gender, and environmental concern. Environ Behav 25(3):322–348

Sweeney JC, Soutar GN (2001) Consumer perceived value: the development of a multiple item scale. J Retail 77(2):203–220

Thompson SCG, Barton MA (1994) Ecocentric and anthropocentric attitudes toward the environment. J Environ Psychol 14(2):149–157

Tregear A, Dent JB, McGregor MJ (1994) The demand for organically grown produce. Brit Food J 96(4):21–25

Tsai SP (2005) Impact of personal orientation on luxury-brand purchase value. Int J Market Res 47 (4):429–454

Van der Werff E, Steg L, Keizer K (2013) The value of environmental self-identity: the relationship between biospheric values, environmental self-identity and environmental preferences, intentions and behaviour. J Environ Psychol 34:55–63

Van Lange PAM (2000) Cooperation and competition. In: Kazdin AE (ed) Encyclopedia of psychology, 2nd edn. American Psychological Association & Oxford University Press Washington, pp 296–300

Wiedmann KP, Hennigs N, Siebels A (2007) Measuring consumers' luxury value perception. Acad Mark Sci Rev 11(7):1–21

Wiedmann KP, Hennigs N, Siebels A (2009) Value-based segmentation of luxury consumption behaviour. Psychol Market 26(7):625–651

Williams PRD, Hammitt JK (2001) Perceived risks of conventional and organic produce: pesticides, pathogens and natural toxins. Risk Anal 21(2):319–330

Zanoli R, Naspetti S (2002) Consumer motivations in the purchase of organic food: a means-end approach. Brit Food J 104(8):643–653

Zeithaml VA (1988) Consumer perceptions of price, quality, and value: a means-end model and synthesis of evidence. J Mark 52(3):2–22

Lust for Our Better Versions—The Meta Modern Reality's Roots and Implications

Koronaki Eirini and Panigyrakis George

Abstract A self that goes on changing is a self that goes on living, as Virginia Woolf had stated. The same, however, goes for any brand, and especially for luxury brands, whose success is highly related to the connection they form with their customers (Kastanakis and Balabanis in J Bus Res 67(10):2147–2154, 2014). Consumers today demand ethical and socially responsible behavior from luxury brands (Carrigan et al. in J Mark Manage 29(11–12):1277–1307, 2013). Which underlying forces led to this shift of perception? Today's consumer, termed as the meta-modern consumer, seeks to use brands as cultural resources in his/her attempt to build their ideal selves through consumption (Holt in J Consum Res 29(1):70–90, 2002). Brands are, thus, evaluated upon their ability to provide their customers with appropriate subjective intangible benefits to use for the construction of their identities. As a result, a fundamental part of the perception of an ideal self today is one oriented towards a more ethical and responsible consumption. However, which actions lead towards an ideal self? To thoroughly understand its elements, a chronological overview of the concept is crucial. A philosophical overview from Plato to Chomsky will be presented in this chapter, concerning how the concept of ideal self evolved through time to reach its current form. In the next part of the chapter the connection between ideal selves and ethical luxury will be examined. Can luxury brands provide their customers with socially-oriented benefits? Which activities are more efficient in that direction? Two problems emerge, according to relevant research (Torelli et al. in J Consum Res 38(5):948–963, 2012); luxury consumers do not care that much about ethical luxury and an oxymoron is formed between the selfish nature of luxury consumption and the non selfish nature of

K. Eirini (✉) · P. George
Athens University of Economics and Business, Patision 76, 104 34
Athens, Greece
e-mail: eirini_kor@hotmail.com

P. George
e-mail: pgg@aueb.gr

M.A. Gardetti (ed.), *Sustainable Management of Luxury*,
Environmental Footprints and Eco-design of Products and Processes,
DOI 10.1007/978-981-10-2917-2_22

corporate social responsibility activities (Janssen et al. in J Bus Ethics 119(1):45–57, 2014). Can these two issues be solved? Luxury's everlasting attachment to long term quality and craftsmanship can be a catalyst in that direction.

Keywords Ideal self · Philosophy · Ethical luxury

1 Introduction

A self that goes on changing is a self that goes on living, as Virginia Woolf had stated. This urge to adapt to the prevailing circumstances is a prerequisite not only for human beings, but also for brands which wish to stay relevant. Now, imagine how imperative this ability is for luxury brands, which owe their success to the perceived level of intangible benefits they are providing their customers with (Kastanakis and Balabanis 2014). It is the relevance of these benefits to each individual that leads to the creation of strong emotional bonds to luxury brands (see Fig. 1).

A prerequisite for the creation of such bonds is not only constantly responding to the contemporary consumers' needs and desires, but also realizing their shifts and readjusting to them. It is through such readjustments, that the most successful brands manage to constantly stay relevant in their consumers' hearts and minds. During an era of economic turbulence, political instability, shifts in values, luxury brands could not remain unaffected.

It is under such circumstances that the new paradigm of sustainable luxury emerged and continues to gain ground. Throughout this chapter, the underlying reasons for the development of a more ethical approach to luxury will be presented, to continue with how it can be leveraged on by companies, in an attempt to maintain their place in the market space.

2 What Do Consumers See in Consumption?

2.1 Consumer Culture Theory Approach

For a thorough understanding of the prevailing sustainability trend, an in-depth understanding of what consumers see in consumption is a prerequisite. Consumer

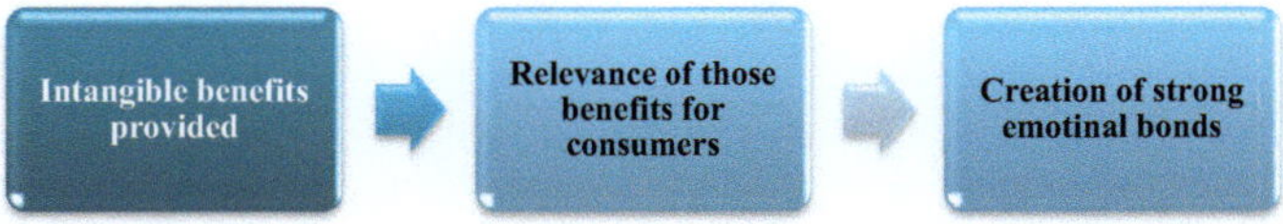

Fig. 1 Building strong luxury brands

Fig. 2 Key interrelated areas according to consumer culture theory

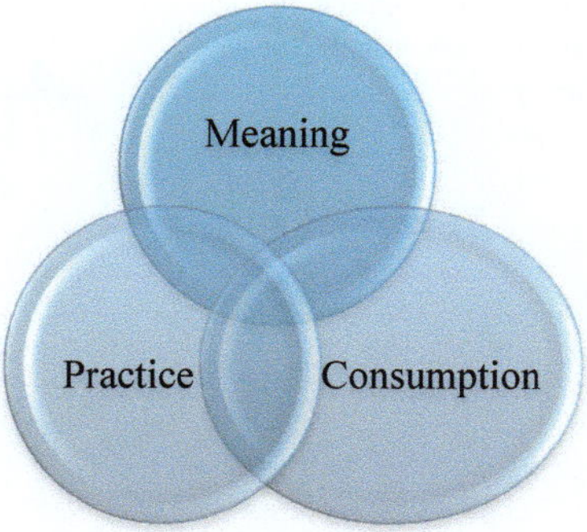

culture theory is an approach that can be applied in that direction. Characterized as one of the most "fascinating contemporary social forces by McDonald and Wearing (2013), consumer culture studies the relations between social and cultural patterns of meaning, practice and consumption. How do the concepts of meaning, practice and consumption interrelate (see Fig. 2)? Looking back into classical consumer research perspectives, scholars tended to consider the consumer as a rational decision maker, whose decisions are an outcome of the objective comparison of his/her choices. However, consumption has come to express more than just the evaluation of a range of products.

Despite the fact that consumption has become a common and unremarkable aspect of everyday life, it expresses an integral aspect of modern life. Through the social practices of consumption the individual can express his/her personality, values and underlying motivation. It can in fact be argued that when consuming, through the use of certain systems and forms of symbolic interactions, individuals have the opportunity to create meaning in their personal and social lives. So what is the meaning that today's consumer seeks to create?

As Baudrillard (1970/1998, p. 217) underlined consumption is an active form of relationship not only to objects, but also to society and to the world, a global response which founds our entire cultural system. Isn't it the case that today's world has been gradually shifting towards the issue of sustainability? How could luxury consumers ignore such a major trend or how could the issue of sustainability not be a part of their ideal selves? The connections among meaning, practice and consumption, thus, emerge, revealing the logic behind consumer's culture theory's approach.

2.2 *Approaches to Consumer Culture Theory*

However, one comes to wonder what this approach means and how it can be used to understand emerging trends in marketing in general. The key here is to understand that there have been developed two ways of examining consumer culture (McDonald and Wearing 2013) (see Fig. 3):

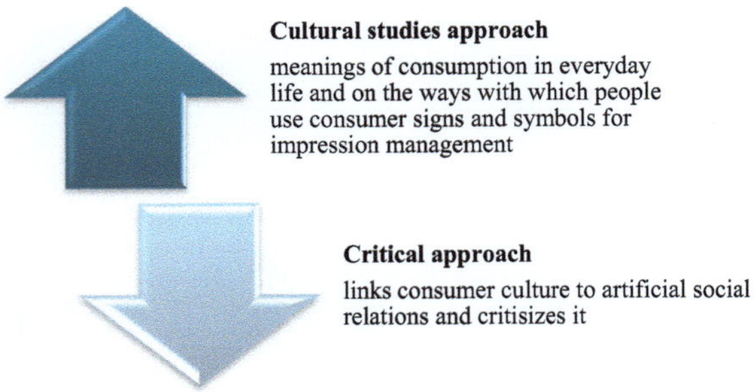

Fig. 3 Two approaches to studying consumer culture theory

(1) The first one is the cultural studies approach. It focuses on the meanings of consumption in everyday life and on the ways with which people use consumer signs and symbols for impression management. (Schor 2007).
(2) The critical approach on the other hand, links consumer culture to superficial social relations from a primarily self centered individual and critically comments on such issues.

Despite the different orientation of these two approaches, they have a common viewpoint. They both see consumption as central in the individual's self-identity formation in the contemporary modern society. According to their approach self-identity is not considered the stable entity that it once was, influenced primarily by fixed institutions such as one's family. The idea emerged that self-identity is in a process of constant transformation, depending on the impressions that one seeks to build. (Elliott and Lemert 2005).

The whole rationale of analysis on why sustainable consumption emerged lies on this argument. Consumers do actually define themselves through their consumption and form what they perceive as ideal self with the assistance of their purchases. This means that changes in the prevailing ideals and beliefs will have the power to affect consumers' behavior, according to consumer culture theory.

2.3 Consumer Culture Theory and Company Practices

Consumer culture is thus basically the rationale that explains what and how people consume, which renders it a key determinant of the marketers' branding activities. The principles defining how firms seek to build their brands has been termed as the branding paradigm. These principles are in accordance with the key assumptions of the extant consumer culture (Holt 2002).

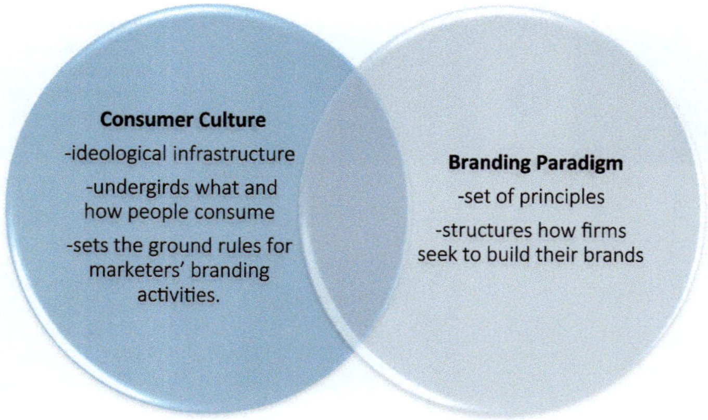

Fig. 4 Consumer culture and branding paradigm (Holt 2002)

That is the relation between the prevailing consumer culture and the respective branding paradigm. An understanding of the contemporary consumer culture will provide a clear idea of what sustainable luxury consists of and aims to achieve (see Fig. 4).

3 Prevailing Consumer Culture and the Construction of the Identity Puzzle

Marketing is dead. Long live marketing! A shift in the nature of marketing is evident lately, with consumers denying the passive consumption of dreams and desires (van Raaij 1993) for an Aristotelian ideal of wealth creation, in which one becomes a better person and the world a better place to live in (Aristotle 1954). The era we are going through, termed as post post-modernsim or meta-modernism (Nealon 2012), denies postmodernism's prevailing consumer culture of creating one's identity through the consumption of certain brands (Firat and Venkatesh 1995).

Contemporary consumer culture demands brands to serve as cultural resources in their attempt to consummation, giving rise to the Citizen-Artist consumer model (Holt 2002). Value is the key criterion for the suitability of brands for this role, in fact consummation-assisting value. Luxury brands have been characterized as the pros in building strong relationships with their customers and providing them with the desirable values.

The questions inevitably emerges; for which reasons and with which processes do consumers create such profound relationships to luxury brands? Today's consumer, termed as the meta-modern consumer, seeks to use brands as cultural resources in his/her attempt to build their ideal selves through consumption (Holt 2002). But how can brands function as resources in the individual's identity-building process?

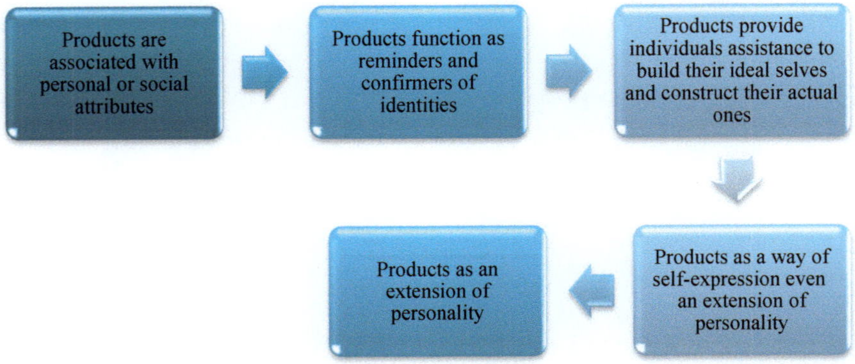

Fig. 5 How can products assist in identity building?

Possessions are by a smaller or bigger extent seen by individuals as part of themselves, a way of self- expression or even an extension of personality (Belk 1988). In fact, products provide individuals assistance to build their ideal selves and construct their actual ones in a much easier and quicker way, considering that they can function as reminders and confirmers of their identities (Hoge and McCarthy 1984). Their ability to function as such is grounded upon the reason that products are always associated with personal or social attributes (Mead 1938). Brands are thus evaluated upon their ability to provide their customers with appropriate subjective intangible benefits (see Fig. 5).

4 Relation to Luxury: Chronologically Revisited

Luxury brands have long been purchased for their symbolic benefits and the meanings they are endowed with. However, for an in-depth understanding of the contemporary relation to luxury, a presentation of how it has been perceived throughout the years is critical. Luxury is a concept which has always existed in human societies, with varying forms and relations to it. Looking back from Ancient Greece, to Ancient Rome, to the Byzantium and the Renaissance, getting to the French Aristocrats to the invention of the haute couture and the later evolution of luxury brands as we know them today, it becomes clear that individuals have been treating themselves with luxurious goods for quite some time now. However, the individual's relation to luxury can be divided into particular and definite eras, with varying attitudes as to whether luxury is accepted or not and for which purposes (see Fig. 6).

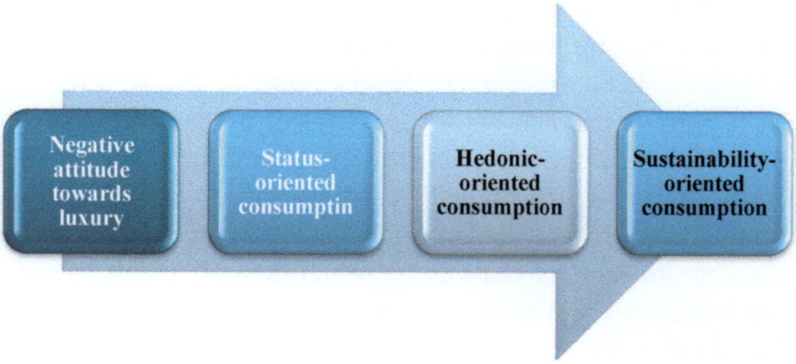

Fig. 6 Chronological evolution of focus in luxury consumption

4.1 Luxury: Hindering Society's Welfare

Up until the Renaissance, luxury was endowed with negative connotations. Despite the fact that it was indeed present in many forms from Ancient Greece up until the 17th century, it was considered to halt the individual from focusing onto the fundamental aspects of life, spirit and society. This indulgence was thus considered as a delay in personal development and it was perceived as a weakness of the upper class. Citizens were expected to care about the welfare of their society, to discuss their moral dispositions and political stance or even to perform their religious obligations. Any interference with expensive and rare objects and, either when consuming them or even when trying to acquire them, was considered as holding them back of the actually important things in life.

4.2 Achieving Status Through Luxury Goods

However, these negative connotations started to fade away, parallel to the Industrial Revolution. This shift was partially due to the increasing number of goods produced by factories and the simultaneous increasing disposable income. Luxury goods were starting to gradually dispose of the negative elements they were connected with, and started to be looked at as elements of social distinction. It is interesting to see that as other elements of social distinction started to fade away, consumers wanted to maintain the use of such products to show their place in society and to differentiate themselves from others.

At the same time, while serving as an assistant for distinction for some, it served as an assistant for belonging to others. Because a number of individuals were purchasing and consuming such goods, to show that they belonged to an aspired reference group. Say for example that someone wanted to be perceived as a traveler or a highly sophisticated person. It seemed much easier to purchase a product with travel-spirit-endowed elements or respectively sophisticated ones, than actually trying to cultivate these particular characteristics. No particular resources, in terms of time or money were required and it definitely was an easier solution.

However, at the time being, a certain level of education was in fact expected for an individual to truly appreciate luxury. A sense of refinement, taste and appreciation of beauty was required to recognize the dimensions rendering a product or good from commodity to luxurious. The ability to pay the high price was not considered sufficient to properly understand what the luxurious product represented.

That is also one of the key reasons for which luxury is characterized by such high levels of subjectivity. What is seen as prestigious in one culture is not necessarily perceived that way in another, and the aspiring reference groups can also significantly differ depending on the context.

4.3 Respective Research

The research referring to this approach to luxury can be divided into the following three categories (see Fig. 7).

4.3.1 Status Motivation

Vigneron and Johnson (2004) termed this effect as the Veblen effect (perceived conspicuous value), an effect which draws on early studies on the influence of reference groups on the consumption of luxury brands (Wiedmann et al. 2009; Christodoulides et al. 2009; Chaudhuri and Majumdar 2006). As Mason (1984) contends:

To the purely conspicuous consumer, the satisfaction derived from any particular purchase comes not from its value in use but from audience reaction to the wealth displayed by the purchaser in being able to secure the product for consumption (p. 26).

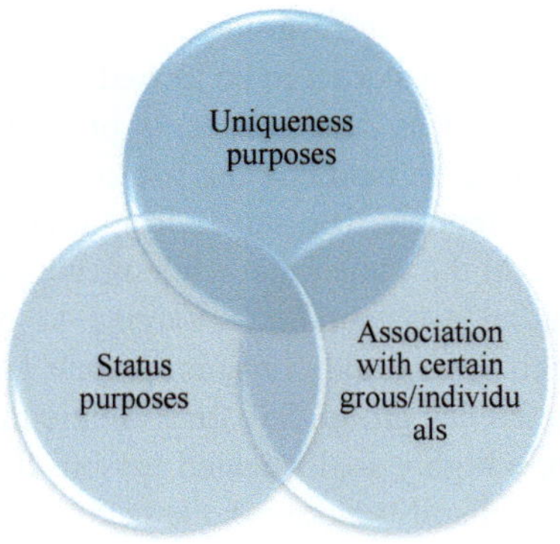

Fig. 7 Luxury goods consumption motivations according to interpersonal/external factors

Veblen coined the term "conspicuous consumption" to refer to public consumption of luxury brands. The consumption of prestige brands is in fact seen as a signal of status and wealth (Nia and Zaichkowsky 2000). Their price, expensive by normal standards, enhances the value of such a signal (Mason 1998).

4.3.2 Uniqueness Motivation

The other underlying motivation for externally-oriented luxury consumption is related to the individuals' need for uniqueness (Snyder and Fromkin 2012), defined in literature as (Tian et al. 2001):

"the individual's pursuit of differentness relative to others achieved through the acquisition, utilization and disposition of consumer goods for the purpose of developing and enhancing one's personal and social identity" (p. 50).

This motivation has been termed as the snob effect (perceived unique value) by Vigneron and Johnson (2004) and is based on the idea that if virtually everyone owns a particular brand it is by definition not prestigious (Groth and McDaniel 1993).

Research shows that consumers' perceptions of limited supply of a brand may lead to increased preference for that brand (Lynn 1991). This is not surprising given consumers' need for uniqueness (Wiedmann et al. 2009). It also addresses personal matters such as the individual's relationship to materialism (Richins and Dawson 1992). This mentality is the outcome of a social comparison process in which the individual's desire is to be perceived as different from other individuals (Snyder and Fromkin 2012). Historically, it has been expressed through the use of valuable materials such as gold, silver or diamonds, which are naturally scarce (Catry 2003).

4.3.3 Association with Aspiration Groups Motivation

The third underlying motivation for external luxury consumption is related to the individual's desire to be associated with certain groups or individuals. The role-playing value and the social value of prestige brands can be instrumental in the decision to buy (Brinberg and Plimpton 1986). Belk's (1988) concept of extended self suggests that our possessions are "a major contributor to and reflection of our identities" (p. 139). Through ownership of luxury brands consumers aspire to be associated with certain prestige groups and at the same time seek to be disassociated from non-prestige groups (Wilcox et al. 2009). This motivation has been termed as the Bandwagon effect-perceived social value or perceived extended self by Vigneron and Johnson (2004).

4.4 Pleasing the Senses Through Luxury

After the proliferation of a shameless approach to conspicuous luxury, another approach started to thrive. Consumers gradually started to perceive more personally oriented benefits to luxurious products. Whether referring to the pleasing of their senses from the superior quality of those goods, or to their general satisfaction of need for beauty, a less conspicuous orientation towards those goods started to gain ground.

This approach to luxury came naturally, since they were considered as objects of superior beauty and taste from their very existence. Consumers were not trying to impress somebody else through the use of such products anymore. What they wanted to do was to indulge into a pleasing-the-senses experience, since they considered that it was something they "deserved". Luxury products became symbols of self gratification from symbols of classification. However, it is worth noticing at this point, that the latter approach does not delude the former. Both approaches can be evident both at a certain point in time in a given society, but also within an individual.

4.5 Respective Research

The research referring to the more personal approach to luxury can be divided into the following two categories (see Fig. 8).

4.5.1 Emotional Motivation

This motivation refers to the consumption of luxury for the emotions provoked, and has been termed as the Hedonic effect/perceived emotional value (Vigneron and Johnson 2004). For a brand which satisfies an emotional desire such as a prestige brand, a product's subjective intangible benefits, such as aesthetic appeal and the pleasure-seeking satisfaction is clearly determining the brand selection (Atwal and Williams 2009; Dubois and Laurent 1994).

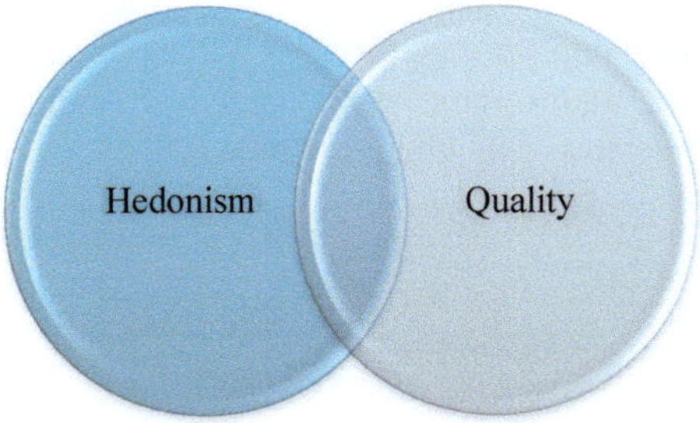

Fig. 8 Luxury goods consumption motivations according to personal/internal factors

Luxury products are in fact predominantly hedonic (Vickers and Renand 2003; Holbrook and Hirschman 1982; Fenigstein et al. 1975), given that their consumption is primarily associated with an affective and sensory experience of aesthetic or sensual pleasure, fantasy and fun (Dhar and Wertenbroch 2000).

4.5.2 Superior Quality Motivation

Prestige is derived partly from the technical superiority and the extreme care that takes place during the production process. Termed as the functional dimension of luxury value by Wiedmann et al. (2009), it refers to such core product benefits and basic utilities as quality, uniqueness, usability, reliability and durability (Sheth et al. 1991). Consumers in fact expect, long before they consumer a luxury brand, that it will exhibit superior product characteristics and performance compared with other non-luxury brands (Garfein 1989). It is therefore, highly unlikely that a luxury brand image can be sustained when product quality is not consistently maintained at a high standard (Christodoulides et al. 2009). This motivation has been termed as the Perfectionism effect (perceived quality value) by Vigneron and Johnson (2004).

4.6 The Ethical Renaissance of Luxury

As aforementioned, the economic and political environment, in line with the prevalence of a more ethical approach to consumption could not have left luxury brands unaffected. So what is it that luxury consumers demand from the respective companies today? Ethicality, ethicality, ethicality (Carrigan et al. 2013). The concept of sustainability is key in contemporary luxury management, whether referring to business practices or key ideas promoted by the brand. Luxury consumers increasingly demand ethical behavior from these companies and companies which wish to stay relevant and appealing have to adapt their practices accordingly.

An oxymoron calling for understanding and resolution seems to be emerging in this sustainability in luxury trend. Research and trends have proven that luxury consumers do demand ethical behavior (Achabou and Dekhili 2013). However, simultaneously, they do not actually care about ethical luxury (Torelli et al. 2012). The selfish nature of luxury consumption and the non selfish nature of corporate social responsibility activities are the ones that basically contradict (Janssen et al. 2014), leaving various question marks both to academics and to practitioners in the field as to what this contradiction means, and whether a company could simultaneously satisfy the traditional ideas on luxury and the sustainability trend (see Fig. 9).

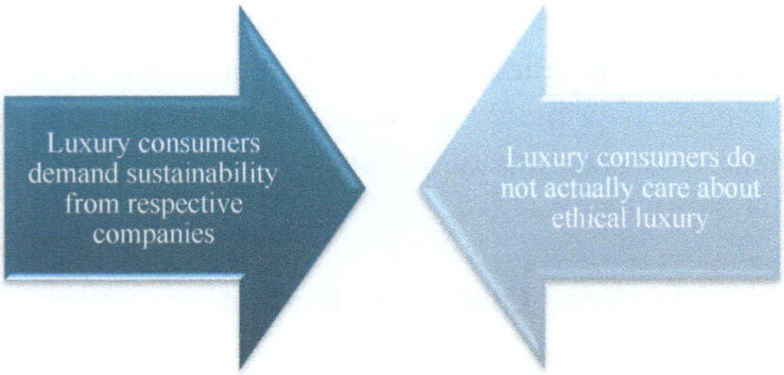

Fig. 9 Oxymoron in sustainable luxury

5 Reaching Personal Consummation: Philosophically Revisited

Luxury brands are key facilitators in the individuals' attempt to construct their ideal selves, as aforementioned. However, the concept of the ideal self is a highly relevant concept, not only in terms of societies but also in terms of eras. What s considered as ideal attributes for an individual and the subsequent ideal behavior depends on the prevailing norms, values and beliefs. So, in order to understand its contemporary form, one needs to understand its evolution through time and the parameters it has included throughout the years.

Undoubtedly, the dimension of sustainability and its attributed importance is a key parameter of the contemporary ideal self. But what else is there to that and which are the other dimensions included? Consumers increasingly demand ethical behavior from the companies whose brands they are using. For a thorough understanding of how sustainability has come to mean that much in contemporary societies, the concept of the ideal self will be presented in the following part of the chapter.

5.1 *The Ideal Self in the Pre-modern Era*

During the pre-modern era, the focus of personal consummation was the equilibrium between personal desires and social or divine orders (Midgley 1991) (see Table 1). Plato is one of the first ones to examine the issue of personal fulfillment in his work Symposium, in the 5th century BC (Plato and Rowe 1998), by underlining the need for more thinking, being open to change, appreciating beauty and changing society. He further pinpoints the importance of the agent, when classifying a need as necessary or not (Irwin 1995). An intellectual person who cared about his surroundings and the society he lived in with a certain level of education was thus the prevailing ideal self at the time.

Further on in the pre-modern era, Aristotle defines "desire" in the 4th century, as an appetite for a pleasant state (Aristotle 1954), by underlining that it is "reason" that classifies the varying types of motives. He further criticized any kind of excess

Table 1 Pre-modern era (B.C.) (I)

Plato	Underlines the need for more thinking, being open to change, appreciating beauty and changing society
Aristotle	Criticizes excess and promotes the idea that harmony comes with moderation
Epicurus	Believes that the purpose of life is about being happy and avoiding pain

whether in consumption or desire and promoted the harmony that comes with moderation. One can thus see the negative attitude towards indulging and exaggerating. The ideal of the time was more about finding a balance between the pleasures and satisfactions in life. A shift, however, occurred later on with the beliefs of philosophers such as Epicurus. For Epicurus, in the 3rd century, the purpose of life is about being happy and avoiding pain (Bailey 1964). It was basically the first form of hedonism, during which enjoying the pleasures of life through the satisfaction of all the senses was not frowned upon anymore and an ideal self included parameters of recognizing such qualities.

Later on, during the Christian years, philosophers in the 11th century such as Pierre Abelard promoted the idea that desires are neutral in themselves. It is the actor's motives that characterize them morally (Haldane 1997). One can thus see that the underlying morals of the individual's behavior started to gain significant ground in what was considered an ideal self. The individual was expected to behave a certain way, according to what was considered ethical and right at the moment.

To contradict this belief, Machiavelli, in the 15th century, promoted the idea that living virtuously does not necessarily lead to happiness (dei Machiavelli 1989). The importance attributed to virtue gradually looses ground to happiness, rendering ethical behavior as less important in the construction of an ideal situation.

During the Age of Enlightenment, philosophers such as Hume proposed the idea that desires are the individuals' ideals about how they want the world to be, whereas philosophers such as Descartes support that reason is the underlying motive for any desire (Smith 1997) One can thus see not only the increasing significance of reason and thinking, but also a desire to think about and pursue and ideal world with certain characteristics (see Table 2).

Table 2 Pre-modern era (B.C.) (II)

Pierre Abelard	Supports the idea that desires are neutral in themselves and that it is the actor's motives that characterize them morally
Machiavelli	Believes that living virtuously does not necessarily lead to happiness
Hume	Underlines that desires are the individuals' ideals about how they want the world to be
Descartes	Promotes the idea that reason is the underlying motive for any desire

5.2 *The Ideal Self in the Modern Era*

During Industrialization the viewpoint prevails that welfare can only be achieved through possessions (van Raaij 1993). It was Adam Smith (1863) who first supports the viewpoint that profit is the main virtue in societies and the road to welfare. Individual motivation as key to happiness is also supported by the Schopenhauer (2012). The acquisition of material goods to reach happiness both as an individual and as a society as a whole was emerging at the time, relating the ideal self to one with an increased number of goods.

This orientation towards wealth creation is hardly criticized by Marx and Engels (1975). They did not accept this belief, and were not willing to accept that the cost that the increased wealth creation would mean in terms of social classes. Parallel to that is the idea of Nietzsche's value-creating Ubermensch/Super-human, who overcomes the traditional views on morality (Loewer 2015) One can thus identify a significant change into the role that objects, materialism and wealth creation played in the construction of an ideal self (see Table 3).

Psychologists come into play in the 20th century, supporting the viewpoint that products can satisfy the individual's underlying motives, if endowed with certain benefits. The ideas of the symbol and its power, and their connection to the ego-ID and the superego, are promoted by Freud (1957). Influencing the masses through their underlying ideals is facilitated with work such as that of Carl Jung, who categorized archetypes to which people can relate to (Carr 2002) and Edward Bernays, who pinpoints how the public opinion can be influenced in support of certain lifestyles (Curtis and Lambert 2002). One can thus see how products at the time are not only considered as a significant parameter of an ideal self, but they are expected to include certain intangible benefits and values, which individuals are expected to use for the creation of their ideal selves (Table 4).

5.3 *The Ideal Self in the Post-modern Era*

The Revolution of the 60s comes in line with ideas on self-liberation and personal expression, (Cahoone 1996), denying imposed materialism. However, it is noteworthy that despite this denial on materialism, consumers widely use "rebellious

Table 3 Modern era philosophers

Adam Smith	Thinks that profit is the main virtue in societies and the road to welfare
Schopenhauer	Supports that individual motivation as key to happiness
Marx	Considers class struggle a negative element of contemporary societies
Engels	Believes that the material world determined our ideas not the other way round
Nietsche	Describes the value-creating Ubermensch/Super-human, who overcomes the traditional views on morality

Table 4 Modern era psychologists

Freud	Promotes the ideas of the symbol and its power, and their connection to the ego-ID and the superego
Jung	Believes that influencing the masses through their underlying ideals is possible (archetypes)
Bernays	Supports that the public opinion can be influenced in support of certain lifestyles

brands" in their effort of personal expression. It is under this scope that philosophers at the time dealt with the power of consumption in everyday life. Baudrillard and Benedict (2005) underline the power of the signs in the sought of the "Great Virtual Satisfaction". Consumption was now seen as a key parameter of forming relations between individuals within a society.

Foucault (1982) examines the power of marketing as a discourse and as a sexual hiding metaphor, in which the role of subjectivity, as underlined by Chomsky (1959), also arises as a key issue. Having thus gained such significant importance, marketing is seen as a concept with certain levels of power, which the individuals can use to create, maintain or change their relations to others or what they wish to communicate about themselves. A shift is evident, according to which individuals are more interested into promoting a certain ideal self to others, rather than actually reaching that ideal self. Products endowed with certain ideals, values and intangible characteristics to which the consumers' ideal self related to are preferred to activities that would actually bring the individual closer to reaching that ideal self.

Responding to such a problematique, Russell (Loewer 2015) underlines that in a "scientific society", wars would stop and welfare would arise. The question of Being in its finest form is set by Heidegger, who underlines the significance of actually Being (Loewer 2015). Such a positioning stresses out that individuals were gradually starting to realize that their actual being and not just the being that they promoted was in fact more significant, preparing the ground for the contemporary more ethical approach and for what has come to be included in the ideal self today (Table 5).

Table 5 Postmodern philosophers

Baudrillard	Sees the social dimensions of consumption
Benedict	Underlines the power of the signs
Foucault	Underlines the power of marketing as a discourse and as a sexual hiding metaphor
Chomsky	Promotes the importance of subjectivity
Russell	Believes that in a "scientific society", wars would stop and welfare would arise
Heidegger	Stresses the importance of actually "Being"

Fig. 10 Re-harmonizing society and marketing

5.4 The Ideal Self Today

The issue with these modern philosophical approaches was that they were based on the idea that marketing was taking the inside-out approach, whereas its nature calls for an outside-in approach (Burchell et al. 2013). As aforementioned, consumers today seek again for a holistic type of welfare as proposed by Aristotle. Based on this different perspective, the consideration of value as a driving force is crucial for re-harmonizing society and marketing, also termed as mindful consumption (Sheth et al. 2011) (see Fig. 10).

After the presentation of the concept of consummation over the centuries its complexity became even clearer. The concept of reaching one's ideal self is highly related to culture and time in general and cannot be easily specified. However, after having examined how the included elements and ideals evolved over time, its contemporary form can be more easily understood. The key point to keep in mind at this point is that luxury brands are highly related to the environment they belong to, since they aim to represent its individuals. Only when acknowledging the connectedness of philosophy to the innate nature of luxury will one be able to understand its varying axes. And this is what firms are asked to do today; to understand their customers up to that point that they will be able to provide them with the desired resources for building their ideal selves.

6 Companies' Reaction to the New Conditions

Companies are seeking for ways to make themselves more appealing under the aforementioned circumstances and to reign the value creation dynamics (Bruch and Ghoshal 2003). Strategic value drivers such as ethicality and sustainability are thus applied. Adler (2006, p. 487) interestingly reports Harvard Business Professor Rob Austin, who underlines that in this value creation quest, artists are the best ones to be consulted. So how can artists help in this direction?

The arts have the potential to function as a key driver in providing emotion-laden experiences and to increase engagement (Schiuma 2011). Arts in business is thus an efficacious tool for creating value, leading to the emergence of the so-called Arts-Based Initiatives (Schiuma 2009). These are organizational and managerial interventions, based on one or more types of art, which engage people in emotional or rational art experiences within an organizational context, and render art to a business asset. They are experience-oriented mechanisms, whether active or passive in nature.

Although the link between luxury brands and responsible or ethical consumption appears to be counterintuitive (Davies et al. 2012). However, interestingly, recent business practices seem to be showing to another direction. Few brands have shown more interest in systematically promoting consumers' intellectual cultivation through art initiatives than major marketers of luxury goods. Luxury brands are increasingly promoting and organizing a variety of cultural events on the axes of fine arts, letters, savoir-vivre or scholarly excellence. Such orchestrated attempts can go beyond the mere role of sponsor, when recognizing the transformational power of the arts (Berthoin Antal and Strauß 2013).

In the following table (see Table 6), a variety of tactics towards the sustainability trend are presented (Koronaki and Panigyrakis 2014).

Table 6 Implications for marketers

Arts and aesthetics	Knowledge and academic excellence	Savoir faire and good taste
• Opening of art centers • Sponsoring museums, exhibitions or artists • Funding or co-funding governmental artistic or archeological initiatives • Investing in architecture in their premises • Infusing artistic elements in various minor points-of-interaction with the consumer	• Hosting literary events for customers • Hosting literary events for anyone interested • Promoting renowned contemporary writers • Offering scholarships for talents in artistic majors • Supporting academic excellence no matter the field • Organizing literary competitions	• Styling seminars • Travelling seminars • Supporting sporting activities • Organizing one's own sporting activities both actual and potential customers • Wine-knowledge seminars • Introducing applications promoting these ideas

7 Conclusions

For the majority of luxury brands, an orientation towards cultural issues has been considered a natural extension of their nature (Dubois and Duquesne 1993). While this tendency shifted with the democratization of luxury and the idea that no particular cultural background was needed to fully appreciate luxury proliferated (Twitchell 2002), since the global recession, luxury brands have reshaped their strategies, in their attempt to maintain their position in the market. Chanel, Armani, Prada, Tod's, Louis Vuitton, Hermes, Gucci, Burberry, Rolex and Moet and Chandon are only some of the luxury brands which have initiated an integrated approach towards Arts-Based Initiatives for their clientele.

This basically means that despite the existence of a certain contradiction between luxury and sustainability, sustainable luxury is here to stay. Considered a paradigm shift, luxury companies cannot afford to ignore and are called to systematically leverage on the fact that a key parameter in their clientele's consummation is ethicality, rendering issues of sustainability as a crucial parameter when evaluating a luxury brand. Luxury consumers might not always care about parameters of ethicality when consuming, however, they do demand it from luxury companies. Those brands which wish to stay relevant and appealing need to adapt to the new paradigm and behave accordingly. Because it is the inner and deeper needs of their consumers that luxury brands want to satisfy, and such a significant shift cannot but be acknowledged.

References

Achabou MA, Dekhili S (2013) Luxury and sustainable development: is there a match? J Bus Res 66(10):1896–1903

Adler NJ (2006) The arts and leadership. Now that we can do anything, what will we do? Acad Manag Learn Educ 5(4): 486–499

Aristotle R (1954) Basic works (trans: Rhys Roberts W)

Atwal G, Williams A (2009) Luxury brand marketing: The experience is everything! J Brand Manage 18(5/6):338–346

Bailey C (1964) The greek atomists and epicurus

Baudrillard J (1970/1998) The consumer society: Myths and structures. London, Sage.

Baudrillard J, Benedict J (2005) Le système des objets. vol 3. Verso

Belk RW (1988) Possessions and the Extended Self. J Consum Res 15(2):139–168

Berthoin Antal A, Strauß A (2013) Artistic interventions in organisations: finding evidence of values-added. Creative Clash Report, Berlin

Brinberg D, Plimpton L (1986) Self monitoring and product conspicuousness on reference group influence. Adv Consum Res, vol 13

Bruch H, Ghoshal S (2003) Unleashing organizational energy. MIT Sloan Manage Rev 45(1):45

Burchell K, Rettie R, Patel K (2013) Marketing social norms: social marketing and the 'social norm approach'. J Consum Behav 12(1):1–9

Cahoone L (1996) From modernism to postmodernism: an anthology. Blackwell Publishers, Malden, MA, USA

Carr A (2002) Jung, archetypes and mirroring in organizational change management: lessons from a longitudinal case study. J Organ Change Manage 15(5):477–489

Carrigan M, Moraes C, McEachern M (2013) From conspicuous to considered fashion: a harm-chain approach to the responsibilities of luxury-fashion businesses. J Mark Manage 29 (11–12):1277–1307

Catry B (2003) The great pretenders: magic of luxury goods. Bus Strategy Rev. 14(3):11–7

Chaudhuri A, Holbrook MB (2001) The chain of effects from brand trust and brand affect to brand performance: the role of brand loyalty. J Mark 81–93

Christodoulides G, Michaelidou N, Li CH (2009) Measuring perceived brand luxury: an evaluation of the BLI scale. J Brand Manage 16(5–6):395–405

Chomsky N (1959) A review of BF Skinner's verbal behavior. Language 35(1):26–58

Curtis A, Lambert S (2002) The century of the self. BBC, UK

Davies IA, Lee Z, Ahonkhai I (2012) Do consumers care about ethical-luxury? J Bus Ethics 106 (1):37–51

dei Machiavelli NDB (1989) Machiavelli: the chief works and others. vol 3. Duke University Press, USA

Dhar R, Wertenbroch K (2000) Consumer choice between hedonic and utilitarian goods. J Mark Res 37:60–71

Dubois B, Duquesne P (1993) The market for luxury goods: income versus culture. Eur J Mark 27 (1):35–44

Dubois B, Laurent G (1994) Attitudes toward the concept of luxury: An Exploratory Analysis. Asia-Pac Adv Consum Res 1(2):273–278

Elliott A, Lemert C (2005) The new individualism: The emotional costs of globalization. Routledge, London

Fenigstein A, Scheier MF et al (1975) Public and private self-consciousness: assessment and theory. J Consul Clin Psychol 43(4):522

Firat AF, Venkatesh A (1995) Liberatory postmodernism and the reenchantment of consumption. J Consum Res 22(3):239

Foucault M (1982) The subject and power. Crit Inquiry 8(4):777–795

Freud S (1957) A connection between a symbol and a symptom. In: The standard edition of the complete psychological works of sigmund freud, volume XIV (1914–1916): on the history of the psycho-analytic movement, Papers on metapsychology and other works, pp 339–340

Garfein RT (1989) Cross-cultural perspectives on the dynamics of prestige. J Serv Mark 3:17–24

Groth JC, McDaniel SW (1993) The exclusive value principle: The basis for prestige racing. J Consum Mark 10(1):10–16

Haldane J (1997) Medieval and renaissance ethics. In: Singer P (1991) (ed) A companion to ethics, Blackwell, Oxford

Hoge DR, McCarthy JD (1984) Influence of individual and group identity salience in the global self-esteem of youth. J Pers Soc Psychol 47:403–414

Holbrook MB, Hirschman EC (1982) The experiential aspects of consumption: consumer fantasies, feelings, and fun. J Consum Res 9(2):132–140

Holt DB (2002) Why do brands cause trouble? a dialectical theory of consumer culture and branding. J Consum Res 29(1):70–90

Irwin TH (1995) Plato's objections to the sophists. Greek World 568–587

Janssen C, Vanhamme J, Lindgreen A, Lefebvre C (2014) The catch-22 of responsible luxury: effects of luxury product characteristics on consumers' perception of fit with corporate social responsibility. J Bus Ethics 119(1):45–57

Kapferer JN (2010) All that glitters is not green: The challenge of sustainable luxury. Eur Bus Rev (Nov-Dec), 40–45

Kapferer JN, Valette-Florence P (2016) Is luxury sufficient to create brand desirability? A cross-cultural analysis of the relationship between luxury and dreams. Luxury Res J 1(2):110–127

Kastanakis MN, Balabanis G (2014) Explaining variation in conspicuous luxury consumption: an individual differences' perspective. J Bus Res 67(10):2147–2154

Koronaki E, Panigyrakis G (2014) Luxury as societal mentor. In: Berghaus B, Mueller-Stewens G, Reinecke S (ed) The management of Luxury Book on Management of Organizations in the Luxury Market

Loewer B (2015) Philosophie in 30 Sekunden. Librero IPB, Netherlands

Lynn M (1991) Scarcity effects on value: A quantitative review of the commodity theory literature. Psychol Mark 8:43–57

Marx K, Engels F (1975) Collected works. vol 10, Marx and Engels: 1849–1851. Lawrence & Wishart, UK

Mason RS (1984) Conspicuous consumption: a literature review. Eur J Mark 18(3):26–39

Mason RS (1998) The economics of conspicuous consumption. Books

Matsuyama K (2002) The rise of mass consumption societies. J Polit Econ 110(5):5–1070

Mead GH (1938) The philosophy of the act. University of Chicago Press, Chicago

Midgley M (1991) The origin of ethics. In: Singer P (ed) (2013) A companion to ethics, Blackwell Reference, London, pp 3–13

Nealon J (2012) Post-postmodernism: or, the cultural logic of just-in-time capitalism. Stanford University Press, USA

Nia A, Zaichkowsky J (2000) Do counterfeits devalue the ownership of luxury brands? J Prod Brand Manage 9(7):485–497

Pantzalis I (1995) Exclusivity strategies in pricing and brand extension. Doctoral dissertation, University of Arizona, Tucson, AZ.

Plato, Rowe CJ (1998) Plato, symposium: edited with an introduction, translation and commentary by. Aris & Phillips

Richins ML, Dawson S (1992) A consumer values orientation for materialism and its measurement: scale development and validation. J Consum Res 19(3):303–316

Schiuma G (2009) The value of arts-based initiatives. Arts & Business, London

Schiuma G (2011) The value of arts for business. Cambridge University Press, Cambridge and New York

Schopenhauer A (2012) The world as will and representation. vol 1. Courier Corporation, USA

Schor J (2007) In defence of consumer critique: Revisiting the consumption debates of the 20th century. Ann Am Acad Polit Soc Sci 611:16–29

Sheth JN, Newman BI et al (1991) Why we buy what we buy: a theory of consumption values. J Bus Res 22(2):159–170

Sheth JN, Sethia NK, Srinivas S (2011) Mindful consumption: a customer-centric approach to sustainability. J Acad Mark Sci 39(1):21–39

Smith A (1863) An inquiry into the nature and causes of the wealth of nations: with a life of the author, an introductory discourse, notes, and supplemental dissertations by JR M'Culloch. Ad. and Ch. Black

Smith M (1977) Realism. In: Singer P (ed) (2013) A companion to ethics, Blackwell Reference, London, pp 3–13

Snyder CR, Fromkin HL (2012) Uniqueness: The human pursuit of difference. Springer Science & Business Media

Tian KT, Bearden WO, Hunter GL (2001) Consumers' need for uniqueness: Scale development and validation. J Cons Res 28:50–66

Torelli CJ, Monga AB, Kaikati AM (2012) Doing poorly by doing good: corporate social responsibility and brand concepts. J Consum Res 38(5):948–963

Twitchell JB (2002) Living it up: our love affair with luxury. Columbia University Press, USA
Van Raaij WF (1993) Postmodern consumption. J Econ Psychol 14(3):541–563
Vigneron F, Johnson LW (2004) Measuring perceptions of brand luxury. J Brand Manage 11 (6):484–506
Wearing SL, McDonald M, Wearing M (2013) Consumer culture, the mobilisation of the narcissistic self and adolescent deviant leisure. Leisure Stud 32(4):367–381
Wiedmann KP, Hennigs N et al (2007) Measuring consumers' luxury value perception: a cross-cultural framework. Acad Mark Sci Rev 7(1):1–21
Wiedmann KP, Hennigs N, Siebels A (2009) Value-based segmentation of luxury consumption behavior. Psychol Market 26(7):625–651
Wilcox K, Kim HM, Sen S (2009) Why do consumers buy counterfeit luxury brands? J Mark Res 46(2):247–59

Moral Responsibility for Sustainable Luxury Consumption and Production

Duane Windsor

Abstract Sustainable development requires that both planetary sustainability and economic development for material welfare of individuals be positive. A tradeoff in which material welfare improves by damaging planetary sustainability is risky, given increasing deterioration in the world's climate. Embedded in sustainable development are the two issues of whether poverty is being relieved progressively and whether economic inequality is decreasing or increasing. The Paris agreement of December 2015, entering into force in 2020, is the newest framework for addressing climate change. The action plan tries to halt rise in global emissions, with a delay afforded to developing countries, and then undertake rapid reductions from the peak. Environmental change, negative and positive, involves many small effects accumulating. This chapter develops the proposition that luxury consumption and production generate special moral responsibility for sustainability, responsibility that holds for both buyers and sellers. While all consumption and production increasingly require sustainability strategies, there is a key distinction between necessities and luxuries. Sustainability is a regressive tax on necessities and a progressive tax on luxuries. Luxury consumption might be reduced in order to transfer wealth to the poor. The voluntary billionaire's club is a step in this direction. The focus of this chapter is on the responsibility imposed by wealth to make luxury consumption and production as environmentally friendly as possible. For instance, the wealthy can afford a premium price to obtain carbon neutral and even more desirably carbon reducing consumption and production. Capacity generates responsibility in this context. Climate change imposes a greater moral responsibility on the wealthy than on the non-wealthy, above a minimum standard for everyone. This conception of moral responsibility has important implications for how firms manage strategic relationship with sustainable luxury stakeholders. The methodology of the chapter is to identify relevant key literature on its central

D. Windsor (✉)
Jesse H. Jones Graduate School of Business, Rice University, MS-531,
PO Box 2932, Houston, TX 77252-2932, USA
e-mail: odw@rice.edu
URL: http://www.business.rice.edu/person/oliver-duane-windsor

M.A. Gardetti (ed.), *Sustainable Management of Luxury*,
Environmental Footprints and Eco-design of Products and Processes,
DOI 10.1007/978-981-10-2917-2_23

proposition, and to develop the logic of special responsibility and the implications of that special responsibility for stakeholder management by luxury-oriented firms.

Keywords Cap-and-trade · Carbon taxation · Exotic tourism · Luxury taxation · Moral responsibility · New middle class · Sustainable development · Sustainable luxury

1 Introduction

The contribution of this chapter is to address the question of the moral responsibility of consumers and producers for sustainable luxury consumption and production. These consumers and producers are best positioned to move consumption and production to a more sustainable basis. Whether luxury consumption and production should be sharply curtailed is a different policy matter.

Sustainable development is the context for examining the question of the moral responsibility of consumers and producers for sustainable luxury consumption and production (Eriksson and Svensson 2016). I argue that this responsibility is shared equally by consumers and producers; and that each side of a luxury market has equal responsibility to influence the other side. This moral responsibility includes the consequences of luxury consumption and production for future generations (van der Dussen 2016). This responsibility is due to the environmental impacts of luxury consumption and production, and to the greater wealth of luxury consumers and producers which means greater capability and competence to address the economic burdens of increased sustainability.

Everyone regardless of wealth arguably has a shared responsibility for planetary sustainability, but I argue that the responsibility rises with luxury consumption and production linked to environmental impact and to wealth. Responsibility rises above some minimum for all individuals in relationship to wealth and luxury activities. Shared responsibility further involves the question of whether the wealthy should transfer resources to the poor in compensation for the economic burden of the minimum responsibility. However, this distribution or distributive justice question is not the topic addressed in this chapter.

The wealthier are more likely to engage in luxury consumption; and the wealthier have greater capacity and competence to bear economic burdens of the increased costs that might occur from improvements in sustainability of luxury consumption. The wealthier are also better able to transfer resources to the poor for the narrow purpose of improving planetary sustainability. Nevertheless, the special responsibility is due to luxury consumption and production rather than to wealth as such: it is simply that the former is associated with the latter. It may also be the empirical case that luxury producers are better positioned to undertake sustainability improvements relative to necessities producers.

There is a long history of concern about the relationship between luxury and responsibility (Windsor 2014). Even so, luxury consumption grows rapidly and globally and markedly in some emerging or developing countries. The luxury market arguably has shifted from a relatively homogeneous group of a few consumers to multiple and heterogeneous market segments (Chandon et al. 2016), which in aggregate are much larger.

Luxury has the general meaning of unusually comfortable lifestyle or sumptuous consumption. The basic distinction is between luxury and necessity. A luxury is not necessary but is desirable, and at the same time is either relatively expensive or relatively rare. In central London or central Paris, an automobile may not be necessary due either to walking proximity or government investment in public transport (buses and subway system). In Los Angeles or Houston, an automobile may be necessary due to the absence of walking proximity or adequate government investment in public transport. But while a bottom-of-the-line Fiat will serve, a top-of-the-line Maserati (owned by Fiat since 1993) is an expensive luxury.

Luxury covers a broad scope of goods and services. One instance is exotic tourism, meaning expensive tourism to environmentally and socially fragile locations (Windsor 2016). Another instance is luxury hotels and resorts and restaurants, independently of exotic tourism locations (Kim and Kim 2005). Vehicles, clothing, travel, and so forth are also within the luxury category. While the emphasis has been on lifestyles of the super-rich (Davison 2016), a new middle class is arguably discernible in various countries, including at least some emerging markets such as Brazil, China, and India (Kravets and Sandikci 2014). The term "new middle class" conveys both an increase in size of the "class" and also a shift toward luxury consumption.

The remainder of this chapter is organized as follows. The second section explains the author's conception of moral responsibility for sustainable luxury consumption and production. The third section explains the relationship between carbon taxation and luxury taxation and the difference between taxation and voluntary moral responsibility for sustainability. The fourth section addresses the likely effects of sustainable luxury for the financial performance of luxury producers. The fifth section examines luxury consumer willingness to pay for sustainability improvements. The conclusion summarizes the arguments of the chapter and the implications for scholars and practitioners.

2 Moral Responsibility in the Luxury Sector

This chapter concerns moral responsibility for sustainable luxury consumption and production. The author's thesis is that consumers have a moral responsibility to prefer sustainable luxury goods and services and also to influence producers concerning their preferences. Similarly, producers have a moral responsibility to make and sell sustainable luxury goods and services and also to influence consumers concerning the appropriate options.

A moral responsibility is voluntary in the sense that a person—such as a consumer or a producer—accepts the responsibility out of a personal sense of duty or obligation. Legislation defines compliance with law: that is, legislation regulates behavior through threat of penalty. Legislation sets standards for conduct and misconduct. Environmental impact is of course widely legislated. Such legislation would be superfluous if individuals widely practiced voluntary moral responsibility. Carbon taxation and/or direct luxury taxation reflect the absence or inadequacy of voluntary moral responsibility. Absence suggests that individuals act contrary to moral responsibility; inadequacy suggests that even if individuals act in accord with moral responsibility, such action is not sufficient to address the problem of environmental deterioration. If voluntary moral responsibility were sufficient, then carbon taxation and/or direct luxury taxation would not be necessary.

Ethical analysis typically involves consideration of motives, actions, and consequences (or impacts). Wood (2010) developed this approach into modern corporate social performance (CSP) theory and measurement. The approach can be separated into consumer and producer sides of the market. Consumers have responsibilities; producers have responsibilities. Ideally, these two responsibilities match in the marketplace. This conception of dual and mutually reinforcing responsibilities is quite different from *caveat emptor* (buyer beware). Today, customers and suppliers can co-create value through the shift from firm- and product-centric orientation to personalized brand experiences (Tynan et al. 2010). These personalized brand experiences are highly heterogeneous (Vigneron and Johnson 2004). Customers may be willing to pay for such luxury experiences.

It has been argued that there is general confusion about moral responsibility (Feinberg 1970, p. 37). In this chapter, moral responsibility is a voluntary moral recognition and acceptance of desirable actions. A moral actor tries to take into account consequences of her or his actions or inactions on others, and accepts responsibility for her or his contributions to those consequences (Carroll 1991). This conception of responsibility may apply to those individuals acting within organizations as well as to individuals acting on their own behalf (Sepinwall 2016; Wolf 1985). Responsibility may be linked to moral competence to form judgments (Sekerka 2016; Wolf 2016).

Banfield (1985, p. 337) points out that any maximizing model must include goals (what is sought), time horizon (over what time period), and voluntary moral constraints. The latter are simply prohibitions against specified actions. An actor freely accepts the prohibitions from a sense of duty or obligation. A moral constraint is different from an externally imposed prohibition, which is essentially a "cost." In this approach, a carbon tax or a luxury tax is an external pricing intended to discourage certain consumption and production activities. A consumer may voluntarily prefer sustainable luxury goods over traditional luxury goods as a moral choice.

The circumstance driving sustainable consumption and production most generally is worsening climate change. February 2016 is now viewed as the warmest February on record and 2015 as the warmest year on record. A 2016 study casts significant doubt on whether governments can in reality achieve the goal of the Paris Accord to limit global warming to below 2 °C, or 3.6 °F (Jones and Warner 2016). There are three basic reasons. First, world population will continue to increase. Current projections expect world population to reach 10.9 billion by 2100. About one-fifth of the current population does not have electricity yet. One can anticipate rising consumption of limited resources of various kinds, including per capita energy consumption. Second, the global energy mix necessary to address population and climate change in combination requires extraordinary and timely investment. Fossil fuels, emitting carbon dioxide, must be replaced by various renewable energy sources such as solar and wind power, hydropower, and biofuels. Renewable energy production was 9% in 2014. Jones and Warner estimate that renewable energy production must be 87–94% of global energy consumption by 2100. Using two different scenarios for when at least half (50%) of these renewable energy needs will occur, at 2028 (to hold global warming to 2 °C) and 2054 (to hold global warming to 2.5–3 °C), the authors conclude that meeting the target of less than 2 °C is simply unlikely. Third, governments have never met such a challenge, and may face pressures to increase energy supply rather than address global warming effectively. The authors recommend expanding renewable energy sources as rapidly as possible, simply to limit warming effects to between 2.5 and 3 °C.

Figure 1 is a conceptual sketch of the basic ideas rather than a formal modeling approach. The vertical axis differentiates among wealth (including income and assets) in terms of low, middle, and wealthy. The horizontal axis separates aggregate consumption level into necessities, luxuries, and sumptuous luxuries.

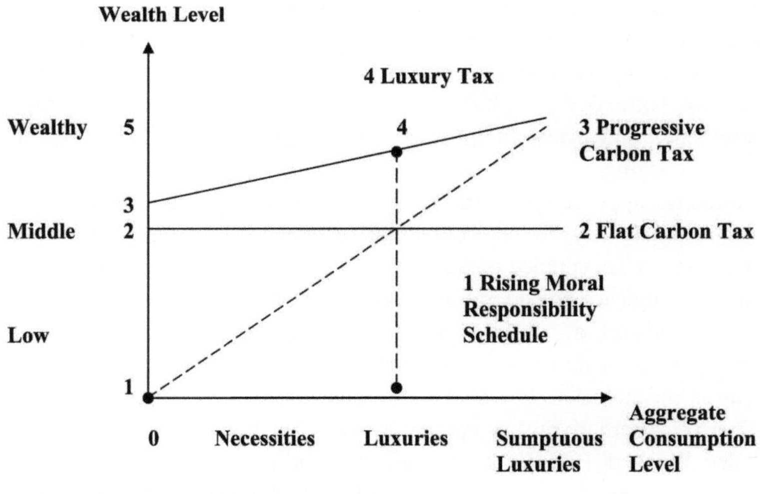

1: rising moral responsibility schedule

2: flat carbon tax on all items

3: progressive (rising rate) carbon tax on all items

4: luxury tax can be either a flat rate or a progressive rate tax on luxury and sumptuous luxury items and exempts necessities

5: wealth tax on wealth holders

Fig. 1 Luxury consumption and carbon taxation

Historically, wealthy was a relatively small proportion of a population and emphasized sumptuous luxuries and hence rarity; both middle and low wealth groups emphasized necessities. There was thus a separation between wealthy and everyone else, and particularly between sumptuous consumption and necessities. The empirical evidence indicates increasing proportions of wealthy and middle populations; and expansion of consumption and production into luxuries much more widely available.

Labeled #1 in Fig. 1, the dashed line originating at the zero point, at approximately a 45-degree angle, is a rising moral responsibility schedule. The argument is that moral responsibility rises with both increased consumption of luxuries (due to rising environmental impacts) and with increasing wealth (due to capability to bear the economic burden of environmental improvements). In Fig. 1, there is no exemption on moral responsibility. One might introduce an argument for an exemption at poverty level of wealth. Labeled #2, a flat carbon tax, shown by a solid line parallel to the horizontal axis, attempts to shift consumption and production toward less carbon-intensive substitutes. The level of the flat carbon tax is given by location along the vertical axis; that level can be moved up or down the vertical axis as a public policy decision. A flat carbon tax is regressive, and there is no exemption depicted in Fig. 1. Labeled #3, a progressive rate carbon tax, shown by a rising (positive slope) solid line not parallel to the horizontal axis, attempts to improve the shift by wealthier people and more sumptuous luxury consumers. Sumptuous luxury has the sense or meaning of luxury consumption for the sole purpose of status or prestige. Labeled #4, a luxury tax begins only in the center of Fig. 1 at the vertical line, which indicates a point at which an exemption for necessities ends and a tax on luxuries and sumptuous luxuries—whether flat or progressive—begins.

The three tax approaches are the subject of the next section. This section focuses on moral responsibility for environmental sustainability. Moral responsibility can arise in voluntary acceptance of duty or obligation; or moral responsibility can be asserted by another for consideration. A consumer may voluntarily accept fair trade pricing for bananas as a way of helping small producers and perhaps also encouraging more sustainable production techniques (Wilson and Jackson 2016). Fair trade is an increasingly important dimension of sustainable luxury (Schmidt et al. 2016). Alternatively, a campaign such as the one by the Coalition of Immokalee Workers (C.I.W.) may seek to mobilize consumer pressure to create higher compensation for employees (Tammelleo and Lombardi 2014). Either fair trade or higher compensation campaigns concern distribution, unless in some way a resulting effect is increased sustainability.

For instance, one might receive benefits from a common resource (Vanderheiden 2016). One can argue that such receipt generates an obligation to contribute toward maintenance of the common resource. This obligation might operate for a specific common resource—such as fish stocks or the global climate—or for the whole of common resources in the form of the planet. Vanderheiden develops an argument that individual contributions to climate change might be regarded as wrongful free riding on the planetary system: the individual benefits from the contribution but refuses to contribute to costs of maintenance of the planetary system. This line of reasoning leads to a fair share conception of burden. However, remedial action might be achieved better through collective action than through allegation of individual harm resulting in bad environmental effects (Vanderheiden 2016). For example, climate change and rising sea level will stimulate migration, whether undocumented or permitted, from developing countries to developed countries (Nawrotzki et al. 2016). Since the developed countries contributed to climate change and rising sea level, the question of moral responsibility of the developed countries arises for consideration.

The moral responsibility approach is different from the question of the international justice of climate change control treaties. Shue (1993) suggests four questions concerning international justice in relationship to subsistence emissions and luxury emissions. The first question is the fair allocation of the costs of preventing avoidable global warming. The second question is the fair allocation of the costs of coping with the social effects of non-avoidable global warming. The third question concerns the distribution of wealth necessary for a fair process of international bargaining over the first and second questions. The fourth question is the fair allocation of greenhouse gases both during transition to the long-term allocation and at the long-term allocation itself. These four questions affect sources of transfers and recipients of those transfers. Shue points out that answering these four questions involves a choice between fault-based principles and no-fault-based principles. A fault-based principle directs attention to who caused global warming effects. A no-fault-based principle directs attention to who can afford to pay for transfers and who is harmed by the effects of global warming.

Moral responsibility of consumers and producers has important management implications in the luxury sector. Three key implications can be readily identified. First, managers should work to reduce the environmental effects of production and consumption. This responsibility can occur independently of consumers; however, strict independence requires constant customer prices and constant consumption experience. Managers should be proactive in reducing environmental consequences. Second, managers should seek to influence customers to reflect their moral responsibility in relationship to willingness to pay and willingness to alter consumption experience. Managers should be proactive in influencing customers, and also in responding to customers seeking to influence producers toward greater environmental responsibility. Third, one can extend M.K. Gandhi's idea of trusteeship from a focus on the wealthy helping the poor (Gopinath 2005; Shah 2014) to a broader conception including the wealthy assuming responsibility for environmental sustainability (Jose 2016). Trusteeship is a mindset, one quite different from agency on behalf of a company's ownerships or stewardship on behalf of the organization itself defined as the organization's key stakeholders. Trusteeship views society and environment as the primary responsibilities of a business or the wealthy that a business serves.

3 Carbon Taxation and Luxury Taxation

As noted above, taxation is an alternative to voluntary moral responsibility. Society can levy taxes to compel mandatory sustainability improvements. There are two basic approaches relevant here and depicted in Fig. 1: carbon taxation and luxury taxation. Carbon taxation, which might be made progressive in some way, seeks to reduce rising carbon levels in the natural environment by raising the effective price of carbon. Luxury taxation seeks to reduce luxury consumption and production directly. Luxury taxation combines issues of environmental sustainability and equitable distribution of wealth.

Two general approaches to reduce carbon emissions have been proposed: cap-and-trade and carbon tax. Cap-and-trade (or emissions trading) is a partly flexible regulatory mechanism in which a government (or set of governments) issues a fixed set of allowances for emissions, and then permits trading among affected parties such that parties achieving efficiencies below an allowance can sell to parties not achieving efficiencies below an allowance. Trading provides an incentive for improvements, while subjecting overall emissions to a cap. A carbon tax charges parties for consumption and production that generates carbon emissions (Zhang et al. 2016). The taxed parties should have an incentive to reduce carbon emissions in order to reduce tax liability.

There is considerable debate over whether a cap-and-trade or carbon tax approach is superior. One recent study of this question incorporates the real-world condition that carbon intensive energy sources may be heavily subsidized (Lui 2016). Subsidy deters renewable energy investments and decreases effectiveness of carbon taxation (Lui 2016). At a policy level, which system should be introduced may be considerably less important than immediate action—as either system will help deter carbon emissions (Keohane and Wagner 2015).

There is a vital difference between a flat tax (or flat rate tax) and a progressive tax (or rising rate tax). A flat tax is a fixed percentage (or proportion) of a taxed base, such as consumption, income, sales, or wealth. A flat tax is thus a constant marginal rate of taxation. (Each additional dollar is taxed at the same rate.) Any set of deductions, exemptions, or caps to the taxed base can generate variation in the marginal rate of taxation. A true flat tax is generally regressive in the sense that the economic burden of a flat tax rate is heavier on the poorer taxpayers. One can add a certain degree of progressivity to a flat tax by exempting poorer taxpayers, basically resulting in a two-rate tax system: zero tax below some cutoff, and a proportional tax above that cutoff. A progressive tax has a rising marginal rate of taxation, meaning that the economic burden of the tax rises with consumption, incomes, sales, or wealth. The fair tax proposal of the Americans for Fair Tax (AFT) organization would shift from income to consumption as a tax base (Santoro and Fuller 2016), ideally on a revenue neutral basis. As a sales tax, the fair tax is inherently regressive (Santoro and Fuller 2016).

A flat rate carbon tax would be regressive in economic burden. One could think of a progressive carbon tax system in which there is some exemption at low income or consumption and the tax rate rises with carbon emission. When parties must consumer carbon-intensive goods above a subsistence level, the carbon tax is regressive unless revenues are used for uniform lump-sum transfers which redistribution then introduces progressivity (Klenert and Mattauch 2016). A difficulty in assessing tax systems is the possibility raised by some scholars that the analysis is as much ideology as science (Wagner 2016).

In Fig. 1, a flat carbon tax falls on necessities as well as luxuries, and as discussed above is thereby regressive in burden. A progressive carbon tax rises with carbon generation and/or consumption; but it also falls on necessities as well as luxuries unless there is some exemption at the bottom of the income distribution. With an exemption that excludes all necessities, a progressive carbon tax becomes a subset of a luxury one—one focused on carbon generation and consumption associated with luxury goods. Technically, a luxury tax would have a different target. By definition, a luxury tax is levied on goods or services that are deemed by public policy to be non-essential. A luxury tax is in effect a progressive tax on luxury production and/or consumption; all necessities—as so deemed by public policy—are exempt from luxury taxation. The difference is as follows. A carbon tax —flat or progressive—is levied on carbon generation and/or consumption. A luxury tax is levied on identified items: for example, expensive vehicles and vessels, artwork, luxury housing, and so forth. Sumptuous luxuries are basically luxuries consumed for the sole purpose of status or prestige (Rokka 2016), and thus might be subjected to a higher rate of taxation—creating effectively two luxury tax rates. Typically, a luxury tax is levied as a sales tax or value added tax (VAT) in some form. The purpose of a luxury tax is to reduce production and consumption of the targeted items. There may be somewhat different effects than a progressive carbon tax, but a luxury tax might influence environmental sustainability more broadly than a carbon tax. Systematic comparison of progressive carbon taxation and luxury taxation is desirable. Frank (2005) presents the social welfare case against reducing taxes for the wealthy. Luxury taxation has tended to be used for purposes of raising revenue. The idea considered here is that luxury taxation might be an approach to encouraging environmental sustainability and other social benefits by raising the price of luxury goods and services (Briggs et al. 2016).

A luxury tax is different from the 2012 UN proposal for a 1% global tax on billionaires to help fund development finance; that proposal is a wealth tax which would be levied on the upper end of the vertical axis classified as "wealthy" in Fig. 1, labeled #5. The UN proposal report also included a global carbon tax, a transfer (or exchange) tax on key currencies, and earnings from allocation of International Monetary Fund (IMF) Special Drawing Rights (SDRs)—all proposals being directed to funding of development finance (UN 2012).

As an alternative to carbon taxation or luxury taxation, voluntary moral responsibility raises two questions. One question, examined in the next section, concerns the effects of more sustainable luxury on the financial performance of luxury producers. These effects should be considered in light of the financial effects of either carbon taxation or luxury taxation. The other question, examined in the subsequent section, concerns the consumers' willingness to pay for sustainable goods and services.

4 Luxury Producer Financial Performance

This section examines the likely effects of sustainability on luxury producers' financial performance. There is a developing movement in which luxury producers are financing foundations to accomplish good in the public interest. One example is the Louis Vuitton Foundation, established 2014 in Paris (Herrero 2016). Higher profits might be shifted into such activities. Such actions signal that profitability may not be at severe risk from sustainability improvements. The evidence below supports a finding that luxury producers can benefit from sustainability initiatives.

This chapter's specification of moral responsibility is different from simple responsiveness to market demand or from manipulation of market demand. One can formulate a straightforward theory of business supply of responsible goods and services in which some stakeholder group pays for the responsible production, thus holding constant the profit of the business (McWilliams and Siegel 2001). The business simply responds to a form of market demand, no different economically from any other form of market demand. The stakeholder group might be consumers or employees or even donors. Alternatively, a business might manipulate demand in such a way as to increase profit margin—as through advertising or deception. The business might engage in so-called "greenwashing" for the sole purpose of increasing profit. The authors argue that cost-benefit analysis will result in an "ideal" level of CSR supply, maintaining a neutral relationship between firm CSR and firm financial performance.

In theory, socially responsible production likely reduces monopolistic profit and is unlikely to emerge from such a supplier. However, with supplier competition socially responsible production may positively affect consumers' buying decisions and increase supplier profits—within a limit concerning pricing not being too high for the market (Pigors and Rockenbach 2016). Whether this view works similarly in luxury consumption and production should be established directly. In luxury markets, pricing is not the dominant consideration.

This view is reinforced by an empirical finding in another study that stakeholder management—building better relationships with primary stakeholders including communities, customers, employees, and suppliers—can strengthen competitive advantage and improve shareholder value, while social issue participation has the opposite effect (Hillman and Keim 2001). It may be critically important whether sustainable luxury consumption and production reflect stakeholder management or social issues participation.

In luxury fashion, there is some evidence that company design might be viewed more positively than user design (Fuchs et al. 2013). The reason is that consumers view user design as lower in quality and thus not signaling high status. If such evidence is valid, then companies may have the opportunity to influence consumers toward green products. User design works better when luxury fashion products do not involve status signaling.

Barnett and Salomon (2006, 2012) report empirical evidence, using two different data sources, that the relationship between social responsibility or performance and financial performance may be curvilinear. While firms can be profitable through low responsibility or irresponsibility, firms can be somewhat more profitable through higher responsibility. This evidence suggests that the relationship is not dichotomous, meaning either that bad is unprofitable and good is profitable or bad is profitable and good is unprofitable.

A kind of greenwashing may be involved in the following information. International firms may face contradictory incentives concerning whether to reduce environmental impacts and provide transparent environmental information (Aragón-Correa et al. 2016). A comparison of the world's 100 most international nonfinancial firms to 16,023 firms in their industries and also a group of selected similar firms finds a significant gap between environmental performance and voluntary disclosure for the 100. That set of firms has a much better record of environmental disclosure than their industries and the matched firms. However, that set of firms has a worse financial performance. The authors conclude that the 100 seek legitimacy through voluntary disclosure.

The conventional conception of corporate social responsibility (CSR) has emphasized a pyramidal construct in which the foundation is economic performance including profitability within a context of legal compliance and ethical responsibility, the latter extending to voluntary (or discretionary) philanthropy (Carroll 1991). Within the perspective of this conventional CSR conception, moral managers behave differently toward obligations and stakeholders than do amoral managers or immoral managers (Carroll 1991). Amoral managers act in accord with a cost-benefit calculus, while immoral managers are simply concerned with profitability. The conventional conception embeds a tension or contest among ideas of economic responsibility, ethical responsibility, and corporate citizenship responsibility (Windsor 2006). Economic responsibility theory advocates wealth creation in a context of minimally regulated markets. Ethical responsibility theory advocates corporate duties of self-restraint and altruism in a context of much stronger regulations and stakeholder rights. Corporate citizenship responsibility theory tends to be a political metaphor emphasizing either in one of two conflicting variants that either (1) philanthropy is a strategic lever for increasing corporate reputation and market opportunities, or (2) firms should accept an increased emphasis on universal human rights.

A recent development in the literature distinguishes between CSR and corporate social irresponsibility (Windsor 2013, 2015). The point of the differentiation is that irresponsibility—illustrated by damaging the natural environment, harming stakeholders' interests, and so forth—is much more harmful in its effects and thus more desirably to be regulated in some way relative to the much more limited benefits of encouraging corporate altruism or citizenship. Plainly it would be most desirable to reduce irresponsibility and encourage citizenship.

A cross-national study investigated the perceptions of corporate responsibility of Chinese and Finnish managers of small and medium-sized enterprises (SMEs) in the mechanical wood processing industry (Li et al. 2016). The term corporate responsibility is neutral with respect to the role of corporate social responsibility (CSR). The authors conducted semi-structured interviews with 23 line managers (during the period December 2009–December 2010). They report that managers adopt informal CSR strategies and tools in order to meet the specific expectations of their key stakeholders. Interestingly, this finding held across the different cultural and operational contexts for Chinese and Finnish managers. While legal aspects drive corporate responsibility the four aspects of production technology, efficient procurement, customer focus, and competent workplace drive competitive advantage.

The important opportunity is for luxury businesses to make the supply chain and production and distribution process as environmentally friendly as possible. Lean production and total quality management are even more important in luxury production, as illustrated by a case study of an Italian fashion company (Carmignani 2016). The fashion market is an important dimension of luxury consumption and production (Miller and Mills 2012).

Luxury car manufacturers typically engage in relatively low volume production. Such manufacturers may be able to increase production and reduce overall emissions. Green manufacturing technologies and environmental strategies may be easier for luxury car manufacturers than for high volume manufacturers (Nunes et al. 2016).

5 Luxury Consumer Willingness to Pay

The financial performance effects for luxury producers seeking to improve environmental sustainability depend partly on the willingness to pay of luxury consumers. Technically, willingness to pay is measured by the demand function for a good or service. For luxury consumers, sustainable consumption may involve an increase in willingness to pay for reductions in environmental impacts of their consumption. Technically, the product or service would separate into the two dimensions of consumption and reduced environmental impact: consumers pay for both dimensions. The alternative possibility is that luxury producers can generate cost savings to pay for improved environmental sustainability while holding prices and consumption quality constant. Thus, sustainable luxury may involve both increased willingness to pay and greater producer cost reductions. However, it should be emphasized that cost reduction is a complex topic, and that a high degree of cost reduction might result in goods and services that are not in reality luxury. This deterioration of luxury is arguably a strong empirical risk with rising mass consumption and marketing of goods simply labeled luxury. The alternative addressed in this chapter is environmentally-oriented cost reduction holding luxury consumption quality constant—a quite different concept. Thomas (2007) argues that the rise of luxury goods holding companies (or conglomerates) consolidating small family-owned luxury firms has resulted in reduced quality and increased counterfeiting among other negative effects. However, it should be emphasized that Thomas attributes these negative effects to industry consolidation and mass consumption—as distinct from the producer environmental sustainability orientation defined in this chapter.

There is a considerable and increasing literature, especially in marketing, on consumer psychology of sustainable luxury. This section summarizes that literature and its implications. The literature helps support the possibility of luxury consumers preferring sustainable luxury consumption if products and services are made available.

There are studies of consumers' willingness to pay for environmental improvements. An analysis of respondent data from three metropolitan cities (Delhi, Mumbai, and Bangalore) in India finds, using contingent valuation methodology, positive willingness to pay (Gupta 2016). This willingness is affected by environmental awareness (influenced in turn by a number of factors including environmental interest, environmental activities, education, income, and age). Taxation might be used to address environmental externalities (Gupta 2016). A US study addresses consumer response to a bottled water tax (Villas-Boas et al. 2016). Response may include willingness to pay for environmental improvements.

Consumer psychology is a key driver of voluntary acceptance of moral responsibility for the consequences of luxury consumption (He et al. 2016; Sen and Bhattacharya 2001; Sen et al. 2016). Green claims and brand responsibility of producers can influence consumers (Olsen et al. 2014). The presence of other consumers may also influence luxury orientation (Pozharliev et al. 2015). Marketing is thus a key responsibility of luxury producers, and that marketing should be directed to reshaping preferences toward environmentally sustainable options and features (Kumar 2015).

In self-determination theory, consumers' intrinsic values may influence conscious choices in purchasing of goods and services (Ki and Kim 2016). One study of 452 respondents explored whether they chose a "sustainable luxury purchase" or a "conspicuous luxury purchase" and whether this choice was influenced by intrinsic values (such as personal style and social consciousness) or extrinsic values (such as fashion, public self-consciousness, and status consciousness). Additionally, the authors examined whether the choice resulted in positive affect influencing in turn repurchase intention. The authors reported that sustainable luxury purchase predicted positive affect more strongly than did conspicuous luxury purchase. This finding suggests that positive emotional responses toward luxury brands influence repurchase intention.

A study of "brand prominence"—defined as conspicuousness of a brand mark or logo on a product—separated consumers into four groups determined by wealth and need for status (Han et al. 2010). Using field experiments and market data analysis, the study found that across each group preference for conspicuously or inconspicuously branded luxury products depended on desire to associate or dissociate with members of each consumer's own group or other groups. Thus wealthy but low in need for status consumers prefer to associate with others of their own group and pay a premium for non-prominent (or "quiet") goods. In contrast, wealthy but high in need for status consumers prefer prominent (or "loud") goods to signal to less affluent differentiation in status. High in need for status consumers lacking sufficient wealth for true luxury use "loud" counterfeit products to imitate the wealthy.

There is a trend of "downward extensions" in which producers introduce further product lines at more accessible prices (Kapferer and Laurent 2016). The trend expands consumption of luxuries. The authors examined 8376 respondents in seven countries to identify consumer perceptions of minimum price for a luxury good in 21 categories of such goods. The idea is to identify consumer perceptions of where luxury begins. They found an extreme variance across consumers, but a large majority perceived very low minimum prices for defining luxuries and consistently so across the product categories. The study concludes that "expensiveness" and "luxury" are now relative concepts, with extreme heterogeneity across consumers. As a result, the idea of luxury extends from traditional (sumptuous) luxury to new (affordable) luxury. In other language, "rarity" has become "abundant" (Kapferer 2012). Traditional luxury rested on rarity (or scarcity) of availability and limited production and distribution. In effect, traditional luxury involved monopolistic practices yielding relatively high premiums on relatively low volumes. Growth of luxury brands has led to an "abundant rarity" or new luxury in which consumers attach feelings of privilege to the brand. These feelings of privilege rest on experience, pricing, prestige, and life-style symbolism. Kapferer and Valette-Florence (2016) interviewed 1286 luxury consumers about 12 well known luxury brands as the basis for these findings.

There is some experimental evidence suggesting that status goods may support higher pricing as follows (Kuksov and Xie 2012). Consumers value status goods because of rarity of consumption by other consumers. In a situation of two competitors, a reduction in the price of one competitor increases demand for the other competitor. Two status products may be imperfect substitutes in the consumer utility function but complements in the producer profit function. A result is that competition can result in higher prices relative to monopoly ownership of each product.

The distinction between wealthy and non-wealthy may cut across countries (Martinez-Alier 1995). In rich countries, there is rising demand for environmental amenities and simultaneously environmental movements against the negative effects of affluence. In poor countries, there are also environmental movements that can be labeled "environmentalism of the poor" (Martinez-Alier 1995).

Luxury typically includes the experience of brand consumption itself (Atwal and Williams 2009). Luxury also includes consumer perception and producer crafting of brand authenticity. A study of the strategies of 26 luxury wine businesses reveals that creating perception of authenticity involves a sincere story blending industrial and rhetorical attributes (Beverland 2005). Sincerity rested on hand crafted techniques, wine uniqueness, place relationship, and passion on the one hand, and disavowal of commercial motives, rational production techniques, and modern marketing techniques. Authenticity status was associated with price premiums and competitive advantage. Sincerity emphasized quality and relevance rather than commercial considerations.

East Asia is the largest volume market for Western luxury and prestige brands; this phenomenon arguably rests on cultural factors including materialism and conspicuous consumption (Schultz and Jain 2016; Wong and Ahuvia 1998; Wu et al. 2016). An empirical study across twenty countries is revealing (Dubois et al. 2005). Cultural variations may have significant impacts on natural environment by location. For instance, wild meat consumption is reportedly significant and impactful around Hue, Vietnam (Sandalj et al. 2016). Demand for luxury seafood is rising in China, with effects on supply chains in the Philippines and North America (Fabinyi 2016). Luxury demand should be distinguished from rising standard for necessities, and such a distinction can be made in seafood.

A study in Singapore found different perceptions of ownership of luxury brands relative to the USA (Phau and Prendergast 2000). In the USA, the rarity principle may hold such that producers maintain prestige through exclusivity resting on high levels of awareness but tightly controlled brand access. In Singapore, Asian consumers are reportedly different. Brand popularity may positively increase the "dream value" of the brand. As a result, increasing awareness positively increases brand preference and purchase intentions. Active marketing works to strengthen brand image and perception of brand benefits.

There is empirical evidence that strong branding is associated with substantially lower executive pay (Tavassoli et al. 2014). The proposed explanation is that executives value association with strong brands and thus accept lower compensation. This effect appears strongest for CEOs and younger executives. The data came from a large and cross-industry sample of executives.

One study argues that a new class of American consumer has arisen (Silverstein and Fiske 2003). This new class—resting on increasingly wide income disparities, higher education levels, and rising awareness of other cultures and those cultures' conceptions of the good life—is consuming new luxury products and services (such as vehicles, clothing, foods, and alcohols including wines, spirits, and beers). Basically, the American middle market is trading up in quality and taste. Silverstein and Fiske point out that (in 2003) the middle market earning $50,000 and above annually have about $1 trillion in disposable income. These consumers are willing to pay significant premiums for preferred products and services. The authors argue that new luxury goods engage consumer emotions and aspirations for a better life. The authors attribute new luxury to both consumers and producers. The result is relatively high sales volumes despite relatively high prices. New luxury businesses are thus profitable and growing. As the authors point out, a Panera sandwich is in the same new luxury market as a lower priced Mercedes.

A study of condominium sales in the Tokyo metropolitan area (Fuerst and Shimizu 2016) found that eco-labeled buildings command a small, but statistically significant, premium on both asking and final transaction prices. The authors report general consistency with results found in other countries. The authors further analyze buyer characteristics to discern sources of demand for eco-labeled real estate. They find that the price premium reflects higher willingness to pay of wealthier purchasers. There is a similar willingness to pay found for less affluent purchasers, but the premium effect is smaller (due to lower wealth). The authors, however, also report a problem difficult to disentangle empirically. Capitalized utility bill savings likely account for much of the premium (including for less affluent purchasers). After accounting for this effect, the authors think that the higher premium for affluent households still indicates some intangible benefits of green building occupancy.

6 Conclusion

This chapter has examined the moral responsibility of luxury consumers and producers for cooperating in improving sustainable luxury. The definition of sustainable luxury is luxury consumption and production in ways that at a minimum reduce negative environmental impacts. Indeed, it would be highly desirable that sustainable luxury increase positive environmental impacts, as well. This consideration of moral responsibility of luxury consumers and producers is placed in the context of sustainable development, which includes improved progress against poverty. However, it is important to separate luxury from issues of wealth redistribution. The focus of the chapter is on the relationship between luxury and sustainability. Luxury is intertwined with wealth: it is the wealthy who can afford luxury consumption. By the same token, it is the wealthy who can afford sustainable luxury. This concluding section discusses below in separate sub-sections the chapter's key findings, and then the implications of those findings for scholars and managers.

6.1 Findings

The methodology of the chapter is to identify relevant key literature on the central proposition of moral responsibility, and to develop firstly the logic of special responsibility and then secondly the implications of that special responsibility for stakeholder management by luxury-oriented firms.

There is a strong case for moral responsibility for luxury consumption and production due to the increasing climate change that is threatening negative changes to planetary sustainability. Sustainability is a regressive tax on the poorer, and a progressive tax on the wealthy. Wealthy consumers have both the capacity and the responsibility to address sustainability through luxury goods and services. The author explained his conception of voluntary moral responsibility arising in luxury consumption and production. This conception is that both consumers and producers can cooperate in improving sustainability.

The principal alternatives to voluntary moral responsibility involve regulation and/or taxation. Cap-and-trade is a regulatory approach to reduction of carbon production in which firms are allocated carbon production rights ideally below their current carbon production levels; firms can then sell (or trade) excess carbon production rights.

The author examined carbon taxation and luxury taxation as alternatives to voluntary moral responsibility. Carbon taxation can be a flat rate or a progressive rate. A flat rate is regressive. A flat rate above an exemption for necessities or poverty becomes at least a mildly progressive rate. A luxury tax—whether a flat rate or a progressive rate—would be levied on luxury goods and services, with the twin effects of reducing luxury consumption and production, and depending on design with the object of redirecting luxury consumption and production toward greater sustainability outcomes. It is conceivable that reasonably low tax rates might redirect luxury toward sustainability without reducing luxury consumption and production as such. The chapter does not propose a wealth tax as such. Rather, a wealth tax is defined so that the reader can see the differences among carbon taxation, luxury taxation, and wealth taxation. The policy design issue is how to focus on luxury consumption and production for purposes of improving environmental sustainability outcomes.

Much depends on how moral responsibility and/or taxation alternatives may influence behavior of consumers and producers. The chapter examined evidence on the effects of sustainability initiatives on luxury producer financial performance, and the willingness to pay of luxury consumers. The available evidence supports the idea of consumer and producer willingness to undertake sustainability efforts. Firms can do well financially, and have equally or even more satisfied consumers. The financial performance effect for luxury production firms is analogous to the wealth of luxury consumers: financial performance like wealth increases responsibility for sustainability—akin to a progressive tax rate.

6.2 Implications for Scholars

Understanding the relationship of luxury consumption and production to sustainable development has important implications for scholars. Research should refine moral responsibility theory, theories and constructs for luxury consumption and production, relationship to carbon, luxury, and wealth taxation, and evidence concerning producer financial incentives and consumer willingness to pay. All these aspects are important elements of sustainable luxury consumption and production. It is particularly important to study the differences between why consumers and producers would adopt voluntary moral responsibility for sustainable development and how external regulation and taxation would influence behaviors of consumers and producers. The most desirable condition for luxury industries is that consumer satisfaction, firm financial performance, and environmental sustainability should all improve jointly. Reductions in consumer satisfaction and/or firm financial performance to improve environmental sustainability may be necessary, but reflect inferior policy alternatives. Consumer satisfaction can include the consumer's concern for the natural environment, such that satisfaction and concern are interdependent.

6.3 Implications for Managers

For practitioners—here meaning luxury industry producers—there are opportunities for environmental leadership and strategic advantages in sustainable initiatives. Detailed case studies of luxury producers are desirable. Guidance for practitioners and consumers is also desirable. Producers should be responsive to consumer expectations of sustainable products and services, and should seek to influence those consumer expectations concerning luxury products, luxury experiences, and pricing. Consumers may be willing to pay more for sustainability; and producers should seek to discover how to hold or improve firm financial performance through sustainability initiatives. As was discussed previously, there has been a development noted by Thomas (2007) in which consolidation of small family firms into larger conglomerates has moved toward mass luxury consumption and production in ways that adversely affect luxury and sustainability. If accurate, the mindset involved is one of profit seeking antithetical to environmental sustainability when that mindset ignores negative environmental effects.

References

Aragón-Correa JA, Marcus A, Hurtado-Torres N (2016) The natural environmental strategies of international firms: old controversies and new evidence on performance and disclosure. Acad Manag Perspect 30(1):24–39. doi:10.5465/amp.2014.0043

Atwal G, Williams A (2009) Luxury brand marketing–the experience is everything! J Brand Manag 16(5):338–346. doi:10.1057/bm.2008.48

Banfield EC (1985) Here the people rule: selected essays. Plenum Press, New York

Barnett ML, Salomon RM (2006) Beyond dichotomy: the curvilinear relationship between social responsibility and financial performance. Strateg Manag J 27(11):1101–1122. doi:10.1002/smj.557

Barnett ML, Salomon RM (2012) Does it pay to be really good? Addressing the shape of the relationship between social and financial performance. Strateg Manag J 33(11):1304–1320. doi:10.1002/smj.1980

Beverland MB (2005) Crafting brand authenticity: the case of luxury wines. J Manag Stud 42 (5):1003–1029. doi:10.1111/j.1467-6486.2005.00530.x

Briggs AD, Kehlbacher A, Tiffin R, Scarborough P (2016) Simulating the impact on health of internalising the cost of carbon in food prices combined with a tax on sugar-sweetened beverages. BMC Public Health 16(1). doi:10.1186/s12889-016-2723-8 (open access)

Carmignani G (2016) Lean supply chain model and application in an Italian fashion luxury company. In: Chiarini A, Found P, Rich N (eds) Understanding the lean enterprise. Springer International Publishing, Cham, Switzerland, pp 203–220. doi:10.1007/978-3-319-19995-5_9

Carroll AB (1991) The pyramid of corporate social responsibility: toward the moral management of organizational stakeholders. Bus Horiz 34(4):39–48. doi:10.1016/0007-6813(91)90005-G

Chandon JL, Laurent G, Valette-Florence P (2016) Pursuing the concept of luxury: introduction to the JBR special issue on “luxury marketing from tradition to innovation”. J Bus Res 69 (1):299–303. doi:10.1016/j.jbusres.2015.08.001

Davison A (2016) The luxury of nature: the environmental consequences of super-rich lives. In: Hay I, Beaverstock JV (eds) Handbook on wealth and the super-rich. Edward Elgar Publishing, Northampton and Cheltenham, pp 339–359

Dubois B, Czellar S, Laurent G (2005) Consumer segments based on attitudes toward luxury: empirical evidence from twenty countries. Mark Lett 16(2):115–128. doi:10.1007/s11002-005-2172-0

Eriksson D, Svensson G (2016) A balance model of theoretical sustainability—framework and propositions. Corp Gov Int J Bus Soc 16(1):21–34. doi:10.1108/CG-01-2015-0006

Fabinyi M (2016) Producing for Chinese luxury seafood value chains: different outcomes for producers in the Philippines and North America. Mar Policy 63:184–190. doi:10.1016/j.marpol.2015.03.024

Feinberg J (1970) Doing and deserving: essays in the theory of responsibility. Princeton University Press, Princeton

Frank RH (2005) Positional externalities cause large and preventable welfare losses. Am Econ Rev 95(2):137–141. doi:10.1257/000282805774670392

Fuchs C, Prandelli E, Schreier M, Dahl DW (2013) All that is users might not be gold: how labeling products as user designed backfires in the context of luxury fashion brands. J Mark 77 (5):75–91. doi:10.1509/jm.11.0330

Fuerst F, Shimizu C (2016) Green luxury goods? The economics of eco-labels in the Japanese housing market. J Japan Int Economies 39:108–122. doi:10.1016/j.jjie.2016.01.003

Gopinath C (2005) Trusteeship as a moral foundation for business. Bus Soc Rev 110(3):331–344. doi:10.1111/j.0045-3609.2005.00019.x

Gupta M (2016) Willingness to pay for carbon tax: a study of Indian road passenger transport. Transp Policy 45:46–54. doi:10.1016/j.tranpol.2015.09.001

Han YJ, Nunes JC, Drèze X (2010) Signaling status with luxury goods: the role of brand prominence. J Mark 74(4):15–30. doi:10.1509/jmkg

He H, Zhu W, Gouran D, Kolo O (2016) Moral identity centrality and cause-related marketing: the moderating effects of brand social responsibility image and emotional brand attachment. Eur J Mark 50(1/2):236–259. doi:10.1108/EJM-10-2014-0613

Herrero M (2016) Cultural foundations and luxury brands: the case of brand philanthropy. Paper presented at the Third ISA Forum of Sociology, Vienna, Austria, 10–14 July 2016 (abstract)

Hillman AJ, Keim GD (2001) Shareholder value, stakeholder management, and social issues: what’s the bottom line? Strateg Manag J 22(2):125–139. doi:10.1002/1097-0266(200101)22:2<125:AID-SMJ150>3.0.CO;2-H

Jones GA, Warner KJ (2016) The 21st century population-energy-climate nexus. Energy Policy 93:206–212. doi:10.1016/j.enpol.2016.02.044

Jose PD (2016) Business and society: creating shared value: in conversation with NR Narayana Murthy, Founder, Infosys. IIMB Manag Rev 28(1):43–51. Retrieved from http://www.sciencedirect.com/science/article/pii/S0970389616000069. doi:10.1016/j.iimb.2016.02.003

Kapferer J-N (2012) Abundant rarity: the key to luxury growth. Bus Horiz 55(5):453–462. doi:10.1016/j.bushor.2012.04.002

Kapferer J-N, Laurent G (2016) Where do consumers think luxury begins? A study of perceived minimum price for 21 luxury goods in 7 countries. J Bus Res 69(1):332–340. doi:10.1016/j.jbusres.2015.08.005

Kapferer J-N, Valette-Florence P (2016) Beyond rarity: the path of luxury desire—how luxury brands grow yet remain desirable. J Prod Brand Manag 25(2) (forthcoming)

Keohane NO, Wagner G (2015) Pluses and minuses of a carbon tax: either a tax or a cap can put a price on climate pollution. The goal is to reduce it. New York Times. Environmental Defense Fund letter. http://www.nytimes.com/2015/06/15/opinion/pluses-and-minuses-of-a-carbon-tax.html

Ki CW, Kim YK (2016) Sustainable versus conspicuous luxury fashion purchase: applying self-determination theory. Fam Consum Sci Res J 44(3):309–323. doi:10.1111/fcsr.12147

Kim HB, Kim WG (2005) The relationship between brand equity and firms' performance in luxury hotels and chain restaurants. Tourism Manag 26(4):549–560. doi:10.1016/j.tourman.2004.03.010

Klenert D, Mattauch L (2016) How to make a carbon tax reform progressive: the role of subsistence consumption. Econ Lett 138:100–103. doi:10.1016/j.econlet.2015.11.019

Kravets O, Sandikci O (2014) Competently ordinary: new middle class consumers in the emerging markets. J Mark 78(4):125–140. doi:10.1509/jm.12.0190

Kuksov D, Xie Y (2012) Competition in a status goods market. J Mark Res 49(5):609–623. doi:10.1509/jmr.11.0005

Kumar V (2015) Evolution of marketing as a discipline: what has happened and what to look out for. J Mark 79(1):1–9. doi:10.1509/jm.79.1.1

Li N, Toppinen A, Lantta M (2016) Managerial perceptions of SMEs in the wood industry supply chain on corporate responsibility and competitive advantage: evidence from China and Finland. J Small Bus Manag 54(1):162–186. doi:10.1111/jsbm.12136

Lui CY (2016) Keeping the carbon buried: cap-and-trade is more efficient than a world carbon tax. MA thesis, Economics, University of Victoria, Canada. http://web.uvic.ca/~chuiying/Lui_quota_tax.pdf

Martinez-Alier J (1995) The environment as a luxury good or "too poor to be green"? Ecol Econ 13(1):1–10. doi:10.1016/0921-8009(94)00062-Z

McWilliams A, Siegel D (2001) Corporate social responsibility: a theory of the firm perspective. Acad Manag Rev 26(1):117–127. doi:10.5465/AMR.2001.4011987

Miller KW, Mills MK (2012) Contributing clarity by examining brand luxury in the fashion market. J Bus Res 65(10):1471–1479. doi:10.1016/j.jbusres.2011.10.013

Nawrotzki RJ, Riosmena F, Hunter LM, Runfola DM (2016) Undocumented migration in response to climate change: the moral responsibility of industrialized nations. Int J Pop Stud 1(1):60–74. doi:http://dx.doi.org/10.18063/IJPS.2015.01.004

Nunes B, Bennett D, Shaw D (2016) Green operations strategy of a luxury car manufacturer. Technol Anal Strateg Manag 28(1):24–39. doi:10.1080/09537325.2015.1068933

Olsen MC, Slotegraaf RJ, Chandukala SR (2014) Green claims and message frames: how green new products change brand attitude. J Mark 78(5):119–137. doi:10.1509/jm.13.0387

Phau I, Prendergast G (2000) Consuming luxury brands: the relevance of the 'rarity principle'. J Brand Manag 8(2):122–138. doi:10.1057/palgrave.bm.2540013

Pigors M, Rockenbach B (2016) Consumer social responsibility. Manag Sci. doi:10.1287/mnsc.2015.2279 (forthcoming)

Pozharliev R, Verbeke WJMI, Van Strien JW, Bagozzi RP (2015) Merely being with you increases my attention to luxury products: using EEG to understand consumers' emotional experience with luxury branded products. J Mark Res 52(4):546–558. doi:10.1509/jmr.13.0560

Rokka J (2016) Champagne: marketplace icon. Consumption Mark Cult. doi:10.1080/10253866.2016.1177990

Sandalj M, Treydte AC, Ziegler S (2016) Is wild meat luxury? Quantifying wild meat demand and availability in Hue, Vietnam. Bio Conserv 194:105–112. doi:10.1016/j.biocon.2015.12.018

Santoro JR, Fuller CS (2016) Reassessing the fair tax. U Pitt Law Rev 77:385–411

Schmidt S, Hennigs N, Behrens S, Karampournioti E (2016) The luxury of sustainability: examining value-based drivers of fair trade consumption. In: Gardetti MA, Muthu SS (eds) Handbook of sustainable luxury textiles and fashion, vol 2, Springer Singapore, Singapore, pp 121–136. doi:10.1007/978-981-287-742-0_7

Schultz D, Jain V (2016) Exploring luxury brand country of origin impact on purchasing behaviors in India: a conceptual framework. Asia Pac J Bus Adm 7(3):216–236. doi:10.1108/APJBA-11-2014-0129

Sekerka LE (2016) Moral competencies. In: Sekerka LE (ed) Ethics is a daily deal. Springer International Publishing, Cham, Switzerland, pp 199–211

Sen S, Bhattacharya CB (2001) Does doing good always lead to doing better? Consumer reactions to corporate social responsibility. J Mark Res 38(2):225–243. doi:10.1509/jmkr.38.2.225.18838

Sen S, Du S, Bhattacharya CB (2016) Corporate social responsibility: a consumer psychology perspective. Curr Opin Psychol 10:70–75. doi:10.1016/j.copsyc.2015.12.014

Sepinwall AJ (2016) Corporate moral responsibility. Philos Compass 11(1):3–12. doi:10.1111/phc3.12293

Shah S (2014) Corporate social responsibility: a way of life at the Tata Group. J Hum Values 20 (1):59–74. doi:10.1177/0971685813515591

Shue H (1993) Subsistence emissions and luxury emissions. Law Policy 15(1):39–60. doi:10.1111/j.1467-9930.1993.tb00093.x

Silverstein MJ, Fiske N (2003) Luxury for the masses. Harvard Bus Rev 81(4):48–57

Tammelleo S, Lombardi LG (2014) Consumer social responsibility? Bus Prof Ethics J 33(1):99–126. doi:10.5840/bpej20145110

Tavassoli NT, Sorescu A, Chandy R (2014) Employee-based brand equity: why firms with strong brands pay their executives less. J Mark Res 51(6):676–690. doi:10.1509/jmr.13.0435

Thomas D (2007) Deluxe: how luxury lost its luster. Penguin Press, New York

Tynan C, McKechnie S, Chhuon C (2010) Co-creating value for luxury brands. J Bus Res 63 (11):1156–1163. doi:10.1016/j.jbusres.2009.10.012

UN Department of Economic and Social Affairs, World Economic and Social Survey (2012) In search of new development finance: overview 2012. Retrieved from http://www.un.org/en/development/desa/policy/wess/wess_current/2012wess_overview_en.pdf

van der Dussen J (2016) Responsibility for future generations. In: van der Dussen J (ed) Studies on collingwood, history and civilization. Springer International Publishing, Cham, Switzerland, pp 299–314

Vanderheiden S (2016) Climate change and free riding. J Moral Philos 13(1):1–27. doi:10.1163/17455243-4681046

Vigneron F, Johnson LW (2004) Measuring perceptions of brand luxury. J Brand Manag 11 (6):484–506. doi:10.1057/palgrave.bm.2540194

Villas-Boas SB, Berck P, Stevens A, Moe-Lange J (2016) Measuring consumer responses to a bottled water tax policy. Am J Agric Econ. Available from https://escholarship.org/uc/item/3qw4h1pf#page-2. Accessed 26 Mar 2016 (forthcoming)

Wagner RE (2016) The language of taxation: ideology masquerading as science. GMU working paper in economics no. 16-01, George Mason University, Department of Economics. doi:10.2139/ssrn.2731508. Accessed 25 Mar 2016

Wilson M, Jackson P (2016) Fair trade bananas in the Caribbean: towards a moral economy of recognition. Geoforum 70:11–21. doi:10.1016/j.geoforum.2016.01.003

Windsor D (2006) Corporate social responsibility: three key approaches. J Manag Stud 43(1):93–114. doi:10.1111/j.1467-6486.2006.00584.x

Windsor D (2013) Corporate social responsibility and irresponsibility: a positive theory approach. J Bus Res 66(10):1937–1944. doi:10.1016/j.jbusres.2013.02.016

Windsor D (2014) Heritage of luxury and responsibility. In: Berghaus B, Müller-Stewens G, Reinecke S (eds) The management of luxury: a practitioner's handbook. Kogan Page, London, pp 395–407

Windsor D (2015) Identifying reasons why some firms maximize corporate social irresponsibility and some firms minimize corporate social responsibility. In: Stachowicz-Stanusch A (ed) Corporate social performance: paradoxes, pitfalls, and pathways to the better world. IAP—Information Age Publishing, Charlotte, NC, pp 67–89

Windsor D (2016) Exotic tourism in very fragile locations: sustainable value creation in environmentally and socially fragile locations. In: Gardetti MA, Torres AL (eds) Sustainability in hospitality: how innovative hotels are transforming the industry. Greenleaf Publishing, Sheffield, UK, pp 61–78

Wolf S (1985) The legal and moral responsibility of organizations. In: Pennock R, Chapman J (eds) Nomos XVII: criminal justice. New York University Press, New York, pp 267–286

Wolf S (2016) Character and responsibility. J Philos 112(7):356–372. doi:10.5840/jphil2015112722

Wood DJ (2010) Measuring corporate social performance: a review. Int J Manag Rev 12(1):50–84. doi:10.1111/j.1468-2370.2009.00274.x

Wong NY, Ahuvia AC (1998) Personal taste and family face: luxury consumption in confucian and western societies. Psychol Mark 15(5):423–441. doi:10.1002/(SICI)1520-6793(199808)15:5<423:AID-MAR2>3.0.CO;2-9

Wu M-SS, Chen C-HS, Nguyen B (2016) Luxury brand purchases and the extended self: a cross-cultural comparison of young female consumers in Taiwan and the UK. Asia Pac J Bus Adm 7(3):153–173. doi:10.1108/APJBA-05-2015-0046

Zhang K, Wang Q, Liang QM, Chen H (2016) A bibliometric analysis of research on carbon tax from 1989 to 2014. Renew Sustain Energy Rev 58:297–310. doi:10.1016/j.rser.2015.12.089

MIX
Papier aus verantwortungsvollen Quellen
Paper from responsible sources
FSC® C105338

If you have any concerns about our products,
you can contact us on
ProductSafety@springernature.com

In case Publisher is established outside the EU,
the EU authorized representative is:
Springer Nature Customer Service Center GmbH
Europaplatz 3, 69115 Heidelberg, Germany

Printed by Libri Plureos GmbH
in Hamburg, Germany